Environmental Influences on the Immune System

Charlotte Esser
Editor

Environmental Influences on the Immune System

Springer

Editor
Charlotte Esser
IUF – Leibniz Research Institute for Environmental Medicine at the Heinrich-Heine University of Düsseldorf
Düsseldorf
Germany

ISBN 978-3-7091-1888-7 ISBN 978-3-7091-1890-0 (eBook)
DOI 10.1007/978-3-7091-1890-0

Library of Congress Control Number: 2015958356

Springer Vienna Heidelberg New York Dordrecht London

Printed on acid-free paper

Springer-Verlag GmbH Wien is part of Springer Science+Business Media (www.springer.com)

To my mother and father

Preface

The poetry of earth is never dead.
John Keats (1795–1821)

Environment (\in-ˈvī-rə(n)-mənt) is a noun that refers to the sum of physical, chemical, and biological factors that act upon an organism or an ecological community, but can also refer to all social and cultural conditions that influence the life of an individual or community. Thus, the environment comprises factors as diverging as climate, microbes, lifestyle, stress, diet, sun exposure, chemical pollution, and much more. Ultimately, the environment determines how we can live.

The term "immune system" appeared first in the early twentieth century and describes the many interacting and specialized body functions that protect from disease and infections. The immune system has evolved as the response of animals against bacteria, viruses, and fungi and it is most highly developed in vertebrate animals. It also serves to detect and eliminate cancer cells and contributes to epithelial integrity, thus protecting our barriers to the environment. Without our immune system, we cannot survive. Manipulating the immune status by vaccination and drugs and dietary compounds and supporting immune competence by hygiene measurements have saved countless human lives.

Looking at the interaction and communication of the immune system with the environment, especially the chemical environment in the broadest sense, is an interdisciplinary exercise. In the last decades, research was limited often to niches such as immunotoxicology and immunopharmacology. This is currently changing. There is a growing awareness in mainstream immunology that environmental conditions and environmental factors far beyond infections can influence the immune system. The influence can strengthen or weaken the immune system, including vaccination success, or give relevant cues for the adaptive direction an effective immune response should take.

There is a great need to understand how this communication between the environment and the immune system works. In a modern world we must understand how chemicals and the environment affect health: people suffering from allergies or autoimmunity, cancer, or immune-related morbidity want answers. Better knowledge will open new avenues for preventive or therapeutic strategies, informed policy decisions, or changes towards a healthier personal lifestyle.

This book wants to address this need and thus close an important gap. The book is divided into three parts, which cover human factors such as age, stress and diet, environmental factors, and important natural and man-made factors (UV light and chemicals). As a final section, a chapter looks at the gaps and challenges, and at the human rights perspective and the obligations coming with it. I have invited leading experts in the field for their contributions, and I am honored and grateful that so many have taken up the challenge. Thereby the book will serve as an excellent and up-to-date source of information for scholars from immunology, toxicology, allergy, and other fields. It will serve both scientists and those who make decisions in the field of public health to better understand the breadth and importance of understanding the influence of our environment on the immune system and thereby our health.

I am indebted to my current and former colleagues in immunology and toxicology for enthusiastic discussion and great science. Finally, I am grateful for the generous financial support by the Deutsche Forschungsgemeinschaft towards my lab over the years.

Düsseldorf, Germany — Charlotte Esser
July 2015

Contents

Contributors

Diana Boraschi Institute of Protein Biochemistry, National Research Council, Naples, Italy

Lisbeth A. Boule Department of Environmental Medicine, University of Rochester School of Medicine & Dentistry, Rochester, NY, USA

Angela Ceribelli Division of Rheumatology and Clinical Immunology, Humanitas Research Hospital, Rozzano, Milan, Italy

BIOMETRA Department University of Milan, Milan, Italy

Charlotte Esser IUF – Leibniz Research Institute for Environmental Medicine at the University of Düsseldorf, Düsseldorf, Germany

Geraint Florida-James School of Life, Sport and Social Sciences, Edinburgh Napier University, Edinburgh, UK

Remo Frei Molecular Immunology, Swiss Institute of Allergy and Asthma Research, University of Zurich, Zürich, Switzerland

Christine Kühne-Center for Allergy Research and Education, Davos, Switzerland

Renee M. Gardner, PhD Department of Public Health Sciences, Karolinska Institute, Stockholm, Sweden

Department of Pathology, Microbiology and Immunology University of South Carolina School of Medicine, Columbia, SC, USA

Elena Generali Division of Rheumatology and Clinical Immunology, Humanitas Research Hospital, Rozzano, Milan, Italy

Elisa Couto Gomes School of Life, Sport and Social Sciences, Edinburgh Napier University, Edinburgh, UK

Philippe Grandjean, MD, DMSc Department of Environmental Medicine, University of Southern Denmark, Odense, Denmark

Department of Environmental Health Harvard School of Public Health, Boston, MA, USA

Carsten Heilmann, MD, DMSc Department of Paediatric and Adolescent Medicine, Copenhagen University Hospital Rigshospitalet, Copenhagen, Denmark

Christian Herder Institute for Clinical Diabetology, German Diabetes Center, Leibniz Center for Diabetes Research at Heinrich Heine University Düsseldorf, Düsseldorf, Germany

German Center for Diabetes Research (DZD e.V.), Düsseldorf, Germany

Paola Italiani Institute of Biomedical Technologies, National Research Council, Segrate, Italy

Katrine Kielsen, MD Institute for Inflammation Research and Department of Paediatric and Adolescent Medicine, Copenhagen University Hospital Rigshospitalet, Copenhagen, Denmark

Ursula Krämer IUF – Leibniz Research Institute for Environmental Medicine, Düsseldorf, Germany

Roger Lauener Children's Hospital of Eastern Switzerland, St. Gallen, Switzerland

Christine Kühne-Center for Allergy Research and Education, Davos, Switzerland

B. Paige Lawrence, PhD Department of Environmental Medicine, University of Rochester School of Medicine & Dentistry, Rochester, NY, USA

Julia N. Mälzer Regenerative Immunology and Aging, Berlin-Brandenburg Center for Regenerative Therapies (BCRT), Charité Universitätsmedizin Berlin, Berlin, Germany

Jennifer F. Nyland, PhD Department of Pathology, Microbiology and Immunology, University of South Carolina School of Medicine, Columbia, SC, USA

Charles L. Raison Division of Family Studies and Human Development, Norton School of Family and Consumer Sciences, College of Agriculture and Life Sciences, University of Arizona, Tucson, AZ, USA

Department of Psychiatry College of Medicine, University of Arizona, Tucson, AZ, USA

Rebecca G. Reed Division of Family Studies and Human Development, Norton School of Family and Consumer Sciences, College of Agriculture and Life Sciences, University of Arizona, Tucson, AZ, USA

Department of Psychiatry College of Medicine, Norton School of Family and Consumer Sciences, College of Agriculture and Life Sciences, University of Arizona, Tucson, AZ, USA

Caroline Roduit Department of Allergy/Immunology, Children's Hospital, University of Zurich, Zürich, Switzerland

Christine Kühne-Center for Allergy Research and Education, Davos, Switzerland

Lars P. Ryder, MSc Department of Clinical Immunology, Copenhagen University Hospital Rigshospitalet, Copenhagen, Denmark

Axel R. Schulz Regenerative Immunology and Aging, Berlin-Brandenburg Center for Regenerative Therapies (BCRT), Charité Universitätsmedizin Berlin, Berlin, Germany

Carlo Selmi, MD, PhD Division of Rheumatology and Clinical Immunology, Humanitas Clinical and Research Center, Rozzano, Milan, Italy

Zaiba Shamim, MSc, PhD Institute for Inflammation Research and Department of Paediatric and Adolescent Medicine, Copenhagen University Hospital Rigshospitalet, Copenhagen, Denmark

Dinah Shelton The George Washington University Law School, Washington, DC, USA

Tom Teichert Institute for Clinical Diabetology, German Diabetes Center, Leibniz Center for Diabetes Research at Heinrich Heine University Düsseldorf, Düsseldorf, Germany

German Center for Diabetes Research (DZD e.V.), Düsseldorf, Germany

Andreas Thiel Regenerative Immunology and Aging, Berlin-Brandenburg Center for Regenerative Therapies (BCRT), Charité - Universitätsmedizin Berlin, Berlin, Germany

Stephen E. Ullrich, PhD Department of Immunology, The University of Texas MD Anderson Cancer Center, Houston, TX, USA

Hans-Werner Vohr Department of Toxicology, Bayer Pharma AG Pharmaforschungszentrum, Wuppertal, Germany

Erika von Mutius Head, Department of Asthma and Allergy, Dr von Hauner Children's Hospital, Ludwig Maximilians University Munich, Munich, Germany

Abbreviations

4-n NP	4-n-nonylphenol
ACD	Allergic contact dermatitis
ACh	Acetylcholine
ACTH	Adrenocorticotropic hormone
AD	Atopic dermatitis
AfCHPR	African Charter on Human and Peoples' Rights
AHR/AhR	Aryl hydrocarbon receptor
ANA	Antinuclear autoantibodies
ANoA	Anti-nucleolar autoantibodies
ANS	Autonomic nervous system
APC	Antigen-presenting cells
ATSDR	Agency for Toxic Substances and Disease Registry (of the USA)
AU	African Union
AVP	Arginine vasopressin
BALF	bronchoalveolar lavage fluid
BPA	Bisphenol A
BSA	Bovine serum albumin
C57BL/6	An inbred mouse strain used in immunology
CD	Crohn's disease
CDC	Centers for Disease Control and Prevention (of the USA)
CHS	Contact hypersensitivity
CLR	C-type lectin receptors
CMV	Cytomegalovirus
CNS	Central nervous system
CNT	Carbon nanotubes
CO	Carbon monoxide
COPD	Chronic obstructive pulmonary disease
COX-2	Cyclooxygenase-2
CRH	Corticotropin-releasing hormone
CRP	C-reactive protein
CRS	Cytokine release syndrome, also called "cytokine storm"
CVB	Coxsackievirus B3
CVD	Cardiovascular disease
dACC	Dorsal anterior cingulate cortex

DAMP	Damage-associated molecular patterns
DC	Dendritic cells
DES	Diethylstilbestrol
DIT	Developmental immunotoxicity
DNMT	DNA methyltransferase
DOTC	di-*n*-octyltin dichloride
DPRA	Direct Peptide Reactivity Assay
dsRNA	Double-stranded ribonucleic acid
DTH	Delayed-type hypersensitivity
EAC	Emotional approach coping
EBV	Epstein-Barr virus
ECHR	European Convention for the Protection of Human Rights and Fundamental Freedoms
EDCs	Endocrine disrupting chemicals
ELISA	Enzyme-linked immunosorbent assay
EOGRTS	Extended one-generation reproductive toxicity study
EPA	United States Environmental Protection Agency
ERK	Extracellular signal-regulated kinase
EWAS	Environment-wide association
FICZ	6-Formylindolo[3,2-b]carbazole
FOXP3	Forkhead box P3
GABA-BDZ	γ-aminobutyric acid-benzodiazepines
GD	Graves' disease
GerES	German Environmental Survey
GI	Gastrointestinal
GM-CSF	Granulocyte macrophage colony-stimulating factor
GSE	Gluten-sensitive enteropathy or celiac disease
GSH	Glutathione
HBV	Hepatitis B virus
HCB	Hexachlorobenzene
HCV	Hepatitis C virus
Hg	Mercury
HIV	Human immunodeficiency virus
HLA	Human leukocyte antigen
HPA	Hypothalamic-pituitary-adrenal
HSC	Hematopoietic stem cells
HT	Hashimoto's thyroiditis
IACHR	Inter-American Commission on Human Rights
ICCPR	International Covenant on Civil and Political Rights
ICESCR	International Covenant on Economic, Social and Cultural Rights
IDF	International Diabetes Federation
IFN	Interferon
Ig	Immunoglobulin (can be of M, D, G, E, A type)
IL	Interleukin
ILC	Innate lymphoid cells

IVDK	Information Network of Departments of Dermatology for recording and scientific analysis of contact allergies
I-κB	Inhibitor of κB
JAK-STAT	Janus kinase/signal transducers and activators of transcription
JP-8	Jet propulsion 8
kDa	Kilodalton
LAL	Limulus amoebocyte lysate
LEDS	Life Events and Difficulties Schedule
LLNA	Local lymph node assay
LN	Lymph node
LPS	Lipopolysaccharide
MeHg	Methylmercury
MHC	Major histocompatibility complex
miRNA	MicroRNA
MPS	Mononuclear phagocyte system
MRL	Minimal risk levels
MS	Multiple sclerosis
MWCNT	Multi-walled carbon nanotubes
NAG	N-acetyl-β-d-glucosaminidase
NET	Neutrophil extracellular traps
NFκB	Nuclear factor kappa B
NHANES	National Health and Nutrition Examination Survey (of the USA)
NHP	Nonhuman primates
NIEHS	National Institute of Environmental Health Sciences (of the USA)
NIOHS	National Institute for Occupational Safety and Health (of the USA)
NK	Natural killer cell
NKT	Natural killer T cell
NLR	NOD-like receptors
NLRP3	NLR-related protein 3
NO_x	Nitrogen oxides
NP	Engineered nanoparticles
NPY	Neuropeptide Y
NSAID	Nonsteroidal anti-inflammatory drugs
OAS	Organization of American States
OAU	Organization of African Unity
OECD	Organisation for Economic Co-operation and Development
OVA	Ovalbumin
PAF	Platelet-activating factor (1-alkyl-2-acetyl-*sn*-glycero-3-phosphocholine)
PAMP	Pathogen-associated molecular patterns
PATHOS-D	Pathogen Host Defense
PBC	Primary biliary cirrhosis
PBMC	Peripheral blood mononuclear cells
PCBs	Polychlorinated biphenyls
PCR	Polymerase chain reaction

PEG	Polyethylene glycol
PFAS	Polyfluorinated alkylate substances
PFCA	Plaque forming cell assay
PFCs	Perfluorinated compounds
PFOA	Perfluorooctanoic acid
PFOS	Perfluorooctane sulfonic acid
PGE_2	Prostaglandin E2
PHENX	Phenotypes and exposures
PM	Particulate matter
PMN	Polymorphonuclear leukocytes
PNS	Parasympathetic nervous system
PPARs	Peroxisome proliferator-activated receptors
PPV23	Pneumococcal vaccine
PRR	Pattern-recognition receptors
PSS	Perceived Stress Scale
PUVA	Psoralen plus UVA
PVN	Paraventricular nucleus
RA	Rheumatoid arthritis
RANK	Receptor activator of NF-κB
RANKL	Receptor activator of NF-κB ligand
REACH	Registration, Evaluation, Authorisation and Restriction of Chemicals (a European regulation)
RF	Rheumatoid factor
RfD	Reference dose
ROS	Reactive oxygen species
SCF	Stem cell factor
SCFA	Short-chain fatty acids
sICAM-1	Soluble intercellular adhesion molecule-1
SIgA	Secretory immunoglobulin A
SLE	Systemic lupus erythematosus
SMOL	Small molecules
SNS	Sympathetic nervous system
SOD	Superoxide dismutase
SO_x	Sulfur oxides
SRBC	Sheep red blood cells
sTNF-RII	Tumor necrosis factor receptor type-II
T1D	Type 1 diabetes
T2D	Type 2 diabetes
TAS-20	Toronto Alexithymia Scale
TBTO	Tributyltin oxide
Tc	Cytotoxic T cells
TCDD	2,3,7,8-tetrachlorodibenzo-*p*-dioxin
TCR	T cell receptor
Tdap	Tetanus, diphtheria, and pertussis vaccine
TGF	Transforming growth factor

Th/T_H	T helper
TLR	Toll-like receptors
TNF-α	Tumor necrosis factor-α
TREC	T cell receptor excision circles
Treg	Regulatory T cell(s)
TSP	Total suspended particles
TSST	Trier Social Stress Test
UC	Ulcerative colitis
UFP	Ultrafine particles
UN	United Nations
URTI	Upper respiratory tract infections
UV	Ultraviolet
UVA	Ultraviolet light, wavelength 320–400 nm
UVB	Ultraviolet light, wavelength 290–320 nm
UVC	Ultraviolet light, wavelength 100–290 nm
VIP	Vasoactive intestinal polypeptide
VZV	Varicella zoster virus
WHO	World Health Organization
WoE	Weight of evidence

Principles of the Immune System: Players and Organization

1

Charlotte Esser

Contents

1.1 Introduction

The immune system of vertebrate organisms is an organ of enormous complexity (Abbas et al. 2015; Paul 2013). The immune system is necessary for survival, yet its dysfunction can lead to great morbidity or even mortality. The immune system enables the organism to cope with pathogenic microorganisms and their toxins, detect and kill cancer cells, and contribute to epithelial integrity at the barriers with the environment. These broad functions are reflected in a staggering variety of functionally diverse cell subsets and effector molecules. Indeed, new immune cell subsets and signaling molecules continue to be discovered, and for nonimmunologists the terminology is often daunting.[1]

[1] Cell surface proteins are often used to name or characterize immune cells. Cell surface proteins were first distinguished serologically in various laboratories by raising antibodies against immune cells, and if the same molecule was discovered in parallel, several names existed. Eventually, to bring order into chaos, the highly useful nomenclature of "clusters of differentiation," or CD molecules, was developed. CD molecules are assigned a number in chronological order (Zola et al. 2007). Currently, there are more than 350 CD known. For many CD molecules, the underlying proteins and their function(s) have been characterized by now.

C. Esser
IUF – Leibniz Research Institute for Environmental Medicine at the University of Düsseldorf, Düsseldorf, Germany
e-mail: charlotte.esser@uni-duesseldorf.de

C. Esser (ed.), *Environmental Influences on the Immune System*,
DOI 10.1007/978-3-7091-1890-0_1

The immune system has a number of special characteristics not shared by other organs. First, the immune system functions via spatial interactions. Thus, immune cells and lymphoid tissues and organs are found across the entire body (Fig. 1.1). Many immune cells are migratory and capable to shuttle between tissues, interstitial tissue fluids, the bloodstream, lymph nodes, and the lymphatics; indeed, many immune cells circulate continuously. Migration can be either random or guided by chemokines, which attract cells to a site of inflammation. Meeting points, such as the lymph nodes, enable different immune cell subsets with different functions to communicate and interact closely in a coordinate fashion. Second, the immune system is characterized by comprehensive adaptability of responses against the insult. This concerns the participating cell subsets and thus the direction a response may take (e.g., tailored specifically to the type of pathogen), its intensity, its duration, or its spread. Usually the immune cells are classified into two big groups, namely, as belonging to the "innate" or "adaptive" immune system. The latter term is used for immunity based on the reaction of T and B cells. These cells undergo during their development a genetic process which results in each cell having a gene coding for a unique antigen-binding molecule (the antibodies in B cells, the T-cell receptor in T cells). Soluble antibodies or the surface-bound T-cell receptor recognize molecular structures (antigens) in the serum or on other cells and mount a humoral (=antibody based) or cellular immune response accordingly. T cells only recognize peptide antigens, whereas B cells can generate antibodies against proteins, lipids, carbohydrates, or any other molecule with minimal size requirements. The "gene rearrangement" to generate high affinity and high specificity antigen receptors has developed in vertebrates only (Hirano et al. 2011). T or B cells are not present in invertebrates. Cells of the innate immune system can detect antigen via their pathogenic molecular pattern (PAMP) receptors (Janeway and Medzhitov 2002). Such receptors (Toll-like receptors [TLR], the mannose receptor, to name important ones) detect only a limited number of molecular structures such as double-stranded RNA, lipopolysaccharides (LPS), unmethylated CpG, or flagellin; all such molecules are typical for evolutionary old molecular patterns of bacteria or viruses and are not made by higher organisms. Innate immune cells are highly diverse; they include granulocytes (which are the majority of white blood cells), dendritic cells, natural killer cells, macrophages, or mucosal-specific innate lymphoid cells. Upon recognition of such structures, innate immune cells immediately fight the infection by, e.g., phagocytizing the bacteria and oxidative burst or by inflammatory cytokine secretion. Moreover, some innate immune cells have the additional capacity to digest pathogen proteins and display it as small peptide pieces on their surface. This instructs and directs antigen-specific T cells that danger is at hand and thus starts the adaptive immune response. Specialists for this "antigen presentation" are the dendritic cells. T cells need the interaction with antigen-presenting cells to mature into effector cells. In turn, B cells need help from T cells to function. Only a subset of B produces antibodies without T-cell help. Finally, the immune system is continuously renewing itself from hematopoietic stem cells. All lineages are derived from common hematopoietic stem cells, which are found in the fetal liver before birth and in the

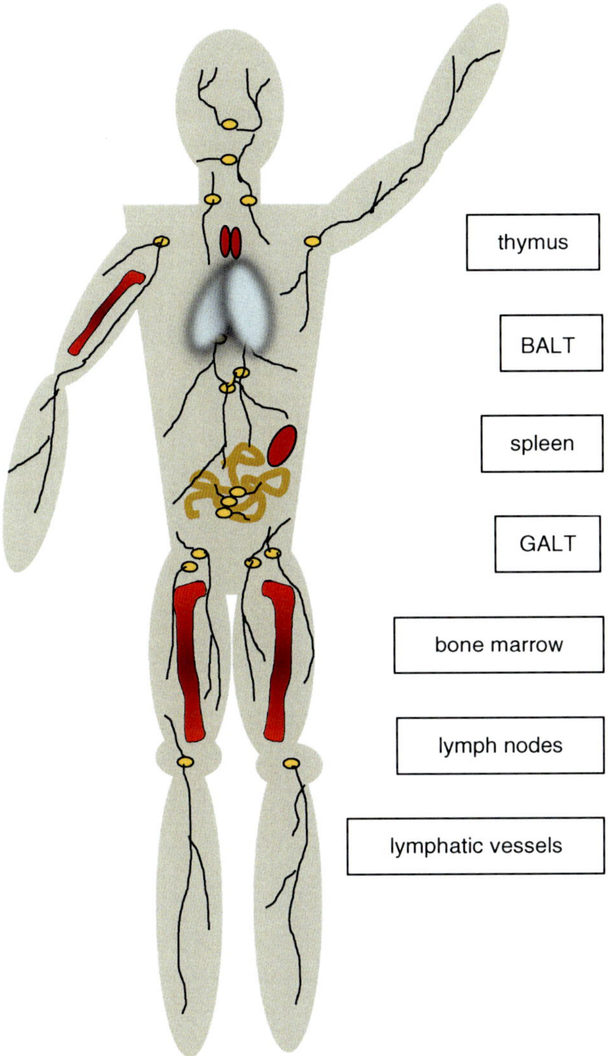

Fig. 1.1 The immune system as an organ across the body. Schematic presentation of major immune structures in the body. Primary lymphoid organs are the (1) sites of hematopoiesis, i.e., bone marrow of all hollow bones, and (2) the thymus which selects T cells that are neither autoreactive nor nonresponsive to foreign antigen presented to them. Secondary lymphoid organs are the spleen and the lymph nodes (LNs); LNs contain T cells, B cells, and dendritic cells; tissue fluid and cells drain via lymph vessels to the nodes; cells can also enter from the blood stream via high endothelial venule cells. LNs are coordination sites for T-cell differentiation upon antigen presentation; note that lymphatic vessels and lymph nodes are much more numerous than presented in this scheme. Lymph vessels empty into the bloodstream from the thoracic duct lymph vessel at the clavicular vein (not shown here). *BALT* (bronchoalveolar-associated lymphoid tissue). *GALT* (gut-associated lymphoid tissue: intraepithelial lymphocytes, Peyer's patches)

bone marrow after birth. The stem cells divide, and their descendants differentiate further, using intrinsic and extrinsic clues. The differentiation is irreversible. Many immune cells are short-lived and replaced by newly differentiated ones. However, some immune cells, in particular the memory cells and long-living plasma cells of the adaptive immune system, can stay around in niches of the bone marrow for years or decades. This allows the immune system to fight a new response faster and more vigorously than at the first encounter. Memory can last lifelong; thus, one gets some infectious diseases only once. The measles are well known for this. On the flip side, also autoimmune disorders and allergies can last as long as the specific memory T cells and long-lived plasma cells survive; allergies will flare up again and again upon renewed contact with antigen. This poses the major challenge for any therapy beyond treating symptoms. Last but not least, it is also a feature of the immune system that it can switch off immune responses (while preserving memory). This is vital of course as an ongoing inflammation or tissue destruction can lead to devastating health consequences. It requires again a complicated signaling network and involves cytokines, immunosuppressive enzymes, and surface molecules to give negative feedback signals to immune cells and stop their activity.

The basic principles of the immune system are *recognition* followed by *response*. What is “recognized” by the immune system, in other words: what are antigens? The answer appears trivial: pathogens, infective organisms, cancer cells, and toxins. However, the T- and B-cell receptors do not bind and react to (in immunology language: recognize) complete organisms, but they bind to organic molecules: lipids, carbohydrates, and proteins/peptides. Small molecular weight chemicals such as some toxins can be recognized by T cells only when bound to a protein. Not all of these molecules are specific for harmful pathogens; they also exist in harmless bacteria, in plants, in pollen, in food, or in one’s own body. It is vital that the immune system does not use its destructive potential when there is no infectious risk or possible harm to the body or when the antigen is indeed a molecule of one’s own body. Immunologists have coined the term “self” for this. Recognition thus must be able to distinguish between harmless and harmful organisms and molecules and between self and non-self. One way to ensure correct recognition are the pathogenic molecular pattern recognition receptors mentioned above. Another one is the requirement of several signals at the same time. One signal is the antigen recognition itself; the other is a costimulatory signal on the cell surface, usually provided by antigen-presenting cells. In addition, soluble cytokines contribute to the start and direction of an immune response.

Once started, an immune response leads to a response on a cellular level, where cells begin to migrate, produce, and secrete effector molecules such as cytokines and chemokines or differentiate to become effector cells themselves, e.g., able to kill infected cells via direct cell contact. The arsenal of responses possibilities is huge: phagocytosis and intracellular killing by oxidative burst; cytokine and mediator secretion to change the tissue micro-milieu and permeability, rendering immune cells more sensitive; direct killing of infected cells by T cells or natural killer (NK) cells; antibody production by B cells; complement activation; cell proliferation; and more.

As always, there is a prize to pay for such high sophistication, and thus, the immune system can go awry and become dangerous. When *recognition* goes wrong, the immune system can attack its own body cells and tissues (resulting in autoimmunity) or mount responses against harmless proteins or drugs such as food proteins, inhaled pollen, penicillin, etc. (resulting in allergies, asthma, eczema, etc.). In both cases, what the immune system recognizes as "danger" cannot be resolved and eliminated by the immune response. Thus, the immune response results in chronic, sterile inflammation and tissue damage. Figure 1.2 illustrates this. In general, therapy of the alerted immune system cannot revert it to a naive state. Rather, therapeutic options aim at treating the symptoms, dampening the immune response, or simply avoiding the antigen or inducing chemical (if that is possible).

On the other hand, when the *response* goes wrong, by whatever causes, the immune system can create an equally crippling outcome. A wrong response could be either a heightened inflammation by secretion of the wrong cytokines or a state of immunosuppression, where, for instance, B cells do not secrete enough antibodies or T cells remain unresponsive despite receiving proper signals. As a result, an infection might be resolved too slowly or not at all and eventually overwhelm the body. Chemicals such as environmental pollutants or drugs can cause or contribute to a response going wrong. Examples for immunosuppressive chemicals are diethylstilbestrol or 2,3,7,8-tetrachloro-dibenzo-p-dioxin. Other immunosuppressive drugs are cyclophosphamide or cortisol, which indeed are important therapeutics for autoimmune or allergic immune responses or in cases of transplantation.

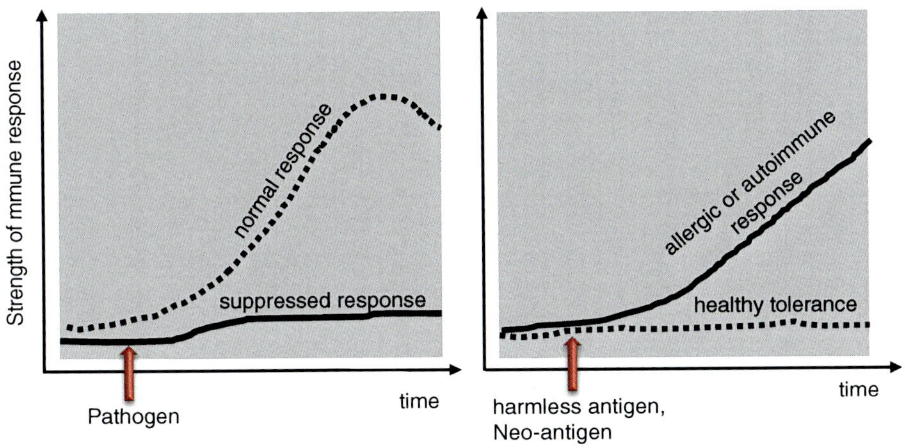

Fig. 1.2 Adverse immune reactions. Schematic representations of immune responses in an immuno-suppressive situation or an autoimmune/allergic situation. *Arrow*: exposure event; *dotted line*: normal/healthy response kinetics; *straight line*: adverse immune response; (**a**) immunosuppressive scenario, (**b**) allergic or autoimmune situation

1.2 Innate and Adaptive Immunity

The immune system is made up of various cell types (see Table 1.1), which have highly diverse functions and interact with each other at short or long distance. Communication is via cell–cell contact using receptor–ligand surface structures or over considerable distances via cytokines and chemokines. Lymph nodes situated throughout the body along the lymph vessels provide relevant spatial structures as meeting points for direct communication of immune cells.

There are two main arms of the immune system, the so-called innate immunity (which is evolutionary older) and the adaptive immune system (which evolved only in vertebrates). Macrophages, granulocytes, and dendritic cells (and many cell types; see below) belong to the former and T cells and B cells to the latter. Cells of the innate immune system mainly protect by phagocytosis (followed by intracellular destruction) of pathogens and by secreting cytokines, which generate and orchestrate an inflammatory tissue milieu unfavorable for pathogens. To detect pathogens, cells of the innate immune system have so-called pathogen recognition receptors for structures produced or found exclusively in bacteria, fungi, or viruses, such as lipopolysaccharides, flagellins, unmethylated CpG, or dsRNA. Upon recognition of such structures, innate immune cells immediately fight the infection by, e.g., phagocytosing the bacteria and oxidative burst. Secretion of various cytokines generates either a tolerogenic or inflammatory micro-milieu, adapted to the type of pathogen and the immunological situation. Moreover, some innate immune cells have the additional capacity to digest pathogen proteins and display it as small peptide pieces on special surface molecules (the major histocompatibility complex class I or class II, MHC-I or MHC-II). Specialists for this "antigen presentation" are the dendritic cells, but also macrophages and other innate immune cells, or even B cells can present antigen. Antigen presentation starts and directs the adaptive immune response by activating or suppressing T cells. T cells can recognize peptides on the MHC with their T-cell receptor. Naive T cells differentiate upon recognition of their cognate antigen and costimulatory signals provided by antigen-presenting cells (APC). APC activities range from ensuring immune tolerance against dietary antigens to the initiation of a potent immune response upon entering bacteria into skin wounds. Cytotoxic CD8+ T cells are capable of killing infected cells or cancer cells. T helper (Th) cells, on the other hand, orchestrate adaptive and innate immune responses by secretion of cytokines; for instance, they help B cells to differentiate and undergo immunoglobulin class switching and provide proinflammatory or immunosuppressive cytokines for other immune cells. Differentiation from naive CD4+ T cells into T helper (Th) 1, Th2, or Th17 cells is driven by combinations of cytokines in the micro-milieu, which are also provided by APC. Only a subset of B cells, the so-called CD5 B cells, produces antibodies without T-cell help.

The generation of the T-cell receptor (always cell surface bound) and the B-cell receptor (which can be surface bound or soluble and is then called antibody or immunoglobulin) requires a fascinating genetic process called gene rearrangement. Millions of different antigens exist, and the repertoire of B cells and T cells matches this high number. It is beyond the scope of this article to go into details, but the

Table 1.1 Cells of the immune system (excluding precursor and progenitor cells)

Cell name	Subsets	Function	Remark
Innate immune cells			
Macrophages	M1	Produce proinflammatory cytokines (e.g., TNF-α), promote Th1 response; strongly microbicidal	Monocytes in the circulation develop into macrophages in tissues under the influence of macrophage-stimulating factor
	M2	Produce immunosuppressive cytokines (e.g., IL-10) Promote tissue remodeling	
Granulocytes	Neutrophils Basophils Eosinophils Mast cells (see below)	Phagocytosis	Neutrophils constitute 40–70% of total white blood cells; first line of defense against infection. Granulocytes are also called polymorphonuclear cells (PMN) because of the lobed shape of their nuclei
Mast cells		Secretion of allergic mediators	Located at boundary of tissue with environment, cooperation with T cells due to their selective expression of cytokines upon stimulation
Dendritic cells	Lymphoid DC Myeloid DCs Langerhans cell Plasmacytoid DC CD103+ gut DC CD103– gut DC	Antigen recognition, transport into lymph nodes and presentation to T cells	Dendritic cells can be named and categorized in parallel for function (e.g., "tolerogenic" or "inflammatory," depending on how they act on T cells), for differentiation stage (as immature/mature), by motile behavior (migratory DC), or by lineage (lymphoid/myeloid). These subtypes overlap and can be very confusing at times. DCs are usually identified by various combinations of surface markers such as CD8, CD11b, CD11c, CD103, or DEC-205. It should be noted that the function of many CD molecules is not precisely known

(continued)

Table 1.1 (continued)

Cell name		Subsets	Function	Remark
Innate lymphoid cells (Spits et al. 2013) Derived from a common precursor Have a lymphoid morphology Classification is done on phenotype and function		Natural killer cells	Killing of virus-infected cells or cancer cells	Belong to group 1 ILC
		ILC1	IFN-γ secretion	Tissue development and remodeling, early immune response against microorganisms
		ILC2 (including nuocytes and natural helper cells)	IL-4, IL-5 secretion, amphiregulin	Resistance against nematodes, repair of lung tissue after viral infection, tissue development and remodeling, early immune response against microorganisms
		$CCR6^{low}$ILC3 (also called NK22 cells)	IL-17, IL-22 secretion immune response	Tissue development and remodeling, early immune response against microorganisms
		$CCR6^{+}$ ILC3, /lymphoid tissue inducer cells (LTi)		Formation of secondary lymphoid organs during embryogenesis
Adaptive immune cells				
αβ T cell receptor (TCR) T cells	CD4 helper T cells	Th1 Th2 Th17	Cytokine production IFN-γ (and others) IL-4 (and others) IL-17, IL-22 Differentiate after antigen contact	All T cells undergo processes of selection in the thymus, which only a few percent of incoming precursor cell survive Naive T cells refer to T cells which have not recognized antigen
	Regulatory T cells (Treg)	CD4+CD24+FoxP3+ natural regulatory T cells (nTreg)	Down-modulation of immune response	Leave thymus as Treg
		Inducible CD4+ regulatory T cells (iTreg): CD4+Foxp3+ Treg CD4+Tr1 Treg CD4+ Th3 cells CD8+ inducible Treg	Suppression and limitation of immune responses, by cell–cell contact or secretion of cytokines, in particular IL-10	Are induced from naive T cells in the periphery
	Killer cells	CD8+ killer T cells NK T cells	Cytotoxicity Cytotoxicity	

Table 1.1 (continued)

Cell name	Subsets	Function	Remark
γδ TCR	Invariant TCR (skin) Invariant TCR (gut) Invariant TCR (vaginal) Variant TCR	Epithelial integrity, other functions unclear	γδ T cells are considered an intermediate between adaptive and innate immune responses Invariant γδ T cells are formed exclusively in the fetal thymus
B cells	Conventional B cells	Humoral immune response	The terminal differentiation stage of B cells is plasma cells, which secrete as many as 10,000 antibody molecules per second (Hibi and Dosch 1986)
	CD5 B cells	T-independent antibody production against certain bacterial antigens	

process uses stochastic assembly of gene segments from the T cell or immunoglobulin locus to generate individually in each cell a new full gene, which codes either for the T-cell receptor or the B-cell receptor. As a result, every T cell and every B cell has its own unique receptor, which is specific for a single antigen. The process generates also T or B cells with specificity for self-proteins, but these autoimmune cells are deleted in a complex series of selection processes. In addition, regulatory T cells are made both in the thymus and during immune responses to keep in check possible autoreactive responses.

A third group of immune cells, termed "innate lymphoid cells" (ILCs), has been discovered recently. They do not have specific, rearranged receptors like T or B cells, but are derived from the same precursor cells in hematopoiesis (the common lymphoid progenitor cells). ILCs are either cytotoxic (NK cells) or helper cells; the latter ILC can be subdivided according to the transcription factors needed to generate them and, similar to T helper cells, according to their typical patterns of secreted cytokines. For instance, ILC1 produce large amounts of IFN-γ and are needed for fast response against intracellular parasites. ILC2 secrete IL-4 and IL-13 and have special roles in fighting worm infections and contribute to allergies. ILC3 produce IL-17 and IL-22, are needed to fight certain bacterial infections, and subsets have lymphoid tissue inducer functions.

All cells of the immune system are generated throughout life from the common hematopoietic stem cell. They have individual life spans ranging from a few days to many years. Immune cells follow their intrinsic programs and/or adapt to external cues, relayed into the cells by surface receptors coupled to signal transduction pathways. Thus, transcription factors are pivotal in shaping the immune response, as all immune cells pass at some point through the executive steps of up- or downregulation of genes. Major pathways in immune cells are G protein-coupled receptors, the MAP kinases, NF-κB, or the Janus kinase (JAK)–STAT pathways. Another is direct activation by ligand of latent transcription factors (such as glucocorticoid receptors). As pointed out below, these signaling pathways can become targets of immune interfering molecules, which may result in toxicity or exploited pharmacologically.

1.3 Cytokines as Players

As discussed above, immune cells produce and react to soluble mediators, which are essential in any immune response. These mediators include cytokines, chemokines, and mediators such as histamine or antibacterial peptides. They act over long ranges or locally and are fundamental in communication between immune cells and information exchange of tissues. Whereas chemokines attract and direct immune cells spatially, cytokines take center stage in orchestrating immune responses. Cytokines are small glycoproteins secreted by many cells, not exclusively immune cells (the old name lymphokines or interleukins (IL) suggested wrongly that they are specific for lymphocyte communication), and act via binding to cognate receptors, which leads to changes in the receiving cells. Cytokines are pleiotropic and redundant and act synergistically and anergistically. Expression of cytokine

receptors allows the integration of a variety of cytokine signals. Their role in immune responses cannot be overstated. Changes in cytokine efficacy, regulation, or production – by environmental factors or by medical interventions – affect the immune response and thus can have unwanted adverse or desired beneficial effect. Originally cytokines were identified and named for function; by now most of them are simply numbered (currently IL-1 to IL-38). Cytokines operate at every stage in inflammation, drive lymphocyte proliferation and differentiation, regulate growth and repair of epithelial cells, maintain memory, trigger immunoglobulin class switching, start the acute phase response, cause fever, cause migratory behavior, act as pro- or antiapoptotic, signal lipogenesis, are proangiogenic, and more.

1.4 Signaling

Cells communicate with the "outer world" via receptors, either cell surface bound or intracytoplasmic. Once a ligand (which can be growth factors, cytokines, hormones, or small molecular weight chemicals) has bound to its receptor, a series of intracellular events is triggered, which eventually lead to changes in gene expression and an activity of the cell adapted to the changed environment: production of enzymes (or growth factors, chemokines, cytokines, etc.), proliferation, differentiation, apoptosis, and more. There are several major signaling pathways, some specific for immune cells. The activity of signaling molecules can be restricted to cell–cell contacts (where one cell has the ligand on its surface, the other the receptor) and be limited to nearby cells (called paracrine signaling) or across the body (endocrine signaling). Characteristic for cell signaling are response thresholds, the amplification of the signal within the cell, and the possibility to integrate signals derived from different pathways synergistically or antagonistically in a tightly controlled network. Finally, the outcome of signaling is cell specific and may vary according to, e.g., the target gene(s) accessibility at a distinct cellular stage. Immune cells evolved to sense changes in the environment, especially with respect to the living environment; from the complexity of response and communication described in the paragraphs above, it is clear that immune signaling must be very sophisticated to do its job. Again, cellular signaling can be a target of immune interference. Table 1.2 lists major signaling pathways in immune cells.

1.5 Immunotoxicology and Environmental Immunology

The term adverse or unwanted immune response concerns immune responses which go wrong. These are (1) autoimmunity, where the antigens are own body proteins; (2) allergies, where the antigens are noninfectious (harmless) proteins of the environment such as pollen or food antigens; and (3) immunosuppression, leading to weak or absent immune response against infections (Fig. 1.2).

Adverse immune responses are mostly studied in pharmacology or immunotoxicology, both of which disciplines look at the role of chemicals (Table 1.3).

Table 1.2 Major signaling pathways in immune cells

Receptor	Receptor kinase	G protein-coupled receptors (>1000 members in this family)	Latent cytoplasmic	Immunoglobulin superfamily
Typical ligand (immune system relevant)	Cytokines	Chemokines	Small molecular weight chemicals (e.g., glucocorticoids, dioxins)	Peptide on MHC molecule (T cell) Any other molecule (B cell)
Effects	Phosphorylation of tyrosines on key signaling molecules, e.g., STATs	G protein activation, generation of second messengers (cAMP; di-acyl-glycerat (DAG); inositol-3 phosphate; cyclic GMP; nitric oxide)	Transformation of receptor, which becomes transcription factor; gene induction	Antigen sensing
Example	EGF, TGF, erythropoietin, cytokine; IL-2, IL-4, IL-6, IL-7, IL-7, IL-10; interferons	CCR5, CXC3R	Arylhydrocarbon (Ah) receptor, glucocorticoid receptor, estrogen receptor	T-cell receptor, B-cell receptor

Immunotoxicology became a field of interest and hard science along with growing concerns about environmental pollution, in the search for causes of cancer and allergies and the advent of better tools and basic concepts in immunology (Kerkvliet 2012). The chapters in this book look at the environment in a broad sense: not only man-made factors such as (toxic) chemicals but also endogenous factors and lifestyle facts: foods, stress, exercise, and how being young or old affects our immune competence (Fig. 1.3).

1.5.1 Immunosuppression

Chemicals and drugs can interfere with cellular physiological functions of all cells, including cells of the immune system. Immunosuppression by chemicals can be caused by (a) killing of cells of the immune system, (b) changes in cell differentiation leading to fewer or incapacitated immune cells, and (c) changes in typical cell

Table 1.3 Some known immunotoxic chemicals causing…

Immunosuppression	Allergy	Autoimmunity (disease)
2,3,7,8-tetrachlorodibenzo-*p*-dioxin and other polycyclic aromatic hydrocarbons	Urushiol (from poison ivy)	Gold salts (rheumatoid arthritis)
Mercury salts	Penicillin	D-Penicillamine (Pemphigus vulgaris)
Organotin compounds (antifungal ship paint)	Formaldehyde	Procainamide (systemic lupus)
Arsenic (salts and organic arsenic)	Nickel (jewelry, dental braces, implants)	Vinyl chloride (sclerodermia)
Asbestos	Mercury salts	D-Methyldopa (hemolytic anemia)
Various insecticides	Volatile organic compounds	
Benzene	p-Phenylenediamine (compound in many hair dyes)	
Cyclophosphamide (drug)		
FK506 (drug)		
Glucocorticoids (stress)		
UV irradiation[b]		

[a]It is estimated that there are more than 100,000 chemicals on the market (Fischetti M. Sci. Am. 303:92 (2019); Foth and Hayes, Hum. Exp. Toxicol. 27:5 (2008). Only for a minority, immunotoxic properties have been systematically tested. Novel legislation in the European Union, known as the Registration, Evaluation, Authorisation and Restriction of Chemicals (REACH), which entered into force on 1 June 2007, applies to all chemical substances produced at more than 1 ton/year and requires the provision of information for chemicals from companies

[b]UV irradiation is not a chemical, of course. However, the energy of UV radiation is high and can modify DNA bases. Moreover, UV radiation generates metabolites from intracellular compounds such as tryptophan, which can serve as signaling molecules and modulate Treg differentiation (Navid et al. 2013)

function such as cytokine secretion or expression of costimulatory molecules. Unfortunately, the exact mechanisms of interference by a given chemical are often unclear, and such lack of knowledge cripples treatment, prevention or optimizing drug usage in medicine. Immune interference may be caused by blockade or activation of intracellular enzymes, cell signaling molecules, transcription factors, or other proteins. The actions of dioxins, furans, or glucocorticoids are very good examples of this.

Immunosuppression is an operational term, which refers to an immune system operating with less efficiency than normal. Immune responses may start later, or they might be weaker. Immunosuppression might be general or restricted to certain pathogens. Whether caused intentionally by pharmacotherapy, unintentionally by environmental chemicals, or caused by aging processes, immunosuppressed individuals are more susceptible to infections and the development of spontaneous cancers. Immunosuppression can also cause inefficient vaccination. Epidemiologically, immunosuppression can be measured by comparing the average healthy population

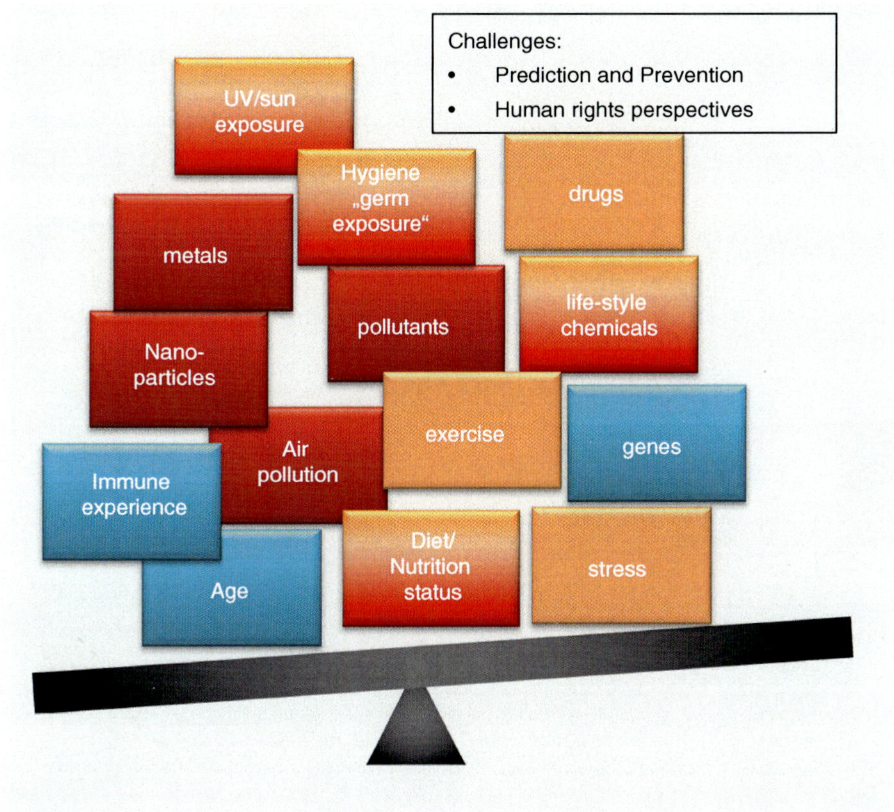

Fig. 1.3 Environmental influences. Factors influencing immune competence. Many factors contribute to the immune competence or induce adverse immune responses. Some set overall thresholds or capacities, such as the genetic predisposition, and cannot be influenced (*turquoise boxes*). Some factors can be influenced by oneself (*orange*), such as use of lifestyle chemicals, exercise, or a healthy diet. Other factors can be subject to changes in global health perspectives and political intervention (*red boxes*) such as extent of pollution or chemical exposure in consumer products, including food

with a particular subgroup, e.g., workers exposed to a toxic substance. However, no easily accessible and universally accepted markers for immunosuppression have been identified in vivo, and the considerable functional reserve of the immune system must be exceeded before immunosuppression becomes clinically relevant (Descotes 2005; Putman et al. 2003). Loss of immune cells (unless very high, like in AIDS) or a shift in proportion of cells in the blood or in lymphoid organs is of limited diagnostic value in humans. Standard immunotoxicity tests to detect immunosuppression or potentiation have been developed and validated (Luster et al. 1988, 1993). It is worth to keep in mind, though, that both acute clinical illness and small shifts in the susceptibility to normal infections can be of economic relevance on the population scale.

1.5.2 Autoimmunity and Allergy: Immunotoxicity Caused by Responses against Autoantigens or Harmless Antigens

Tolerance is the immunological term for the fact that the immune system does not react to its own body proteins, lipids, DNA, or other molecules, although they are potential antigens. One type of adverse immune reaction, often found with drugs, is a loss of tolerance, i.e., failure of the immune system to keep up the distinction between self and non-self. Tolerance can be lost in several ways. One way could be an unspecific impairment or loss of the cells important for tolerance, such as regulatory T cells. Another way is the drug-induced generation of self-peptides which are normally not made by standard protein degradation in the body or are not exposed to the immune system. To such self-antigens, the immune system has never acquired tolerance. T cells respond to peptide antigens presented to their T-cell receptors by MHC molecules on other cells. Importantly, also the body's own cellular proteins are continuously presented on MHC by all body cells. This is a way of tissue cells to signal "I am healthy" to the immune system. During protein catabolism in the cells, a typical range of peptides are generated and become presented on the surface. T cells do not react toward such normal peptides derived from its own body proteins, either because T cells with the respective specificity were eliminated in the thymus (termed "central tolerance"), or because they were silenced by regulatory T cells, or because they became unresponsive (termed "anergic") after antigen contact in the periphery (termed "peripheral tolerance"). During infection or in the case of cancer, the normal pattern of presented peptides changes, and different or additional peptides appear on the cell surface, tagging the cell as dangerous and in trouble. Notably, also chemicals can change the normal range of presented peptides of a cell, which appear as neoantigens. In other words, chemicals can render healthy cells recognizable by T cells. Chemicals might do this in several ways. First, they can covalently bind to self-peptides on MHC molecules (as "haptens") and be presented along with this peptide piggyback-like. In this case, a cell looks "foreign" or "infected," and there will be T cells in the body which consequently might attack it. In another scenario, some chemicals might interfere with the normal antigen processing and presentation of body proteins, leading to the presentation of normally cryptic self-antigens, against which no central T-cell tolerance exists. Examples for this mechanism are drugs containing gold salts (Griem et al. 1996). A T-cell reaction ensues, which, however, can calm down once the chemical or drug is removed (although memory cells persist). It is not always easy to predict whether a chemical will bind to cellular proteins and cause the formation of haptenated neoantigens or expose cryptic antigens. One indicator or risk factor is the presence of reactive groups in the chemical which can lead to the formation of protein adducts by chemicals. What type of an adverse immune reaction results from chemical exposure, and whether a reaction develops at all, depends on the chemical itself but also on auxiliary circumstances, the genetic predisposition, and exposure regimen and site. The outcome of chemical-induced antigen distortion can be autoimmunity or allergies. Predicting the immunotoxic potential of a given chemical is therefore still a great challenge. Allergies are overacting immune responses against harmless antigens such as pollen or food proteins or chemicals. In the latter case, the binding to self-proteins as described above can be relevant. Penicillin allergy is a famous example for this. Allergic responses can be based on IgE secretion and mast cell degranulation or involve dominantly T cells which secrete proinflammatory cytokines leading to tissue destruction. Whether or not an allergy will develop is hard to predict; genetic disposition plays an important role and other auxiliary circumstances as well. Clinical manifestation often occurs only after several contacts with the antigen ("sensitization" phase).

1.6 Summary and Conclusion

The immune system is vital for well-being, health, and survival. Increasingly it is understood that environmental factors can become a risk or a benefit for the immune system, impairing or boosting its function. It is important to understand these environmental influences on a public health scale on the one hand and their underlying molecular mechanisms on the other hand. Together the knowledge will enable individuals, the science and medical community, and governments to take the necessary steps to discover, develop, and implement therapeutic and preventive opportunities and ensure a healthier environment for all.

References

Abbas A, Lichtmann AHH, Pillai S (2015) Cellular and molecular immunology. Elsevier, Philadelphia

Descotes J (2005) Immunotoxicology: role in the safety assessment of drugs. Drug Saf 28:127–136

Griem P, Panthel K, Kalbacher H, Gleichmann E (1996) Alteration of a model antigen by Au(III) leads to T cell sensitization to cryptic peptides. Eur J Immunol 26:279–287

Hibi T, Dosch HM (1986) Limiting dilution analysis of the B cell compartment in human bone marrow. Eur J Immunol 16:139–145

Hirano M, Das S, Guo P, Cooper MD (2011) The evolution of adaptive immunity in vertebrates. Adv Immunol 109:125–157

Janeway CA, Medzhitov R (2002) Innate immune recognition. Annu Rev Immunol 20:197–216

Kerkvliet NI (2012) TCDD: an environmental immunotoxicant reveals a novel pathway of immunoregulation – a 30-year odyssey. Toxicol Pathol 40:138–142

Luster MI, Munson AE, Thomas PT, Holsapple MP, Fenters JD, White KL Jr, Lauer LD, Germolec DR, Rosenthal GJ, Dean JH (1988) Development of a testing battery to assess chemical-induced immunotoxicity: national toxicology program's guidelines for immunotoxicity evaluation in mice. Fundam Appl Toxicol 10:2–19

Luster MI, Portier C, Pait DG, Rosenthal GJ, Germolec DR, Corsini E, Blaylock BL, Pollock P, Kouchi Y, Craig W (1993) Risk assessment in immunotoxicology. II. Relationships between immune and host resistance tests. Fundam Appl Toxicol 21:71–82

Navid F, Bruhs A, Schuller W, Fritsche E, Krutmann J, Schwarz T, Schwarz A (2013) The aryl hydrocarbon receptor is involved in UVR-induced immunosuppression. J Invest Dermatol 133:2763–2770

Paul W (2013) Fundamental immunology, 7th edn. Lippincott, Williams & Wilkins, Philadelphia

Putman E, van der Laan JW, van Loveren H (2003) Assessing immunotoxicity: guidelines. Fundam Clin Pharmacol 17:615–626

Spits H, Artis D, Colonna M, Diefenbach A, Di Santo JP, Eberl G, Koyasu S, Locksley RM, McKenzie AN, Mebius RE, Powrie F, Vivier E (2013) Innate lymphoid cells – a proposal for uniform nomenclature. Nat Rev Immunol 13:145–149

Zola H, Swart B, Banham A, Barry S, Beare A, Bensussan A, Boumsell L, Buckley D, Buhring HJ, Clark G, Engel P, Fox D, Jin BQ, Macardle PJ, Malavasi F, Mason D, Stockinger H, Yang X (2007) CD molecules 2006 – human cell differentiation molecules. J Immunol Methods 319:1–5

Part I

Age and Lifestyle

Influence of Early-Life Environmental Exposures on Immune Function Across the Life Span

2

Lisbeth A. Boule and B. Paige Lawrence

Contents

2.1 Receptor Binding Chemicals

Cell surface and intracellular receptors provide targets by which exogenous chemicals alter the function of the immune system. Indeed, this principle underlies the mechanism of action of numerous pharmaceutical agents and is exploited to design new drugs. However, these receptors also bind pollutants to which we are regularly exposed. Yet, how these exposures, particularly when they occur during

L.A. Boule • B.P. Lawrence, PhD (✉)
Department of Environmental Medicine, University of Rochester School of Medicine & Dentistry, 575 Elmwood Avenue, Rochester, NY 14642, USA
e-mail: Paige_Lawrence@urmc.rochester.edu

C. Esser (ed.), *Environmental Influences on the Immune System*,
DOI 10.1007/978-3-7091-1890-0_2

development, lead to changes in the function of the immune system later in life is, for the most part, unknown. Nevertheless, evidence derived from human cohorts and animal models provides a compelling database that supports the idea that early-life exposures to several common environmental agents that bind specific cellular receptors have an impact on the development and function of the immune system (Table 2.1). In the following sections, we draw from this database and provide several examples. Relevant literature for receptor-binding chemicals that are not discussed below can be found in Table 2.1.

2.1.1 Aryl Hydrocarbon Receptor

One receptor that has been implicated in causing persistent changes in the function of the immune system after developmental activation is the aryl hydrocarbon receptor (AHR). Numerous environmentally derived chemicals bind this receptor, including dioxins, some polychlorinated biphenyls (PCBs), and polycyclic aromatic hydrocarbons. In addition, naturally derived molecules such as certain tryptophan metabolites, indoles and bioflavonoids, bind to the AHR (Nguyen and Bradfield 2008). Many studies have examined the consequences of activating the AHR on the immune system of adult animals. AHR activation alters multiple aspects of the immune system and subsequent disease outcome, with a range of outcomes observed depending upon the AHR ligand used (Lawrence and Vorderstrasse 2013).

While the metabolism, distribution, and excretion of all the ligands for the AHR have not been comprehensively studied, humans are regularly exposed to environmental contaminants that bind the AHR, primarily via our diet (Institute of Medicine 2003). Dioxins and PCBs cross the placenta and are excreted in breast milk (Gasiewicz et al. 1983). In fact, it is estimated that infants are exposed to considerably higher levels of these AHR ligands than adults due to bioaccumulation (Domingo and Bocio 2007). This information raises the question of whether activating the AHR inappropriately during development leads to persistent changes in the immune system. Summarized below are several studies that indicate that the answer to this question is yes: developmental exposure to anthropogenic AHR ligands elicits long-term changes in the function of the immune system.

The first studies that examined this were published several decades ago and showed that AHR activation during development altered function of the offspring's immune system (Faith and Moore 1977; Vos and Moore 1974; Thomas and Hinsdill 1979; Luster et al. 1980). Specifically, these studies showed that offspring of dams treated with the prototypical ligand 2,3,7,8-tetrachlorodibenzo-*p*-dioxin (TCDD) exhibited impairment in hallmark immune responses later in life, including the antibody response to sheep red blood cells (SRBC) and delayed-type hypersensitivity (DTH) responses (Vos and Moore 1974; Faith and Moore 1977; Thomas and Hinsdill 1979; Walker et al. 2004; Gehrs et al. 1997; Gehrs and Smialowicz 1997, 1999). Other groups have studied this phenomenon, showing that AHR activation during development fundamentally changes disease processes later in life, including increased susceptibility to bacterial and tumor challenge and decreased antiviral

Table 2.1 Reported immune consequences of developmental exposures to agents that bind cellular receptors

Chemical	Receptor binding	Animal models	Human studies
Polycyclic aromatic hydrocarbons	AHR	Increased cancer incidence; reduced resistance to tumor challenge (Rodriguez et al. 1999; Urso and Gengozian 1982, 1984; Urso and Johnson 1987) Reduction in thymocyte and peripheral T-cell populations (Rodriguez et al. 1999) DNA adducts in thymocytes (Rodriguez et al. 2002)	Altered fetal IgE levels (Herr et al. 2011) Alerted cord blood leukocyte proportions (Herr et al. 2010) Increased incidence of infection (Jedrychowski et al. 2005) Altered gene expression in immune pathways in cord blood cells (Sram et al. 2013)
Organochlorines (including dioxins and PCBs)	AHR	Increased PMN[1] phagocytosis (Bilrha et al. 2004) Decreased antibody response to bacterial challenge (Bilrha et al. 2004) Decreased adaptive immune response to IAV infection (Vorderstrasse et al. 2004, 2006; Hogaboam et al. 2008) Increased susceptibility to bacterial infection and tumor challenge (Luster et al. 1980; Sugita-Konishi et al. 2003; Thomas and Hinsdill 1979) Decreased DTH and CHS responses (Gehrs et al. 1997; Gehrs and Smialowicz 1997, 1999; Walker et al. 2004; Thomas and Hinsdill 1979; Faith and Moore 1977) Reduced skin graft rejection (Vos and Moore 1974) Decreased T-dependent antibody response (Thomas and Hinsdill 1979; Faith and Moore 1977) Increased autoantibody levels and immune complex deposition (Mustafa et al. 2008, 2011a, b)	Increased incidence of respiratory infections (Sunyer et al. 2010; Jedrychowski et al. 2005; Dewailly et al. 2000; Kimbrough and Krouskas 2001; Guo et al. 2004; Dallaire et al. 2004; Dallaire et al. 2006; Miyashita et al. 2011; Jedrychowski et al. 2005; Glynn et al. 2008; Stolevik et al. 2013) Decreased antibody response to childhood vaccinations (Heilmann et al. 2006, 2010; Weisglas-Kuperus et al. 2000, 2004; Stolevik et al. 2013; Hochstenbach et al. 2012) Decreased thymus size (Park et al. 2008)
Atrazine	Androgen receptor	Sex-dependent changes in MLR and antibody levels (Rowe et al. 2008) Suppressed DTH response (Rooney et al. 2003)	
Nonylphenol	Androgen receptor	Decreased T-cell number (Karrow et al. 2004) Increased NK cell killing (Karrow et al. 2004)	

(continued)

Table 2.1 (continued)

Chemical	Receptor binding	Animal models	Human studies
BPA	Estrogen receptor	Increased T-dependent antibody responses and cytokine production (Yoshino et al. 2004) Altered oral tolerance development (Ohshima et al. 2007) Increased effector CD4$^+$ T cells and reduced T regulatory cells after infection (Yan et al. 2008) Faster onset of type 1 diabetes (Bodin et al. 2014) Only minor changes in allergic airway disease symptoms (Bauer et al. 2012) Increased allergen sensitization, but not pulmonary inflammation, in allergic model (O'Brien et al. 2014) Increased bronchopulmonary inflammation and airway hyperresponsiveness in allergic model (Midoro-Horiuti et al. 2010; Nakajima et al. 2012) Decreased innate responses, but no change in adaptive immune response, to IAV infection (Roy et al. 2012) Only minor changes in intestinal inflammation in mouse model of IBD (Roy et al. 2013)	Increased rate of asthma and childhood wheeze incidence (Donohue et al. 2013; Spanier et al. 2012; Vaidya and Kulkarni 2012)
DES	Estrogen receptor	Decreased T-dependent antibody responses (Luster et al. 1978, 1979; Kalland 1980b) Reduced NK cell activity (Kalland 1980d) Reduced antibody responses in female offspring (Kalland 1980a) Decreased lymphocyte proliferation (Kalland 1980c)	Increased incidence of autoimmune disease in women (Noller et al. 1988) Increased cancer incidence (Herbst et al. 1971)
Organophosphates	GABA and acetylcholinic receptors	Decreased cytokine production and worsened lung immunopathology after RSV infection (Watanabe et al. 2013) Reduced antibody response to SRBC (Singh et al. 2013a) Reduced cytokine and proliferation responses from lymphocytes (Singh et al. 2013a)	

Toluene	GABA and acetylcholinic receptors	Dose-dependent altered antibody production (Yamamoto et al. 2009a, b)	
Corticosteroids	Glucocorticoid receptor	Decreased thymic output and T-cell migration (Bakker et al. 1995) Decreased DTH reaction (Dietert et al. 2003) Increased antibody responses (Dietert et al. 2003)	
Organotins	PPARs	Reduced proliferation of T cells to mitogen (Smialowicz et al. 1988) Decreased antibody responses (Tonk et al. 2011a, b) Decreased MLR (Smialowicz et al. 1989) Decreased lymphocyte proliferation (Smialowicz et al. 1989) Decreased NK cytotoxic activity (Smialowicz et al. 1989) Decreased host resistance to infection (Vos et al. 1990) No change in DTH response or NK cell killing ability (DeWitt et al. 2007)	
Perfluorinated compounds	PPARs	Reduction in NK cytotoxicity (Keil et al. 2008) No change in T-cell number or cytokine production (Hu et al. 2012) Reduced antibody responses (Peden-Adams et al. 2009; Keil et al. 2008; Hu et al. 2012)	Reduced antibody responses to childhood vaccinations (Grandjean et al. 2012) Positive correlation between PFC exposure and IgE levels (Wang et al. 2011)

Abbreviations: *CHS* contact hypersensitivity, *DTH* delayed-type hypersensitivity, *IAV* influenza A virus, *IBD* inflammatory bowel disease, *MLR* mixed lymphocyte reaction, *NK* natural killer, *PMN* polymorphonuclear cell, *RSV* respiratory syncytial virus, *SRBC* sheep red blood cells immunity (Luster et al. 1980; Sugita-Konishi et al. 2003; Vorderstrasse et al. 2004, 2006; Hogaboam et al. 2008). In addition, AHR activation during development increases autoimmune-like symptoms in wild type and autoimmune-prone mouse strains (Mustafa et al. 2008, 2011a, b; Holladay et al. 2011). Although AHR represents one of the more heavily studied receptors when it comes to examining relationships between early-life exposures and changes in immune function later in life, the mechanism by which these persistent changes arise remains unknown. Recent studies in other organ systems suggest that one of consequences of AHR activation during development is alteration of epigenetic profiles, but how the AHR is mediating these changes has yet to be elucidated (Manikkam et al. 2012; Papoutsis et al. 2013). Nevertheless, these reports suggest that in addition to acting as a transcription factor, early-life AHR signaling may influence immune responses later in life via a mechanism that involves modulation of epigenetic regulatory pathways.

2.1.2 Peroxisome Proliferator-Activated Receptors

Peroxisome proliferator-activated receptors (PPARs) are transcription factors whose normal physiological function includes metabolic regulation. PPARs bind fatty acids and lipid-derived moieties, leading to transcription of genes that regulate processes such as adipogenesis and inflammation (Ferre 2004). PPARs also bind some anthropogenic chemicals, including di-n-octyltin dichloride (DOTC), perfluorinated compounds (PFCs), and tributyltin oxide (TBTO). DOTC is used as a stabilizer in polyvinyl chloride plastics, and the main route of human exposure is through drinking water (World Health Organization 2004). PFCs are used as stain and water repellents, food wrappings, fire-fighting foams, and in numerous other household goods (Giesy and Kannan 2002). Humans are exposed to PFCs through the diet and through contamination of groundwater, as these compounds leach out of commercial products and bioaccumulate (Giesy and Kannan 2002). TBTO is used as a marine biocide, and humans are exposed mainly through the diet, although some studies suggest that TBTO exposure occurs through non-dietary sources, such as via rubber gloves and baking sheets (World Health Organization 1999). Importantly, evidence indicates these compounds are transferred to developing fetus and neonate through the placenta and during lactation, leading to early-life exposure within the human population (World Health Organization 1999, 2004).

Several studies have examined the effect of activating PPARs by exogenous chemicals during development and have shown that the immune system is altered later in life. One of the more striking among these is an epidemiological study showed that early-life PFC levels correlated with reduced antibody responses to routine childhood vaccinations (Grandjean et al. 2012). Pre- and early postnatal PFC exposure levels also positively correlated with atopy in a separate birth cohort study (Wang et al. 2011). Animal studies show dose-dependent effects of perfluorinated compound exposure during development on antibody responses later in life (Keil et al. 2008; Hu et al. 2012; Peden-Adams et al. 2009). While point sources of PFCs are decreasing in the some countries, such as the United States, PFC exposure sources and levels are increasing in other countries, such as China and Southeast

Asia (Webster 2010). Given these results of studies to date, and likelihood for continued global exposure, more studies of the developmental immunotoxicity of PFCs are clearly needed to better characterize their impact on immune function later in life (Corsini et al. 2014).

Other PPAR ligands, such as the organotins TBTO and DOTC, show conflicting changes in immune responses after developmental exposure. Offspring of dams treated with TBTO have persistent decreases in ex vivo assays, such as mixed lymphocyte reactions and mitogen-induced proliferation (Smialowicz et al. 1989; Vos et al. 1990). On the other hand, developmental exposure to DOTC increased DTH responses in male offspring (Smialowicz et al. 1989). Differences in the doses used, route of exposure, and ligand-specific downstream consequences of PPAR activation are possible explanations for these disparate results. In addition to directly affecting immune cells, organotins may influence the function of the immune system indirectly, by altering the developmental programming of other physiological systems that interact with the immune system. For example, developmental exposure to TBTO skews the differentiation pattern of mesenchymal stem cells (Kirchner et al. 2010), which may have consequences for immune function later in life.

In summary, given that PPAR activation by a broad spectrum of ligands exerts a variety of effects on the immune and other organ systems, comprehensive and systematic studies that examine the long-term consequences of exposure to specific PPAR ligands are necessary. Studies should include a wide range of environmentally relevant doses, examine sex-specific differences, and include multiple aspects of immune function. These approaches should be integrated with studies of human cohorts, as this assimilation will be critical for determining the likelihood that developmental exposure will modulate the immune system and influence human health and disease, and for pinpointing the mechanisms by which they do so.

2.1.3 Hormone Receptors

Chemicals that bind to hormone receptors or modulate cellular signaling mediated by hormone receptors are referred to as endocrine-disrupting chemicals (EDCs). EDCs can be hormone mimetics, act as receptor antagonists, or alter hormone-mediated signaling. Sometimes whether they act as receptor agonists or antagonists is dose-dependent (Vandenberg et al. 2012). Numerous chemicals have been defined as EDCs and include PCBs, bisphenol A (BPA), and various phytoestrogens (National Institute of Environmental Health Sciences 2010). Many EDCs are abundant in our environment and are detected readily in human urine, blood, and other tissues. Moreover, pregnant and lactating women are exposed to EDCs, and they are found in cord blood, placenta, and breast milk (Bonfanti et al. 2009; Domingo and Bocio 2007; Calafat et al. 2008; Bushnik et al. 2010; Kurzer and Xu 1997). Examples of receptors bound by EDCs include the estrogen, androgen, thyroid hormone, and glucocorticoid receptors. To date, research on the pathophysiological effects of most EDCs has focused on the reproductive and nervous system, as well as regulation of central metabolism (Newbold et al. 2007; Gore 2008). However, given the appreciation that the endocrine and immune systems influence each other, attention is slowly turning toward the effects of EDCs on the developing immune system. To date, the majority of these studies have examined the effect of estrogenic EDCs on immune function. This makes it clear that there are many opportunities to investigate the effects of EDCs that act via other hormone-regulated pathways.

Two of the compounds that bind the estrogen receptor, and for which there is some evidence that developmental exposure impacts the function of the immune system, are bisphenol A (BPA) and diethylstilbestrol (DES) (Chapin et al. 2008). BPA is used in plastics, is transferred through the placenta, and is secreted in breast milk (Calafat et al. 2008; Bushnik et al. 2010). Additionally, compared to adults, infants have the highest measured levels of BPA to date (Geens et al. 2012). Therefore, BPA is a candidate developmental immunotoxicant. Correlations between early-life BPA exposure and childhood asthma and the odds of developing wheeze in early life have been reported (Donohue et al. 2013; Spanier et al. 2012; Vaidya and Kulkarni 2012). Depending on the study, in rodent models prenatal or perinatal exposure to BPA has enhanced or repressed immune responses (Yoshino et al. 2004; Ohshima et al. 2007; Yan et al. 2008). In mouse models of allergic asthma, developmental BPA has had mixed effects on immune parameters and overall metrics of pulmonary inflammation (Midoro-Horiuti et al. 2010; Bauer et al. 2012; O'Brien et al. 2014; Nakajima et al. 2012). In other studies, maternal exposure to BPA enhanced aspects of pulmonary innate immune responses to infection with influenza A virus, but had no measurable effect on adaptive immune parameters or viral clearance (Roy et al. 2012). While findings from these studies present some inconsistencies due to species differences, route of exposure, dosing paradigms, and immunological model systems, they provide evidence that developmental exposure to BPA modulates the function of the immune system later in life.

In contrast to unintended exposures to BPA, the synthetic nonsteroidal estrogen DES was administered deliberately for therapeutic purposes. Unfortunately, under the misguided impression that it would reduce preterm birth and alleviate pregnancy complications, DES was given to pregnant women for over 30 years. The developmental toxicity of DES was first revealed in daughters, who developed a high incidence of what had previously been a rare vaginal cancer. Further research revealed that maternal treatment with DES alters the development of the both male and female reproductive systems (Newbold et al. 2006; Couse et al. 2001; Prins et al. 2001). Moreover, DES exposure has persistent immunomodulatory effects (Luster et al. 1978, 1979; Kalland 1980a, b, c, d). Given these consequences, the administration of DES to pregnant women and its presence in animal feed were eliminated; however, much continues to be learned about the effects of developmental exposure to DES. For example, it is unknown whether the changes in the immune system after developmental DES exposure are direct effects of estrogen receptor activation in immune cells or arise indirectly due to alterations in other organ systems. Additionally, because hormones, especially estrogens, have been implicated as one reason for why women have a higher incidence of autoimmune diseases than men, further studies examining developmental exposure to the extensive cadre of estrogenic compounds to which we are exposed on the incidence and progression of autoimmune diseases are important.

It is important to bear in mind that estrogenic compounds are not the only type of EDCs. For example, the herbicide atrazine binds to the androgen receptor (US Environmental Protection Agency 2007). Atrazine is a drinking and groundwater contaminant that has gained increasing interest in environmental and toxicological studies (US Environmental Protection Agency 2007). Although banned in some parts of the world, atrazine levels remain high in some areas of the United States, where its use as an herbicide is still common (Breckenridge et al. 2010). While only a few studies have examined whether developmental exposure to atrazine affects the immune system, the evidence from these studies is compelling. Consistent with what one would predict for an EDC, maternal exposure to atrazine causes sex-specific differences in the immune system of the offspring. Depending on the study, female offspring were either unaffected by developmental atrazine exposure or their immune response was suppressed (Rowe et al. 2008). On the other hand, male offspring exhibited enhancement of immune responses, such as cytokine and antibody production after antigen exposure (Rowe et al. 2008). While still showing sex-specific effects, a different study showed decreased immune responses in male mice that were developmentally exposed to atrazine, including suppression of DTH (Rooney et al. 2003). Despite the contradictory direction of changes, these data suggest that developmental exposure to atrazine appears to influence the function of the immune system.

There are numerous other receptors to which chemicals from the environment can bind, acting as either agonists or antagonists. Moreover, there are other chemicals, not reviewed herein, for which there is some evidence that they interact with receptors. The advantage of studying receptor-binding chemicals is that the research questions regarding immune modulation build on what is already known about the pathways regulated by these receptors, and there are numerous pharmacological, biochemical, and genetic tools available to study many receptors and receptor families. Future studies should take advantage of these tools and reagents to determine the mechanism by which unintended receptor activation during development leads to permanent changes in immune function. Furthermore, it is worth also considering that activating a particular receptor during development may lead to events that are not part of its canonical signaling pathway, but cause durable changes in immune function that are not revealed until later in life.

2.2 Smoke

Smoke is one of the most ubiquitous forms of environmental exposure. Point sources for smoke include biomass fuels burned for cooking and heating, commercial and backyard waste incineration, as well as direct and secondhand exposure to smoke from tobacco products. Exposure to various types of smoke during pregnancy is associated with health complications for the developing fetus and neonate (Knopik et al. 2012; DiFranza et al. 2004). Understanding how exposures to these assorted types of smoke, which are complex chemical mixtures, influence immune function of the offspring has profound implications for improving public health. In the sections below, it will be evident that cigarette smoke exposures have been more extensively studied than biomass fuels; yet, the proportion of the population, especially of women and children, exposed to biomass smoke is much larger (Martin et al. 2011). While there are likely many differences in composition and downstream pathophysiological consequences of exposure, cigarette and biomass fuel smoke exposure lead to a similar enhancement in inflammatory responses in human pulmonary cell cultures and in mouse models (Mehra et al. 2012). This suggests that some of the findings generated studying how maternal smoking impacts the development and function of the immune system may inform thinking about how biomass fuel smoke exposure during development may lead to persistent changes in the immune system later in life.

2.2.1 Cigarette Smoke

It has long been known that cigarette smoke is unhealthy, yet worldwide smoking rates continue to be upwards of 15 % of the global population (World Health Organization 2014). There continue to be reports of pregnant women actively smoking, and many pregnant women are exposed to secondhand smoke. Tobacco smoke exposure during development has many health complications for children later in life (some of which implicate altered immune function). These studies have been the focus of many comprehensive reviews (Doherty et al. 2009; Hofhuis et al. 2003; Cupul-Uicab et al. 2012). A recent meta-analysis of epidemiological data shows that prenatal passive smoke exposure increases the incidence of asthma at least 20 % and in some studies upwards of 80 % (Burke et al. 2012). Studies in animal models and with other human cohorts support this analysis, demonstrating that either direct mainstream smoke or passive secondhand smoke exposure during pregnancy positively correlates with increased asthma symptoms and incidence (Wu et al. 2009; Herr et al. 2011; Jaakkola et al. 2006; DiFranza et al. 2004; Cheraghi and Salvi 2009; Selgrade et al. 2013; Barber et al. 1996). Other epidemiological studies associate exposure to cigarette smoke during development with higher incidence of respiratory infection and other pulmonary complications (Jaakkola et al. 2006; DiFranza et al. 2004; Bradley et al. 2005; Cheraghi and Salvi 2009; Jedrychowski et al. 2005; Hylkema and Blacquiere 2009). Animal studies corroborate and extend these observations, showing altered responses to immune challenges later in life after developmental exposure to cigarette smoke (Ng and Zelikoff 2007; Ng et al. 2006; Ng and Zelikoff 2008).

Developmental exposure to cigarette smoke also affects lung structure (Hylkema and Blacquiere 2009; Singh et al. 2013b; Ji et al. 1998; Manoli et al. 2012), which can lead to further complications after secondary respiratory insults later in life, such as additional environmental exposures (e.g., ozone) (Han et al. 2011). It has been posited that altered retinoic acid signaling during lung development may be responsible for some of the structural changes seen in the lung after cigarette smoke during development (Manoli et al. 2012). Other cellular mediators that contribute to exacerbated pulmonary responses to allergens and infections also appear to result from early-life exposure to cigarette smoke (in both animal and human studies), including reduced T-cell responses to stimulation (Singh et al. 2006; Ng et al. 2006; Tebow et al. 2008), altered cytokine responses to TLR ligands (Noakes et al. 2006), and a predisposition of immune cells in offspring to Th2-type responses (Noakes et al. 2003; Devereux et al. 2002; Penn et al. 2007). Thus, while it remains to be determined which of these observed effects are due to changes in immune cells versus effects of early-life exposure on non-hematopoietic cells, these studies collectively support that cigarette smoke exposure during development affects the function of the immune system later in life.

The mechanisms by which cigarette smoke causes these effects have yet to be fully elucidated. However, recent studies point to epigenetic changes as one mechanism that may drive some of these observed changes in immune function after perinatal cigarette smoke exposure. Cord blood mononuclear cells and placental cells from smoking mothers have altered patterns of DNA methylation (Maccani et al. 2013; Murphy et al. 2012; Martino and Prescott 2011). Another study showed increased expression of a specific microRNA associated with immune function positively correlated with increased cigarette exposure during to development (Herberth et al. 2014). As we learn more about how cigarette smoke exposures during development alter the immune system, strategies to reverse or treat these consequences must be evaluated. Two recent studies, one in mice and one in humans, suggest potential therapeutic targets or interventions that may be successful (Singh et al. 2013b; McEvoy et al. 2014). These are intriguing and encouraging reports and emphasize the need for future efforts to delineate the cellular and molecular mechanisms by which early-life exposure to smoke leads to alterations in immune function later in life.

2.3 Biomass Fuel Smoke

Biomass fuels are derived of biological materials (animal and vegetable) and are used for cooking, heating, and light in many areas of the world (Torres-Duque et al. 2008). The most common form of biomass fuel is wood, although dung, crop residues, corncobs, and grass are also used (Torres-Duque et al. 2008; Babalik et al. 2013). Biomass fuel smoke, as with cigarette smoke, is composed of a mixture of chemicals, and the actual composition varies depending upon the type of biomass fuel used (Torres-Duque et al. 2008). While only a few studies have examined exposure to biomass fuel smoke during early childhood and risk for disease, the evidence is strong and consistent. Early-life exposure to biomass fuel leads to higher risks of acute lower respiratory tract infections and altered rates of asthma (Torres-Duque et al. 2008; Diette et al. 2012). Moreover, there is growing evidence that exposure to biomass fuel smoke across the life span increases disease incidence, as women exposed to biomass fuel smoke have increased risk for respiratory infections, asthma, and lung cancer (Trevor et al. 2014; Guarnieri et al. 2014). Further support of the idea that biomass fuels are likely to affect the developing immune system, and lead to poorer health later in life, can be gleaned from studies examining developmental exposures to chemicals that are constituents of biomass fuel smoke, including AHR ligands, oxidants, and particulate matter. Given the correlation between these exposures and disease incidence in adults, and that a large portion of the global population us exposed and the compelling evidence from the studies examining early-life exposures, biomass fuel smoke exposure most likely impacts the development and function of the immune system. Further studies examining the specific impact of biomass fuel smoke exposure during development on immunological outcomes later in life are necessary to determine whether proposed intervention strategies, such as changing cooking methods or using different heat sources, are likely to improve long-term health.

2.4 Heavy Metals

Exposure to heavy metals is often considered relevant only in the context of heavily industrialized regions. However, heavy metals are ubiquitous in the environment due to a combination of anthropogenic and natural sources (Bjermo et al. 2013; Mahaffey et al. 1975; Jarup 2003). In addition to soils, heavy metals are found in ground and drinking water, as well as in commonly consumed foods, such as rice and other grains (Bjermo et al. 2013). As such, human exposure to heavy metals is a common global health issue. Furthermore, some of the most widely studied developmental exposures that have documented effects the immune system are heavy metals, such as lead, mercury, arsenic, and cadmium (Bjermo et al. 2013). The immunological consequences of early-life exposure to lead have been comprehensively reviewed by others and will therefore not be presented here (Luebke et al. 2006; Dietert et al. 2004).

2.4.1 Mercury

Mercury is found in three distinct forms: elemental, inorganic, and methylmercury. Exposure to mercury, particularly methylmercury, occurs mainly through diets rich in fish (Bjermo et al. 2013). Exposure to inorganic mercury occurs through dental amalgams, fungicides, skin lightening creams, paints, and some tattoo dyes (US Department of Health and Human Services 1999). It has been reported that 5 % of women of childbearing age have blood mercury levels that are higher than the oral reference dose (Schober et al. 2003). Globally, pregnant women have detectable, blood mercury levels, although many have levels that are lower than the oral reference dose (Davidson et al. 2006; Stern et al. 2001; Razzaghi et al. 2014; Bjermo et al. 2013; Al-Saleh et al. 2011; Sanders et al. 2012). It has not been established whether lower doses of mercury have immunotoxic effects. However, studies examining adults exposed to the three forms of mercury show an association with higher incidence of infection and an increase in autoimmune nephritis symptoms (Holmes et al. 2009). Developmental mercury exposure has long been associated with neurological defects in children, and some studies now associate alterations in the immune system as a partial cause for these defects (Zhang et al. 2011). Yet, few studies of developmental exposures to mercury have probed whether it leads to persistent changes in immune function. Some studies support a correlation between mercury levels in pregnant women with a higher incidence of autoimmune biomarkers, yet others show no correlation (Karagas et al. 2012).

Animal studies examining effects of the different forms of mercury on the mature (adult) immune system show an increase in autoimmune disease symptoms, such as autoimmune nephritis associated with systemic lupus erythematosus (SLE) (Abedi-Valugerdi et al. 1999; Hu et al. 1999; Pollard et al. 2001; Via et al. 2003). Examination of the effects of developmental exposure to mercury in all of its forms has primarily focused on ex vivo cellular assays and metrics indicative of autoimmune disease phenotypes in mouse models, although the nature of the changes is not consistent across studies. The specific type of leukocytes altered by developmental exposure to mercury remains understudied. For instance, one study showed that methylmercury exposure during development increased B-cell proliferative responses and antibody secretion (Thuvander et al. 1996). A separate study showed that developmental exposure to methylmercury dose-dependently decreased cytokine production (Tonk et al. 2010). Studies of inorganic mercury have reported increases in immune responses in offspring exposed during development. In particular, offspring of dams treated with inorganic mercury had higher levels of autoantigen-specific $CD4^+$ T cells and B cells, as well as increased cytokine production by these cells (Pilones et al. 2007, 2009). These studies did not demonstrate gender-specific differences in developmentally exposed offspring, yet a study in a different strain of mice, which are not prone to autoimmunity, showed that males had enhanced immune responses after developmental exposure to inorganic mercury, while female offspring had reduced responses (Silva et al. 2005). Given the multitude of cell types involved in an immune response, it is important to define cellular targets of developmental exposure to methylmercury and inorganic and elemental mercury. This would greatly improve our understanding of disease pathogenesis after developmental exposure to mercury and help explain some of the seemingly contradictory literature. Other future studies need to elucidate gender-specific differences in the immune responses after developmental exposure to mercury.

Interestingly, there have not been any studies examining whether developmental exposure to mercury alters host responses to infection. If developmental exposure to mercury skews the immune system of offspring toward a more hyperreactive state, then it is possible there will be increased immune-mediated pathology after infection. Another possibility is that developmental exposure to mercury alters central and/or peripheral tolerance mechanisms, which may not have consequences for fighting infections, but will manifest in autoimmune symptoms upon aging. Longitudinal studies in both animals and exposed human populations will be informative and help to define the consequences of developmental exposure to mercury much later in life. Finally, given that all forms of mercury have different toxicological profiles (Holmes et al. 2009), it is important to systematically compare the effects of early-life exposure to each type of mercury on multiple aspects of immune function.

2.4.2 Arsenic

Exposure to arsenic occurs mainly through drinking water contamination and the diet. Bedrock in certain areas contains high levels of arsenic, which dissolves in water and contaminates drinking supplies, particularly in areas that primarily use untreated well water. Arsenic is readily passed through the placenta and breast milk (Vahter 2008). Epidemiological studies support the idea that developmental exposure to arsenic has immunotoxic effects later in life. Arsenic exposure during development is correlated with higher risk of lung cancer and nonmalignant lung disease, a higher incidence of respiratory infections, and reduced thymus size (Smith et al. 2006; Moore et al. 2009; Raqib et al. 2009). Additionally, a correlation between arsenic exposure levels during development and a reduction in T-cell receptor gene rearrangement in the thymus has been reported, possibly due to alterations in oxidative stress and apoptosis pathways (Ahmed et al. 2012). Other epidemiological studies examining early-life arsenic exposure have shown a correlation with altered patterns of DNA methylation in cord blood leukocytes (Kile et al. 2014; Koestler et al. 2013; Intarasunanont et al. 2012). Changes in DNA methylation may be responsible for altered expression of genes in pathways involved with inflammation and stress responses. Another study examined gene expression profiles in cord blood leukocytes after developmental exposure to arsenic and showed increased expression levels of genes involved in inflammatory pathways (Fry et al. 2007). Although no studies have directly correlated the changes in DNA methylation with altered gene expression in general or in a specific type of leukocyte, these reports suggest one potential mechanism by which arsenic exposure during development leads to persistent changes in immune function.

Animal studies support the reports based on human populations. For instance, mice exposed to arsenic during development have decreased lung function, whereas adult exposure to the same levels of arsenic did not lead to these changes (Lantz et al. 2009). This suggests that the developmental period may be particularly susceptible to arsenic. Another study showed that developmental exposure to arsenic led to worse outcomes after respiratory infection (Ramsey et al. 2013). In parallel with human studies, epigenetic changes after arsenic exposure during development have been reported in animal models. For example, histone acetylation is altered after developmental arsenic exposure. This particular study did not examine the immune system per se, but did correlate the changes in histone acetylation with functional changes in the neurological system (Cronican et al. 2013). Other organ systems, such as the immune system, most likely also experience changes in epigenetic profiles after developmental exposure to arsenic, but this remains a testable hypothesis. In addition to epigenetics, studies that identify the other mechanisms by which arsenic causes persistent changes in immune function are needed, particularly animal studies that utilize a wider variety of immune challenges after developmental exposure to environmentally relevant levels of arsenic. These types of studies will help connect observed epigenetic changes with arsenic to associated alterations in the ability of the immune system to respond to insult, maintain homeostasis, or prevent disease.

2.4.3 Cadmium

Cadmium exposure occurs through the diet, via contaminated soil and foods, and through cigarette smoke exposure (US Department of Health and Human Services 2009). Most recently, the revelation that rice in southern China is tainted with high levels of cadmium has raised further concern about the potential long-term health effects of exposure (Zhang et al. 2014; Zhai et al. 2008). Cadmium is transferred through the placenta; therefore the developing fetus is exposed to cadmium (Jarup and Akesson 2009). There are very few epidemiological studies that have examined effects of cadmium exposure during development on the immune system later in life. However, one study showed that developmental cadmium exposure altered DNA methylation profiles of maternal and cord blood leukocytes (Sanders et al. 2014), which suggests that changes in leukocyte function might occur as a result of cadmium exposures during development.

Animal studies examining the immune system after developmental exposure to cadmium support the idea that early-life cadmium exposure has persistent effects on immune function. Initial studies revealed a confusing picture: developmental exposure increased proliferative responses of lymphocytes, yet decreased DTH responses (Soukupova et al. 1991). More recent studies affirm that developmental exposure to cadmium can have both immunostimulatory and immunosuppressive effects. Lactational exposure to cadmium leads to decreased proliferation of lymphocytes (Pillet et al. 2005), whereas exposure to cadmium throughout development leads to increased T-dependent and T-independent antibody responses (Holaskova et al. 2012; Hanson et al. 2012). Cadmium exposure in utero disrupts key signaling pathways in the thymus early in life (Hanson et al. 2010). Collectively, these studies suggest that differences in the timing, dose, and slate of endpoints tested may influence whether cadmium is perceived as enhancing or suppressing immune function.

Similar to other recent studies of metal toxicity, developmental exposure to cadmium may alter the epigenome of the offspring. However, data are only just emerging; therefore the precise nature of changes remains to be fully defined. One study showed a sex-specific change in DNA methylation and expression of DNA methyltransferase (DNMT) enzymes. Specifically, male offspring had higher levels of DNMT3a and hypermethylation of DNA at specific sites, whereas female offspring had lower DNMT3a levels and hypomethylated DNA. Additionally, this study correlated the changes in DNA methylation with altered expression of mRNA and protein levels of the glucocorticoid receptor (Castillo et al. 2012). While this study did not examine immune function specifically, it suggests that cadmium has the potential to alter the DNA methylation profiles of exposed mice, leading to durable functional changes in many organ systems. To understand the long-term impact of cadmium, further research needs to examine the specific hematopoietic cellular targets of developmental exposure to cadmium. In addition, the use of in vivo systems that test multiple aspects of immune function, including infectious, hypersensitivity, and autoimmune models, will be needed to understand how developmental exposure to cadmium alters immune responses later in life.

2.4.4 Pharmaceuticals

Given that the human immune system continues to develop well after birth, the potential for long-lasting effects of pharmaceuticals on immune function across the life span is an important consideration. Yet, how developmental exposure to pharmaceuticals affects the immune system is largely understudied. Although many pharmaceuticals are tested for overt toxicity to the fetus before, the ability of these drugs to cause persistent changes in the immune system of those exposed early in life is rarely considered. Further understanding of this issue may lead to changes in the therapeutic range of use of certain drugs or additional monitoring of subjects exposed to particular drugs in utero or shortly after birth. In this section we discuss three specific pharmaceuticals for which there has been consideration of the effects of developmental exposures on immune function later in life. The first example is cyclosporine A, a drug for which there exist several studies examining the effects of developmental exposure on the immune system. The second example is diazepam, a compound for which there is also evidence that developmental exposure alters the immune system. Lastly, we point out a single study that has examined the effects of a broadly used pharmaceutical family, nonsteroidal anti-inflammatory drugs (NSAIDs), and the potential impact of developmental exposure to NSAIDs on the immune system.

Information obtained thus far suggests that developmental exposure to pharmaceuticals can alter the lifelong function of the immune system. For example, cyclosporine A is used to induce immunosuppression after organ transplantation and in people with autoimmune diseases such as rheumatoid arthritis and psoriasis. Cyclosporine A is one of the most well-studied pharmaceuticals when examining the effects of developmental exposure on the immune system. However, there are still unanswered questions, because epidemiological studies present conflicting results as to whether or not exposure during development has an effect on the immune system later in life. One study shows exposure may alter T-cell development, but others show no effect (Baarsma and Kamps 1993; Motta et al. 2007, 2008; Pilarski et al. 1994). Several rodent studies show that cyclosporine A exposure during development impacts the immune system of offspring born to treated dams (Hussain et al. 2005; Barrow et al. 2006). Some studies show that the offspring have reduced adaptive immune responses, such as decreased DTH reactions, T-dependent antibody production, and cytokine production (Hussain et al. 2005). Other studies show an increase in autoantibody levels in offspring of cyclosporine-treated dams (Sakaguchi and Sakaguchi 1989; Classen and Shevach 1991). Additionally, studies have examined the durability of the effects of developmental exposure to cyclosporine A and suggest that changes in the immune system do not persist into adulthood (Padgett and Seelig 2002; Barrow et al. 2006). While the conclusiveness of these studies is not firm, when considered together, they support the need to methodically determine whether early-life exposure to cyclosporine A, and related immunosuppressive drugs, has a sustained impact on immune functions later in life, with particular attention paid to host defenses against infection and the incidence or progression of autoimmune diseases.

Another example of a commonly used drug for which developmentally immune effects have been reported is diazepam. It, along with other benzodiazepines, is prescribed to treat a multitude of conditions. Diazepam was the focus of several studies of developmental immunotoxicity in the 1980s and 1990s. These studies showed that animals exposed to diazepam during development had decreased resistance to infection (Ugaz et al. 1999; Schlumpf et al. 1994b; Laschi et al. 1983). Furthermore, mice developmentally exposed to diazepam had increased tumor incidence, which correlated with decreased T-cell responses to antigen and cytokine production (Schlumpf et al. 1989, 1993, 1994a; Livezey et al. 1986; Schreiber et al. 1993a, b; Dostal et al. 1995). These reports, combined with numerous other studies, led to warnings for women taking diazepam during pregnancy and lactation.

Nonsteroidal anti-inflammatory drugs (NSAIDs) are widely used as an over-the-counter treatment for numerous symptoms; however, extensive examination of the developmental immunotoxicity of the compounds in this drug family has not been carried out. The metabolism of NSAIDs has been studied, and it is known that NSAIDs can cross the placenta and can be excreted in breast milk (Antonucci et al. 2012). Yet, they are worth noting here because single study looked at developmental exposure to multiple NSAID doses and effects on immune responses. At the highest maternal dose used, there were a reduction in the cellularity of immune organs and a decrease in antigen-specific immune responses in female offspring (Kushima et al. 2007). Given the wide use and availability of NSAIDs, further studies determining the immunotoxic potential of developmental exposure to NSAIDs will help to determine if pregnant women need to monitor their NSAID use.

In summary, for most pharmaceutical agents, there is scant information regarding potential ability for developmental exposure to alter immune function and contribute to disease later in life. Presented here are three examples, showing the range from firm evidence, to some evidence, to a single study that suggests developmental immunotoxicity. Importantly, there is growing appreciation for the need to design and harmonize strategies to evaluate the potential for early-life exposure to in-use and emerging pharmaceuticals to not only be overtly toxic to the developing immune system but to look for subtle changes which, over the life span, could manifest in enhanced susceptibility to disease (Collinge et al. 2012).

2.5 Maternal Diet

The profound influence of maternal nutrition on the overall health of her offspring has become widely studied recently, especially considering the current obesity epidemic in many developed countries. However, the effects of maternal diet during pregnancy are not isolated to high-fat diets and maternal obesity. Restrictive diets and dietary supplements also contribute to effects on fetal development and have implications for offspring later in life. In short, understanding the durable effects of maternal diet on the health of her offspring is an emerging area of research, with new attention being paid to the immune system. There are two possible contributions to immune responses in offspring of mothers whose diet has been altered: (1)

maternal metabolic changes and (2) direct effects of the diet on the offspring. The work presented here does not separate these effects, but future studies should address how each of these effects may contribute to lasting changes in immune function. In this section, we discuss several studies that demonstrate effects of maternal diet on the immune system later in life and provide insight into at least some possible mechanisms by which these changes may be occurring.

2.5.1 Nutritional Restriction

Lack of certain important food groups, such as protein, and general reduction in intake of nutrients have implications for fetal growth and development. In addition, there may be long-lasting effects on multiple organ systems, including the immune system. Very few studies have addressed how maternal dietary restrictions led to persistent immunological changes in her offspring, and none has addressed the mechanism by which these changes may be occurring. While there is insufficient epidemiological evidence to define the precise impact of a restrictive maternal diet on the immune system of her children, overall children born to malnourished mothers are not as healthy as controls (Victora et al. 2008). A study by Silva et al. showed that a protein-free diet during pregnancy reduced leukocyte migration from the bone marrow and increased global inflammation in the offspring (Silva et al. 2010). General nutritional restriction, not just restriction of one specific component of the diet, reduced thymic and splenic cellularity in offspring, as well as the number of antibody producing cells in the spleen (Carney et al. 2004). Clearly, restricting aspects of the maternal diet can have persistent impacts on offspring health, including immunological parameters, warranting further study of how dietary restrictions lead to persistent changes in immune function. In addition, it will be interesting to define the implications of maternal dietary restriction on disease outcomes, such as susceptibility to infection and tumor incidence.

2.5.2 High-Fat Diet

The obesity epidemic has led to many studies examining the effects of maternal obesity and high-fat diet on her offspring. Obesity during pregnancy can lead to metabolic alterations in children, but less is known about how maternal obesity alters development of the immune system of children (Schmatz et al. 2010). A few studies have examined the effects of maternal high-fat diet on persistent changes in immune function in offspring. Overall, it seems that most studies support an increase in inflammation in the offspring of dams on a high-fat diet. For example, offspring of mothers that consumed a high-fat diet had higher levels of proinflammatory cytokines and increased incidence and symptom severity of nonalcoholic fatty liver disease (Odaka et al. 2010; Mouralidarane et al. 2013). One comprehensive study of many different immune outcomes showed that offspring of dams on a high-fat diet had increased disease severity in multiple models, including a worse disease score in an experimental autoimmune encephalomyelitis model, increased lesion size after methicillin-resistant *Staphylococcus aureus*, and increased mortality in a model of bacterial sepsis (Myles et al. 2013). Interestingly, the authors in this study went on to explore a potential mechanism by which these changes might be occurring in the offspring. They found that the microbiome of dams on the high-fat diet, and subsequently their offspring, was different than normal chow controls and their offspring. This change could be rescued by cohousing offspring of dams on the normal chow diet and offspring of dams on the high-fat diet (Myles et al. 2013). This finding suggests that the microbiome should be incorporated into mechanistic studies that link effects of maternal diet to the health of her offspring. Additional studies examining strategies to mitigate effects of maternal high-fat diets will help to determine possible means to reverse these effects. It is equally important to define whether and how other dietary changes during pregnancy influence immune function in the offspring. A better understanding of this is important when considering ways to prevent or reduce the immunological consequences of maternal obesity.

2.5.3 Dietary Supplementation

Changes in maternal diet during pregnancy are not always unhealthy for the fetus. In fact, supplementing the diet during pregnancy is suggested for certain nutrients not found in typical diets of a particular region, such as the recommendation to supplement omega-3 fatty acids and folic acid in Western diets to reduce neural tube defects. Epidemiological evidence suggests that docosahexaenoic acid (DHA, one type of omega-3 fatty acid) supplementation is correlated with a reduction in allergy incidence and a lower incidence of respiratory infections (Shek et al. 2012). Another study in patients showed that infants of mothers who were on a DHA-supplemented diet had a higher proportion of circulating naïve $CD4^+$ T cells, and when $CD4^+$ T cells were isolated from these infants and stimulated ex vivo, they produced less of the cytokine interferon gamma (Granot et al. 2011). While these data suggest these infants may have a lower overall "inflammatory state," they also imply that infants of mothers on a DHA-supplemented diet may have hyporesponsive T cells. To our knowledge, no animal studies have been published to evaluate the effects of maternal DHA supplementation on the immune system of offspring later in life. Further causal studies need to be done to support the correlative human work and to reveal the mechanisms by which the immune system is altered by maternal dietary supplementation. In addition, other popular supplements taken during pregnancy, including herbal remedies, should be examined for potential immunomodulatory effects after developmental exposures.

2.6 Conclusions and Future Directions

Whether considering environmental pollutants, therapeutic agents, or dietary factors, we have summarized findings from numerous studies demonstrating that the developing immune system is susceptible to being altered by early-life exposures

(Fig. 2.1). Moreover, changes are often not obvious at the time of exposure and may occur at doses that do not affect the function of the fully mature immune system. Instead, the current body of evidence suggests that many developmental exposures do not cause overt toxicity (e.g., obvious loss of cells); rather they affect the integrated function of the immune system, such that there is an increased or decreased response to challenge later in life. Many diseases, particularly autoimmune and inflammatory diseases, do not become clinically symptomatic until much later in life. Therefore, allowing subjects that were exposed during development to age before examining immune-related endpoints will permit measurement of defined immunological parameters implicated in disease pathogenesis.

Although there has been an effort to do more research in silico and in vitro, determining whether developmental exposures lead to durable changes in the immune system will require research conducted using animal models that mirror human diseases, and when possible, longitudinal human cohorts. Especially informative will be

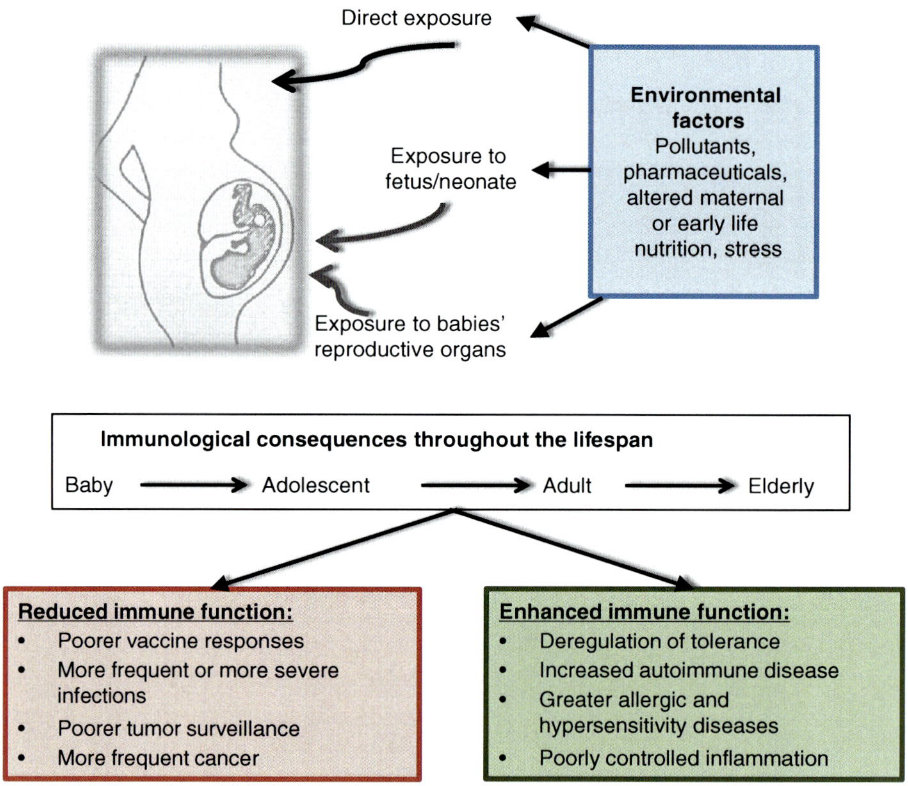

Fig. 2.1 Early life exposures influence immune function over the lifespan—and maybe beyond. A conceptualization of how early life exposures may manifest in persistent alterations in immune function. Developmental exposures have the potential to change the responsive capacity of the immune system to challenge later in life, leading to detrimental consequences for the host at many life stages

epidemiological studies that document early-life exposures and monitor immune function and disease occurrence much later in life. These types of studies will improve our general understanding of whether a particular exposure has long-lasting immunological consequences. Moreover, they also form the critical foundation for mechanistic studies, as it remains to be determined how exposure during development changes the function of the immune system later in life. Another understudied aspect of how developmental exposures change the immune system is the idea of the duration across the full life span. Rodent students are often conducted using young adult animals, and human studies rarely have been able to connect documented developmental exposure with diseases that do not appear until maturity. Thus, it is important to leverage animal models to define how developmental exposures influence the immune system throughout the life span (i.e., immature, young adult, and aged populations). Additionally, multigenerational and transgenerational studies, which examine the immune system of the F2 generation (grandchildren) and F3 generation (great grandchildren), are needed to determine if the changes observed after developmental exposures are transmitted to subsequent generations, a phenomenon reported to occur in other organ systems (Nilsson et al. 2012).

The research reviewed in this chapter, for the most part, examined the consequences developmental exposure to a single agent. These types of studies, particularly those performed in animal models, provide extraordinarily important insight into the mechanism by which environmental agents affect the developmental programming and function of the immune system. However, future work needs to start considering developmental studies that include mixtures. It is estimated that humans are regularly exposed to 80,000–200,000 different chemicals. Many of these chemicals have not been examined for potentially immunomodulatory effects, either directly or as a result of exposure during development. In addition to expanding the repertoire of environmental agents studied as single exposures (i.e., one at a time), designing approaches to examine how mixtures of environmental factors (beneficial and detrimental) during development alter the immune system will provide critical insight into how our environment truly shapes the development and function of the immune system.

References

Abedi-Valugerdi M, Hu H, Moller G (1999) Mercury-induced anti-nucleolar autoantibodies can transgress the membrane of living cells in vivo and in vitro. Int Immunol 11(4):605–615

Ahmed S, Ahsan KB, Kippler M, Mily A, Wagatsuma Y, Hoque AM, Ngom PT, El Arifeen S, Raqib R, Vahter M (2012) In utero arsenic exposure is associated with impaired thymic function in newborns possibly via oxidative stress and apoptosis. Toxicol Sci 129(2):305–314. doi:10.1093/toxsci/kfs202

Al-Saleh I, Shinwari N, Mashhour A, Mohamed Gel D, Rabah A (2011) Heavy metals (lead, cadmium and mercury) in maternal, cord blood and placenta of healthy women. Int J Hyg Environ Health 214(2):79–101. doi:10.1016/j.ijheh.2010.10.001

Antonucci R, Zaffanello M, Puxeddu E, Porcella A, Cuzzolin L, Pilloni MD, Fanos V (2012) Use of non-steroidal anti-inflammatory drugs in pregnancy: impact on the fetus and newborn. Curr Drug Metab 13(4):474–490

Baarsma R, Kamps WA (1993) Immunological responses in an infant after cyclosporine A exposure during pregnancy. Eur J Pediatr 152(6):476–477

Babalik A, Bakirci N, Taylan M, Bostan L, Kiziltas S, Basbug Y, Calisir HC (2013) Biomass smoke exposure as a serious health hazard for women. Tuberk Toraks 61(2):115–121

Bakker JM, Schmidt ED, Kroes H, Kavelaars A, Heijnen CJ, Tilders FJ, van Rees EP (1995) Effects of short-term dexamethasone treatment during pregnancy on the development of the immune system and the hypothalamo-pituitary adrenal axis in the rat. J Neuroimmunol 63(2):183–191

Barber K, Mussin E, Taylor DK (1996) Fetal exposure to involuntary maternal smoking and childhood respiratory disease. Ann Allergy Asthma Immunol 76(5):427–430. doi:10.1016/S1081-1206(10)63459-X

Barrow PC, Horand F, Ravel G (2006) Developmental immunotoxicity investigations in the SD rat following pre- and post-natal exposure to cyclosporin. Birth Defects Res B Dev Reprod Toxicol 77(5):430–437. doi:10.1002/bdrb.20093

Bauer SM, Roy A, Emo J, Chapman TJ, Georas SN, Lawrence BP (2012) The effects of maternal exposure to bisphenol A on allergic lung inflammation into adulthood. Toxicol Sci 130(1):82–93. doi:10.1093/toxsci/kfs227

Bilrha H, Roy R, Wagner E, Belles-Isles M, Bailey JL, Ayotte P (2004) Effects of gestational and lactational exposure to organochlorine compounds on cellular, humoral, and innate immunity in swine. Toxicol Sci 77(1):41–50. doi:10.1093/toxsci/kfg240

Bjermo H, Sand S, Nalsen C, Lundh T, Enghardt Barbieri H, Pearson M, Lindroos AK, Jonsson BA, Barregard L, Darnerud PO (2013) Lead, mercury, and cadmium in blood and their relation to diet among Swedish adults. Food Chem Toxicol 57:161–169. doi:10.1016/j.fct.2013.03.024

Bodin J, Bolling AK, Becher R, Kuper F, Lovik M, Nygaard UC (2014) Transmaternal bisphenol A exposure accelerates diabetes type 1 development in NOD mice. Toxicol Sci 137(2):311–323. doi:10.1093/toxsci/kft242

Bonfanti P, Colombo A, Villa S, Comelli F, Costa B, Santagostino A (2009) The effects of accumulation of an environmentally relevant polychlorinated biphenyl mixture on cytochrome P450 and P-glycoprotein expressions in fetuses and pregnant rats. Chemosphere 75(5):572–579. doi:10.1016/j.chemosphere.2009.01.063

Bradley JP, Bacharier LB, Bonfiglio J, Schechtman KB, Strunk R, Storch G, Castro M (2005) Severity of respiratory syncytial virus bronchiolitis is affected by cigarette smoke exposure and atopy. Pediatrics 115(1):e7–e14. doi:10.1542/peds.2004-0059

Breckenridge BCSJ, Eldridge JC, Stevens JT (eds) (2010) Symmetrical triazine herbicides: a review of regulatory endpoints. Handbook of pesticide toxicology: agents, 3rd edn. Academic, New York

Burke H, Leonardi-Bee J, Hashim A, Pine-Abata H, Chen Y, Cook DG, Britton JR, McKeever TM (2012) Prenatal and passive smoke exposure and incidence of asthma and wheeze: systematic review and meta-analysis. Pediatrics 129(4):735–744. doi:10.1542/peds.2011-2196

Bushnik T, Haines D, Levallois P, Levesque J, Van Oostdam J, Viau C (2010) Lead and bisphenol A concentrations in the Canadian population. Health Rep 21(3):7–18

Calafat AM, Ye X, Wong LY, Reidy JA, Needham LL (2008) Exposure of the U.S. population to bisphenol A and 4-tertiary-octylphenol: 2003–2004. Environ Health Perspect 116(1):39–44. doi:10.1289/ehp.10753

Carney EW, Zablotny CL, Marty MS, Crissman JW, Anderson P, Woolhiser M, Holsapple M (2004) The effects of feed restriction during in utero and postnatal development in rats. Toxicol Sci 82(1):237–249. doi:10.1093/toxsci/kfh249

Castillo P, Ibanez F, Guajardo A, Llanos MN, Ronco AM (2012) Impact of cadmium exposure during pregnancy on hepatic glucocorticoid receptor methylation and expression in rat fetus. PLoS One 7(9), e44139. doi:10.1371/journal.pone.0044139

Chapin RE, Adams J, Boekelheide K, Gray LE Jr, Hayward SW, Lees PS, McIntyre BS, Portier KM, Schnorr TM, Selevan SG, Vandenbergh JG, Woskie SR (2008) NTP-CERHR expert panel report on the reproductive and developmental toxicity of bisphenol A. Birth Defects Res B Dev Reprod Toxicol 83(3):157–395. doi:10.1002/bdrb.20147

Cheraghi M, Salvi S (2009) Environmental tobacco smoke (ETS) and respiratory health in children. Eur J Pediatr 168(8):897–905. doi:10.1007/s00431-009-0967-3

Classen JB, Shevach EM (1991) Evidence that cyclosporine treatment during pregnancy predisposes offspring to develop autoantibodies. Transplantation 51(5):1052–1057

Collinge M, Burns-Naas LA, Chellman GJ, Kawabata TT, Komocsar WJ, Piccotti JR, Shenton J, Wierda D (2012) Developmental immunotoxicity (DIT) testing of pharmaceuticals: current practices, state of the science, knowledge gaps, and recommendations. J Immunotoxicol 9(2):210–230. doi:10.3109/1547691X.2012.661486

Corsini E, Luebke RW, Germolec DR, Dewitt JC (2014) Perfluorinated compounds: Emerging POPs with potential immunotoxicity. Toxicol Lett. doi:10.1016/j.toxlet.2014.01.038

Couse JF, Dixon D, Yates M, Moore AB, Ma L, Maas R, Korach KS (2001) Estrogen receptor-alpha knockout mice exhibit resistance to the developmental effects of neonatal diethylstilbestrol exposure on the female reproductive tract. Dev Biol 238(2):224–238. doi:10.1006/dbio.2001.0413

Cronican AA, Fitz NF, Carter A, Saleem M, Shiva S, Barchowsky A, Koldamova R, Schug J, Lefterov I (2013) Genome-wide alteration of histone H3K9 acetylation pattern in mouse offspring prenatally exposed to arsenic. PLoS One 8(2), e53478. doi:10.1371/journal.pone.0053478

Cupul-Uicab LA, Skjaerven R, Haug K, Melve KK, Engel SM, Longnecker MP (2012) In utero exposure to maternal tobacco smoke and subsequent obesity, hypertension, and gestational diabetes among women in the MoBa cohort. Environ Health Perspect 120(3):355–360. doi:10.1289/ehp.1103789

Dallaire F, Dewailly E, Muckle G, Vezina C, Jacobson SW, Jacobson JL, Ayotte P (2004) Acute infections and environmental exposure to organochlorines in Inuit infants from Nunavik. Environ Health Perspect 112(14):1359–1365

Dallaire F, Dewailly E, Vezina C, Muckle G, Weber JP, Bruneau S, Ayotte P (2006) Effect of prenatal exposure to polychlorinated biphenyls on incidence of acute respiratory infections in preschool Inuit children. Environ Health Perspect 114(8):1301–1305

Davidson PW, Myers GJ, Weiss B, Shamlaye CF, Cox C (2006) Prenatal methyl mercury exposure from fish consumption and child development: a review of evidence and perspectives from the Seychelles Child Development Study. Neurotoxicology 27(6):1106–1109. doi:10.1016/j.neuro.2006.03.024

Devereux G, Barker RN, Seaton A (2002) Antenatal determinants of neonatal immune responses to allergens. Clin Exp Allergy 32(1):43–50

Dewailly E, Ayotte P, Bruneau S, Gingras S, Belles-Isles M, Roy R (2000) Susceptibility to infections and immune status in Inuit infants exposed to organochlorines. Environ Health Perspect 108(3):205–211

DeWitt JC, Copeland CB, Luebke RW (2007) Immune function is not impaired in Sprague-Dawley rats exposed to dimethyltin dichloride (DMTC) during development or adulthood. Toxicology 232(3):303–310. doi:10.1016/j.tox.2007.01.017

Dietert RR, Lee JE, Olsen J, Fitch K, Marsh JA (2003) Developmental immunotoxicity of dexamethasone: comparison of fetal versus adult exposures. Toxicology 194(1–2):163–176

Dietert RR, Lee JE, Hussain I, Piepenbrink M (2004) Developmental immunotoxicology of lead. Toxicol Appl Pharmacol 198(2):86–94. doi:10.1016/j.taap.2003.08.020

Diette GB, Accinelli RA, Balmes JR, Buist AS, Checkley W, Garbe P, Hansel NN, Kapil V, Gordon S, Lagat DK, Yip F, Mortimer K, Perez-Padilla R, Roth C, Schwaninger JM, Punturieri A, Kiley J (2012) Obstructive lung disease and exposure to burning biomass fuel in the indoor environment. Glob Heart 7(3):265–270. doi:10.1016/j.gheart.2012.06.016

DiFranza JR, Aligne CA, Weitzman M (2004) Prenatal and postnatal environmental tobacco smoke exposure and children's health. Pediatrics 113(4 Suppl):1007–1015

Doherty SP, Grabowski J, Hoffman C, Ng SP, Zelikoff JT (2009) Early life insult from cigarette smoke may be predictive of chronic diseases later in life. Biomarkers 14(Suppl 1):97–101. doi:10.1080/13547500902965898

Domingo JL, Bocio A (2007) Levels of PCDD/PCDFs and PCBs in edible marine species and human intake: a literature review. Environ Int 33(3):397–405. doi:10.1016/j.envint.2006.12.004

Donohue KM, Miller RL, Perzanowski MS, Just AC, Hoepner LA, Arunajadai S, Canfield S, Resnick D, Calafat AM, Perera FP, Whyatt RM (2013) Prenatal and postnatal bisphenol A exposure and asthma development among inner-city children. J Allergy Clin Immunol 131(3):736–742. doi:10.1016/j.jaci.2012.12.1573

Dostal M, Benesova O, Tejkalova H, Soukupova D (1995) Immune response of adult rats is altered by administration of diazepam in the first postnatal week. Reprod Toxicol 9(2):115–121

Faith RE, Moore JA (1977) Impairment of thymus-dependent immune functions by exposure of the developing immune system to 2,3,7,8-tetrachlorodibenzo-p-dioxin (TCDD). J Toxicol Environ Health 3(3):451–464. doi:10.1080/15287397709529578

Ferre P (2004) The biology of peroxisome proliferator-activated receptors: relationship with lipid metabolism and insulin sensitivity. Diabetes 53(Suppl 1):S43–S50

Fry RC, Navasumrit P, Valiathan C, Svensson JP, Hogan BJ, Luo M, Bhattacharya S, Kandjanapa K, Soontararuks S, Nookabkaew S, Mahidol C, Ruchirawat M, Samson LD (2007) Activation of inflammation/NF-kappaB signaling in infants born to arsenic-exposed mothers. PLoS Genet 3(11), e207. doi:10.1371/journal.pgen.0030207

Gasiewicz TA, Geiger LE, Rucci G, Neal RA (1983) Distribution, excretion, and metabolism of 2,3,7,8-tetrachlorodibenzo-p-dioxin in C57BL/6J, DBA/2J, and B6D2F1/J mice. Drug Metab Dispos 11(5):397–403

Geens T, Aerts D, Berthot C, Bourguignon JP, Goeyens L, Lecomte P, Maghuin-Rogister G, Pironnet AM, Pussemier L, Scippo ML, Van Loco J, Covaci A (2012) A review of dietary and non-dietary exposure to bisphenol-A. Food Chem Toxicol 50(10):3725–3740. doi:10.1016/j.fct.2012.07.059

Gehrs BC, Smialowicz RJ (1997) Alterations in the developing immune system of the F344 rat after perinatal exposure to 2,3,7,8-tetrachlorodibenzo-p-dioxin I. [correction of II]. Effects on the fetus and the neonate. Toxicology 122(3):219–228

Gehrs BC, Smialowicz RJ (1999) Persistent suppression of delayed-type hypersensitivity in adult F344 rats after perinatal exposure to 2,3,7,8-tetrachlorodibenzo-p-dioxin. Toxicology 134(1):79–88

Gehrs BC, Riddle MM, Williams WC, Smialowicz RJ (1997) Alterations in the developing immune system of the F344 rat after perinatal exposure to 2,3,7,8-tetrachlorodibenzo-p-dioxin: II. Effects on the pup and the adult. Toxicology 122(3):229–240

Giesy JP, Kannan K (2002) Perfluorochemical surfactants in the environment. Environ Sci Technol 36(7):146A–152A

Glynn A, Thuvander A, Aune M, Johannisson A, Darnerud PO, Ronquist G, Cnattingius S (2008) Immune cell counts and risks of respiratory infections among infants exposed pre- and postnatally to organochlorine compounds: a prospective study. Environ Health 7:62. doi:10.1186/1476-069X-7-62

Gore AC (2008) Developmental programming and endocrine disruptor effects on reproductive neuroendocrine systems. Front Neuroendocrinol 29(3):358–374. doi:10.1016/j.yfrne.2008.02.002

Grandjean P, Andersen EW, Budtz-Jorgensen E, Nielsen F, Molbak K, Weihe P, Heilmann C (2012) Serum vaccine antibody concentrations in children exposed to perfluorinated compounds. JAMA 307(4):391–397. doi:10.1001/jama.2011.2034

Granot E, Jakobovich E, Rabinowitz R, Levy P, Schlesinger M (2011) DHA supplementation during pregnancy and lactation affects infants' cellular but not humoral immune response. Mediators Inflamm 2011:493925. doi:10.1155/2011/493925

Guarnieri MJ, Diaz JV, Basu C, Diaz A, Pope D, Smith KR, Smith-Sivertsen T, Bruce N, Solomon C, McCracken J, Balmes JR (2014) Effects of woodsmoke exposure on airway inflammation in rural Guatemalan women. PLoS One 9(3), e88455. doi:10.1371/journal.pone.0088455

Guo YL, Lambert GH, Hsu CC, Hsu MM (2004) Yucheng: health effects of prenatal exposure to polychlorinated biphenyls and dibenzofurans. Int Arch Occup Environ Health 77(3):153–158. doi:10.1007/s00420-003-0487-9

Han SG, Bhoopalan V, Akinbiyi T, Gairola CG, Bhalla DK (2011) In utero tobacco smoke exposure alters pulmonary responses of newborn rats to ozone. J Toxicol Environ Health A 74(10):668–677. doi:10.1080/15287394.2011.539133

Hanson ML, Brundage KM, Schafer R, Tou JC, Barnett JB (2010) Prenatal cadmium exposure dysregulates sonic hedgehog and Wnt/beta-catenin signaling in the thymus resulting in altered thymocyte development. Toxicol Appl Pharmacol 242(2):136–145. doi:10.1016/j.taap.2009.09.023

Hanson ML, Holaskova I, Elliott M, Brundage KM, Schafer R, Barnett JB (2012) Prenatal cadmium exposure alters postnatal immune cell development and function. Toxicol Appl Pharmacol 261(2):196–203. doi:10.1016/j.taap.2012.04.002

Heilmann C, Grandjean P, Weihe P, Nielsen F, Budtz-Jorgensen E (2006) Reduced antibody responses to vaccinations in children exposed to polychlorinated biphenyls. PLoS Med 3(8), e311. doi:10.1371/journal.pmed.0030311

Heilmann C, Budtz-Jorgensen E, Nielsen F, Heinzow B, Weihe P, Grandjean P (2010) Serum concentrations of antibodies against vaccine toxoids in children exposed perinatally to immunotoxicants. Environ Health Perspect 118(10):1434–1438. doi:10.1289/ehp.1001975

Herberth G, Bauer M, Gasch M, Hinz D, Roder S, Olek S, Kohajda T, Rolle-Kampczyk U, von Bergen M, Sack U, Borte M, Lehmann I (2014) Maternal and cord blood miR-223 expression associates with prenatal tobacco smoke exposure and low regulatory T-cell numbers. J Allergy Clin Immunol 133(2):543–550. doi:10.1016/j.jaci.2013.06.036

Herbst AL, Ulfelder H, Poskanzer DC (1971) Adenocarcinoma of the vagina. Association of maternal stilbestrol therapy with tumor appearance in young women. N Engl J Med 284(15):878–881. doi:10.1056/NEJM197104222841604

Herr CE, Dostal M, Ghosh R, Ashwood P, Lipsett M, Pinkerton KE, Sram R, Hertz-Picciotto I (2010) Air pollution exposure during critical time periods in gestation and alterations in cord blood lymphocyte distribution: a cohort of livebirths. Environ Health 9:46. doi:10.1186/1476-069X-9-46

Herr CE, Ghosh R, Dostal M, Skokanova V, Ashwood P, Lipsett M, Joad JP, Pinkerton KE, Yap PS, Frost JD, Sram R, Hertz-Picciotto I (2011) Exposure to air pollution in critical prenatal time windows and IgE levels in newborns. Pediatr Allergy Immunol 22(1 Pt 1):75–84. doi:10.1111/j.1399-3038.2010.01074.x

Hochstenbach K, van Leeuwen DM, Gmuender H, Gottschalk RW, Stolevik SB, Nygaard UC, Lovik M, Granum B, Namork E, Meltzer HM, Kleinjans JC, van Delft JH, van Loveren H (2012) Toxicogenomic profiles in relation to maternal immunotoxic exposure and immune functionality in newborns. Toxicol Sci 129(2):315–324. doi:10.1093/toxsci/kfs214

Hofhuis W, de Jongste JC, Merkus PJ (2003) Adverse health effects of prenatal and postnatal tobacco smoke exposure on children. Arch Dis Child 88(12):1086–1090

Hogaboam JP, Moore AJ, Lawrence BP (2008) The aryl hydrocarbon receptor affects distinct tissue compartments during ontogeny of the immune system. Toxicol Sci 102(1):160–170. doi:10.1093/toxsci/kfm283

Holaskova I, Elliott M, Hanson ML, Schafer R, Barnett JB (2012) Prenatal cadmium exposure produces persistent changes to thymus and spleen cell phenotypic repertoire as well as the acquired immune response. Toxicol Appl Pharmacol 265(2):181–189. doi:10.1016/j.taap.2012.10.009

Holladay SD, Mustafa A, Gogal RM Jr (2011) Prenatal TCDD in mice increases adult autoimmunity. Reprod Toxicol 31(3):312–318. doi:10.1016/j.reprotox.2010.08.001

Holmes P, James KA, Levy LS (2009) Is low-level environmental mercury exposure of concern to human health? Sci Total Environ 408(2):171–182. doi:10.1016/j.scitotenv.2009.09.043

Hu H, Moller G, Abedi-Valugerdi M (1999) Mechanism of mercury-induced autoimmunity: both T helper 1- and T helper 2-type responses are involved. Immunology 96(3):348–357

Hu Q, Franklin JN, Bryan I, Morris E, Wood A, DeWitt JC (2012) Does developmental exposure to perflurooctanoic acid (PFOA) induce immunopathologies commonly observed in neurodevelopmental disorders? Neurotoxicology 33(6):1491–1498. doi:10.1016/j.neuro.2012.10.016

Hussain I, Piepenbrink MS, Fitch KJ, Marsh JA, Dietert RR (2005) Developmental immunotoxicity of cyclosporin-A in rats: age-associated differential effects. Toxicology 206(2):273–284. doi:10.1016/j.tox.2004.08.019

Hylkema MN, Blacquiere MJ (2009) Intrauterine effects of maternal smoking on sensitization, asthma, and chronic obstructive pulmonary disease. Proc Am Thorac Soc 6(8):660–662. doi:10.1513/pats.200907-065DP

Institute of Medicine (2003) Dioxins and dioxin-like compounds in the food supply. National Academy of Sciences, Washington, D.C.

Intarasunanont P, Navasumrit P, Waraprasit S, Chaisatra K, Suk WA, Mahidol C, Ruchirawat M (2012) Effects of arsenic exposure on DNA methylation in cord blood samples from newborn babies and in a human lymphoblast cell line. Environ Health 11:31. doi:10.1186/1476-069X-11-31

Jaakkola JJ, Kosheleva AA, Katsnelson BA, Kuzmin SV, Privalova LI, Spengler JD (2006) Prenatal and postnatal tobacco smoke exposure and respiratory health in Russian children. Respir Res 7:48. doi:10.1186/1465-9921-7-48

Jarup L (2003) Hazards of heavy metal contamination. Br Med Bull 68:167–182

Jarup L, Akesson A (2009) Current status of cadmium as an environmental health problem. Toxicol Appl Pharmacol 238(3):201–208. doi:10.1016/j.taap.2009.04.020

Jedrychowski W, Galas A, Pac A, Flak E, Camman D, Rauh V, Perera F (2005) Prenatal ambient air exposure to polycyclic aromatic hydrocarbons and the occurrence of respiratory symptoms over the first year of life. Eur J Epidemiol 20(9):775–782. doi:10.1007/s10654-005-1048-1

Ji CM, Royce FH, Truong U, Plopper CG, Singh G, Pinkerton KE (1998) Maternal exposure to environmental tobacco smoke alters Clara cell secretory protein expression in fetal rat lung. Am J Physiol 275(5 Pt 1):L870–L876

Kalland T (1980a) Alterations of antibody response in female mice after neonatal exposure to diethylstilbestrol. J Immunol 124(1):194–198

Kalland T (1980b) Decreased and disproportionate T-cell population in adult mice after neonatal exposure to diethylstilbestrol. Cell Immunol 51(1):55–63

Kalland T (1980c) Ovarian influence on mitogen responsiveness of lymphocytes from mice neonatally exposed to diethylstilbestrol. J Toxicol Environ Health 6(1):67–74. doi:10.1080/15287398009529831

Kalland T (1980d) Reduced natural killer activity in female mice after neonatal exposure to diethylstilbestrol. J Immunol 124(3):1297–1300

Karagas MR, Choi AL, Oken E, Horvat M, Schoeny R, Kamai E, Cowell W, Grandjean P, Korrick S (2012) Evidence on the human health effects of low-level methylmercury exposure. Environ Health Perspect 120(6):799–806. doi:10.1289/ehp.1104494

Karrow NA, Guo TL, Delclos KB, Newbold RR, Weis C, Germolec DR, White KL Jr, McCay JA (2004) Nonylphenol alters the activity of splenic NK cells and the numbers of leukocyte subpopulations in Sprague-Dawley rats: a two-generation feeding study. Toxicology 196(3):237–245. doi:10.1016/j.tox.2003.11.009

Keil DE, Mehlmann T, Butterworth L, Peden-Adams MM (2008) Gestational exposure to perfluorooctane sulfonate suppresses immune function in B6C3F1 mice. Toxicol Sci 103(1):77–85. doi:10.1093/toxsci/kfn015

Kile ML, Houseman EA, Baccarelli AA, Quamruzzaman Q, Rahman M, Mostofa G, Cardenas A, Wright RO, Christiani DC (2014) Effect of prenatal arsenic exposure on DNA methylation and leukocyte subpopulations in cord blood. Epigenetics 9(5):774–782. doi:10.4161/epi.28153

Kimbrough RD, Krouskas CA (2001) Polychlorinated biphenyls, dibenzo-p-dioxins, and dibenzofurans and birth weight and immune and thyroid function in children. Regul Toxicol Pharmacol 34(1):42–52. doi:10.1006/rtph.2001.1484

Kirchner S, Kieu T, Chow C, Casey S, Blumberg B (2010) Prenatal exposure to the environmental obesogen tributyltin predisposes multipotent stem cells to become adipocytes. Mol Endocrinol 24(3):526–539. doi:10.1210/me.2009-0261

Knopik VS, Maccani MA, Francazio S, McGeary JE (2012) The epigenetics of maternal cigarette smoking during pregnancy and effects on child development. Dev Psychopathol 24(4):1377–1390. doi:10.1017/S0954579412000776

Koestler DC, Avissar-Whiting M, Houseman EA, Karagas MR, Marsit CJ (2013) Differential DNA methylation in umbilical cord blood of infants exposed to low levels of arsenic in utero. Environ Health Perspect 121(8):971–977. doi:10.1289/ehp.1205925

Kurzer MS, Xu X (1997) Dietary phytoestrogens. Annu Rev Nutr 17:353–381. doi:10.1146/annurev.nutr.17.1.353

Kushima K, Oda K, Sakuma S, Furusawa S, Fujiwara M (2007) Effect of prenatal administration of NSAIDs on the immune response in juvenile and adult rats. Toxicology 232(3):257–267. doi:10.1016/j.tox.2007.01.012

Lantz RC, Chau B, Sarihan P, Witten ML, Pivniouk VI, Chen GJ (2009) In utero and postnatal exposure to arsenic alters pulmonary structure and function. Toxicol Appl Pharmacol 235(1):105–113. doi:10.1016/j.taap.2008.11.012

Laschi A, Descotes J, Tachon P, Evreux JC (1983) Adverse influence of diazepam upon resistance to Klebsiella pneumoniae infection in mice. Toxicol Lett 16(3–4):281–284

Lawrence BP, Vorderstrasse BA (2013) New insights into the aryl hydrocarbon receptor as a modulator of host responses to infection. Semin Immunopathol 35(6):615–626. doi:10.1007/s00281-013-0395-3

Livezey GT, Marczynski TJ, McGrew EA, Beluhan FZ (1986) Prenatal exposure to diazepam: late postnatal teratogenic effect. Neurobehav Toxicol Teratol 8(5):433–440

Luebke RW, Chen DH, Dietert R, Yang Y, King M, Luster MI (2006) The comparative immunotoxicity of five selected compounds following developmental or adult exposure. J Toxicol Environ Health B Crit Rev 9(1):1–26. doi:10.1080/15287390500194326

Luster MI, Faith RE, McLachlan JA (1978) Alterations of the antibody response following in utero exposure to diethylstilbestrol. Bull Environ Contam Toxicol 20(4):433–437

Luster MI, Faith RE, McLachlan JA, Clark GC (1979) Effect of in utero exposure to diethylstilbestrol on the immune response in mice. Toxicol Appl Pharmacol 47(2):279–285

Luster MI, Boorman GA, Dean JH, Harris MW, Luebke RW, Padarathsingh ML, Moore JA (1980) Examination of bone marrow, immunologic parameters and host susceptibility following pre- and postnatal exposure to 2,3,7,8-tetrachlorodibenzo-p-dioxin (TCDD). Int J Immunopharmacol 2(4):301–310

Maccani JZ, Koestler DC, Houseman EA, Marsit CJ, Kelsey KT (2013) Placental DNA methylation alterations associated with maternal tobacco smoking at the RUNX3 gene are also associated with gestational age. Epigenomics 5(6):619–630. doi:10.2217/epi.13.63

Mahaffey KR, Corneliussen PE, Jelinek CF, Fiorino JA (1975) Heavy metal exposure from foods. Environ Health Perspect 12:63–69

Manikkam M, Tracey R, Guerrero-Bosagna C, Skinner MK (2012) Dioxin (TCDD) induces epigenetic transgenerational inheritance of adult onset disease and sperm epimutations. PLoS One 7(9), e46249. doi:10.1371/journal.pone.0046249

Manoli SE, Smith LA, Vyhlidal CA, An CH, Porrata Y, Cardoso WV, Baron RM, Haley KJ (2012) Maternal smoking and the retinoid pathway in the developing lung. Respir Res 13:42. doi:10.1186/1465-9921-13-42

Martin WJ 2nd, Glass RI, Balbus JM, Collins FS (2011) Public health. A major environmental cause of death. Science 334(6053):180–181. doi:10.1126/science.1213088

Martino D, Prescott S (2011) Epigenetics and prenatal influences on asthma and allergic airways disease. Chest 139(3):640–647. doi:10.1378/chest.10-1800

McEvoy CT, Schilling D, Clay N, Jackson K, Go MD, Spitale P, Bunten C, Leiva M, Gonzales D, Hollister-Smith J, Durand M, Frei B, Buist AS, Peters D, Morris CD, Spindel ER (2014) Vitamin C supplementation for pregnant smoking women and pulmonary function in their newborn infants: a randomized clinical trial. JAMA 311(20):2074–2082. doi:10.1001/jama.2014.5217

Mehra D, Geraghty PM, Hardigan AA, Foronjy R (2012) A comparison of the inflammatory and proteolytic effects of dung biomass and cigarette smoke exposure in the lung. PLoS One 7(12), e52889. doi:10.1371/journal.pone.0052889

Midoro-Horiuti T, Tiwari R, Watson CS, Goldblum RM (2010) Maternal bisphenol a exposure promotes the development of experimental asthma in mouse pups. Environ Health Perspect 118(2):273–277. doi:10.1289/ehp.0901259

Miyashita C, Sasaki S, Saijo Y, Washino N, Okada E, Kobayashi S, Konishi K, Kajiwara J, Todaka T, Kishi R (2011) Effects of prenatal exposure to dioxin-like compounds on allergies and infections during infancy. Environ Res 111(4):551–558. doi:10.1016/j.envres.2011.01.021

Moore SE, Prentice AM, Wagatsuma Y, Fulford AJ, Collinson AC, Raqib R, Vahter M, Persson LA, Arifeen SE (2009) Early-life nutritional and environmental determinants of thymic size in infants born in rural Bangladesh. Acta Paediatr 98(7):1168–1175. doi:10.1111/j.1651-2227.2009.01292.x

Motta M, Ciardelli L, Marconi M, Tincani A, Gasparoni A, Lojacono A, Chirico G (2007) Immune system development in infants born to mothers with autoimmune disease, exposed in utero to immunosuppressive agents. Am J Perinatol 24(8):441–447

Motta M, Tincani A, Meroni PL, Cimaz R (2008) Follow-up of children exposed antenatally to immunosuppressive drugs. Rheumatology (Oxford) 47 Suppl 3:iii32–iii34. doi:10.1093/rheumatology/ken149

Mouralidarane A, Soeda J, Visconti-Pugmire C, Samuelsson AM, Pombo J, Maragkoudaki X, Butt A, Saraswati R, Novelli M, Fusai G, Poston L, Taylor PD, Oben JA (2013) Maternal obesity programs offspring nonalcoholic fatty liver disease by innate immune dysfunction in mice. Hepatology 58(1):128–138. doi:10.1002/hep.26248

Murphy SK, Adigun A, Huang Z, Overcash F, Wang F, Jirtle RL, Schildkraut JM, Murtha AP, Iversen ES, Hoyo C (2012) Gender-specific methylation differences in relation to prenatal exposure to cigarette smoke. Gene 494(1):36–43. doi:10.1016/j.gene.2011.11.062

Mustafa A, Holladay SD, Goff M, Witonsky SG, Kerr R, Reilly CM, Sponenberg DP, Gogal RM Jr (2008) An enhanced postnatal autoimmune profile in 24 week-old C57BL/6 mice developmentally exposed to TCDD. Toxicol Appl Pharmacol 232(1):51–59. doi:10.1016/j.taap.2008.04.015

Mustafa A, Holladay S, Witonsky S, Zimmerman K, Manari A, Countermarsh S, Karpuzoglu E, Gogal R (2011a) Prenatal TCDD causes persistent modulation of the postnatal immune response, and exacerbates inflammatory disease, in 36-week-old lupus-like autoimmune SNF1 mice. Birth Defects Res B Dev Reprod Toxicol 92(1):82–94. doi:10.1002/bdrb.20285

Mustafa A, Holladay SD, Witonsky S, Sponenberg DP, Karpuzoglu E, Gogal RM Jr (2011b) A single mid-gestation exposure to TCDD yields a postnatal autoimmune signature, differing by sex, in early geriatric C57BL/6 mice. Toxicology 290(2–3):156–168. doi:10.1016/j.tox.2011.08.021

Myles IA, Fontecilla NM, Janelsins BM, Vithayathil PJ, Segre JA, Datta SK (2013) Parental dietary fat intake alters offspring microbiome and immunity. J Immunol 191(6):3200–3209. doi:10.4049/jimmunol.1301057

Nakajima Y, Goldblum RM, Midoro-Horiuti T (2012) Fetal exposure to bisphenol A as a risk factor for the development of childhood asthma: an animal model study. Environ Health 11:8. doi:10.1186/1476-069X-11-8

National Institute of Environmental Health Sciences (2010) Endocrine disruptors. National Institutes of Health, Research Triangle Park

Newbold RR, Padilla-Banks E, Jefferson WN (2006) Adverse effects of the model environmental estrogen diethylstilbestrol are transmitted to subsequent generations. Endocrinology 147(6 Suppl):S11–S17. doi:10.1210/en.2005-1164

Newbold RR, Padilla-Banks E, Snyder RJ, Phillips TM, Jefferson WN (2007) Developmental exposure to endocrine disruptors and the obesity epidemic. Reprod Toxicol 23(3):290–296. doi:10.1016/j.reprotox.2006.12.010

Ng SP, Zelikoff JT (2007) Smoking during pregnancy: subsequent effects on offspring immune competence and disease vulnerability in later life. Reprod Toxicol 23(3):428–437. doi:10.1016/j.reprotox.2006.11.008

Ng SP, Zelikoff JT (2008) The effects of prenatal exposure of mice to cigarette smoke on offspring immune parameters. J Toxicol Environ Health A 71(7):445–453. doi:10.1080/15287390701839281

Ng SP, Silverstone AE, Lai ZW, Zelikoff JT (2006) Effects of prenatal exposure to cigarette smoke on offspring tumor susceptibility and associated immune mechanisms. Toxicol Sci 89(1):135–144. doi:10.1093/toxsci/kfj006

Nguyen LP, Bradfield CA (2008) The search for endogenous activators of the aryl hydrocarbon receptor. Chem Res Toxicol 21(1):102–116. doi:10.1021/tx7001965

Nilsson E, Larsen G, Manikkam M, Guerrero-Bosagna C, Savenkova MI, Skinner MK (2012) Environmentally induced epigenetic transgenerational inheritance of ovarian disease. PLoS One 7(5), e36129. doi:10.1371/journal.pone.0036129

Noakes PS, Holt PG, Prescott SL (2003) Maternal smoking in pregnancy alters neonatal cytokine responses. Allergy 58(10):1053–1058

Noakes PS, Hale J, Thomas R, Lane C, Devadason SG, Prescott SL (2006) Maternal smoking is associated with impaired neonatal toll-like-receptor-mediated immune responses. Eur Respir J 28(4):721–729. doi:10.1183/09031936.06.00050206

Noller KL, Blair PB, O'Brien PC, Melton LJ 3rd, Offord JR, Kaufman RH, Colton T (1988) Increased occurrence of autoimmune disease among women exposed in utero to diethylstilbestrol. Fertil Steril 49(6):1080–1082

O'Brien E, Bergin IL, Dolinoy DC, Zaslona Z, Little RJ, Tao Y, Peters-Golden M, Mancuso P (2014) Perinatal bisphenol A exposure beginning before gestation enhances allergen sensitization, but not pulmonary inflammation, in adult mice. J Dev Orig Health Dis 5(2):121–131. doi:10.1017/S204017441400004X

Odaka Y, Nakano M, Tanaka T, Kaburagi T, Yoshino H, Sato-Mito N, Sato K (2010) The influence of a high-fat dietary environment in the fetal period on postnatal metabolic and immune function. Obesity (Silver Spring) 18(9):1688–1694. doi:10.1038/oby.2009.513

Ohshima Y, Yamada A, Tokuriki S, Yasutomi M, Omata N, Mayumi M (2007) Transmaternal exposure to bisphenol a modulates the development of oral tolerance. Pediatr Res 62(1):60–64. doi:10.1203/PDR.0b013e3180674dae

Padgett EL, Seelig LL Jr (2002) Effects on T-cell maturation and proliferation induced by lactational transfer of cyclosporine to nursing pups. Transplantation 73(6):867–874

Papoutsis AJ, Selmin OI, Borg JL, Romagnolo DF (2013) Gestational exposure to the AhR agonist 2,3,7,8-tetrachlorodibenzo-p-dioxin induces BRCA-1 promoter hypermethylation and reduces BRCA-1 expression in mammary tissue of rat offspring: preventive effects of resveratrol. Mol Carcinog. doi:10.1002/mc.22095

Park HY, Hertz-Picciotto I, Petrik J, Palkovicova L, Kocan A, Trnovec T (2008) Prenatal PCB exposure and thymus size at birth in neonates in Eastern Slovakia. Environ Health Perspect 116(1):104–109. doi:10.1289/ehp.9769

Peden-Adams MM, Stuckey JE, Gaworecki KM, Berger-Ritchie J, Bryant K, Jodice PG, Scott TR, Ferrario JB, Guan B, Vigo C, Boone JS, McGuinn WD, DeWitt JC, Keil DE (2009) Developmental toxicity in white leghorn chickens following in ovo exposure to perfluorooctane sulfonate (PFOS). Reprod Toxicol 27(3–4):307–318. doi:10.1016/j.reprotox.2008.10.009

Penn AL, Rouse RL, Horohov DW, Kearney MT, Paulsen DB, Lomax L (2007) In utero exposure to environmental tobacco smoke potentiates adult responses to allergen in BALB/c mice. Environ Health Perspect 115(4):548–555. doi:10.1289/ehp.9780

Pilarski LM, Yacyshyn BR, Lazarovits AI (1994) Analysis of peripheral blood lymphocyte populations and immune function from children exposed to cyclosporine or to azathioprine in utero. Transplantation 57(1):133–144

Pillet S, Rooney AA, Bouquegneau JM, Cyr DG, Fournier M (2005) Sex-specific effects of neonatal exposures to low levels of cadmium through maternal milk on development and immune functions of juvenile and adult rats. Toxicology 209(3):289–301. doi:10.1016/j.tox.2004.12.007

Pilones K, Lai ZW, Gavalchin J (2007) Prenatal HgCl(2) exposure alters fetal cell phenotypes. J Immunotoxicol 4(4):295–301. doi:10.1080/15476910701680178

Pilones K, Tatum A, Gavalchin J (2009) Gestational exposure to mercury leads to persistent changes in T-cell phenotype and function in adult DBF1 mice. J Immunotoxicol 6(3):161–170. doi:10.1080/15476910903084021

Pollard KM, Pearson DL, Hultman P, Deane TN, Lindh U, Kono DH (2001) Xenobiotic acceleration of idiopathic systemic autoimmunity in lupus-prone bxsb mice. Environ Health Perspect 109(1):27–33

Prins GS, Birch L, Couse JF, Choi I, Katzenellenbogen B, Korach KS (2001) Estrogen imprinting of the developing prostate gland is mediated through stromal estrogen receptor alpha: studies with alphaERKO and betaERKO mice. Cancer Res 61(16):6089–6097

Ramsey KA, Foong RE, Sly PD, Larcombe AN, Zosky GR (2013) Early life arsenic exposure and acute and long-term responses to influenza A infection in mice. Environ Health Perspect 121(10):1187–1193. doi:10.1289/ehp.1306748

Raqib R, Ahmed S, Sultana R, Wagatsuma Y, Mondal D, Hoque AM, Nermell B, Yunus M, Roy S, Persson LA, Arifeen SE, Moore S, Vahter M (2009) Effects of in utero arsenic exposure on child immunity and morbidity in rural Bangladesh. Toxicol Lett 185(3):197–202. doi:10.1016/j.toxlet.2009.01.001

Razzaghi H, Tinker SC, Crider K (2014) Blood mercury concentrations in pregnant and non-pregnant women in the United States: National Health and Nutrition Examination Survey 1999–2006. Am J Obstet Gynecol 210(4):357.e351–357.e359. doi:10.1016/j.ajog.2013.10.884

Rodriguez JW, Kirlin WG, Wirsiy YG, Matheravidathu S, Hodge TW, Urso P (1999) Maternal exposure to benzo[a]pyrene alters development of T lymphocytes in offspring. Immunopharmacol Immunotoxicol 21(2):379–396. doi:10.3109/08923979909052769

Rodriguez JW, Kohan MJ, King LC, Kirlin WG (2002) Detection of DNA adducts in developing CD4+ CD8+ thymocytes and splenocytes following in utero exposure to benzo[a]pyrene. Immunopharmacol Immunotoxicol 24(3):365–381. doi:10.1081/IPH-120014723

Rooney AA, Matulka RA, Luebke RW (2003) Developmental atrazine exposure suppresses immune function in male, but not female Sprague-Dawley rats. Toxicol Sci 76(2):366–375. doi:10.1093/toxsci/kfg250

Rowe AM, Brundage KM, Barnett JB (2008) Developmental immunotoxicity of atrazine in rodents. Basic Clin Pharmacol Toxicol 102(2):139–145. doi:10.1111/j.1742-7843.2007.00175.x

Roy A, Bauer SM, Lawrence BP (2012) Developmental exposure to bisphenol A modulates innate but not adaptive immune responses to influenza A virus infection. PLoS One 7(6), e38448. doi:10.1371/journal.pone.0038448

Roy A, Gaylo A, Cao W, Saubermann LJ, Lawrence BP (2013) Neither direct nor developmental exposure to bisphenol A alters the severity of experimental inflammatory colitis in mice. J Immunotoxicol 10(4):334–340. doi:10.3109/1547691X.2012.747231

Sakaguchi S, Sakaguchi N (1989) Organ-specific autoimmune disease induced in mice by elimination of T cell subsets. V Neonatal administration of cyclosporin A causes autoimmune disease. J Immunol 142(2):471–480

Sanders AP, Flood K, Chiang S, Herring AH, Wolf L, Fry RC (2012) Towards prenatal biomonitoring in North Carolina: assessing arsenic, cadmium, mercury, and lead levels in pregnant women. PLoS One 7(3), e31354. doi:10.1371/journal.pone.0031354

Sanders AP, Smeester L, Rojas D, DeBussycher T, Wu MC, Wright FA, Zhou YH, Laine JE, Rager JE, Swamy GK, Ashley-Koch A, Lynn Miranda M, Fry RC (2014) Cadmium exposure and the epigenome: exposure-associated patterns of DNA methylation in leukocytes from mother-baby pairs. Epigenetics 9(2):212–221. doi:10.4161/epi.26798

Schlumpf M, Ramseier H, Lichtensteiger W (1989) Prenatal diazepam induced persisting depression of cellular immune responses. Life Sci 44(7):493–501

Schlumpf M, Lichtensteiger W, Ramseier H (1993) Diazepam treatment of pregnant rats differentially affects interleukin-1 and interleukin-2 secretion in their offspring during different phases of postnatal development. Pharmacol Toxicol 73(6):335–340

Schlumpf M, Butikofer EE, Schreiber AA, Parmar R, Ramseier HR, Lichtensteiger W (1994a) Delayed developmental immunotoxicity of prenatal benzodiazepines. Toxicol In Vitro 8(5):1061–1065

Schlumpf M, Lichtensteiger W, van Loveren H (1994b) Impaired host resistance to Trichinella spiralis as a consequence of prenatal treatment of rats with diazepam. Toxicology 94(1–3):223–230

Schmatz M, Madan J, Marino T, Davis J (2010) Maternal obesity: the interplay between inflammation, mother and fetus. J Perinatol 30(7):441–446. doi:10.1038/jp.2009.182

Schober SE, Sinks TH, Jones RL, Bolger PM, McDowell M, Osterloh J, Garrett ES, Canady RA, Dillon CF, Sun Y, Joseph CB, Mahaffey KR (2003) Blood mercury levels in US children and women of childbearing age, 1999–2000. JAMA 289(13):1667–1674. doi:10.1001/jama.289.13.1667

Schreiber AA, Frei K, Lichtensteiger W, Schlumpf M (1993a) Alterations in interleukin-6 production by LPS- and Con A-stimulated mixed splenocytes, spleen macrophages and lymphocytes in prenatally diazepam-exposed rats. Agents Actions 39(3–4):166–173

Schreiber AA, Frei K, Lichtensteiger W, Schlumpf M (1993b) The effect of prenatal diazepam exposure on TNF-alpha production by rat splenocytes. Agents Actions 38(3–4):265–272

Selgrade MK, Blain RB, Fedak KM, Cawley MA (2013) Potential risk of asthma associated with in utero exposure to xenobiotics. Birth Defects Res C Embryo Today 99(1):1–13. doi:10.1002/bdrc.21028

Shek LP, Chong MF, Lim JY, Soh SE, Chong YS (2012) Role of dietary long-chain polyunsaturated fatty acids in infant allergies and respiratory diseases. Clin Dev Immunol 2012:730568. doi:10.1155/2012/730568

Silva IA, El Nabawi M, Hoover D, Silbergeld EK (2005) Prenatal HgCl2 exposure in BALB/c mice: gender-specific effects on the ontogeny of the immune system. Dev Comp Immunol 29(2):171–183. doi:10.1016/j.dci.2004.05.008

Silva SV, Garcia-Souza EP, Moura AS, Barja-Fidalgo C (2010) Maternal protein restriction during early lactation induces changes on neutrophil activation and TNF-alpha production of adult offspring. Inflammation 33(2):65–75. doi:10.1007/s10753-009-9159-6

Singh SP, Razani-Boroujerdi S, Pena-Philippides JC, Langley RJ, Mishra NC, Sopori ML (2006) Early postnatal exposure to cigarette smoke impairs the antigen-specific T-cell responses in the spleen. Toxicol Lett 167(3):231–237. doi:10.1016/j.toxlet.2006.10.001

Singh AK, Parashar A, Singh R (2013a) Pre-natal/juvenile chlorpyrifos exposure associated with immunotoxicity in adulthood in Swiss albino mice. J Immunotoxicol 10(2):141–149. doi:10.3109/1547691X.2012.700653

Singh SP, Gundavarapu S, Smith KR, Chand HS, Saeed AI, Mishra NC, Hutt J, Barrett EG, Husain M, Harrod KS, Langley RJ, Sopori ML (2013b) Gestational exposure of mice to secondhand cigarette smoke causes bronchopulmonary dysplasia blocked by the nicotinic receptor antagonist mecamylamine. Environ Health Perspect 121(8):957–964. doi:10.1289/ehp.1306611

Smialowicz RJ, Riddle MM, Rogers RR, Rowe DG, Luebke RW, Fogelson LD, Copeland CB (1988) Immunologic effects of perinatal exposure of rats to dioctyltin dichloride. J Toxicol Environ Health 25(4):403–422. doi:10.1080/15287398809531220

Smialowicz RJ, Riddle MM, Rogers RR, Luebke RW, Copeland CB (1989) Immunotoxicity of tributyltin oxide in rats exposed as adults or pre-weanlings. Toxicology 57(1):97–111

Smith AH, Marshall G, Yuan Y, Ferreccio C, Liaw J, von Ehrenstein O, Steinmaus C, Bates MN, Selvin S (2006) Increased mortality from lung cancer and bronchiectasis in young adults after exposure to arsenic in utero and in early childhood. Environ Health Perspect 114(8):1293–1296

Soukupova D, Dostal M, Piza J (1991) Developmental toxicity of cadmium in mice. II. Immunotoxic effects. Funct Dev Morphol 1(4):31–36

Spanier AJ, Kahn RS, Kunselman AR, Hornung R, Xu Y, Calafat AM, Lanphear BP (2012) Prenatal exposure to bisphenol A and child wheeze from birth to 3 years of age. Environ Health Perspect 120(6):916–920. doi:10.1289/ehp.1104175

Sram RJ, Binkova B, Dostal M, Merkerova-Dostalova M, Libalova H, Milcova A, Rossner P Jr, Rossnerova A, Schmuczerova J, Svecova V, Topinka J, Votavova H (2013) Health impact of air pollution to children. Int J Hyg Environ Health 216(5):533–540. doi:10.1016/j.ijheh.2012.12.001

Stern AH, Gochfeld M, Weisel C, Burger J (2001) Mercury and methylmercury exposure in the New Jersey pregnant population. Arch Environ Health 56(1):4–10. doi:10.1080/00039890109604048

Stolevik SB, Nygaard UC, Namork E, Haugen M, Meltzer HM, Alexander J, Knutsen HK, Aaberge I, Vainio K, van Loveren H, Lovik M, Granum B (2013) Prenatal exposure to polychlorinated biphenyls and dioxins from the maternal diet may be associated with immunosuppressive effects that persist into early childhood. Food Chem Toxicol 51:165–172. doi:10.1016/j.fct.2012.09.027

Sugita-Konishi Y, Kobayashi K, Naito H, Miura K, Suzuki Y (2003) Effect of lactational exposure to 2,3,7,8-tetrachlorodibenzo-p-dioxin on the susceptibility to Listeria infection. Biosci Biotechnol Biochem 67(1):89–93

Sunyer J, Garcia-Esteban R, Alvarez M, Guxens M, Goni F, Basterrechea M, Vrijheid M, Guerra S, Anto JM (2010) DDE in mothers' blood during pregnancy and lower respiratory tract infections in their infants. Epidemiology 21(5):729–735. doi:10.1097/EDE.0b013e3181e5ea96

Tebow G, Sherrill DL, Lohman IC, Stern DA, Wright AL, Martinez FD, Halonen M, Guerra S (2008) Effects of parental smoking on interferon gamma production in children. Pediatrics 121(6):e1563–e1569. doi:10.1542/peds.2007-2795

Thomas PT, Hinsdill RD (1979) The effect of perinatal exposure to tetrachlorodibenzo-p-dioxin on the immune response of young mice. Drug Chem Toxicol 2(1–2):77–98. doi:10.3109/01480547908993183

Thuvander A, Sundberg J, Oskarsson A (1996) Immunomodulating effects after perinatal exposure to methylmercury in mice. Toxicology 114(2):163–175

Tonk EC, de Groot DM, Penninks AH, Waalkens-Berendsen ID, Wolterbeek AP, Slob W, Piersma AH, van Loveren H (2010) Developmental immunotoxicity of methylmercury: the relative sensitivity of developmental and immune parameters. Toxicol Sci 117(2):325–335. doi:10.1093/toxsci/kfq223

Tonk EC, de Groot DM, Penninks AH, Waalkens-Berendsen ID, Wolterbeek AP, Piersma AH, van Loveren H (2011a) Developmental immunotoxicity of di-n-octyltin dichloride (DOTC) in an extended one-generation reproductive toxicity study. Toxicol Lett 204(2–3):156–163. doi:10.1016/j.toxlet.2011.04.027

Tonk EC, Verhoef A, de la Fonteyne LJ, Waalkens-Berendsen ID, Wolterbeek AP, van Loveren H, Piersma AH (2011b) Developmental immunotoxicity in male rats after juvenile exposure to di-n-octyltin dichloride (DOTC). Reprod Toxicol 32(3):341–348. doi:10.1016/j.reprotox.2011.08.005

Torres-Duque C, Maldonado D, Perez-Padilla R, Ezzati M, Viegi G (2008) Biomass fuels and respiratory diseases: a review of the evidence. Proc Am Thorac Soc 5(5):577–590. doi:10.1513/pats.200707-100RP

Trevor J, Antony V, Jindal SK (2014) The effect of biomass fuel exposure on the prevalence of asthma in adults in India – review of current evidence. J Asthma 51(2):136–141. doi:10.3109/02770903.2013.849269

U.S. Department of Health and Human Services (1999) Toxicological profile for mercury. Agency for toxic substances and disease registry. Division of Toxicology/Toxicology Information Branch, Atlanta

U.S. Department of Health and Human Services (2009) Toxicological profile for cadmium. Agency for toxic substances and disease registry. Division of Toxicology and Human Health Sciences, Atlanta

Ugaz EM, Pinheiro SR, Guerra JL, Palermo-Neto J (1999) Effects of prenatal diazepam treatment on Mycobacterium bovis-induced infection in hamsters. Immunopharmacology 41(3):209–217

United States Environmental Protection Agency (2007) Atrazine. US EPA, Washington, D.C

Urso P, Gengozian N (1982) Alterations in the humoral immune response and tumor frequencies in mice exposed to benzo[a]pyrene and X-rays before or after birth. J Toxicol Environ Health 10(4–5):817–835. doi:10.1080/15287398209530297

Urso P, Gengozian N (1984) Subnormal expression of cell-mediated and humoral immune responses in progeny disposed toward a high incidence of tumors after in utero exposure to benzo[a]pyrene. J Toxicol Environ Health 14(4):569–584. doi:10.1080/15287398409530606

Urso P, Johnson RA (1987) Early changes in T lymphocytes and subsets of mouse progeny defective as adults in controlling growth of a syngeneic tumor after in utero insult with benzo(a) pyrene. Immunopharmacology 14(1):1–10

Vahter M (2008) Health effects of early life exposure to arsenic. Basic Clin Pharmacol Toxicol 102(2):204–211. doi:10.1111/j.1742-7843.2007.00168.x

Vaidya SV, Kulkarni H (2012) Association of urinary bisphenol A concentration with allergic asthma: results from the National Health and Nutrition Examination Survey 2005–2006. J Asthma 49(8):800–806. doi:10.3109/02770903.2012.721041

Vandenberg LN, Colborn T, Hayes TB, Heindel JJ, Jacobs DR Jr, Lee DH, Shioda T, Soto AM, vom Saal FS, Welshons WV, Zoeller RT, Myers JP (2012) Hormones and endocrine-disrupting chemicals: low-dose effects and nonmonotonic dose responses. Endocr Rev 33(3):378–455. doi:10.1210/er.2011-1050

Via CS, Nguyen P, Niculescu F, Papadimitriou J, Hoover D, Silbergeld EK (2003) Low-dose exposure to inorganic mercury accelerates disease and mortality in acquired murine lupus. Environ Health Perspect 111(10):1273–1277

Victora CG, Adair L, Fall C, Hallal PC, Martorell R, Richter L, Sachdev HS (2008) Maternal and child undernutrition: consequences for adult health and human capital. Lancet 371(9609):340–357. doi:10.1016/S0140-6736(07)61692-4

Vorderstrasse BA, Cundiff JA, Lawrence BP (2004) Developmental exposure to the potent aryl hydrocarbon receptor agonist 2,3,7,8-tetrachlorodibenzo-p-dioxin Impairs the cell-mediated immune response to infection with influenza a virus, but enhances elements of innate immunity. J Immunotoxicol 1(2):103–112. doi:10.1080/15476910490509244

Vorderstrasse BA, Cundiff JA, Lawrence BP (2006) A dose-response study of the effects of prenatal and lactational exposure to TCDD on the immune response to influenza a virus. J Toxicol Environ Health A 69(6):445–463. doi:10.1080/15287390500246985

Vos JG, Moore JA (1974) Suppression of cellular immunity in rats and mice by maternal treatment with 2,3,7,8-tetrachlorodibenzo-p-dioxin. Int Arch Allergy Appl Immunol 47(5):777–794

Vos JG, De Klerk A, Krajnc EI, Van Loveren H, Rozing J (1990) Immunotoxicity of bis(tri-n-butyltin) oxide in the rat: effects on thymus-dependent immunity and on nonspecific resistance following long-term exposure in young versus aged rats. Toxicol Appl Pharmacol 105(1):144–155

Walker DB, Williams WC, Copeland CB, Smialowicz RJ (2004) Persistent suppression of contact hypersensitivity, and altered T-cell parameters in F344 rats exposed perinatally to 2,3,7,8-tetrachlorodibenzo-p-dioxin (TCDD). Toxicology 197(1):57–66. doi:10.1016/j.tox.2003.12.012

Wang IJ, Hsieh WS, Chen CY, Fletcher T, Lien GW, Chiang HL, Chiang CF, Wu TN, Chen PC (2011) The effect of prenatal perfluorinated chemicals exposures on pediatric atopy. Environ Res 111(6):785–791. doi:10.1016/j.envres.2011.04.006

Watanabe W, Yoshida H, Hirose A, Akashi T, Takeshita T, Kuroki N, Shibata A, Hongo S, Hashiguchi S, Konno K, Kurokawa M (2013) Perinatal exposure to insecticide methamidophos suppressed production of proinflammatory cytokines responding to virus infection in lung tissues in mice. Biomed Res Int 2013:151807. doi:10.1155/2013/151807

Webster G (2010) Potential human health effects of perfluorinated chemicals (PFCs). National Collaborating Centre for Environmental Health, Vancouver

Weisglas-Kuperus N, Patandin S, Berbers GA, Sas TC, Mulder PG, Sauer PJ, Hooijkaas H (2000) Immunologic effects of background exposure to polychlorinated biphenyls and dioxins in Dutch preschool children. Environ Health Perspect 108(12):1203–1207

Weisglas-Kuperus N, Vreugdenhil HJ, Mulder PG (2004) Immunological effects of environmental exposure to polychlorinated biphenyls and dioxins in Dutch school children. Toxicol Lett 149(1–3):281–285. doi:10.1016/j.toxlet.2003.12.039

World Health Organization (1999) Tributyltin oxide. World Health Organization, Geneva

World Health Organization (2004) Dialkyltins in drinking-water. World Health Organization, Geneva

World Health Organization (2014) Tobacco fact sheet number 339. http://www.who.int/mediacentre/factsheets/fs339/en/. Accessed May 2014

Wu ZX, Hunter DD, Kish VL, Benders KM, Batchelor TP, Dey RD (2009) Prenatal and early, but not late, postnatal exposure of mice to sidestream tobacco smoke increases airway hyperresponsiveness later in life. Environ Health Perspect 117(9):1434–1440. doi:10.1289/ehp.0800511

Yamamoto S, Tin Tin Win S, Yoshida Y, Kunugita N, Arashidani K, Fujimaki H (2009a) Children's immunology, what can we learn from animal studies (2): Modulation of systemic Th1/Th2 immune response in infant mice after prenatal exposure to low-level toluene and toll-like receptor (TLR) 2 ligand. J Toxicol Sci 34(Suppl 2):SP341–SP348

Yamamoto S, Win-Shwe TT, Yoshida Y, Kunugita N, Arashidani K, Fujimaki H (2009b) Suppression of Th1- and Th2-type immune responses in infant mouse spleen after prenatal and postnatal exposure to low-level toluene and peptidoglycan. Inhal Toxicol 21(9):793–802. doi:10.1080/08958370902798448

Yan H, Takamoto M, Sugane K (2008) Exposure to Bisphenol A prenatally or in adulthood promotes T(H)2 cytokine production associated with reduction of CD4CD25 regulatory T cells. Environ Health Perspect 116(4):514–519. doi:10.1289/ehp.10829

Yoshino S, Yamaki K, Li X, Sai T, Yanagisawa R, Takano H, Taneda S, Hayashi H, Mori Y (2004) Prenatal exposure to bisphenol A up-regulates immune responses, including T helper 1 and T helper 2 responses, in mice. Immunology 112(3):489–495. doi:10.1111/j.1365-2567.2004.01900.x

Zhai L, Liao X, Chen T, Yan X, Xie H, Wu B, Wang L (2008) Regional assessment of cadmium pollution in agricultural lands and the potential health risk related to intensive mining activities: a case study in Chenzhou City, China. J Environ Sci (China) 20(6):696–703

Zhang Y, Gao D, Bolivar VJ, Lawrence DA (2011) Induction of autoimmunity to brain antigens by developmental mercury exposure. Toxicol Sci 119(2):270–280. doi:10.1093/toxsci/kfq334

Zhang WL, Du Y, Zhai MM, Shang Q (2014) Cadmium exposure and its health effects: a 19-year follow-up study of a polluted area in China. Sci Total Environ 470–471:224–228. doi:10.1016/j.scitotenv.2013.09.070

3 Environmental Influences on the Immune System: The Aging Immune System

Julia N. Mälzer, Axel R. Schulz, and Andreas Thiel

Contents

3.1 Introduction

Immunological competence progressively declines with age, resulting in increased susceptibility of the elderly to infection (Gavazzi and Krause 2002) and higher incidences of certain malignant and autoimmune diseases (Goronzy and Weyand 2012; Fulop et al. 2007; Jemal et al. 2011). Impaired responses to vaccines as well as increased inflammatory activity are further key features of immunological deregulations at higher age (Weinberger and Grubeck-Loebenstein 2012; Vadasz et al. 2013). The underlying biological changes for this gradual deterioration are complex, and a multitude of components of the immune system are affected. Only some of the mechanisms have been identified so far at the molecular, cellular, or systemic level. In the context of a rapidly aging human population, an enhanced understanding of

J.N. Mälzer • A.R. Schulz
Regenerative Immunology and Aging, Berlin-Brandenburg Center for Regenerative Therapies (BCRT), Charité Universitätsmedizin Berlin, Berlin, Germany

A. Thiel (✉)
Regenerative Immunology and Aging, Berlin-Brandenburg Center for Regenerative Therapies (BCRT), Charité Universitätsmedizin Berlin, Campus Virchow-Klinikum, Berlin, Germany
e-mail: andreas.thiel@charite.de

C. Esser (ed.), *Environmental Influences on the Immune System*,
DOI 10.1007/978-3-7091-1890-0_3

the mechanisms driving "immunosenescence" is of great interest to promote healthy aging as well as to identify potential threats to it.

Accelerated population aging results from increasing life expectancy as well as declining fertility and can be classified as one of the major global challenges of the twenty-first century. Since many physiological alterations can be attributed to aging and potentially lead to age-related clinical syndromes (Stanziano et al. 2010), it may be considered as a pandemic affecting the world's population. In 2013, the global share of people over the age of 60 was 11.7 % and is estimated to reach 21.1 % by 2050. That means that the population stratum of people aged 60 years and older will more than double in size from 841 million in 2013 to over 2 billion in 2050 (United Nations and World Population 2013). The declining ratio between working-age and elderly adults resulting from this demographic shift affects both social and economical levels of society, particularly challenging health systems (Mathers and Loncar 2006). Importantly, population aging is clearly not only affecting industrialized countries, where societies have already reached a status of being considerably aged. On the contrary, even nowadays, two thirds of all people over the age of 60 are living in less developed countries, a share which is expected to rise up to 80 % by 2050 (United Nations and World Population 2013).

Especially in the developing regions of the world, infectious diseases are contributing to overall morbidity and mortality (WHO and The top 10 causes of death 2014). But also in industrialized countries, pneumonia and infections with seasonal influenza are ranked under the leading causes of death in persons over the age of 65. More than 85 % of fatal outcomes due to influenza infection fall into the age group above the age 65 years (Thompson et al. 2009). An analysis of an outbreak of West Nile virus in New York in 1999 also identified higher age as the most prominent risk factor associated with death even when a correction for comorbidities was made (Nash et al. 2001).

Certain autoimmune diseases such as giant cell arthritis only start to manifest in patients during late adulthood, while the incidence of others such as rheumatoid arthritis is gradually increasing with age. The underlying mechanisms are not well understood. Age-related impairments in B-cell development have been discussed as an underlying mechanism. Several studies demonstrated elevated levels of autoantibodies in older adults. However, it has been demonstrated that these antibodies are not necessarily associated with the clinical manifestation of autoimmunity (Ramos-Casals et al. 2004). The immune system also plays a fundamental role in the detection and control of aberrant tumor cells, arising in every individual. Cancer incidence and mortality markedly increase after the age of 65, leveling off around age 85–90 (Azar and Ballas 2014). This suggests that together with the accumulation of mutational defects through intrinsic genetic defects, toxic environmental influences, deficient reparation, and clearing potentials, an age-related altered immune competence might well foster cancer formation in the elderly (Malaguarnera et al. 2010).

Altered responses to various vaccinations are another essential feature of age-related changes in human immunity. As a matter of principle, vaccinations can be highly efficient in the prevention of infectious diseases with relatively low medical effort and costs and tolerable side effects for the individual. Unfortunately, the responsiveness to many vaccines is often curtailed in the elderly, comprising recombinant, inactivated, as well as attenuated live vaccines. In developing countries on the one hand, access to medical resources can be limited, and therefore, efficient disease prevention by successful vaccination is vital. In industrialized countries on the other hand, the group of "very old" people over the age of 80 years (United Nations and World Population 2013), being particularly vulnerable to infections and their complications, is rapidly growing (Matias et al. 2014). Finding strategies to overcome this impairment is therefore an important goal in preventing diseases and in preserving a good quality of life in the elderly worldwide.

Investigating senescence changes of the human immune system is highly complex. It can be difficult or even impossible to determine what comes first: Is the general process of biological aging causing immunosenescence, which is then resulting in reduced immunocompetence? Or is the decline in immunological functions observed in the elderly driving the general aging process by making the individual more susceptible for infections and other diseases? In the end, these two statements may not be separated, and the underlying pathways may rather be highly connected. Another major issue is that a lot of the present knowledge on immunosenescence has been obtained from studies in mice. The transferability of the results found in murine studies to the human context is however limited, and we would argue it as highly questionable. Not only is the average life span of a laboratory mouse (1–2 years) approximately 40 times shorter than that of a human (60–90 years), but also environmental conditions like pathogen load and chronic viral infections as well as other aspects such as nutrition, climate, and exercise differ substantially. Laboratory mice are usually kept under rather sterile conditions in comparison with humans who are constantly being challenged with billions of microbes and viruses throughout their life. Furthermore, various species-specific immunological differences in both innate and adaptive immune functions have been discovered, emphasizing the need of finding alternative strategies to understand human immunosenescence (Seok et al. 2013; Goronzy and Weyand 2013).

3.2 The Evolutionary Perspective

When discussing immunosenescence, it can be of interest to view the occurring changes from an evolutionary perspective. The progressive loss of organ function with increasing age accompanied by decreasing fertility and increasing mortality affects most species and most organ systems. On an individual level, these changes are generally unfavorable. On a population level however, "programmed" senescence could be understood as an evolutionary strategy to control population size and facilitate better adaption to changing environments by consistent generation turnover. Humans as well as domesticated animals however, living in a relatively risk-free environment nowadays, can outlive their reproductive period by far, allowing senescence changes to manifest themselves. Since those are revealed only after reproduction, late-acting deleterious mutations driving senescence may escape the process of selection and can therefore be passed on to future generations (Kirkwood and Austad 2000).

The senescence process of the immune system is a phenomenon affecting not only humans but a broad variety of vertebrate and invertebrate species. Thymic involution, for example, meaning the progressive loss of thymus function with advancing age, is evolutionary conserved and occurs in all species possessing a thymus (birds, mammals, and other vertebrates) (Torroba and Zapata 2003). Given the rather uniform and early onset of thymic involution in humans, it does not form part of the gradual loss of organ functions at higher age classically being referred to as senescence. It can rather be understood as a mechanism to optimize the use of external and internal resources (e.g., energy intake, elimination of waste products, damage repair mechanisms) to ensure maximum evolutionary success in the sense of undisturbed reproduction and general fitness. In terms of the immune system, the energetic investment made into its establishment and maintenance needs to be sufficient to protect the organism against pathogens before and during the reproductive period (Kirkwood 1977). Secondly, it needs to be adequately regulated to promote stable self-tolerance and to prevent overshooting immune responses, being both energy consuming and tissue damaging. Lastly, an efficient regulation of immune functions also comprises the aspect of risk reduction for a potential development of malignancies, a hazard inherent to all proliferating systems (Shanley et al. 2009). While a programmed reduction of thymic function once might have been beneficial for the individual, the recent rise of human life expectancy far beyond the reproductive period may now unmask negative effects of thymic involution associated with reduced immunocompetence in the elderly. As previously discussed for the general aging process, also genetic variants promoting immunosenescence escape into the post-reproductive "selection shadow" and can therefore persist throughout generations despite their negative consequences for the individual.

With advancing age, there is not only an increased susceptibility to infection but also to the development of chronic low-grade inflammation, being referred to as "inflamm-aging" by some sources (Franceschi et al. 2000). The underlying subclinical and mostly asymptomatic inflammatory circuits have been suspected to be highly influential in the induction and/or chronification of a variety of age-related diseases (Ferrucci et al. 1999; Jeffery 2008; Bruunsgaard et al. 2003; Teixeira et al. 2014). The drastic improvements of hygiene standards and enhanced means to prevent and control infectious diseases have greatly contributed to reduction of concurrent infections and contact with inflammation inducing or maintaining agents. These changes certainly played a major role in the extension of life span, particularly during the last few centuries. Before that, it was vital to possess the ability to actively fight infections by mounting inflammatory responses, whereas now, pathogen loads are lower and highly effective treatments such as antibiotics are available, and prophylactic vaccinations protect against major pathogens (Shanley et al. 2009).

One has to argue that from an evolutionary view, the immune system has not evolved to provide protection for almost centenarians traveling from one continent to another. Successful long-term aging happened in ancient times too. However, "very old individuals" were the exception and maybe similar to today profiting from coordinated pro- and anti-inflammatory profiles. A balanced control of inflammation may be the most important contributing factor to enable longevity (Franceschi and Bonafe 2003).

3.3 Effects on Hematopoietic Stem Cells and Telomeres

Hematopoietic stem cells (HSCs) residing in the bone marrow give rise to both myeloid and lymphoid progenitor cells as well as to erythrocytes. During aging, bone marrow mass and cellularity decline. However, red cell life span, blood volume, and total white blood cell counts remain relatively stable in healthy elderly individuals (Bagnara et al. 2000). Other hematopoietic branches are more affected. Since regenerative capacities are limited, the hematopoietic system of old individuals is more vulnerable to various challenges like illness, bleedings, malignancies, and malnutrition but also medical treatment like radiation and chemotherapy (Geiger and Rudolph 2009).

One of the restricting elements in many of these processes are telomeres. Telomeres are repetitive nucleotide sequences at each end of a chromatid, protecting the ends of the chromosome from deterioration or from fusion with neighboring chromosomes. During chromosome replication, the enzymes duplicating DNA cannot entirely replicate the full length of the chromosome. This usually affects the telomeres, which serve as a buffer to prevent the actual chromosome from being shortened (Levy et al. 1992). Studies suggest that shortened telomeres are associated with cells entering replicative senescence and undergoing apoptosis (Zhang et al. 1999), but they also have been found in connection with certain cancers (Djojosubroto et al. 2003). The progressive loss of telomeric DNA with an increasing number of divisions can also be observed in HSCs (Vaziri et al. 1994). Cell culture experiments have shown that the shortening of telomeric DNA is associated with a rise in cells undergoing apoptosis. The question whether telomeric shortening also affects the functionality of HSCs in an in vivo situation has not yet been answered.

Besides telomere shortening, there are further age-related mechanisms affecting HSC functioning at higher age. HSCs are subject to an age-associated accumulation of defects in mitochondrial and genomic DNA, also observed in other cell types. Transfer experiments in mice using HSCs of young and old animals showed that when an age-mixed sample was transferred, bone marrow reconstitution in the host was largely driven by "young" stem cells (Albright and Makinodan 1976). In old humans, the proliferative capacity of HSCs is up to four times lower than those derived from young donors (Geiger and Rudolph 2009). Aside from such cell-intrinsic defects, also age-related changes occurring in the bone marrow environment have been reported. A reduction in the local production of stimulating cytokines (such as stem cell factor (SCF), granulocyte macrophage colony-stimulating factor (GM-CSF) and interleukin-3) has been found in elderly adults, potentially explaining protracted periods of reconstitution after stem cell transplantation in older recipients (Small et al. 1999). With regard to this finding, it should be emphasized that alterations in the bone marrow microenvironment may particularly change the immune systems' memory: both antibody-secreting plasma cells, accounting for reactive humoral memory, and memory T cells, representing long-term cellular memory, reside there. Differentiation pathways become affected too. Differentiation toward the lymphoid lineage is impaired in aged subjects, providing one potential explanation to the question why the functions of the adaptive immune system seem to be impaired to a larger extent than innate immunity (Rossi et al. 2007). However, the various lymphoid lineages are not equally affected: production of precursor B cells is found to be reduced, resulting in lower numbers of B cells exiting from the bone marrow, whereas the generation of T-cell precursors remains relatively unchanged (Cancro et al. 2009).

3.4 Aging of the Innate Immune System

Age-related changes have been observed for both the innate and the adaptive arm of the immune system, although they are affected to different extents. The innate immune system is phylogenetically more ancient. It consists of natural barriers like skin and mucosal surfaces, various leukocyte subsets, and noncellular components like complement proteins or local antimicrobial peptides. In contrast to the adaptive immune system, there is no specialization of effector functions or memory formation for individual pathogens (Medzhitov and Janeway 2000). Innate immune cells generically recognize pathogens and modulate their activity through a multitude of invariable, germline-encoded receptors, such as Toll-like receptors (TLRs) and RIG-I-like receptors (Kumar et al. 2011). For a variety of these components, age-related changes have been described (Shaw et al. 2013).

While there is evidence that innate effector mechanisms tend to show declining function in the old, there are also strong hints that with higher age, a state of persistent inflammation can set in (Franceschi et al. 2000). Pro-inflammatory cytokines, such as interleukin-6 (IL-6), interleukin-1-beta (IL-1β), and tumor necrosis factor alpha (TNFα), are present at elevated levels in elderly persons systemically and locally at sites including the brain, blood vessels, and bones (Bagnara et al. 2000). At first and for decades, such micro-inflammations may be asymptomatic and harmless. After years however, they may pose potentially harmful consequences for body tissues and may even be causal for diseases like atherosclerosis (Libby 2002), diabetes (Dandona et al. 2004), or Alzheimer's disease (Akiyama et al. 2000). Undoubtedly, the genetic background and lifestyle choices of an individual represent influencing factors too. The exact underlying mechanisms are still incompletely understood and highly complex (Morrisette-Thomas et al. 2014). However, there is a clear inflammatory component inherent to most of age-associated diseases accounting for major shares of overall morbidity and mortality in the elderly (Ferrucci et al. 1999; Ershler and Keller 2000; Cohen et al. 2003). This can result in a dilemma: an inflammatory disposition, which may be beneficial for establishing a good level of immune protection in the young, may later in life result in a harmful state of chronic inflammation (Licastro et al. 2005).

Macrophages residing in various body tissues as well as monocytes, circulating in the blood, possess the ability to initiate and modulate immune responses by recognition and uptake of pathogens followed by presentation of antigenic peptides to T cells and the release of various cytokines (Murray and Wynn 2011). Numbers of macrophage precursors in bone marrow decline with age (Ogawa et al. 2000). TLRs, present on the surfaces of macrophages, further leukocyte populations, as well as mucosal epithelial cells, are able to recognize molecular patterns inherent to bacterial, fungal, parasitical, and viral pathogens (Kumar et al. 2009). Whether TLR expression decreases with advancing age remains unclear, but the cytokine response following activation of certain TLRs has found to be reduced in monocytes at higher age (van Duin et al. 2007a). In studies in mice and rats, macrophages of old animals were releasing lower quantities of reactive oxygen species crucial for cytotoxic functions (Plowden et al. 2004). Also in aged humans, macrophage activity tends to be reduced, providing a potential explanation for prolonged duration and asymptomatic clinical presentation of infections in the elderly (Lloberas and Celada 2002). Neutrophils, making up the largest share of peripheral blood leukocytes, do not significantly change in numbers with age. However, their functional and chemotactic capacity is compromised in old humans (Tseng and Liu 2014). There is not much data on the other granulocyte populations concerning age-related changes. Eosinophils show a functional decline with age (Mathur et al. 2008), while histamine release of basophils upon stimulation increases (Marone et al. 1986). Absolute numbers of basophils however tend to fall with age (Cohen et al. 2013).

Monocytes perform functions such as phagocytosis, antigen presentation, and cytokine production and are precursors for tissue macrophages and dendritic cells (DCs) circulating in peripheral blood. In humans, three major monocyte subsets can be distinguished: classical $CD14^{++}/CD16^{-}$, intermediate $CD14^{++}/CD16^{+}$, and nonclassical $CD14^{+}/CD16^{+}$ monocytes (Guilliams et al. 2014), with the latter two being more activated and capable at secreting pro-inflammatory cytokines (Tacke and Randolph 2006; Zimmermann et al. 2010). Elderly individuals show a significant shift toward nonclassical monocytes in their monocyte compartment (Sadeghi et al. 1999; Seidler et al. 2010). A deregulation in monocyte function, such as decreased phagocytosis and increased basal TNFα levels is likely contributing to the phenotype of inflamm-aging (Hearps et al. 2012).

DCs form a heterogeneous group of professional antigen-presenting cells (APC), which can be of both myeloid and lymphoid origin. While they undoubtedly play an important role in conveying self-tolerance, initiating immune responses, and interlinking innate and adaptive immunity (Reis e Sousa 2006), the effect of age on DCs remains unclear. Plasmacytoid DCs (pDCs) are important in the defense against viral infection through their production of large amounts of type I interferon (IFN) (Liu 2004). Their numbers seem to decrease in elderly humans (Jing et al. 2009). Also on a functional level, there are reports that pDCs show age-related alterations such as decreased expression of TLR7 and 9 (Jing et al. 2009; Garbe et al. 2012) and diminished secretion of activation-induced type I and III IFNs and TNFα (Panda et al. 2010;

Qian et al. 2011; Sridharan et al. 2011). There is conflicting data whether numbers of myeloid dendritic cells (mDCs) are falling at higher age or remain unchanged. They are powerful APCs and capable of priming naive T cells with antigens they phagocytized. Data suggest that these two features are preserved also at higher age (Castle et al. 1999; Steger et al. 1997). For other functions however, mDCs of aged donors show impaired functionality represented by reduced expression of certain TLRs and lower secretion of cytokines after stimulation (Panda et al. 2010; Qian et al. 2011; Della Bella et al. 2007). With respect to inflamm-aging, a low-level expression of pro-inflammatory cytokines such as IL6 and TNFα has been observed in unstimulated pDCs and mDCs from elderly donors (Panda et al. 2010).

Natural killer (NK) cells and natural killer T (NKT) cells are derived from lymphoid progenitors. NK-cell activity is controlled by a set of activating and inhibitory receptors and plays an important role in the elimination of virus-infected and malignant cells (Lanier 1998). They exert functions such as the direct lysis of target cells as well as release of pro-inflammatory cytokines such as IFNγ. In contrast to other lymphocyte populations, NK-cell numbers tend to increase with advancing age, potentially conveying an important role in infection control at later stages in life (Borrego et al. 1999). Within the NK-cell population, two major subsets of $CD56^{bright}$ $CD16^{dim}$ and $CD56^{dim}$ $CD16^{bright}$ NK cells can be distinguished (Cooper et al. 2001). $CD56^{dim}$ NK cells make up around 90 % of the total population and act highly cytotoxic while only producing low levels of cytokines. Their number tends to increase with advancing age, conserving NK-cell-mediated cytotoxicity, both direct and antibody mediated, up to higher age. In contrast to that, $CD56^{bright}$ NK cells have only minimal cytotoxic activity but are able to produce cytokines like IFNγ. Their quantity reduces with age, but it remains controversial whether cytokine production and cytotoxicity are affected (Mariani et al. 2002; Hayhoe et al. 2010). In conclusion, senescence changes of innate immunity are reducing the ability to properly respond to infections and to initiate the adaptive immune response. Reduced neutrophils and NK-cell activity in the elderly are predictive of increased mortality (Niwa et al. 1989; Ogata et al. 2001), and age-impaired TLR function can be associated with reduced vaccine responsiveness (van Duin et al. 2007b). Figure 3.1 summarizes the major changes with respect to innate immunity cells and their functions.

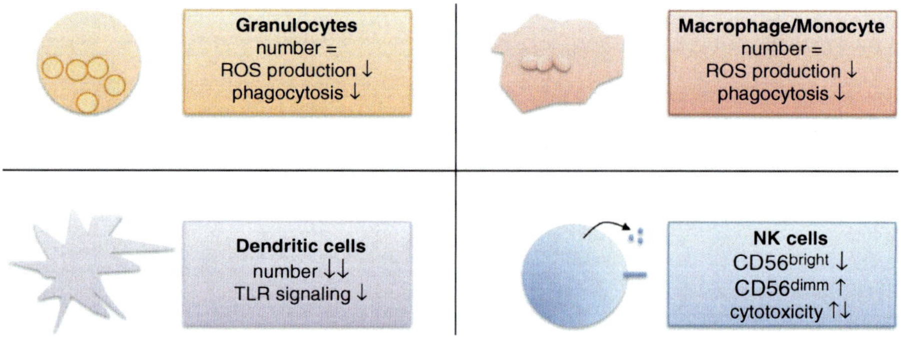

Fig. 3.1 Age-related changes in innate immunity cells and their functions

3.5 Age-Related Changes to Adaptive Immunity

In contrast to innate immunity, adaptive responses are not readily available after birth but are being acquired and refined through a process of selection and adaption. Both the T- and the B-cell compartments are strongly affected by aging. Age-associated changes in T-cell function manifest in declining numbers of T cells exiting from the thymus, a shrinking diversity of the T-cell receptor (TCR) repertoire and a reduced potential of the cells to expand and differentiate upon activation (Naylor et al. 2005; Douek et al. 1998; Goronzy and Weyand 2005). After exiting from the bone marrow, T-cell precursors enter the thymus and undergo further steps of maturation and selection. Thymic involution, meaning a decline in tissue mass and loss of tissue structure which results in a reduced output of naive T cells, has been one of the main features described in the context of immunosenescence (Boehm and Swann 2013). While some authors continue to cite puberty as the onset of thymic involution in humans, there is strong evidence that it sets in a lot earlier (Shanley et al. 2009). Maximum size and peak of activity is reached around the age of 1. By the age of 7, only 10 % of active thymic tissue is left, while the rest has been replaced by fat (Flores et al. 1999). These changes are almost complete by around 50 years of age, although studies have shown that this process can also be modified by external influences: stress, illness, malnutrition, or also pregnancy can lead to transient thymic regression, potentially influenced by inflammatory cytokines and regulated by hormones (Gruver and Sempowski 2008). On the other hand, there has been evidence that under certain conditions, for example, after bone marrow or stem cell transplantation (Alexander et al. 2008; Haynes et al. 2000) or blocking of androgen hormone signaling (Sutherland et al. 2005), a reset with respect to thymic T-cell production can occur. The thymus is an energetically costly organ, with 90 % of the entering thymocytes not passing thymic selection (George and Ritter 1996). Early in life, high thymic activity is needed to build a robust and diverse peripheral T-cell pool, vital for fighting infections and staying healthy and consequently enabling successful reproduction. Once that has been achieved, the longevity of memory T cells as well as their capability to undergo homeostatic proliferation in the periphery allows the maintenance of a stable peripheral antigen-experienced T-cell pool at lower activity levels of the thymus (den Braber et al. 2012). Another beneficial aspect of reducing thymic output might be lowering the risk of lymphatic malignancies. Lastly, the generation of T cells always contains the risk of cellular autoreactivity. Once released into the periphery, autoreactive T cells having escaped negative selection in the thymus (central tolerance) can usually be controlled by various other peripheral mechanisms. However, any inflammatory conditions can alter the various peripheral checkpoints and tolerance mechanisms and result in the activation of autoreactive effector and memory cells. Once generated, autoreactive memory cells can hardly be controlled and pose the risk of inducing and maintaining chronic auto-inflammation with potentially lethal outcome if not adequately treated. It is therefore plausible to maintain thymic activity only throughout a limited time frame in life, in order to provide a sufficiently diverse repertoire of T-cell specificities and in parallel to keep the risk of malignancies and increased autoreactivity low. At steady state, the age-related loss of thymic function apparently can be tolerated without negative clinical side effects, but when confronted with external challenges such as chronic infections, tumors, or very high age, the missing potential to quickly generate fresh, diverse T cells can leave an individual more vulnerable for disease.

One option to assess thymic activity in humans is the measurement of the content of T-cell receptor excision circles (TRECs) in naive peripheral T cells. TRECs are circular DNA fragments created in T cells during TCR gene rearrangement in the thymus. During cell division, TRECs are not replicated and therefore being diluted over generations of descendant cells, rendering them as an indicator of replicative history (Al-Harthi et al. 2000). In one study, a decline in TRECs by 95 % could be observed between two cohorts of 25-year-old or 60-year-old individuals. However, baseline proliferation of $CD4^+$ T cells assessed by Ki-67 expression as well as TCR-β-chain diversity remained relatively stable between the two groups. After the age of 70 years, TRECs declined only slightly further, while homeostatic proliferation doubled. In this group, a drastic drop of TCR diversity was detected, giving a potential cause for reduced vaccine efficacy in the old (Naylor et al. 2005). Thymic activity and homeostatic proliferation of human naive T cells can also be evaluated by using CD31 expression to distinguish two subsets of naive CD4+ T cells with distinct TREC content in the peripheral blood of healthy humans (Kimmig et al. 2002). Existing data so far suggests that homeostatic proliferation of human naive T cells results in TCR repertoire restrictions (Kohler et al. 2005). However, applying the now-available technologies of next-generation sequencing of the entire TCR repertoires from defined samples will enable a more detailed view on human naive T-cell homeostasis on the basis of single T-cell clones. There are further hints from mice and nonhuman primate experiments that low numbers of naive T cells and a constricted TCR repertoire negatively influence primary immune responses against various pathogens (Al-Harthi et al. 2000; Blackman and Woodland 2011; Čičin-Šain et al. 2010). The naive compartments of $CD4^+$ and the $CD8^+$ T-cell compartment seem to be differently affected during aging, with $CD4^+$ cells being generally less compromised than $CD8^+$ T cells (Koch et al. 2008).

A striking feature of the adaptive system is its capability to generate long-lived memory B and T cells, which can convey protection upon rechallenge with a pathogen that has been previously encountered. Usually, this second response is faster and functionally adapted, leading to a fast clearance of the pathogen or antigenic challenge. Like naive T cells, also memory and effector T cells experience age-related changes. An individual's history of immunological exposure to acute infections as well as to persisting pathogens such as human cytomegalovirus (CMV) drives the differentiation and expansion of the respective specific T-cell clones. Especially in the setting of chronic viral infection, repetitive antigen stimulation can give rise to an accumulating population of terminally differentiated T cells which are less functional and replication senescent (Ouyang et al. 2004) (Fletcher et al. 2005) but can make up a significant share of the repertoire (Sylwester et al. 2005). This naturally takes place at the expense of other T cells competing for the same resources and space. Memory inflation therefore needs to be considered as a further crucial key feature of immunosenescence with skewed responses to vaccination and infection in affected individuals (Saurwein-Teissl et al. 2002; Goronzy et al. 2001). A study assessing the naive/memory subset composition in centenarians, which are often considered to be an example of successful aging, showed that these elite donors did not display high numbers of terminally differentiated T cells, but remained within the range of young and middle-aged donors (Nasi et al. 2006).

Age-associated alterations have also been found in signaling pathways of naive T cells, suggesting that not only quantitative but also qualitative changes might affect primary T-cell responses in elderly individuals. In mouse experiments, naive T cells from aged mice showed defects in T-cell synapses, early TCR signaling events, as well as lower levels of cytokines produced by the generated effector cells (Sadighi Akha and Miller 2005). In elderly humans, the TCR-induced extracellular signal-regulated kinase (ERK) phosphorylation in naive $CD4^{+}$ T cells was reduced, caused by increased protein expression of the dual specific phosphatase 6 (DUSP6) due to falling levels of miR-181a. Interestingly, the reconstitution of miR-181a lowered DUSP6 expression in naive $CD4^{+}$ T cells in elderly individuals and improved T-cell responses, making DUSP6 a rare potential therapeutic target for rescuing impaired T-cell responses, for example, in elderly persons (Li et al. 2012).

Many other studies have reported on multiple other signaling defects in T cells isolated from the elderly. However, most studies did not analyze naive T cells but rather different effector cells. One has to emphasize that some of the reported signaling defects are therefore not age related but rather associated with distinct cell types that result from chronic inflammation. They might therefore not be regarded as intrinsic age-related changes, since such effector-type T cells can be high as well in healthy younger adults when they suffer from diseases such as chronic infection or allergy.

There are divergent results about age-related shifts in the cytokine profile of T cells. Some studies have found Th2 cytokines to be increased in the elderly (Mansfield et al. 2012), while others found increased levels of Th1 cytokines (Sakata-Kaneko et al. 2000). However, it is undoubted that a fine balance between a generally pro-inflammatory Th1 profile, needed for effective responses toward bacterial and viral infections, on the one hand and the rather anti-inflammatory Th2 polarization is required for healthy aging (Sandmand et al. 2002). In both in vivo responses to natural infection as well as with regard to vaccination, polyfunctional T cells, being capable of producing different cytokines simultaneously, have been shown to be indicative for a competent acute reaction as well as for an efficient generation of immunological memory (Seder et al. 2008). The number of antigen-specific polyfunctional cells after primary vaccination with attenuated yellow fever virus is lower in old individuals (own unpublished data) but also in the setting of noncontrolled chronic viral infection with HIV (Betts et al. 2006).

B cells are capable of producing antibodies that bind pathogens and neutralize them or make them detectable for other immune cells. As mentioned above, numbers of B-cell precursors in the bone marrow decrease with age and also numbers of naive peripheral B cells tend to modestly decline with age. Alongside with that, a gradual rise in antigen-experienced memory B cells can be observed (Johnson and Cambier 2004). However, these changes are less drastic compared to the decrease in recent thymic emigrants in the T-cell compartment. Levels of immunoglobulins do not significantly fall at higher age (Listì et al. 2006), although amounts of vaccination-induced specific antibodies tend to be lower in the elderly (Goodwin et al. 2006; Stiasny et al. 2012). Similar to T cells, aged B-cell repertoires display a more restricted diversity with age (Dunn-Walters and Ademokun 2010; Howard et al. 2006). A rise in autoantibodies can be observed in the elderly. However, the clinical relevance of this finding remains unclear, since elevated titers do not necessarily correlate with the presence of symptoms of autoimmunity (Mariotti et al. 1992; Hallgren et al. 1973). On the molecular level, a decreased expression of activation-induced cytidine deaminase (AID) leading to an age-associated reduction in class switch recombination could be observed in vitro (Frasca et al. 2008). Finally, a study comparing the B-cell repertoire of old and young adults reported a correlation between old individuals having a highly contracted repertoire and a reduced general health status, higher morbidity and mortality in those (Gibson et al. 2009).

Once established, immunological memory seems to be relatively stable throughout the aging process and clearly less affected than primary activation of naive B and T cells (Sallusto et al. 2004). Interestingly, at least in mice, there is evidence that a T-cell memory for a certain pathogen established at higher age is less robust than the one that has been established early in life, possibly due to a deficient primary response (Kapasi et al. 2002). A good example of immunological memory being resilient to immunosenescence could be observed during the 2009 H1N1 influenza pandemic. In 1957, a very similar influenza strain was circulating and older adults, who had developed a protective memory back then, did now possess higher avidity antibodies than middle-aged adults who had not yet been born back then (Hancock et al. 2009).

One interesting theory when thinking about memory formation and immunosenescence is the concept of the "original antigenic sin," also known as Hoskins effect (Francis 1960). Once there is an established immunological memory for a certain pathogen, in case of a new encounter with a similar but not identical second pathogen, the induced response will be driven by memory cells specific for the first pathogen (Kim et al. 2009). While still executing some protective functions, they do not reach the full potential of a primary response with naive B and T cells of high affinity being optimally selected (Klenerman and Zinkernagel 1998). In theory, this is not only a troubling topic in vaccine development but also for elderly adults, who dispose of and also rely on an established immunological memory for a broad variety of antigens and might therefore be hindered in the establishment of efficient primary responses. But whether this results in an ultimately impaired level of protection has not yet been proven.

3.6 The Role of Persistent Viral Infections

Certain viruses such as CMV, Epstein-Barr virus (EBV), varicella-zoster virus (VZV), hepatitis B and C virus, or human immunodeficiency virus (HIV) can establish persistent viral infections in humans, constantly challenging the host's immune system. Chronic exposure to viral antigens results in the clonal expansion of memory/effector T cells, a phenomenon, which has been particularly well examined for CMV persistence (Pawelec 2014). CMV reaches high seroprevalences of 60 % in the general population in developed countries and above 90 % in older adults (Staras et al. 2006) and in developing countries (Cannon et al. 2010). Besides the clonally expanded memory T-cell pool that can be found also in young infected individuals, chronic CMV infection has been discussed to promote an inflammatory phenotype by higher levels of interleukin-6 and TNFα found in seropositive persons (Trzonkowski et al. 2003; Roberts et al. 2010) and to negatively impact vaccination outcomes as it has been observed for flu vaccinations (Moro-García et al. 2012; Derhovanessian et al. 2013).

EBV is another widespread persistent virus, reaching seroprevalences of over 90 % in the general populations (Cohen 2000). The influence of EBV on the T-cell compartment is milder if compared to CMV infection. However, up to 10 % of $CD8^+$ T cells in healthy EBV-seropositive individuals can be specific for the virus (Hislop et al. 2002; Tan et al. 1999). The impact on the B-cell compartment, with memory B cells being the main target cells of EBV (Babcock et al. 1998), is however by far more pronounced: clonal B-cell expansion is driven by EBV infection resembling typical senescence alterations (Wang et al. 2014).

3.7 Clinical Assessment of Immunosenescence

Without doubt, the various changes associated with immunosenescence do not affect all individuals equally, and the numerical age of a person is not a sufficient indicator for his or her immunological competence. It is therefore of scientific but also clinical interest to establish tools for assessing individual *immune-age* risk profiles. In times where patient-adapted treatment becomes increasingly important, this can, for example, help in choosing optimized vaccination schemes and personalized therapeutic strategies, thus preventing potential risks connected with diseases but also with treatment toxicity. Telomere length in leukocytes can be measured and serve as an indicator of cell replication history. Telomeres can shorten due to oxidative stress and tissue damage (Cawthon et al. 2003). As stated before, thymic activity can be assessed by PCR quantification of TRECs. An alternative option to indirectly assess thymic activity is the analysis of CD31 expression on naive $CD4^+$ T cells by flow cytometry. CD31 is expressed on a subset of naive $CD4^+$ T cells, which have only recently exited the thymus and have not undergone further substantial peripheral proliferation and antigen selection (Kimmig et al. 2002). There have been proposals of "immune risk profiles" focusing on B- and T-cell numbers, their proliferative capacities, the $CD4^+/CD8^+$ ratio, as well as seropositivity for chronic viral infections (Pawelec et al. 2002; Boren and Gershwin 2004). Individuals who showed a high-risk profile had a higher mortality during the 2–6 years of follow-up. In contrast, centenarians usually display risk profiles that are lower than what is predicted for their numerical age (Strindhall et al. 2007). There is diverging data on the role of chronic viral infections in driving immunosenescence. Parallels in terms of increased susceptibility for infectious diseases as well as for cancer can be found between aged humans and individuals suffering from noncontrolled HIV infection (Douek et al. 1998). Also persistent infections with herpes viruses have been associated with promoting immunosenescence. While it is undoubted that the prevalence of infection with cytomegalovirus (CMV) increases with age and that CMV-specific T-cell clones can make up a large share of the repertoire in seropositive individuals (Pawelec et al. 2009), there is ongoing controversy about whether it actually drives immunosenescence and influences morbidity and mortality (Solana et al. 2012).

3.8 Strategies to Overcome Immunosenescence

With an enhanced understanding of immunosenescence, there is a growing interest in finding strategies to overcome its negative aspects in order to facilitate or at least enable optimal health until late in life. The commonly shared belief that a balanced diet and moderate physical exercise play an important role in maintaining health also applies for immunological functions, and various vitamins and trace elements have been shown to be beneficial (Chandra 2004). When deficient, nutritional supplementation can help but there is no evidence, further supplementation beyond the recommended daily amounts can "boost" immune function. Recombinant interleukin-7 (IL-7) is being studied as a therapeutic option to increase thymic output of T cells. Successful reversion of thymic atrophy and increased thymic output could be observed in both mice and macaques after the administration of IL-7 (Faltynek et al. 1992; Moniuszko et al. 2004). In humans, there is data on the use of IL-7 in adult patients with refractory cancer or HIV infection. Here, IL-7 was able to induce increased numbers of naive and central memory cells, while effector T-cell numbers were not affected. $CD8^+$ T-cell repertoire diversity also increased. These effects were age independent and persisted after therapy (for the study duration of 28 days) (Sportès et al. 2008; Levy et al. 2009). While this is a promising data, long-term studies with larger cohorts will have to prove whether IL 7 could also be a target for the enhancement of thymic functions in other clinical settings or even in healthy old adults in the context of disease prevention. Statin drugs, which are classically being used in patients with hypercholesterolemia and cardiovascular disease, were found to have beneficial effects for the process of aging. Studies are preliminary and have demonstrated their anti-inflammatory potential and a potential reduction in telomere shortening by lowering oxidative stress (Ruiz-Limon et al. 2015; Olivieri et al. 2012).

3.9 Vaccinations in the Elderly

Special attention is given to vaccinations in the elderly. Being an excellent mean of disease prevention at relatively low costs, licensed standard vaccines are usually very safe to administer. However, a reduction in vaccination efficacy has been demonstrated in old adults for various live, inactivated, and recombinant vaccines. The previously mentioned age-related immunological alterations such as innate dysfunction, a shrinking naive T-cell repertoire, humoral defects, and coinfection with persistent viruses can definitely influence successful vaccinations (Weinberger and Grubeck-Loebenstein 2012). Immunization recommendations for people over the age of 65 years differ slightly between countries. Health authorities usually recommend the annual administrations of seasonal influenza vaccine, a one-time vaccination against pneumococcal infection and the combined vaccination against tetanus, diphtheria, and pertussis at 10-year intervals (U.S. Department of Health 2014). In some countries, varicella-zoster virus (VZV) vaccine is recommended to prevent the reactivation of the virus in the form of herpes zoster. Various studies have collected data on the reduced efficacy of influenza vaccine in the elderly: while providing protection in 65–90 % of young vaccinees, seasonal influenza vaccine only achieves a rate of 30–50 % of elderly individuals being protected after vaccination (Jefferson et al. 2007). For the 23-valent polysaccharide pneumococcal vaccine (PPV23), some studies have found a drastic reduction in antibody production and protection levels in people above the age of 80, while it showed acceptable efficacy in individuals aged 75 and younger (Andrews et al. 2012). The combined Tdap vaccine (tetanus, diphtheria, and pertussis vaccine) appears to give a satisfactory response in elderly individuals (Weston et al. 2012). For this vaccine, it could be demonstrated that pre-vaccination antibody titers positively influenced vaccination outcome, suggesting a beneficial effect of carefully timed booster immunizations in the elderly (Kaml et al. 2006). By administering vaccines at younger age, the defective memory generation observed in old adults may be avoided. The live viral VZV vaccine showed reduced efficiency in the elderly. It can reduce the incidence but not fully prevent herpes zoster reactivation (Levin 2012). For primary immune responses to vaccination, an increase of non-responsiveness to hepatitis B vaccine could be found in older vaccinees (Fisman et al. 2002). Aside from booster immunizations, there are further clinical strategies to overcome this inability to respond adequately to and enhance the level of protection. Simply higher-dose vaccines may improve vaccination success and however may also lead to decreased efficiencies and/or unwanted side effects due to over-activation. A study of patients with asthma suggested that patients aged 60 and older produced adequate levels of sero-protective antibody to H1N1 vaccine in response to a 30 mcg dose, but not a 15 mcg dose (Busse et al. 2011). For the seasonal influenza vaccination in 2010, a vaccine containing four times as much antigen than the conventional flu shot was approved by the FDA for persons over the age of 65 (Schubert 2010). Further strategies include new vaccine formulations and the use of adjuvants. For pneumococcal vaccination, trials have been made with using 7-valent protein-conjugated vaccine instead of the standard 23-valent polysaccharide formula with satisfactory results (de Roux et al. 2008). Lastly, also the route of vaccine administration is worth considering. One study demonstrated that a better influenza-specific response could be achieved by intradermal application compared with standard intramuscular injection (Holland et al. 2008).

Considering the rapid population aging in developing countries as well as the increasing numbers of older adults traveling to tropical regions, vaccination in travel and tropical medicine for the elderly have become an increasingly important topic. Infections like dengue fever, Japanese encephalitis, or yellow fever threaten the life of the local population as well as that of travelers. However, there is so far only little data on the efficiency of travel vaccines in elderly adults (Ericsson et al. 2001), hallmarking it as an urgent focus for future research.

3.10 Conclusion and Outlook

Gradual alterations of immune functions developing with advancing age are encompassed by the term "immunosenescence." These changes have been demonstrated to account for the distorted immune competence in the elderly, resulting in augmented susceptibility to infection and cancer as well as reduced vaccination efficacies and an altered ability of the immune system to control autoimmunity and chronic inflammation. Immunosenescence occurs with physiological aging: however, it contributes to numerous immunopathological age-related disorders and modifications. In spite of intensive research work, the underlying mechanisms have remained largely obscure so far. A widespread detailed analysis of genomic and post-genomic signatures on the cellular level will be needed in the future in order to understand the basis of these systemic changes of the immune system.

References

Akiyama H et al (2000) Inflammation and Alzheimer's disease. Neurobiol Aging 21(3):383–421

Albright JW, Makinodan T (1976) Decline in the growth potential of spleen-colonizing bone marrow stem cells of long-lived aging mice. J Exp Med 144(5):1204–1213

Alexander T et al (2008) Depletion of autoreactive immunologic memory followed by autologous hematopoietic stem cell transplantation in patients with refractory SLE induces long-term remission through de novo generation of a juvenile and tolerant immune system. Blood 113(1):214–223

Al-Harthi L et al (2000) Detection of T cell receptor circles (TRECs) as biomarkers for de novo T cell synthesis using a quantitative polymerase chain reaction–enzyme linked immunosorbent assay (PCR–ELISA). J Immunol Methods 237(1–2):187–197

Andrews NJ et al (2012) Impact and effectiveness of 23-valent pneumococcal polysaccharide vaccine against invasive pneumococcal disease in the elderly in England and Wales. Vaccine 30(48):6802–6808

Azar A, Ballas ZK (2014) Immune function in older adults. UpToDate. Cited 29 Nov 2014

Babcock GJ et al (1998) EBV persistence in memory B cells in vivo. Immunity 9(3):395–404

Bagnara GP, Bonsi L, Strippoli P, Bonifazi F, Tonelli R, D'Addato S, Paganelli R, Scala E, Fagiolo U, Monti D, Cossarizza A, Bonafé M, Franceschi C (2000) Hemopoiesis in healthy old people and centenarians: well-maintained responsiveness of CD34+ cells to hemopoietic growth factors and remodeling of cytokine network. J Gerontol A Biol Sci Med Sci 55(2):B61–B66

Betts MR et al (2006) HIV nonprogressors preferentially maintain highly functional HIV-specific CD8+ T cells. Blood 107(12):4781–4789

Blackman MA, Woodland DL (2011) The narrowing of the CD8 T cell repertoire in old age. Curr Opin Immunol 23(4):537–542

Boehm T, Swann JB (2013) Thymus involution and regeneration: two sides of the same coin? Nat Rev Immunol 13(11):831–838

Boren E, Gershwin ME (2004) Inflamm-aging: autoimmunity, and the immune-risk phenotype. Autoimmun Rev 3(5):401–406
Borrego F et al (1999) NK phenotypic markers and IL2 response in NK cells from elderly people. Exp Gerontol 34(2):253–265
Bruunsgaard H et al (2003) Elevated levels of tumor necrosis factor alpha and mortality in centenarians. Am J Med 115(4):278–283
Busse WW et al (2011) Vaccination of patients with mild and severe asthma with a 2009 pandemic H1N1 influenza virus vaccine. J Allergy Clin Immunol 127(1):130–137.e3
Cancro MP et al (2009) B cells and aging: molecules and mechanisms. Trends Immunol 30(7):313–318
Cannon MJ, Schmid DS, Hyde TB (2010) Review of cytomegalovirus seroprevalence and demographic characteristics associated with infection. Rev Med Virol 20(4):202–213
Castle SC et al (1999) Antigen presenting cell function is enhanced in healthy elderly. Mech Ageing Dev 107(2):137–145
Cawthon RM et al (2003) Association between telomere length in blood and mortality in people aged 60 years or older. Lancet 361(9355):393–395
Chandra RK (2004) Impact of nutritional status and nutrient supplements on immune responses and incidence of infection in older individuals. Ageing Res Rev 3(1):91–104
Čičin-Šain L et al (2010) Loss of naïve T-cells and repertoire constriction predict poor response to vaccination in old primates. J Immunol 184(12):6739–6745
Cohen JI (2000) Epstein–Barr virus infection. N Engl J Med 343(7):481–492
Cohen HJ, Harris T, Pieper CF (2003) Coagulation and activation of inflammatory pathways in the development of functional decline and mortality in the elderly. Am J Med 114(3):180–187
Cohen AA et al (2013) A novel statistical approach shows evidence for multi-system physiological dysregulation during aging. Mech Ageing Dev 134(3–4):110–117
Cooper MA, Fehniger TA, Caligiuri MA (2001) The biology of human natural killer-cell subsets. Trends Immunol 22(11):633–640
Dandona P, Aljada A, Bandyopadhyay A (2004) Inflammation: the link between insulin resistance, obesity and diabetes. Trends Immunol 25(1):4–7
de Roux A et al (2008) Comparison of pneumococcal conjugate polysaccharide and free polysaccharide vaccines in elderly adults: conjugate vaccine elicits improved antibacterial immune responses and immunological memory. Clin Infect Dis 46(7):1015–1023
Della Bella S et al (2007) Peripheral blood dendritic cells and monocytes are differently regulated in the elderly. Clin Immunol 122(2):220–228
den Braber I et al (2012) Maintenance of peripheral naive T cells is sustained by thymus output in mice but not humans. Immunity 36(2):288–297
Derhovanessian E et al (2013) Cytomegalovirus-associated accumulation of late-differentiated CD4 T-cells correlates with poor humoral response to influenza vaccination. Vaccine 31(4): 685–690
Djojosubroto MW et al (2003) Telomeres and telomerase in aging, regeneration and cancer. Mol Cells 15(2):164–175
Douek DC et al (1998) Changes in thymic function with age and during the treatment of HIV infection. Nature 396(6712):690–695
Dunn-Walters DK, Ademokun AA (2010) B cell repertoire and ageing. Curr Opin Immunol 22(4):514–520
Ericsson CD et al (2001) Travel vaccines and elderly persons: review of vaccines available in the United States. Clin Infect Dis 33(9):1553–1556
Ershler WB, Keller ET (2000) Age-associated increased interleukin-6 gene expression, late-life diseases, and frailty. Annu Rev Med 51(1):245–270
Faltynek CR et al (1992) Administration of human recombinant IL-7 to normal and irradiated mice increases the numbers of lymphocytes and some immature cells of the myeloid lineage. J Immunol 149(4):1276–1282
Ferrucci L et al (1999) Serum IL-6 level and the development of disability in older persons. J Am Geriatr Soc 47(6):639–646

Fisman DN, Agrawal D, Leder K (2002) Effect of age on immunologic response to recombinant hepatitis B vaccine: a meta-analysis. Clin Infect Dis 35(11):1368–1375
Fletcher JM et al (2005) Cytomegalovirus-specific CD4+ T cells in healthy carriers are continuously driven to replicative exhaustion. J Immunol 175(12):8218–8225
Flores KG et al (1999) Analysis of the human thymic perivascular space during aging. J Clin Invest 104(8):1031–1039
Franceschi C, Bonafe M (2003) Centenarians as a model for healthy aging. Biochem Soc Trans 31(2):457–461
Franceschi C et al (2000) Inflamm-aging: an evolutionary perspective on immunosenescence. Ann N Y Acad Sci 908(1):244–254
Francis T (1960) On the doctrine of original antigenic sin. Proc Am Philos Soc 104:572–578
Frasca D et al (2008) Aging down-regulates the transcription factor E2A, activation-induced cytidine deaminase, and Ig class switch in human B cells. J Immunol 180(8):5283–5290
Fulop T et al (2007) Immunosupportive therapies in aging. Clin Interv Aging 2:33–54
Garbe K et al (2012) Plasmacytoid dendritic cells and their Toll-like receptor 9 expression selectively decrease with age. Hum Immunol 73(5):493–497
Gavazzi G, Krause K-H (2002) Ageing and infection. Lancet Infect Dis 2(11):659–666
Geiger H, Rudolph KL (2009) Aging in the lympho-hematopoietic stem cell compartment. Trends Immunol 30(7):360–365
George AJT, Ritter MA (1996) Thymic involution with ageing: obsolescence or good housekeeping? Immunol Today 17(6):267–272
Gibson KL et al (2009) B-cell diversity decreases in old age and is correlated with poor health status. Aging Cell 8(1):18–25
Goodwin K, Viboud C, Simonsen L (2006) Antibody response to influenza vaccination in the elderly: a quantitative review. Vaccine 24(8):1159–1169
Goronzy JJ, Weyand CM (2005) T cell development and receptor diversity during aging. Curr Opin Immunol 17(5):468–475
Goronzy J, Weyand C (2012) Immune aging and autoimmunity. Cell Mol Life Sci 69(10): 1615–1623
Goronzy JJ, Weyand CM (2013) Understanding immunosenescence to improve responses to vaccines. Nat Immunol 14(5):428–436
Goronzy JJ et al (2001) Value of immunological markers in predicting responsiveness to influenza vaccination in elderly individuals. J Virol 75(24):12182–12187
Gruver AL, Sempowski GD (2008) Cytokines, leptin, and stress-induced thymic atrophy. J Leukoc Biol 84(4):915–923
Guilliams M et al (2014) Dendritic cells, monocytes and macrophages: a unified nomenclature based on ontogeny. Nat Rev Immunol 14(8):571–578
Hallgren HM et al (1973) Lymphocyte phytohemagglutinin responsiveness, immunoglobulins and autoantibodies in aging humans. J Immunol 111(4):1101–1107
Hancock K et al (2009) Cross-reactive antibody responses to the 2009 pandemic H1N1 influenza virus. N Engl J Med 361(20):1945–1952
Hayhoe RPG et al (2010) Variation of human natural killer cell phenotypes with age: identification of a unique KLRG1-negative subset. Hum Immunol 71(7):676–681
Haynes BF et al (2000) The role of the thymus in immune reconstitution in aging, bone marrow transplantation, and HIV-1 infection. Annu Rev Immunol 18(1):529–560
Hearps AC et al (2012) Aging is associated with chronic innate immune activation and dysregulation of monocyte phenotype and function. Aging Cell 11(5):867–875
Hislop AD et al (2002) Epitope-specific evolution of human CD8(+) T cell responses from primary to persistent phases of Epstein-Barr virus infection. J Exp Med 195(7): 893–905
Holland D et al (2008) Intradermal influenza vaccine administered using a new microinjection system produces superior immunogenicity in elderly adults: a randomized controlled trial. J Infect Dis 198(5):650–658
Howard WA, Gibson KL, Dunn-Walters DK (2006) Antibody quality in old age. Rejuvenation Res 9(1):117–125

Jefferson T et al (2007) Efficacy and effectiveness of influenza vaccines in elderly people: a systematic review. Lancet 366(9492):1165–1174

Jeffery R (2008) South and Inflammatory and coagulation biomarkers linked to mortality in large treatment interruption trial. P. Clayden (ed.) HIV Treatment Bulletin South. 8/2008. ISSN 20104-1450

Jemal A et al (2011) Global cancer statistics. CA Cancer J Clin 61(2):69–90

Jing Y et al (2009) Aging is associated with a numerical and functional decline in plasmacytoid dendritic cells, whereas myeloid dendritic cells are relatively unaltered in human peripheral blood. Hum Immunol 70(10):777–784

Johnson SA, Cambier JC (2004) Ageing, autoimmunity and arthritis: senescence of the B cell compartment – implications for humoral immunity. Arthritis Res Ther 6(4):131–139

Kaml M et al (2006) Booster vaccination in the elderly: their success depends on the vaccine type applied earlier in life as well as on pre-vaccination antibody titers. Vaccine 24(47–48): 6808–6811

Kapasi ZF, Murali-Krishna K, McRae ML, Ahmed R (2002) Defective generation but normal maintenance of memory T cells in old mice. Eur J Immunol 32(6):1567–1573

Kim JH et al (2009) Original antigenic sin responses to influenza viruses. J Immunol 183(5):3294–3301

Kimmig S et al (2002) Two subsets of naive T helper cells with distinct T cell receptor excision circle content in human adult peripheral blood. J Exp Med 195(6):789–794

Kirkwood TBL (1977) Evolution of ageing. Nature 270(5635):301–304

Kirkwood TBL, Austad SN (2000) Why do we age? Nature 408(6809):233–238

Klenerman P, Zinkernagel RM (1998) Original antigenic sin impairs cytotoxic T lymphocyte responses to viruses bearing variant epitopes. Nature 394(6692):482–485

Koch S et al (2008) Multiparameter flow cytometric analysis of CD4 and CD8 T cell subsets in young and old people. Immun Ageing 5(1):6

Kohler S et al (2005) Post-thymic in vivo proliferation of naive CD4+ T cells constrains the TCR repertoire in healthy human adults. Eur J Immunol 35(6):1987–1994

Kumar H, Kawai T, Akira S (2009) Toll-like receptors and innate immunity. Biochem Biophys Res Commun 388(4):621–625

Kumar H, Kawai T, Akira S (2011) Pathogen recognition by the innate immune system. Int Rev Immunol 30(1):16–34

Lanier LL (1998) NK Cell receptors. Annu Rev Immunol 16(1):359–393

Levin MJ (2012) Immune senescence and vaccines to prevent herpes zoster in older persons. Curr Opin Immunol 24(4):494–500

Levy MZ et al (1992) Telomere end-replication problem and cell aging. J Mol Biol 225(4): 951–960

Levy Y et al (2009) Enhanced T cell recovery in HIV-1–infected adults through IL-7 treatment. J Clin Invest 119(4):997–1007

Li G et al (2012) Decline in miR-181a expression with age impairs T cell receptor sensitivity by increasing DUSP6 activity. Nat Med 18(10):1518–1524

Libby P (2002) Inflammation in atherosclerosis. Nature 420(6917):868–874

Licastro F et al (2005) Innate immunity and inflammation in ageing: a key for understanding age-related diseases. Immun Ageing 2(1):8

Listì F et al (2006) A study of serum immunoglobulin levels in elderly persons that provides new insights into B cell immunosenescence. Ann N Y Acad Sci 1089(1):487–495

Liu Y-J (2004) IPC: professional type 1 interferon-producing cells and plasmacytoid dendritic cell precursors. Annu Rev Immunol 23(1):275–306

Lloberas J, Celada A (2002) Effect of aging on macrophage function. Exp Gerontol 37(12): 1325–1331

Malaguarnera L, Cristaldi E, Malaguarnera M (2010) The role of immunity in elderly cancer. Crit Rev Oncol Hematol 74(1):40–60

Mansfield AS et al (2012) Normal ageing is associated with an increase in Th2 cells, MCP-1 (CCL1) and RANTES (CCL5), with differences in sCD40L and PDGF-AA between sexes. Clin Exp Immunol 170(2):186–193

Mariani E et al (2002) Chemokine production by natural killer cells from nonagenarians. Eur J Immunol 32(6):1524–1529
Mariotti S et al (1992) Thyroid and other organ-specific autoantibodies in healthy ceritenarians. Lancet 339(8808):1506–1508
Marone G et al (1986) Human basophil releasability: I. Age-related changes in basophil releasability. J Allergy Clin Immunol 77(2):377–383
Mathers CD, Loncar D (2006) Projections of global mortality and burden of disease from 2002 to 2030. PLoS Med 3(11):e442
Mathur SK et al (2008) Age-related changes in eosinophil function in human subjects. Chest 133(2):412–419
Matias G et al (2014) Estimates of mortality attributable to influenza and RSV in the United States during 1997–2009 by influenza type or subtype, age, cause of death, and risk status. Influenza Other Respi Viruses 8(5):507–515
Medzhitov R, Janeway C (2000) Innate immunity. N Engl J Med 343(5):338–344
Moniuszko M et al (2004) Recombinant interleukin-7 induces proliferation of naive macaque CD4+ and CD8+ T cells in vivo. J Virol 78(18):9740–9749
Moro-García MA et al (2012) Relationship between functional ability in older people, immune system status, and intensity of response to CMV. Age 34(2):479–495
Morrisette-Thomas V et al (2014) Inflamm-aging does not simply reflect increases in pro-inflammatory markers. Mech Ageing Dev 139:49–57
Murray PJ, Wynn TA (2011) Protective and pathogenic functions of macrophage subsets. Nat Rev Immunol 11(11):723–737
Nash D et al (2001) The outbreak of West Nile virus infection in the New York city area in 1999. N Engl J Med 344(24):1807–1814
Nasi M et al (2006) Thymic output and functionality of the IL-7/IL-7 receptor system in centenarians: implications for the neolymphogenesis at the limit of human life. Aging Cell 5(2):167–175
Naylor K et al (2005) The influence of age on T cell generation and TCR diversity. J Immunol 174(11):7446–7452
Niwa Y et al (1989) Neutrophil chemotaxis, phagocytosis and parameters of reactive oxygen species in human aging: cross-sectional and longitudinal studies. Life Sci 44(22):1655–1664
Ogata K et al (2001) Association between natural killer cell activity and infection in immunologically normal elderly people. Clin Exp Immunol 124(3):392–397
Ogawa T, Kitagawa M, Hirokawa K (2000) Age-related changes of human bone marrow: a histometric estimation of proliferative cells, apoptotic cells, T cells, B cells and macrophages. Mech Ageing Dev 117(1–3):57–68
Olivieri F et al (2012) Telomere/telomerase system: a new target of statins pleiotropic effect? Curr Vasc Pharmacol 10(2):216–224
Ouyang Q et al (2004) Dysfunctional CMV-specific CD8+ T cells accumulate in the elderly. Exp Gerontol 39(4):607–613
Panda A et al (2010) Age-associated decrease in TLR function in primary human dendritic cells predicts influenza vaccine response. J Immunol 184(5):2518–2527
Pawelec G (2014) Immunosenenescence: role of cytomegalovirus. Exp Gerontol 54:1–5
Pawelec G et al (2002) Is human immunosenescence clinically relevant? Looking for 'immunological risk phenotypes'. Trends Immunol 23(7):330–332
Pawelec G et al (2009) Cytomegalovirus and human immunosenescence. Rev Med Virol 19(1):47–56
Plowden J et al (2004) Innate immunity in aging: impact on macrophage function. Aging Cell 3(4):161–167
Qian F et al (2011) Impaired interferon signaling in dendritic cells from older donors infected in vitro with West Nile virus. J Infect Dis 203(10):1415–1424
Ramos-Casals M et al (2004) Systemic autoimmune diseases in elderly patients: atypical presentation and association with neoplasia. Autoimmun Rev 3(5):376–382
Reis e Sousa C (2006) Dendritic cells in a mature age. Nat Rev Immunol 6(6):476–483
Roberts ET et al (2010) Cytomegalovirus antibody levels, inflammation, and mortality among elderly Latinos over 9 years of follow-up. Am J Epidemiol 172(4):363–371

Rossi DJ, Bryder D, Weissman IL (2007) Hematopoietic stem cell aging: mechanism and consequence. Exp Gerontol 42(5):385–390
Ruiz-Limon P et al (2015) Atherosclerosis and cardiovascular disease in systemic lupus erythematosus: effects of in vivo statin treatment. Ann Rheum Dis 74:1450–1458
Sadeghi HM et al (1999) Phenotypic and functional characteristics of circulating monocytes of elderly persons. Exp Gerontol 34(8):959–970
Sadighi Akha AA, Miller RA (2005) Signal transduction in the aging immune system. Curr Opin Immunol 17(5):486–491
Sakata-Kaneko S et al (2000) Altered Th1/Th2 commitment in human CD4+ T cells with ageing. Clin Exp Immunol 120(2):267–273
Sallusto F, Geginat J, Lanzavecchia A (2004) Central memory and effector memory T cell subsets: function, generation, and maintenance. Annu Rev Immunol 22(1):745–763
Sandmand M et al (2002) Is ageing associated with a shift in the balance between Type 1 and Type 2 cytokines in humans? Clin Exp Immunol 127(1):107–114
Saurwein-Teissl M et al (2002) Lack of antibody production following immunization in old age: association with CD8+CD28− T cell clonal expansions and an imbalance in the production of Th1 and Th2 cytokines. J Immunol 168(11):5893–5899
Schubert C (2010) New vaccine tailored to the weakened elderly immune system. Nat Med 16(2):137
Seder RA, Darrah PA, Roederer M (2008) T-cell quality in memory and protection: implications for vaccine design. Nat Rev Immunol 8(4):247–258
Seidler S et al (2010) Age-dependent alterations of monocyte subsets and monocyte-related chemokine pathways in healthy adults. BMC Immunol 11:30
Seok J et al (2013) Genomic responses in mouse models poorly mimic human inflammatory diseases. Proc Natl Acad Sci 110(9):3507–3512
Shanley DP et al (2009) An evolutionary perspective on the mechanisms of immunosenescence. Trends Immunol 30(7):374–381
Shaw AC, Goldstein DR, Montgomery RR (2013) Age-dependent dysregulation of innate immunity. Nat Rev Immunol 13(12):875–887
Small TN et al (1999) Comparison of immune reconstitution after unrelated and related T-cell–depleted bone marrow transplantation: effect of patient age and donor leukocyte infusions. Blood 93:467–480
Solana R et al (2012) CMV and immunosenescence: from basics to clinics. Immun Ageing 9(1):23
Sportès C et al (2008) Administration of rhIL-7 in humans increases in vivo TCR repertoire diversity by preferential expansion of naive T cell subsets. J Exp Med 205(7):1701–1714
Sridharan A et al (2011) Age-associated impaired plasmacytoid dendritic cell functions lead to decreased CD4 and CD8 T cell immunity. Age 33(3):363–376
Stanziano DC et al (2010) A review of selected longitudinal studies on aging: past findings and future directions. J Am Geriatr Soc 58:S292–S297
Staras SAS et al (2006) Seroprevalence of cytomegalovirus infection in the United States, 1988–1994. Clin Infect Dis 43(9):1143–1151
Steger MM, Maczek C, Grubeck-Loebenstein B (1997) Peripheral blood dendritic cells reinduce proliferation in in vitro aged T cell populations. Mech Ageing Dev 93(1–3):125–130
Stiasny K et al (2012) Age affects quantity but not quality of antibody responses after vaccination with an inactivated flavivirus vaccine against tick-borne encephalitis. PLoS One 7(3):e34145
Strindhall J et al (2007) No immune risk profile among individuals who reach 100 years of age: findings from the Swedish NONA immune longitudinal study. Exp Gerontol 42(8):753–761
Sutherland JS et al (2005) Activation of thymic regeneration in mice and humans following androgen blockade. J Immunol 175(4):2741–2753
Sylwester AW et al (2005) Broadly targeted human cytomegalovirus-specific CD4(+) and CD8(+) T cells dominate the memory compartments of exposed subjects. J Exp Med 202(5):673–685
Tacke F, Randolph GJ (2006) Migratory fate and differentiation of blood monocyte subsets. Immunobiology 211(6–8):609–618
Tan LC et al (1999) A re-evaluation of the frequency of CD8+ T cells specific for EBV in healthy virus carriers. J Immunol 162(3):1827–1835

Teixeira BC et al (2014) Inflammatory markers, endothelial function and cardiovascular risk. J Vasc Bras 13(2):108–115

Thompson WW et al (2009) Estimating influenza-associated deaths in the United States. Am J Public Health 99(S2):S225–S230

Torroba M, Zapata AG (2003) Aging of the vertebrate immune system. Microsc Res Tech 62(6):477–481

Trzonkowski P et al (2003) Association between cytomegalovirus infection, enhanced proinflammatory response and low level of anti-hemagglutinins during the anti-influenza vaccination—an impact of immunosenescence. Vaccine 21(25–26):3826–3836

Tseng CW, Liu GY (2014) Expanding roles of neutrophils in aging hosts. Curr Opin Immunol 29:43–48

United Nations, Department of Economic and Social Affairs, Population Division (2013). World Population Ageing 2013. ST/ESA/SER.A/348

Vadasz Z et al (2013) Age-related autoimmunity. BMC Med 11:94

van Duin D et al (2007a) Age-associated defect in human TLR-1/2 function. J Immunol 178(2):970–975

van Duin D et al (2007b) Prevaccine determination of the expression of costimulatory B7 molecules in activated monocytes predicts influenza vaccine responses in young and older adults. J Infect Dis 195(11):1590–1597

Vaziri H et al (1994) Evidence for a mitotic clock in human hematopoietic stem cells: loss of telomeric DNA with age. Proc Natl Acad Sci 91(21):9857–9860

Wang C et al (2014) Effects of aging, cytomegalovirus infection, and EBV infection on human B cell repertoires. J Immunol 192(2):603–611

Weinberger B, Grubeck-Loebenstein B (2012) Vaccines for the elderly. Clin Microbiol Infect 18:100–108

Weston WM et al (2012) Vaccination of adults 65 years of age and older with tetanus toxoid, reduced diphtheria toxoid and acellular pertussis vaccine (Boostrix®): results of two randomized trials. Vaccine 30(9):1721–1728

WHO (2014) World Health Statistics. The top 10 causes of death. Fact sheet N°310, ISBN 978924 1564885; http://www.who.int/mediacentre/factsheets/fs310/en/

Zhang X et al (1999) Telomere shortening and apoptosis in telomerase-inhibited human tumor cells. Genes Dev 13(18):2388–2399

Zimmermann HW et al (2010) Functional contribution of elevated circulating and hepatic non-classical CD14(+)CD16(+) monocytes to inflammation and human liver fibrosis. PLoS One 5(6):e11049

The Hygiene Hypothesis

4

Caroline Roduit, Remo Frei, Erika von Mutius, and Roger Lauener

Contents

C. Roduit (✉)
Department of Allergy/Immunology, Children's Hospital, University of Zurich, Zurich, Switzerland

Department of Allergy/Immunology, Christine Kühne-Center for Allergy Research and Education, Davos, Switzerland
e-mail: caroline.roduit@kispi.uzh.ch

R. Frei
Christine Kühne-Center for Allergy Research and Education, Davos, Switzerland

Molecular Immunology, Swiss Institute of Allergy and Asthma Research, University of Zurich, Zurich, Switzerland

E. von Mutius
Department of Asthma and Allergy, Dr von Hauner Children's Hospital, Ludwig Maximilians University Munich, Munich, Germany

R. Lauener
Christine Kühne-Center for Allergy Research and Education, Davos, Switzerland

Children's Hospital of Eastern Switzerland, St. Gallen, Switzerland

C. Esser (ed.), *Environmental Influences on the Immune System*,
DOI 10.1007/978-3-7091-1890-0_4

4.1 Introduction

A global rise in the prevalence of allergic diseases has been observed over the past decades with a higher prevalence of these conditions in Western countries than in developing countries. This increase is especially problematic in children with currently more than 30 % of children affected by an allergic disease.

First described by the epidemiologist David Strachan in 1989, the so-called hygiene hypothesis was based on the observation that children with an increased number of siblings had less allergic rhinitis and atopic dermatitis (Strachan 1989). The protective effect was attributed to more frequent infections during childhood. Since then, the hygiene hypothesis has been extensively studied in the field of allergy research and proposed to play a role in the rapid increase of the prevalence of allergic diseases over the past decades (The International Study of Asthma and Allergies in Childhood (ISAAC) Steering Committee 1998), which was observed in parallel to a large decrease in infectious diseases (Bach 2002).

A large number of epidemiological studies support the concept of the hygiene hypothesis and even extended it to other immune-mediated diseases, such as type 1 diabetes and inflammatory bowel disease (Brown et al. 2013). Nevertheless, not only infections seemed to protect children against allergies; in addition, environmental exposures with high levels of microbial components, such as farming, have been suggested as one of the major preventive factors for allergy development (Braun-Fahrlander et al. 1999).

The immune mechanisms underlying the hygiene hypothesis are not yet fully elucidated, although several mechanisms have been described. One suggested immunological basis of the hygiene hypothesis is that a reduced microbial stimulation of receptors of the innate immune system in early life leads to a lack of shift toward T-helper cells type 1 (T_H1), away from T_H2 responses (Romagnani 2004a). In fact, researches on the molecular mechanisms of the hygiene hypothesis have reported that receptors of the innate immunity have the ability to modulate allergic responses (Vercelli 2006).

This notion has been debated and it was shown that the molecular basis of allergic disorders cannot be fully explained by the T_H2 paradigm (Mrabet-Dahbi and Maurer 2010; Eyerich and Novak 2013). These observations include that IFN-γ, IL-17, and neutrophils are found in the lungs of asthma patients, and treatments targeting T_H2 cells failed to be effective as hoped in many clinical trials (Kim et al. 2010). Moreover, not only T_H2-associated diseases have increased over the past decades in parallel with elevated hygiene conditions but also T_H1-associated inflammatory and autoimmune diseases (Marshall et al. 2004; Feillet and Bach 2004; Loftus 2004). Furthermore, regulation of T_H1/T_H2 balance was suggested to contribute to the development of allergic and autoimmune conditions (Robinson et al. 2004). Therefore, another suggested mechanism for the hygiene hypothesis is the reduced immune regulation caused by decreased infection stress with an important role of T-regulatory cell activity and the cytokine IL-10 (Yazdanbakhsh et al. 2002; Clemente et al. 2012).

4.2 Infection

As mentioned previously, the first description of the hygiene hypothesis was made on the observation that siblings, with increased exposure to infectious disease through contact with other children, suffer less from allergies (Strachan 1989). Since then, these results were supported by a large number of studies (Strachan et al. 1997a, b; Ball et al. 2000; Bodner et al. 1998). In line with these findings, day care attendance in early life, associated as well with increased microbial exposures in infancy, was shown as a protective factor against allergic diseases (Kramer et al. 1999; Celedon et al. 2003; Hoffjan et al. 2005). Moreover, Italian military students with antibodies to hepatitis A virus showed a lower prevalence of atopy and atopic respiratory diseases (Matricardi et al. 1997).

It was suggested that certain infections that predominantly induce T_H1 responses might limit allergic T_H2 responses. Mycobacterial lipoproteins bind to toll-like receptors (TLRs), and this interaction leads to the prominent synthesis of IL-12, and thus prominent switching toward T_H1 responses (Brightbill et al. 1999). A Japanese study reported that children with positive tuberculin responses had a lower level of IgE and lower levels of T_H2 cytokines, as well as higher levels of IFN-γ, a T_H1 cytokine (Shirakawa et al. 1997). Further, vaccination with bacillus Calmette-Guerin (BCG) that contains attenuated mycobacteria has been shown to be negatively associated with allergic diseases. A prospective international study has shown a protective effect of the BCG vaccine at birth against the development of allergic symptoms at the age of 2 and 5 years, but not against IgE sensitization (Townley et al. 2004). Furthermore, results from the International Study of Asthma and Allergies in Childhood (ISAAC) study showed an inverse relation between tuberculosis notification rates and the prevalence of wheeze (von Mutius et al. 2000a). However, another study only among children with atopic heredity found no association between BCG vaccination and atopy (Alm et al. 1997). Recently, a study could show an increase of IFN-γ and IL-10 production among children receiving BCG vaccination at birth (Akkoc et al. 2010) (Table 4.1).

Further, several studies have reported a negative relation between early childhood infections and the development of allergic diseases, although some findings are inconsistent. A German birth cohort showed an inverse association between repeated viral infections in early life and asthma (Illi et al. 2001). Results from another prospective study revealed that higher levels of infectious episodes in the first 6 months of life were significantly associated with reduced levels of IgE in adolescence (McDade et al. 2004).

These findings are in line with results from animal models that have shown that bacterial species, such as heat-killed *Mycobacteria*, *Listeria monocytogenes*, and *Bordetella pertussis*, may induce a protective effect against the development of allergic diseases (Tukenmez et al. 1999; Li et al. 2003; Kim et al. 2004).

On the contrary, studies analyzing specific childhood infections, such as chickenpox, mumps, whooping cough, or measles have mainly shown positive or no associations with allergic diseases (Bodner et al. 1998; Paunio et al. 2000; Olesen et al. 2003; McKeever et al. 2002). However, a recent US study suggested a risk

Table 4.1 Immune mechanisms underlying the hygiene hypothesis

Environmental factors	Immune system
Helminth infections	Cross-reactive IgE Regulatory B and T cells Regulatory DC IL-10 and TGF-β
Infections/vaccination	IFN-γ IL-10
Microbes	TLR SOCS IRAK IL-10 Less IFN-γ and IL-4
Nutrition (farm milk)	BSA, α-lactalbumin, β-lactoglobulin
Nutrition (fibers, SCFA)	Less inflammation
Probiotics	Regulatory T cells (RALDH)

reduction of asthma, allergic rhinitis, and atopic dermatitis with reported chickenpox infection (Silverberg et al. 2012).

Allergic diseases are chronic inflammatory disorders, and high-sensitivity C-reactive protein (hsCRP) is a sensitivity marker of chronic low-grade inflammation. Interestingly, studies in developing countries have shown low levels of chronic inflammation, measured by C-reactive protein (CRP), despite higher burdens of infectious disease (McDade 2012). In contrast, in high-income countries with reduced levels of infectious disease, a chronic low-grade inflammation has been shown, which has been associated with noncommunicable chronic diseases, such as asthma (Michelson et al. 2009).

On the other hand, no clear association was found between low-grade inflammation and farming environment or atopic sensitization (Mustonen et al. 2012).

Antibiotic use in early life was shown to be associated with an increased risk of asthma and atopic dermatitis (Ong et al. 2014). A German prospective cohort study found that the early exposure to broad-spectrum antibiotics increases the risk of developing atopic dermatitis (Schmitt et al. 2010).

A possible mechanism underlying this increased risk associated with antibiotics is the changes in the host microbiota, leading to an altered development of the infant's immune system. In the same way, murine models have shown that the administration of broad-spectrum antibiotics causes major disruptions to the gut microbiota and reduces T_H1 responses (Schumann et al. 2005; Noverr et al. 2005).

4.3 Farming Environment

4.3.1 Background

After an empiric observation of a Swiss doctor, Markus Gassner, several epidemiological studies across the world have shown that children growing up on a farm suffer less from asthma, allergic rhinitis, and allergic sensitization (Braun-Fahrlander et al. 1999; von Mutius and Vercelli 2010). The first study that supported Gassner's observation was the Swiss Study on Childhood Allergy and Respiratory Symptoms with Respect to Air Pollution (SCARPOL) showing that children growing up on a farm had less sneezing attacks during pollen season (adjusted odds ratio (OR): 0.34, 95 % CI 0.12–0.89) and less atopic sensitization (adjusted OR: 0.31, 95 % CI 0.13–0.73) compared to nonfarmer children (Braun-Fahrlander et al. 1999). These findings were reproduced in various European epidemiological studies such as the Allergy and Endotoxin (ALEX) study (OR in relation to farming status, for asthma: 0.30, 95 % CI 0.15–0.61; for hay fever symptoms: 0.43, 95 % CI 0.24–0.77; for atopic sensitization: 0.61, 95 % CI 0.41–0.92), the Prevention of Allergy Risk Factors for Sensitization in Children Related to Farming and Anthroposophic Life Style (PARSIFAL) study (OR for prevalence of rhinoconjunctivitis symptoms: 0.50 95 % CI 0.38–0.65; for prevalence of atopic sensitization 0.53 95 % CI 0.42–0.67), or the Multidisciplinary Study to Identify the Genetic and Environmental Causes of Asthma in the European Community (GABRIEL) (OR in relation to farming status, for asthma: 0.86; 95 % CI, 0.75–0.99) (Riedler et al. 2001; Alfven et al. 2006; Ege et al. 2011a).

Which factors of this farming environment are responsible for the protective effect on the development of allergic diseases? This question has not been completely elucidated. However, it was shown that farming environment with livestock in the farm stables plays an important role in this protective effect (Von Ehrenstein et al. 2000; Ege et al. 2007; Riedler et al. 2000; Illi et al. 2012). Consumption of unprocessed cow's milk directly from the farm has been often found as being protective against allergies (Riedler et al. 2001; Illi et al. 2012; Bieli et al. 2007; Loss et al. 2011; Sozanska et al. 2013). In a large epidemiologic study, it was shown that neither total viable bacterial counts nor the total fat content of the farm milk were related to asthma or atopy. However, they could show that some whey proteins (bovine serum albumin, alpha-lactalbumin, and beta-lactoglobulin) were associated with a significantly reduced risk of asthma (Loss et al. 2011) (Table 4.1).

This protective "farm effect" was mainly observed with asthma, allergic rhinitis, and allergic sensitization, although the data with atopic dermatitis are inconsistently reported (Riedler et al. 2001; Von Ehrenstein et al. 2000; von Mutius et al. 2000b; Roduit et al. 2011). Nevertheless, two studies showed a protective farm effect on atopic dermatitis, when the exposure occurred during pregnancy (Roduit et al. 2011; Douwes et al. 2008).

The cross-sectional ALEX study first suggested that the timing of exposure plays a crucial role, showing that exposures to stables or consumption of farm milk had a strong protective effect against the development of allergic diseases, especially when those exposures occur during the first year of life (Riedler et al. 2001). Similar results were found in a study conducted in Poland, showing a protective effect of consumption of unpasteurized milk in the first year of life (Sozanska et al. 2013). Moreover, other studies showed that farming exposures induce protection on allergic diseases when occurring already in utero (Douwes et al. 2008; Ege et al. 2006). In the protection against allergy: study in rural environments (PASTURE) birth cohort study, a protective effect on atopic dermatitis up to 2 years of age in children was observed when the mother was working on a farm and had contact to farm animals during pregnancy (Roduit et al. 2011). Interestingly, a dose-response effect was shown with an increasing number of different farm animal species the mother had contact to during pregnancy, resulting in the reduction of the risk of developing atopic dermatitis. Moreover, among children from the same birth cohort study, prenatal exposure to farming activities, especially the contact to different farm animal species and farm dairy products, showed an increase in cord blood cytokine production, such as IFN-γ, resulting in a T_H1-skewed cytokine pattern at birth (Pfefferle et al. 2010). A number of previous studies had reported that decreased levels of IFN-γ at birth predicted the onset of allergies later in life (Kondo et al. 1998; Neaville et al. 2003).

Further, exposure of pregnant mothers to farm stables was also associated with an increase of gene expression of receptors of the innate immunity, TLR2, TLR4, and CD14 (Ege et al. 2006).

These findings suggest that the farming environment may influence the immune system inducing a protective effect on the development of allergic diseases with a critical window of time in early life and even in utero.

Like farm animals, contact with pets has also been extensively studied and has been suggested as a potential protective factor in the development of allergic diseases. A meta-analysis reported some strong evidence of a protective effect of dog exposure on atopic dermatitis, especially when occurred in early life, with an almost uniform effect (Langan et al. 2007). This meta-analysis showed also significant negative association between previous cat exposure and atopic dermatitis. However, another meta-analysis found less risk of childhood asthma associated with cats, but an increased risk with dogs (Takkouche et al. 2008).

In the PASTURE birth cohort study, prenatal contact to cats was shown to have the strongest protective effect on atopic dermatitis (Roduit et al. 2011). A systematic review of longitudinal studies analyzed the association between cat and dog exposure in the prenatal period and allergy (Lodge et al. 2012). They found that for children without a family history of allergy, exposure to dog was protective against the development of allergic diseases.

The increased microbial components of such exposures might play a role in the protective "farm effect" and "pet effect." Exposure to endotoxins (lipopolysaccharides found in the outer cell membrane of gram-negative bacteria) has been suggested as an explanation why pets or farm environmental factors may have a protective effect on allergic diseases. Levels of endotoxin in samples of dust from the child's mattress were inversely related to the occurrence of hay fever, atopic asthma, and atopic sensitization (Braun-Fahrlander et al. 2002). Moreover, a birth cohort study showed a negative association between exposure to high levels of endotoxin and atopic dermatitis, among children with parental history of asthma or allergies (Phipatanakul et al. 2004). Another birth cohort study from Germany, Lifestyle-Related Factors on the Immune System and the Development of Allergies in Childhood (LISA) study, has suggested an up to 50 % reduction of the risk of atopic dermatitis in the first 6 months of life associated with endotoxin exposure measured in dust from mothers' mattresses (Gehring et al. 2001). Further, muramic acid, a component of peptidoglycan found in bacterial cell walls, more abundant in gram-positive bacteria, was shown to be negatively associated with asthma in childhood (van Strien et al. 2004).

All these studies suggest that microbial components play a role in the protective "farm effect" against the development of allergic diseases. Furthermore, recent findings suggested that not only exposure to certain microorganisms or their components but also the exposure to an increased diversity of microbes, in environment such as farming, play an important role. It was shown that farmer children were exposed to a wider range of microbes and that these increased diversity of microbial exposures, bacteria and fungi, had a protective effect on asthma (Ege et al. 2011a).

As underlying mechanism of the hygiene hypothesis, it was suggested that the immune system responds to the microbial burden in the environment and modulates the development of allergic disease. Nevertheless, the exact immune mechanism has not yet been fully elucidated.

The innate immune system constitutes the first line of defense to foreign molecules, such as microbes, and directs the adaptive immune response by T-helper cell activation. Activation of the innate immune system is mediated by pattern recognition receptors (PRR), such as TLRs, CD14, and lectins, which are present on immune cells and recognize pathogen-associated molecular patterns (PAMP). At least ten different TLRs have been described in humans; each TLR is associated with the recognition of certain groups of PAMP (bacterial, fungal, or viral structures) (Takeda and Akira 2005). The development of innate immunity is determined by a combination of genetic and environmental factors. As the hygiene hypothesis suggests that allergic diseases appear because of a reduced microbial exposure and therefore a reduced stimulation of the innate immune system in early life, it has been speculated that alterations in TLR and/or TLR signaling pathway could influence the development of allergy.

Cells of the innate immunity like dendritic cells or macrophages raise the first line of defense against pathogens and initiate and guide the adaptive immune response. This is a tightly regulated process which involves regulation of different critical steps, in order to avoid overboarding inflammation or misleading of the immune system, as it is the case in allergic disorders. Regulatory steps are, besides others, the TLR signaling cascade, secretion of regulatory cytokines IL-10 or TGF-β, or the differentiation of regulatory T cells (T_{REG}). In fact, researches on the molecular mechanisms of the hygiene hypothesis have reported modulations in these processes leading to protection against allergic responses (Vercelli 2006).

4.3.2 Prenatal Exposure and the Immune System

Environmental exposures rich in microbes, such as farming, have been shown to induce an upregulation of innate immunity receptors and this already when exposures occur during pregnancy.

Exposure of pregnant mothers to stables was associated with an increase of gene expression of the innate immunity receptors, TLR2, TLR4, and CD14, measured at school age (Ege et al. 2006). Interestingly, in this study, they found a dose-response effect between the increasing number of farm animal species the mother had contact to during pregnancy and the levels of those receptors measured among their children at school age. These human data are supported by studies with animal models demonstrating that stimulation of TLR2 and TLR4 decreases allergic response (Velasco et al. 2005).

Furthermore, upregulation of those PRRs in association with farm-related exposures was already observed at birth. In the PASTURE birth cohort study, an increase of gene expression of most of the TLRs measured in newborn's white blood cells was associated with consumption of unboiled farm milk during pregnancy (Loss et al. 2012) (Table 4.1).

A direct correlation between the presence of allergic diseases in children and receptors of innate immunity has been shown. It was found that children with a lower expression of TLR5 and TLR9 at birth had an increased risk to develop atopic dermatitis in the first 2 years of age (Roduit et al. 2011). In the same way, a mouse model showed a protective effect of prenatal exposures to farm-derived microbes (*Acinetobacter lwoffii F78*) on asthma among the offspring, and this effect was dependent on TLR signaling pathway (Conrad et al. 2009).

Moreover, it was shown that among farmers, the cord blood mononuclear cells of newborn secrete more IFN-γ and TNF-α in response to phorbol 12-myristate 13-acetate (PMA) compared to nonfarmer. No differences could be demonstrated between the groups regarding the level of the T_H2-associated cytokine, IL-5, the regulatory cytokine, IL-10, and the T_H1-like cytokines, IL-12 (Pfefferle et al. 2010).

As mentioned above, a suggested mechanism for the hygiene hypothesis and the increasing prevalence of allergic diseases is the reduced immune regulation, which is caused by a decrease in microbial exposure and infection stress during childhood resulting in reduced stimulation of T-regulatory cells (T_{REG}) (Yazdanbakhsh et al. 2002; Clemente et al. 2012; Romagnani 2004b).

Children of mothers exposed to farm environment during pregnancy were shown to have an increased T_{REG} cell count in cord blood, as well as an increased level of FOXP3, a transcription factor of T_{REG} cells (Schaub et al. 2009). Moreover, this study showed that T_{REG} cell function was more effective in controlling T_H2 responses in offspring of farming compared with nonfarming mothers. In this study, again the prenatal contact to farm animals showed the strongest effect. The enhanced IFN-γ levels together with more T_{REG} might guide the newborn's immune system in a more regulated and less T_H2 shifted status (von Mutius and Vercelli 2010). Besides T_H2 cells, it was shown that T_H17 cells might also play a role in the pathogenesis of allergies (Kudo et al. 2012; Lluis et al. 2014a). It seems that in early T_H17 differentiation, T_{REG} cells play a promoting role in nonfarmers' children in the presence of endotoxin (Lluis et al. 2014a).

4.3.3 Exposures Later in Life and the Immune System

At school age, among farmers' children, an increased gene expression of CD14, TLR1, TLR2, TLR4, TLR7, and TLR8 in whole-blood cells was measured (Ege et al. 2007; Lauener et al. 2002). The farm-related exposures which were shown to be associated with an increase of gene expression of TLRs are the consumption of farm milk, child's involvement in haying, and use of silage (Ege et al. 2007).

Results from the PASTURE birth cohort study found that children who consumed raw milk in the first year of life had an increased gene expression of those receptors of the innate immunity measured at 1 year of age compared to children who did not consume farm milk (Loss et al. 2012).

The signaling cascade of TLRs is tightly regulated to avoid overinflammation or insufficient defense. Growing up in a farm environment seems to regulate this cascade. Farmers' children have more of the regulatory molecules IRAK-2 and SOCS-4 but also of the kinases IRAK1, TBK1, and RIPK1 (Frei et al. 2014). Moreover, a strong increase of gene expression of regulatory cytokines, IL-10 and TGF-β, was found among farmer's children (Frei et al. 2014). At the same time, farmers' children express less T_H1- and T_H2-associated cytokines, INF-γ and IL-4, in white blood cells. Furthermore, farmers' children have less allergen-specific IgE in serum, and their blood leukocytes secrete less inflammatory cytokines such as TNF-α, INF-γ, IL-10, and IL-12 in response to bacterial components (Braun-Fahrlander et al. 2002). Moreover, exposure to farm or farm-related factors such as farm-milk consumption or staying in the stable leads to increased $CD25^+$/$FOXP3^+$ T_{REG} cells, and this is associated with fewer incidences of asthma and allergen-specific IgE (Lluis et al. 2014b).

All these findings indicate that regulatory mechanisms of the innate and the adaptive immunity might have a critical role in protective farmers' children from the development of allergies (Fig. 4.1).

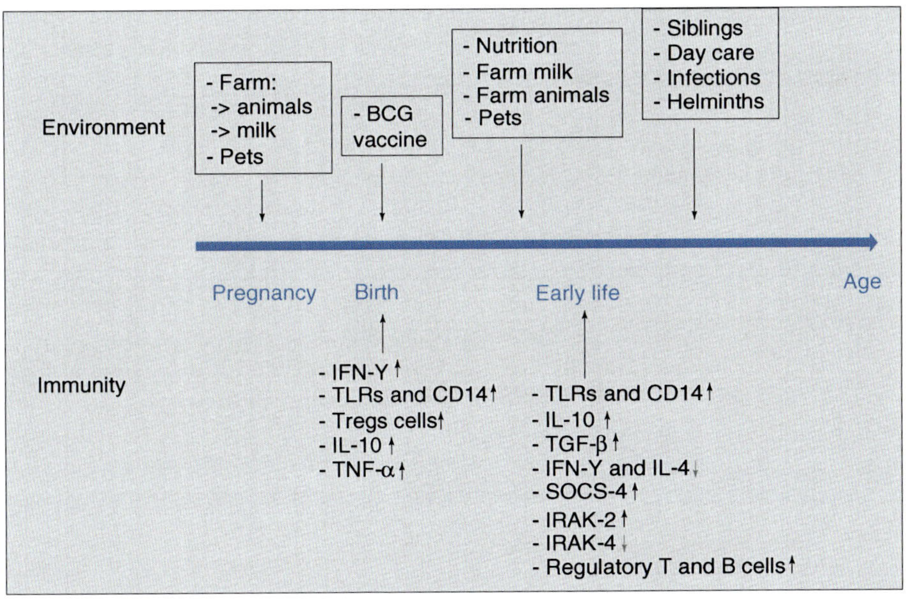

Fig. 4.1 Environment and immune system, time line: the influence of different environmental factors, derived from the hygiene hypothesis, on the immune system over time

4.3.4 Genes, Gene-Environment Interactions, and Epigenetics

Genetic alterations in the innate immunity, especially in receptors of PAMPs, may modify the risk of allergic diseases. Direct associations between single nucleotide polymorphisms (SNP) in PRRs and allergies have been shown (Eder et al. 2004; Fageras Bottcher et al. 2004). *TLR9* and *CD14* promoter polymorphisms were associated with atopic dermatitis (Novak et al. 2007; Litonjua et al. 2005). Also direct associations between single nucleotide polymorphisms (SNP) in *TLRs* and asthma have been shown.

Moreover, several findings indicate that environmental factors associated with high levels of microbial components may interact with existing polymorphisms in *TLRs* on the development of allergic disease, a phenomenon called gene-environment interaction. The concept of gene-environment interaction includes that individuals are only able to benefit from a protective environmental factor if the subject has the susceptible genetic background. It has been shown that two polymorphisms in *CD14* together with exposure to endotoxin, contact to animals, or farm milk lead to lower levels of total and specific IgE or less asthma (Bieli et al. 2007). Further, in the context of a farm environment, an association between a polymorphism in *TLR2* and asthma, hay fever, and specific IgE was reported (Eder et al. 2004). Among children from the PASTURE birth cohort study, an interaction between the same polymorphism in *TLR2* and environment on atopic dermatitis was observed. The prenatal protective effect of contact to cat and farm animal on atopic dermatitis was observed only among children with a specific genotype in a SNP of *TLR2* (Roduit et al. 2011). Another genetic variation in *TLR4* is associated with less specific IgE if endotoxin is present in the environment and a NOD1 polymorphism is associated with less specific IgE, hay fever, and atopic wheeze in the context of farm life and endotoxin (Ege et al. 2011b; Werner et al. 2003; Smit et al. 2009).

Recently, techniques for genome-wide association studies have been developed that will bring up novel interaction candidate genes for asthma and atopy in the context of the hygiene hypothesis (Ege and von Mutius 2013).

Epigenetic mechanisms, mediated in response to the environment or in a heritable fashion, control gene expression by DNA methylation, histone modification, and the expression of noncoding RNA (Kabesch 2014). Farmers' children have at birth hypomethylated regions in promoters of ORMDL1 and STAT6 genes, while RAD50 and IL13 were hypermethylated and therefore not accessible for transcription. Over the age, regions associated with asthma and IgE regulation were changed (Michel et al. 2013). Moreover, prenatal administration of *Acinetobacter lwoffii* F78 to mice prevents the development of an asthmatic phenotype in the progeny of an IFN-γ-dependent way, whose promoter was protected against loss of histone 4 acetylation, which is closely associated with IFN-γ expression (Brand et al. 2011).

4.4 Helminths and Immune Responses

4.4.1 Background

Helminths are known to provoke a strong T_H2 response with their cytokines IL-4, IL-5, IL-9, IL-13, high level of tissue eosinophilia, mucosal mastocytosis, and production of IgE. Furthermore, T_H1 and T_H17 cells can be part of the immune response against parasites. On the other hand, helminth infections cause induction of a whole bunch of anti-inflammatory cells and molecules to avoid hyper-inflammatory responses such as regulatory B and T cells, IL-10, and TGF-β. Moreover, suppressor macrophages and regulatory dendritic cells characterized by expression of IL-10, TGF-β, and indoleamine 2,3-dioxygenase and cyclooxygenase-2 are induced. These regulatory responses are either antigen specific or unrelated to the antigen (Wiria et al. 2012).

4.4.2 Allergy and Asthma

Although helminth infections skew the immune response toward T_H2, they are inversely associated with allergies and asthma. This might be due to the unspecific regulatory response that goes in parallel with a helminth infection meaning that IL-10 and TGF-β might suppress the effector mechanisms that lead to the development of allergies. Nevertheless, the effects seem to be parasite strain and infection intensity dependent. Another mechanism underlying the helminth-mediated reduction in incidence of allergies might be in the cross-reactivity of the IgE. In areas endemic for helminth infections, high total and allergen-specific IgE are rather negatively associated with allergic disorders. The reason may be that helminth IgE antibodies that also bind allergens have low biologic activity and thereby prevent a strong allergic response by occupying the binding site of the real allergen-specific IgE. Development of allergy treatment using helminths is not in use because of insufficient evidence on the efficacy, tolerability, and costs (Wiria et al. 2012).

4.4.3 Autoimmune and Inflammatory Disease

The potential role of helminth infection in reducing severity of multiple sclerosis or inflammatory bowel disease patients has been shown in a number of trials and in mouse models. These T_H1- and T_H17-mediated diseases might be reduced by the immune regulatory potential of a parasite infection (Wiria et al. 2012). Since cardiovascular disease and diabetes are at least partly mediated by inflammation, there is a potential using helminth infection or helminth eggs as therapeutic agents due to their immune regulatory response (Wiria et al. 2012).

4.5 Outlook

Even though a lot of studies showed the interaction between environmental factors with microbial burden and allergic diseases, the exact immune mechanism and factors deriving from the hygiene hypothesis are not yet identified. One reason could be the importance of the gene-environment interaction effect, with different environmental factors having different influences depending on the genetic background, and therefore may differ between populations. On the road to develop effective allergy preventions, these mechanisms have to be further investigated. Additionally, new aspects have been added to the concept of the hygiene hypothesis in the past years of which some are discussed below.

4.5.1 Diversity of Environmental Factors

One new aspect is the role of the diversity of environmental exposures. The important role of the diversity was observed with the prenatal protective effect of the increased number of different farm animal species on atopic dermatitis or recent findings showing that the increased diversity of microbial exposures had a protective effect on asthma (Ege et al. 2011a). Those results support the hypothesis that exposures in early life or even during pregnancy to diverse antigens could increase the maturation of the immune system and induce tolerance networks (Prescott et al. 2008).

A "biodiversity hypothesis" was already proposed by researchers from Finland. They have shown that individuals with atopic sensitization had a lower environmental biodiversity and also a lower diversity of their microbiota (Hanski et al. 2012).

Further, an increased diversity of food introduced within the first year of life was shown to have a protective effect on atopic dermatitis, with an indication of dose-response relationship, and this independently of farming environment or parental history of allergy (Roduit et al. 2012). A reduced risk of asthma, food allergy, and sensitization to food allergens was also be found with an increased exposure to different food antigens in the first year of life, with same dose-response pattern (Roduit et al. 2014; Nwaru et al. 2014).

4.5.2 The Role of Nutrition

With the Western lifestyle, changes in diet have been observed over the past decades and were suggested to also play a role in the increase of the prevalence of allergic diseases.

In Western diets, an increased intake of n-6 long-chain polyunsaturated fatty acid (LC-PUFA) and a reduced intake of n-3 LC-PUFA were observed, and this imbalance was suggested to contribute to the increasing prevalence of allergic diseases. It was shown that n-6 LC-PUFA enhanced pro-inflammatory mediator production by mast cells, while the n-3 LC-PUFA suppressed IL-4 and IL-13 release (van den Elsen et al. 2013).

A reduction of consumption of fiber was also reported in industrialized countries compared to developing countries. Fibers are a wide range of complex oligosaccharides, which are not digested by the host, but enter the colon and provide fermentable substrates for the colonic microbiota (Russell et al. 2011). Soluble fiber, such as prebiotic, was shown to be a major substrate for bacterial growth and increase the number of *Bifidobacteria* (Boehm et al. 2002; Moro et al. 2006). Moreover, it was reported in a mouse model that the dietary fiber content alters the gut microbiota and that a low-fiber diet decreases the diversity of gut microbiota (Trompette et al. 2014).

Further researches are needed to better understand the relationship between nutrition, the gut microbiota, and immune system.

4.5.3 Innate Lymphoid Cells

In the past years, it became clear that a shift in T-helper cell balance toward T-helper cell type 2 is not the main immunological mechanisms underlying allergic disorders. New players such as group 2 innate lymphoid cells (ILC2) could play a role. ILC2 are assigned to the innate immunity, and they secrete large amounts of T_H2 cytokines IL-5 and IL-13 after stimulation with IL-33 and IL-25 released by the epithelium in response to damage by proteases, viruses, or other environmental insults. ILC2 are extremely rare but their ability to secrete T_H2 cytokines is severalfold higher than that of classical T_H2 cells (Vercelli et al. 2014).

4.5.4 The Microbiota

The concept of the hygiene hypothesis has been linked to the composition of the intestinal microbiota. It has been shown that shifts in the composition of the microbiota caused by environmental factors, such as nutrition, infections, or early life antibiotics, may lead to disruption of immune tolerance and therefore to disease (Brown et al. 2013). Bacterial colonization starts at birth and reaches levels until 100 trillion microbes. There are differences in the composition and diversity between individuals living in industrialized countries versus developing countries.

The gut microbiota is responsible for induction of immune tolerance via education of T_{REG} cells out of naive T cells. The T-cell receptor repertoire is increased in lamina propria compared to secondary lymphoid organs. Moreover, the induction of tolerance seems to be dependent on TLR signaling (Brown et al. 2013).

In addition, the host microbiota (which includes commensal and symbiotic microbes) has been demonstrated to be essential for full immunological development. The composition and metabolic activity of the microbiota have profound effects on the induction of immune tolerance. Specific bacterial strains which confer protection from allergic inflammation were shown to be able to induce T_{REG} cells (Lyons et al. 2010). Results from several cross-sectional epidemiologic studies indicate that atopic and nonatopic subjects differ in gut microflora composition

(Bjorksten et al. 2001; Kalliomaki and Isolauri 2003; Watanabe et al. 2003). Moreover, an inverse association between the bacterial diversity of the gut microbiota in the first months of life and the development of atopic dermatitis was reported (Wang et al. 2008). A recent study also showed that a low diversity of the gut microbiota during the first month of life was associated with asthma later in childhood (Abrahamsson et al. 2014). Therefore, perturbations in the gut microbiota, induced by food exposures or other factors, may be involved in the pathogenesis of atopic dermatitis.

Moreover, metabolites produced by intestinal microbiota, such as short-chain fatty acids (SCFAs), were shown to have anti-inflammatory properties (Saemann et al. 2000; Tedelind et al. 2007; Maslowski et al. 2009; Frei et al. 2012). Recently, it was shown in a mouse model that a high-fiber diet increases circulating levels of SCFAs and thereby protects against allergic inflammation in the lung.

These findings led to the development of probiotics or prebiotics as allergy prevention. More studies on composition and immunological effects are necessary to make them effective.

4.6 Conclusion

The hygiene hypothesis includes several and complex environmental factors. New aspects, such as the diversity of the environment and the nutrition, have been shown to play a role. Interestingly, most of these factors do not prevent the production of IgE. Rather, they induce regulatory cells and molecules to minimize the allergic reaction and thereby they reduce the symptoms of the diseases.

References

Abrahamsson TR, Jakobsson HE, Andersson AF, Bjorksten B, Engstrand L, Jenmalm MC (2014) Low gut microbiota diversity in early infancy precedes asthma at school age. Clin Exp Allergy 44:842–850

Akkoc T, Aydogan M, Yildiz A, Karakoc-Aydiner E, Eifan A, Keles S et al (2010) Neonatal BCG vaccination induces IL-10 production by CD4+ CD25+ T cells. Pediatr Allergy Immunol 21:1059–1063

Alfven T, Braun-Fahrlander C, Brunekreef B, von Mutius E, Riedler J, Scheynius A et al (2006) Allergic diseases and atopic sensitization in children related to farming and anthroposophic lifestyle--the PARSIFAL study. Allergy 61:414–421

Alm JS, Lilja G, Pershagen G, Scheynius A (1997) Early BCG vaccination and development of atopy. Lancet 350:400–403

Bach JF (2002) The effect of infections on susceptibility to autoimmune and allergic diseases. N Engl J Med 347:911–920

Ball TM, Castro-Rodriguez JA, Griffith KA, Holberg CJ, Martinez FD, Wright AL (2000) Siblings, day-care attendance, and the risk of asthma and wheezing during childhood. N Engl J Med 343:538–543

Bieli C, Eder W, Frei R, Braun-Fahrlander C, Klimecki W, Waser M et al (2007) A polymorphism in CD14 modifies the effect of farm milk consumption on allergic diseases and CD14 gene expression. J Allergy Clin Immunol 120:1308–1315

Bjorksten B, Sepp E, Julge K, Voor T, Mikelsaar M (2001) Allergy development and the intestinal microflora during the first year of life. J Allergy Clin Immunol 108:516–520

Bodner C, Godden D, Seaton A (1998) Family size, childhood infections and atopic diseases. The Aberdeen WHEASE Group. Thorax 53:28–32

Boehm G, Lidestri M, Casetta P, Jelinek J, Negretti F, Stahl B et al (2002) Supplementation of a bovine milk formula with an oligosaccharide mixture increases counts of faecal bifidobacteria in preterm infants. Arch Dis Child Fetal Neonatal Ed 86:F178–F181

Brand S, Teich R, Dicke T, Harb H, Yildirim AO, Tost J et al (2011) Epigenetic regulation in murine offspring as a novel mechanism for transmaternal asthma protection induced by microbes. J Allergy Clin Immunol 128:618–25.e1–7

Braun-Fahrlander C, Gassner M, Grize L, Neu U, Sennhauser FH, Varonier HS et al (1999) Prevalence of hay fever and allergic sensitization in farmer's children and their peers living in the same rural community. SCARPOL team. Swiss Study on Childhood Allergy and Respiratory Symptoms with Respect to Air Pollution. Clin Exp Allergy 29:28–34

Braun-Fahrlander C, Riedler J, Herz U, Eder W, Waser M, Grize L et al (2002) Environmental exposure to endotoxin and its relation to asthma in school-age children. N Engl J Med 347:869–877

Brightbill HD, Libraty DH, Krutzik SR, Yang RB, Belisle JT, Bleharski JR et al (1999) Host defense mechanisms triggered by microbial lipoproteins through toll-like receptors. Science 285:732–736

Brown EM, Arrieta MC, Finlay BB (2013) A fresh look at the hygiene hypothesis: how intestinal microbial exposure drives immune effector responses in atopic disease. Semin Immunol 25:378–387

Celedon JC, Wright RJ, Litonjua AA, Sredl D, Ryan L, Weiss ST et al (2003) Day care attendance in early life, maternal history of asthma, and asthma at the age of 6 years. Am J Respir Crit Care Med 167:1239–1243

Clemente JC, Ursell LK, Parfrey LW, Knight R (2012) The impact of the gut microbiota on human health: an integrative view. Cell 148:1258–1270

Conrad ML, Ferstl R, Teich R, Brand S, Blumer N, Yildirim AO et al (2009) Maternal TLR signaling is required for prenatal asthma protection by the nonpathogenic microbe Acinetobacter lwoffii F78. J Exp Med 206:2869–2877

Douwes J, Cheng S, Travier N, Cohet C, Niesink A, McKenzie J et al (2008) Farm exposure in utero may protect against asthma, hay fever and eczema. Eur Respir J 32:603–611

Eder W, Klimecki W, Yu L, von Mutius E, Riedler J, Braun-Fahrlander C et al (2004) Toll-like receptor 2 as a major gene for asthma in children of European farmers. J Allergy Clin Immunol 113:482–488

Ege MJ, von Mutius E (2013) Can genes forecast asthma risk? Lancet Respir Med 1:425–426

Ege MJ, Bieli C, Frei R, van Strien RT, Riedler J, Ublagger E et al (2006) Prenatal farm exposure is related to the expression of receptors of the innate immunity and to atopic sensitization in school-age children. J Allergy Clin Immunol 117:817–823

Ege MJ, Frei R, Bieli C, Schram-Bijkerk D, Waser M, Benz MR et al (2007) Not all farming environments protect against the development of asthma and wheeze in children. J Allergy Clin Immunol 119:1140–1147

Ege MJ, Mayer M, Normand AC, Genuneit J, Cookson WO, Braun-Fahrlander C et al (2011a) Exposure to environmental microorganisms and childhood asthma. N Engl J Med 364:701–709

Ege MJ, Strachan DP, Cookson WO, Moffatt MF, Gut I, Lathrop M et al (2011b) Gene-environment interaction for childhood asthma and exposure to farming in Central Europe. J Allergy Clin Immunol 127:138–144, 44.e1–4

Eyerich K, Novak N (2013) Immunology of atopic eczema: overcoming the Th1/Th2 paradigm. Allergy 68:974–982

Fageras Bottcher M, Hmani-Aifa M, Lindstrom A, Jenmalm MC, Mai XM, Nilsson L et al (2004) A TLR4 polymorphism is associated with asthma and reduced lipopolysaccharide-induced interleukin-12(p70) responses in Swedish children. J Allergy Clin Immunol 114:561–567

Feillet H, Bach JF (2004) Increased incidence of inflammatory bowel disease: the price of the decline of infectious burden? Curr Opin Gastroenterol 20:560–564
Frei R, Lauener RP, Crameri R, O'Mahony L (2012) Microbiota and dietary interactions: an update to the hygiene hypothesis? Allergy 67:451–461
Frei R, Roduit C, Bieli C, Loeliger S, Waser M, Scheynius A et al (2014) Expression of genes related to anti-inflammatory pathways are modified among farmers' children. PLoS One 9:e91097
Gehring U, Bolte G, Borte M, Bischof W, Fahlbusch B, Wichmann HE et al (2001) Exposure to endotoxin decreases the risk of atopic eczema in infancy: a cohort study. J Allergy Clin Immunol 108:847–854
Hanski I, von Hertzen L, Fyhrquist N, Koskinen K, Torppa K, Laatikainen T et al (2012) Environmental biodiversity, human microbiota, and allergy are interrelated. Proc Natl Acad Sci U S A 109:8334–8339
Hoffjan S, Nicolae D, Ostrovnaya I, Roberg K, Evans M, Mirel DB et al (2005) Gene-environment interaction effects on the development of immune responses in the 1st year of life. Am J Hum Genet 76:696–704
Illi S, von Mutius E, Lau S, Bergmann R, Niggemann B, Sommerfeld C et al (2001) Early childhood infectious diseases and the development of asthma up to school age: a birth cohort study. BMJ 322:390–395
Illi S, Depner M, Genuneit J, Horak E, Loss G, Strunz-Lehner C et al (2012) Protection from childhood asthma and allergy in Alpine farm environments-the GABRIEL Advanced Studies. J Allergy Clin Immunol 129:1470–1477.e6
Kabesch M (2014) Epigenetics in asthma and allergy. Curr Opin Allergy Clin Immunol 14:62–68
Kalliomaki M, Isolauri E (2003) Role of intestinal flora in the development of allergy. Curr Opin Allergy Clin Immunol 3:15–20
Kim YS, Kwon KS, Kim DK, Choi IW, Lee HK (2004) Inhibition of murine allergic airway disease by Bordetella pertussis. Immunology 112:624–630
Kim HY, DeKruyff RH, Umetsu DT (2010) The many paths to asthma: phenotype shaped by innate and adaptive immunity. Nat Immunol 11:577–584
Kondo N, Kobayashi Y, Shinoda S, Takenaka R, Teramoto T, Kaneko H et al (1998) Reduced interferon gamma production by antigen-stimulated cord blood mononuclear cells is a risk factor of allergic disorders--6-year follow-up study. Clin Exp Allergy 28:1340–1344
Krämer U, Heinrich J, Wjst M, Wichmann HE (1999) Age of entry to day nursery and allergy in later childhood. Lancet 353:450–454
Kudo M, Melton AC, Chen C, Engler MB, Huang KE, Ren X et al (2012) IL-17A produced by alpha beta T cells drives airway hyper-responsiveness in mice and enhances mouse and human airway smooth muscle contraction. Nat Med 18:547–554
Langan SM, Flohr C, Williams HC (2007) The role of furry pets in eczema: a systematic review. Arch Dermatol 143:1570–1577
Lauener RP, Birchler T, Adamski J, Braun-Fahrlander C, Bufe A, Herz U et al (2002) Expression of CD14 and Toll-like receptor 2 in farmers' and non-farmers' children. Lancet 360:465–466
Li XM, Srivastava K, Huleatt JW, Bottomly K, Burks AW, Sampson HA (2003) Engineered recombinant peanut protein and heat-killed Listeria monocytogenes coadministration protects against peanut-induced anaphylaxis in a murine model. J Immunol 170:3289–3295
Litonjua AA, Belanger K, Celedon JC, Milton DK, Bracken MB, Kraft P et al (2005) Polymorphisms in the 5′ region of the CD14 gene are associated with eczema in young children. J Allergy Clin Immunol 115:1056–1062
Lluis A, Ballenberger N, Illi S, Schieck M, Kabesch M, Illig T et al (2014a) Regulation of TH17 markers early in life through maternal farm exposure. J Allergy Clin Immunol 133:864–871
Lluis A, Depner M, Gaugler B, Saas P, Casaca VI, Raedler D et al (2014b) Increased regulatory T-cell numbers are associated with farm milk exposure and lower atopic sensitization and asthma in childhood. J Allergy Clin Immunol 133:551–559

Lodge CJ, Allen KJ, Lowe AJ, Hill DJ, Hosking CS, Abramson MJ et al (2012) Perinatal cat and dog exposure and the risk of asthma and allergy in the urban environment: a systematic review of longitudinal studies. Clin Dev Immunol 2012:176484

Loftus EV Jr (2004) Clinical epidemiology of inflammatory bowel disease: incidence, prevalence, and environmental influences. Gastroenterology 126:1504–1517

Loss G, Apprich S, Waser M, Kneifel W, Genuneit J, Buchele G et al (2011) The protective effect of farm milk consumption on childhood asthma and atopy: the GABRIELA study. J Allergy Clin Immunol 128:766–773.e4

Loss G, Bitter S, Wohlgensinger J, Frei R, Roduit C, Genuneit J et al (2012) Prenatal and early-life exposures alter expression of innate immunity genes: the PASTURE cohort study. J Allergy Clin Immunol 130:523–530.e9

Lyons A, O'Mahony D, O'Brien F, MacSharry J, Sheil B, Ceddia M et al (2010) Bacterial strain-specific induction of Foxp3+ T regulatory cells is protective in murine allergy models. Clin Exp Allergy 40:811–819

Marshall AL, Chetwynd A, Morris JA, Placzek M, Smith C, Olabi A et al (2004) Type 1 diabetes mellitus in childhood: a matched case control study in Lancashire and Cumbria, UK. Diabet Med 21:1035–1040

Maslowski KM, Vieira AT, Ng A, Kranich J, Sierro F, Yu D et al (2009) Regulation of inflammatory responses by gut microbiota and chemoattractant receptor GPR43. Nature 461:1282–1286

Matricardi PM, Rosmini F, Ferrigno L, Nisini R, Rapicetta M, Chionne P et al (1997) Cross sectional retrospective study of prevalence of atopy among Italian military students with antibodies against hepatitis A virus. BMJ 314:999–1003

McDade TW (2012) Early environments and the ecology of inflammation. Proc Natl Acad Sci U S A 109(Suppl 2):17281–17288

McDade TW, Kuzawa CW, Adair LS, Beck MA (2004) Prenatal and early postnatal environments are significant predictors of total immunoglobulin E concentration in Filipino adolescents. Clin Exp Allergy 34:44–50

McKeever TM, Lewis SA, Smith C, Collins J, Heatlie H, Frischer M et al (2002) Early exposure to infections and antibiotics and the incidence of allergic disease: a birth cohort study with the West Midlands General Practice Research Database. J Allergy Clin Immunol 109:43–50

Michel S, Busato F, Genuneit J, Pekkanen J, Dalphin JC, Riedler J et al (2013) Farm exposure and time trends in early childhood may influence DNA methylation in genes related to asthma and allergy. Allergy 68:355–364

Michelson PH, Williams LW, Benjamin DK, Barnato AE (2009) Obesity, inflammation, and asthma severity in childhood: data from the National Health and Nutrition Examination Survey 2001-2004. Ann Allergy Asthma Immunol 103:381–385

Moro G, Arslanoglu S, Stahl B, Jelinek J, Wahn U, Boehm G (2006) A mixture of prebiotic oligosaccharides reduces the incidence of atopic dermatitis during the first six months of age. Arch Dis Child 91:814–819

Mrabet-Dahbi S, Maurer M (2010) Does allergy impair innate immunity? Leads and lessons from atopic dermatitis. Allergy 65:1351–1356

Mustonen K, Keski-Nisula L, Vaarala O, Pfefferle PI, Renz H, Riedler J et al (2012) Few associations between high-sensitivity C-reactive protein and environmental factors in 4.5-year-old children. Pediatr Allergy Immunol 23:522–528

Neaville WA, Tisler C, Bhattacharya A, Anklam K, Gilbertson-White S, Hamilton R et al (2003) Developmental cytokine response profiles and the clinical and immunologic expression of atopy during the first year of life. J Allergy Clin Immunol 112:740–746

Novak N, Yu CF, Bussmann C, Maintz L, Peng WM, Hart J et al (2007) Putative association of a TLR9 promoter polymorphism with atopic eczema. Allergy 62:766–772

Noverr MC, Falkowski NR, McDonald RA, McKenzie AN, Huffnagle GB (2005) Development of allergic airway disease in mice following antibiotic therapy and fungal microbiota increase: role of host genetics, antigen, and interleukin-13. Infect Immun 73:30–38

Nwaru BI, Takkinen HM, Kaila M, Erkkola M, Ahonen S, Pekkanen J et al (2014) Food diversity in infancy and the risk of childhood asthma and allergies. J Allergy Clin Immunol 133:1084–1091

Olesen AB, Juul S, Thestrup-Pedersen K (2003) Atopic dermatitis is increased following vaccination for measles, mumps and rubella or measles infection. Acta Derm Venereol 83:445–450

Ong MS, Umetsu DT, Mandl KD (2014) Consequences of antibiotics and infections in infancy: bugs, drugs, and wheezing. Ann Allergy Asthma Immunol 112:441–445.e1

Paunio M, Heinonen OP, Virtanen M, Leinikki P, Patja A, Peltola H (2000) Measles history and atopic diseases: a population-based cross-sectional study. JAMA 283:343–346

Pfefferle PI, Buchele G, Blumer N, Roponen M, Ege MJ, Krauss-Etschmann S et al (2010) Cord blood cytokines are modulated by maternal farming activities and consumption of farm dairy products during pregnancy: the PASTURE Study. J Allergy Clin Immunol 125:108–15.e1–3

Phipatanakul W, Celedon JC, Raby BA, Litonjua AA, Milton DK, Sredl D et al (2004) Endotoxin exposure and eczema in the first year of life. Pediatrics 114:13–18

Prescott SL, Smith P, Tang M, Palmer DJ, Sinn J, Huntley SJ et al (2008) The importance of early complementary feeding in the development of oral tolerance: concerns and controversies. Pediatr Allergy Immunol 19:375–380

Riedler J, Eder W, Oberfeld G, Schreuer M (2000) Austrian children living on a farm have less hay fever, asthma and allergic sensitization. Clin Exp Allergy 30:194–200

Riedler J, Braun-Fahrlander C, Eder W, Schreuer M, Waser M, Maisch S et al (2001) Exposure to farming in early life and development of asthma and allergy: a cross-sectional survey. Lancet 358:1129–1133

Robinson DS, Larche M, Durham SR (2004) Tregs and allergic disease. J Clin Invest 114:1389–1397

Roduit C, Wohlgensinger J, Frei R, Bitter S, Bieli C, Loeliger S et al (2011) Prenatal animal contact and gene expression of innate immunity receptors at birth are associated with atopic dermatitis. J Allergy Clin Immunol 127:179–185, 85.e1

Roduit C, Frei R, Loss G, Buchele G, Weber J, Depner M et al (2012) Development of atopic dermatitis according to age of onset and association with early-life exposures. J Allergy Clin Immunol 130:130–136.e5

Roduit C, Frei R, Depner M, Schaub B, Loss G, Genuneit J et al (2014) Increased food diversity in the first year of life is inversely associated with allergic diseases. J Allergy Clin Immunol 133:1056–1064

Romagnani S (2004a) Immunologic influences on allergy and the TH1/TH2 balance. J Allergy Clin Immunol 113:395–400

Romagnani S (2004b) The increased prevalence of allergy and the hygiene hypothesis: missing immune deviation, reduced immune suppression, or both? Immunology 112:352–363

Russell DA, Ross RP, Fitzgerald GF, Stanton C (2011) Metabolic activities and probiotic potential of bifidobacteria. Int J Food Microbiol 149:88–105

Saemann MD, Bohmig GA, Osterreicher CH, Burtscher H, Parolini O, Diakos C et al (2000) Anti-inflammatory effects of sodium butyrate on human monocytes: potent inhibition of IL-12 and up-regulation of IL-10 production. FASEB J 14:2380–2382

Schaub B, Liu J, Hoppler S, Schleich I, Huehn J, Olek S et al (2009) Maternal farm exposure modulates neonatal immune mechanisms through regulatory T cells. J Allergy Clin Immunol 123:774–782.e5

Schmitt J, Schmitt NM, Kirch W, Meurer M (2010) Early exposure to antibiotics and infections and the incidence of atopic eczema: a population-based cohort study. Pediatr Allergy Immunol 21:292–300

Schumann A, Nutten S, Donnicola D, Comelli EM, Mansourian R, Cherbut C et al (2005) Neonatal antibiotic treatment alters gastrointestinal tract developmental gene expression and intestinal barrier transcriptome. Physiol Genomics 23:235–245

Shirakawa T, Enomoto T, Shimazu S, Hopkin JM (1997) The inverse association between tuberculin responses and atopic disorder. Science 275:77–79

Silverberg JI, Kleiman E, Silverberg NB, Durkin HG, Joks R, Smith-Norowitz TA (2012) Chickenpox in childhood is associated with decreased atopic disorders, IgE, allergic sensitization, and leukocyte subsets. Pediatr Allergy Immunol 23:50–58

Smit LA, Siroux V, Bouzigon E, Oryszczyn MP, Lathrop M, Demenais F et al (2009) CD14 and toll-like receptor gene polymorphisms, country living, and asthma in adults. Am J Respir Crit Care Med 179:363–368

Sozanska B, Pearce N, Dudek K, Cullinan P (2013) Consumption of unpasteurized milk and its effects on atopy and asthma in children and adult inhabitants in rural Poland. Allergy 68:644–650

Strachan DP (1989) Hay fever, hygiene, and household size. BMJ 299:1259–1260

Strachan DP, Harkins LS, Golding J (1997a) Sibship size and self-reported inhalant allergy among adult women. ALSPAC Study Team. Clin Exp Allergy 27:151–155

Strachan DP, Harkins LS, Johnston ID, Anderson HR (1997b) Childhood antecedents of allergic sensitization in young British adults. J Allergy Clin Immunol 99:6–12

Takeda K, Akira S (2005) Toll-like receptors in innate immunity. Int Immunol 17:1–14

Takkouche B, Gonzalez-Barcala FJ, Etminan M, Fitzgerald M (2008) Exposure to furry pets and the risk of asthma and allergic rhinitis: a meta-analysis. Allergy 63:857–864

Tedelind S, Westberg F, Kjerrulf M, Vidal A (2007) Anti-inflammatory properties of the short-chain fatty acids acetate and propionate: a study with relevance to inflammatory bowel disease. World J Gastroenterol 13:2826–2832

Townley RG, Barlan IB, Patino C, Vichyanond P, Minervini MC, Simasathien T et al (2004) The effect of BCG vaccine at birth on the development of atopy or allergic disease in young children. Ann Allergy Asthma Immunol 92:350–355

Trompette A, Gollwitzer ES, Yadava K, Sichelstiel AK, Sprenger N, Ngom-Bru C et al (2014) Gut microbiota metabolism of dietary fiber influences allergic airway disease and hematopoiesis. Nat Med 20:159–166

Tukenmez F, Bahceciler NN, Barlan IB, Basaran MM (1999) Effect of pre-immunization by killed Mycobacterium bovis and vaccae on immunoglobulin E response in ovalbumin-sensitized newborn mice. Pediatr Allergy Immunol 10:107–111

van den Elsen LW, Nusse Y, Balvers M, Redegeld FA, Knol EF, Garssen J et al (2013) n-3 Long-chain PUFA reduce allergy-related mediator release by human mast cells in vitro via inhibition of reactive oxygen species. Br J Nutr 109:1821–1831

van Strien RT, Engel R, Holst O, Bufe A, Eder W, Waser M et al (2004) Microbial exposure of rural school children, as assessed by levels of N-acetyl-muramic acid in mattress dust, and its association with respiratory health. J Allergy Clin Immunol 113:860–867

Velasco G, Campo M, Manrique OJ, Bellou A, He H, Arestides RS et al (2005) Toll-like receptor 4 or 2 agonists decrease allergic inflammation. Am J Respir Cell Mol Biol 32:218–224

Vercelli D (2006) Mechanisms of the hygiene hypothesis--molecular and otherwise. Curr Opin Immunol 18:733–737

Vercelli D, Gozdz J, von Mutius E (2014) Innate lymphoid cells in asthma: when innate immunity comes in a Th2 flavor. Curr Opin Allergy Clin Immunol 14:29–34

Von Ehrenstein OS, Von Mutius E, Illi S, Baumann L, Bohm O, von Kries R (2000) Reduced risk of hay fever and asthma among children of farmers. Clin Exp Allergy 30:187–193

von Mutius E, Vercelli D (2010) Farm living: effects on childhood asthma and allergy. Nat Rev Immunol 10:861–868

von Mutius E, Pearce N, Beasley R, Cheng S, von Ehrenstein O, Bjorksten B et al (2000a) International patterns of tuberculosis and the prevalence of symptoms of asthma, rhinitis, and eczema. Thorax 55:449–453

von Mutius E, Braun-Fahrlander C, Schierl R, Riedler J, Ehlermann S, Maisch S et al (2000b) Exposure to endotoxin or other bacterial components might protect against the development of atopy. Clin Exp Allergy 30:1230–1234

Wang M, Karlsson C, Olsson C, Adlerberth I, Wold AE, Strachan DP et al (2008) Reduced diversity in the early fecal microbiota of infants with atopic eczema. J Allergy Clin Immunol 121:129–134

Watanabe S, Narisawa Y, Arase S, Okamatsu H, Ikenaga T, Tajiri Y et al (2003) Differences in fecal microflora between patients with atopic dermatitis and healthy control subjects. J Allergy Clin Immunol 111:587–591

Werner M, Topp R, Wimmer K, Richter K, Bischof W, Wjst M et al (2003) TLR4 gene variants modify endotoxin effects on asthma. J Allergy Clin Immunol 112:323–330

Wiria AE, Djuardi Y, Supali T, Sartono E, Yazdanbakhsh M (2012) Helminth infection in populations undergoing epidemiological transition: a friend or foe? Semin Immunopathol 34:889–901

Worldwide variation in prevalence of symptoms of asthma, allergic rhinoconjunctivitis, and atopic eczema: ISAAC. The International Study of Asthma and Allergies in Childhood (ISAAC) Steering Committee (1998) Lancet. 351:1225–1232

Yazdanbakhsh M, Kremsner PG, van Ree R (2002) Allergy, parasites, and the hygiene hypothesis. Science 296:490–494

Stress and the Immune System

5

Rebecca G. Reed and Charles L. Raison

Contents

R.G. Reed (✉)
Division of Family Studies and Human Development, Norton School of Family and Consumer Sciences, College of Agriculture and Life Sciences, University of Arizona, Tucson, AZ 85721-0078, USA

Department of Psychology, University of Kentucky, Lexington, KY, USA
e-mail: rgreed@email.arizona.edu

C.L. Raison
Division of Family Studies and Human Development, Norton School of Family and Consumer Sciences, College of Agriculture and Life Sciences, University of Arizona, Tucson, AZ 85721-0078, USA

Department of Psychiatry, College of Medicine, University of Arizona, Tucson, AZ 85721, USA

C. Esser (ed.), *Environmental Influences on the Immune System*,
DOI 10.1007/978-3-7091-1890-0_5

The association between stress and immune function has received considerable attention in the past several decades (Irwin 2008; Kemeny and Schedlowski 2007; Kiecolt-Glaser et al. 2002). Dysregulation of the neuroendocrine and immune systems, due to chronic stress, is associated with psychological and physiological disorders, including depression, atherosclerosis, asthma, cardiovascular disease, cancers, and the progression of HIV to AIDS (Antoni et al. 2006; Cohen et al. 2007; Dantzer et al. 2008; Irwin 2008). Furthermore, chronic inflammation and other forms of immune dysregulation increase risk for premature all-cause mortality and a variety of diseases including cardiovascular disease, cancer, and metabolic syndrome (Ershler and Keller 2000; Hansson 2005; Hotamisligil 2006; Nabipour et al. 2006; Parkin 2006). Given these significant health outcomes, it therefore seems essential to understand the complex ways in which stress influences immune functioning, as well as the intrapersonal and interpersonal factors that may exacerbate or buffer the effects of stress on immunity.

In this chapter, we provide an overview of how stress affects immune functioning and examine evidence in the literature of various intrapersonal and interpersonal factors that may exacerbate or buffer the health effects of stress. We first review some basic information concerning the immune system to provide the reader with necessary background. We then present the primary pathways by which stress impacts the immune system, including the sympathetic nervous system, the hypothalamic-pituitary-adrenal (HPA) axis, and vagal withdrawal. Next, we discuss how the immune response varies and even goes awry, depending on the nature of the stress (acute versus chronic). Additionally, we discuss how the immune response varies depending upon the individual within whom the stress is occurring. Specifically, we focus on various intrapersonal and interpersonal factors associated with immune functioning. Intrapersonal factors reviewed include rumination, emotion regulation, alexithymia, psychological stress, optimism, and positive affect. Interpersonal factors reviewed include close relationship and family processes such as negative and positive behaviors, ambivalence towards a relationship partner, social rejection and social isolation, and early life adversity. To conclude, we highlight some substantive and methodological considerations relevant to future research on the effects of stress on immunity.

5.1 What Is Stress?

We conceptualize stress to be a constellation of events, beginning with a stressor (stimulus), which precipitates a reaction in the brain (stress perception) that in turn activates a physiological or biological stress response to allow the organism to deal with the threat or opportunity (Dhabhar and McEwen 1997). Psychological stress occurs when events or environmental demands exceed an individual's ability or willingness to cope (Lazarus and Folkman 1984). Being laid off from work, experiencing an argument with a loved one, being diagnosed and living with cancer, or giving a presentation in class are just a few examples of the unexpected obstacles, overwhelming challenges, and uncontrollable events that may be stressful experiences of everyday life. Stress exists on a spectrum—from short-term or acute stress, lasting minutes to hours, to long-term or chronic stress, lasting weeks, months, or years, and the intensity of the stressor is generally linked to its relevance to the survival and reproduction of the organism.

5.2 Overview of the Immune System

Before examining the mechanisms by which psychosocial stressors affect the immune system, we present a brief overview of the immune system as background. The immune system is critical for human health and well-being, as it helps coordinate the body's response to physical injuries and infections that, if left unaddressed, could cause illness or death (Slavich and Irwin 2014). The immune system is composed of two interconnected branches: innate or nonspecific immunity and acquired or specific immunity. Depending on the type of immune response, different components of the immune system may be activated.

The innate response acts immediately (within minutes to hours) when the body is subjected to tissue damage or microbial infection (Medzhitov 2007). The "first line of defense" of innate immunity includes physical barriers such as the skin and mucosal membranes. If these physical barriers are not enough to keep pathogens out, the innate immune response includes neutrophils, monocytes (found in the circulating peripheral blood), and macrophages (found in the tissue) that circulate through the body and use invariant receptors to detect a wide array of pathogens. Upon detecting a pathogen, the cells phagocytize them by engulfing and ingesting them. Additionally, a signaling cascade is activated that results in the activation of nuclear factor-κB (NF-κB) and interferon (IFN) regulatory factors, which are transcription factors that in turn drive the expression of proinflammatory immune-response genes including interleukin (IL)-1 and tumor necrosis factor-α (TNF-α). These genes then produce small protein molecules called cytokines, which are the main actors of the inflammatory response (Raison et al. 2006). Proinflammatory cytokines (e.g., IL-1, IL-6, TNF-α) are those that increase or upregulate inflammation, whereas anti-inflammatory cytokines (e.g., IL-4, IL-10) decrease or downregulate inflammation. The cumulative activities/effects of proinflammatory cytokines are referred to as inflammation. These cytokines initiate a "call to action" and attract other immune cells to the infected area. Another cell involved in innate immunity is the natural killer (NK) cell. NK cells recognize the lack of a self-tissue molecule on the surface of cells (characteristic of many kinds of virally infected cells and some cancerous cells) and lyse the cells by releasing toxic substances on them. The innate immune response is also referred to as a nonspecific response because these mechanisms are not specific to any antigen; rather, this immune response is programmed to recognize features that are shared by groups of foreign substances and will take action to eliminate anything and everything that it deems "foreign" or "not-self."

If a pathogen survives or evades the actions of the innate immune response, then the acquired immune response becomes activated. In contrast to innate immunity, which is nonspecific and does not provide long-lasting protection to the host, acquired immunity involves the proliferation of microbial-specific white blood cells (lymphocytes) that attempt to neutralize or eliminate microbes based on a memory response of having responded to a specific pathogen in the past. The primary cells of the acquired immune response are lymphocytes, including T cells and B cells. T cells include helper T cells (T_H) and cytotoxic T cells (T_C). Helper T cells recognize and interact with an antigen, "raise the alarm" by producing cytokines that call more immune cells to the area, and activate B cells, which produce soluble antibodies. Antibodies are proteins that can neutralize bacterial toxins and bind to free viruses, "tagging" them for elimination and preventing their entry into cells. Cytotoxic T cells recognize antigen expressed by cells that are infected with viruses or otherwise comprised cells (e.g., cancer cells) and lyse those cells. Whereas the innate immune response is rapid, the acquired immune response takes days to fully develop (Barton 2008).

Importantly, acquired immunity in humans is composed of cellular and humoral responses (Elenkov 2008). Cellular immune responses are mounted against intracellular pathogens (e.g., viruses) and are coordinated by a subset of T-helper lymphocytes called *Th1* cells. In the Th1 response, helper T cells produce cytokines, including IL-2, TNF-β, and IFN-γ. These cytokines are associated with the promotion of excessive inflammation and activate macrophages and cytotoxic T cells, which lyse the infected cells. Humoral immune responses are mounted against extracellular pathogens (e.g., parasites, bacteria) and are coordinated by a subset of T-helper lymphocytes called *Th2* cells. In the Th2 response, helper T cells produce different cytokines including IL-4, which stimulate the growth and activation of mast cells and eosinophils, as well as the differentiation of B cells into antibody-secreting B cells. These cytokines also inhibit macrophage activation, T-cell proliferation, and the production of proinflammatory cytokines (Elenkov 2008).

Regulatory T cells (Treg) also play an important role in mediating immune suppression in numerous settings, including, for example, autoimmune disease, allergy, and microbial infection. Treg cells are in the CD4, helper T-cell lineage. They form a subset of cells that also express the cell-surface activation marker CD25, but are best distinguished by the intracellular expression of forkhead box P3 (FOXP3), an important T-cell immunoregulatory transcription factor. Treg cells are an important source of IL-10, once considered a Th2 cytokine but now recognized as being more generally immunoregulatory and anti-inflammatory. Tregs also produce transforming growth factor (TGF)-beta, a cytokine with complex and somewhat contradictory actions but a profile that is generally anti-inflammatory.

Given the general rule that physiological systems in the body have built-in restraining mechanisms, it should perhaps not be surprising that the discovery of

Tregs has prompted the search for regulatory cells in other immune lineages. And indeed, although not as well characterized as Tregs, it is now clear that such cells exist and are important for proper immune functioning. Such cells include regulatory dendritic and B cells and M2-type macrophages. It is increasingly recognized that inflammatory and autoimmune conditions are promoted when these regulatory cells function suboptimally. On the other hand, increasing data suggest that these cells can also pose a risk of inducing patterns of immune suppression that are not always health promoting. For example, regulatory cells have been implicated in vulnerability to cancer development. Increasing evidence also suggests, however, that suboptimal immunoregulatory functioning may be a common feature of major depression and may, in fact, contribute to the proinflammatory state often observed in major depressive disorder.

5.3 Pathways Connecting Stress to Immune Function

Stress can modulate the immune system through various pathways (Fig. 5.1). The first pathway involves the sympathetic nervous system (SNS; adrenergic activation), and the second pathway involves the hypothalamic-pituitary-adrenal (HPA) axis. Both pathways are presented below, and we also discuss evidence suggesting that the parasympathetic nervous system (PNS), specifically vagal withdrawal, affects immune functioning.

5.3.1 Sympathetic Nervous System

Running from a tiger or moving in for a first kiss are various stressful situations, as perceived by the brain, which result in the rapid activation of the autonomic nervous system (ANS). The ANS can be separated into two divisions: the sympathetic nervous system (SNS) and the parasympathetic nervous system (PNS.)

Activation of the SNS rapidly produces many physiological effects evolved to help cope with threat, including increased blood flow to essential organs, such as the brain, heart and lungs, and to skeletal muscles, dilation of lung bronchioles, increased heart rate and contraction strength, and dilation of the pupils to allow more light to enter the eye and enhance far vision. At the same time, SNS activation diverts blood flow away from the gastrointestinal (GI) tract and skin by stimulating vasoconstriction and inhibits GI peristalsis.

The SNS, also referred to as the "fight-or-flight" system, releases mainly norepinephrine (noradrenalin) and epinephrine (adrenaline) from the cells of the adrenal medulla. Once released, these catecholamines act through α- and β-adrenergic receptors to increase production of circulating proinflammatory cytokines including IL-1, IL-6, and TNF-α (Black 2002; Steptoe et al. 2007). In addition, norepinephrine promotes NF-κβ activation, which regulates and increases the gene expression of several proinflammatory mediators, including IL-6 and IL-8 (Fig. 5.1e). These inflammatory mediators, in turn, enhance inflammation.

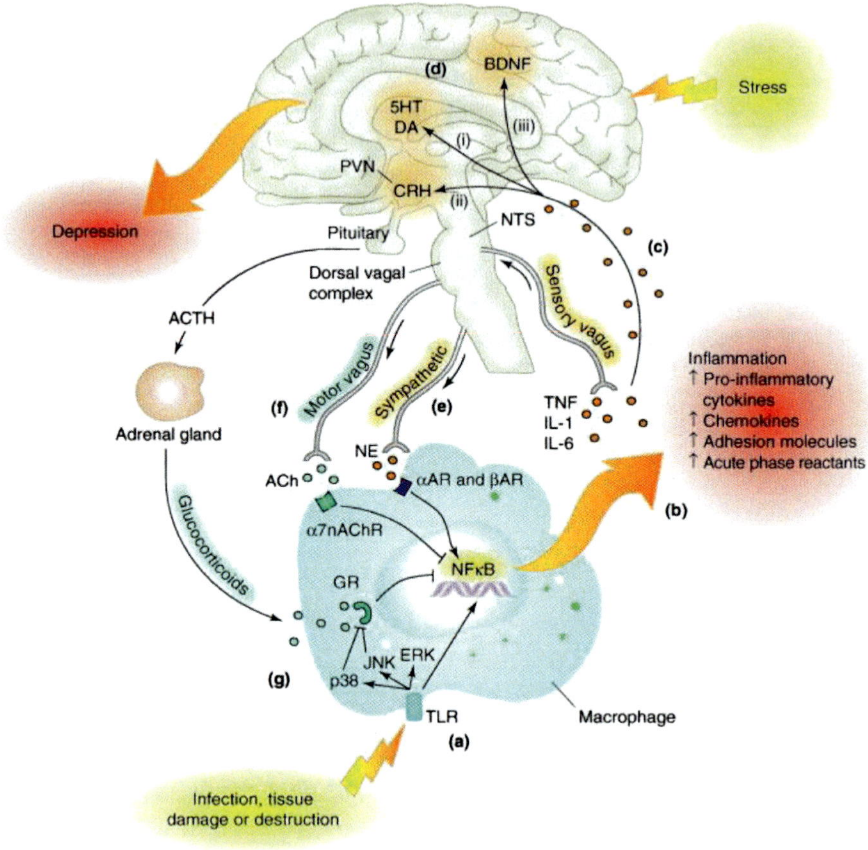

Fig. 5.1 Stress-immune interactions. (*a*) Activation of NF-kB through Toll-like receptors (*TLR*) during immune challenge leads to an inflammatory response including (*b*) the release of proinflammatory cytokines TNF-α, IL-1, and IL-6. (*c*) These cytokines, in turn, access the brain via leaky regions in the blood-brain barrier, active transport molecules, and afferent nerve fibers (e.g., sensory vagus), which relay information through the nucleus tractus solitaries (*NTS*). (*d*) Once in the brain, cytokines participate in various pathways (i, ii, iii) known to be involved in the development of depression [not focused on in this chapter—see Raison et al. 2006]. (*e*) Exposure to environmental stressors promotes activation of inflammatory signaling (NF-κβ) through increased outflow of proinflammatory-sympathetic nervous system responses, including the release of norepinephrine (*NE*), which binds to α- and β-adrenoceptors (αAR and βAR). (*f*) Stressors also induce withdrawal of inhibitory motor vagal input, including the release of acetylcholine (*Ach*), which binds to the α7 subunit of the nicotinic acetylcholine receptor (*α7nAChR*). (*g*) Concurrently with activation of the ANS, stressors induce the production of corticotropin-releasing hormone (*CRH*) in the paraventricular nucleus (*PVN*), which serves to turn on the HPA axis. CRH stimulates the release of adrenocorticotropic hormone (*ACTH*), which then stimulates the release of glucocorticoids (cortisol in humans). Typically, cortisol exerts major suppressive effects on the immune system. However, activation of the mitogen-activated protein kinase pathways (including p38 and Juan amino-terminal kinase [*JNK*]—not discussed here) inhibits the function of glucocorticoid receptors (*GR*), thereby releasing NF-κB from negative regulation by glucocorticoids released as a result of the HPA axis in response to stress (From Raison et al. (2006), with permission)

Neuropeptide Y (NPY) is a co-transmitter of sympathetic nervous innervation and potentiates the actions of norepinephrine. It is considered a stress hormone and mediates many of the cardiovascular effects of stress, including controlling blood pressure and blood flow (Elenkov et al. 2000). NPY can also enhance leukocyte adhesion and together with catecholamines, platelet aggregation, and macrophage activation (Black 2002).

5.3.2 Hypothalamic-Pituitary-Adrenal (HPA) Axis

Concurrently with activation of the ANS, the brain stimulates the production of two closely related neuropeptides in the paraventricular nucleus (PVN) of the hypothalamus via multiple pathways: corticotropin-releasing hormone (CRH) and arginine vasopressin (AVP). Together, these chemicals serve to turn on the HPA axis. CRH is the primary activator of the HPA axis. From the PVN, CRH is transported by a specialized portal circulatory system to the anterior portion of the pituitary gland where it stimulates the release of adrenocorticotropic hormone (ACTH). Importantly, arginine vasopressin (AVP) is a potent synergistic factor with CRH in stimulating ACTH secretion; furthermore, there is a reciprocal positive interaction between CRH and AVP at the level of the hypothalamus, with each neuropeptide stimulating the secretion of the other. ACTH then circulates in the bloodstream and stimulates the outer portion of the adrenal glands (i.e., the zona fasciculate of the adrenal cortex) to release glucocorticoids, mainly cortisol in humans and corticosterone in rats (Fig. 5.1g).

Cortisol is the quintessential stress hormone with multiple effects that enhance the fight-or-flight response. It stimulates the breakdown of amino acids in muscles to be converted into glucose for rapid energy utilization by the body and simultaneously promotes insulin resistance to leave glucose in the bloodstream. It increases blood pressure and enhances the ability of stress-released catecholamines to increase cardiac output, which also increases energy available to the organism for coping with stress. The effects of glucocorticoids on the brain are complex, but in response to acute stress, they narrow and focus attention and enhance memory formation for the circumstances that promoted their release.

Importantly, under normal conditions, cortisol exerts major suppressive effects on the immune system. Cortisol does this by reducing the number and activity of circulating inflammatory cells (including lymphocytes, monocytes, macrophages, neutrophils, eosinophils, mast cells), inhibiting production of proinflammatory mediators (including NF-κβ transcription pathway) and cytokines (IL-1, 2, 3, 6, TNF, interferon gamma), and inhibiting macrophage-antigen presentation and lymphocyte proliferation. Cortisol exerts its effects through cytoplasmic receptors. Activated receptors inhibit, through protein-protein interactions, other transcription factors including NF-κβ.

Additionally, cortisol plays an important negative feedback role on the HPA axis: cortisol binds to glucocorticoid receptors in the hippocampus, which inhibits the production of CRH and ACTH, as well as cortisol, to ultimately turn down or off the activated system. CRH is also negatively regulated by ACTH and itself, as well as by other neuropeptides and neurotransmitters in the brain, such as γ-aminobutyric acid-benzodiazepines (GABA-BDZ) and opioid peptide systems. These mechanisms are critical to ensure that the inflammatory response is appropriately elevated but does not exceed concentrations that would be dangerous for the organism.

5.3.3 How the Immune System "Hears" Changes in the SNS and HPA Axis

Primary and secondary lymphoid organs are innervated by sympathetic noradrenergic nerve fibers (Nance and Sanders 2007). Immune modulation can occur directly through the binding of the hormone to its related receptor at the surface of a cell. Almost all immune cells express receptors for one or more of the stress hormones that are associated with the sympathetic/adrenergic activation and HPA axis (Glaser and Kiecolt-Glaser 2005; Sanders and Kavelaars 2007; Webster et al. 2002). Specifically, T cells, B cells, monocytes, and macrophages express receptors for glucocorticoids, substance P, neuropeptide Y, prolactin, growth hormones, catecholamines (including adrenaline and noradrenaline), and serotonin. T cells also express receptors for corticotropin-releasing hormone. Ultimately, the binding of a stress hormone to a cell-surface receptor triggers a cascade of signals within the cell that can rapidly lead to changes in cell function.

Stress hormones also modulate immune responses indirectly, by altering the production of cytokines, such as IL-1, IL-2, IL-6, and TNF (Glaser and Kiecolt-Glaser 2005). These cytokines have many functions and affect different target cells; thus, dysregulation of these cytokines can cause later downstream effects. Importantly, although not discussed in detail here, these interactions are bidirectional such that cytokines produced by immune cells can feedback and modulate the brain (Fig. 5.1c)—including the SNS and HPA axis (Dantzer et al. 2008; Irwin and Cole 2011; Miller et al. 2009a).

5.3.4 Parasympathetic Activity: Vagal Withdrawal

The sympathetic and parasympathetic nervous systems (PNS) act in tandem to change the state of the body, often by promoting one system and actively withdrawing the other system. The PNS uses primarily the vagus nerve and acetylcholine (cholinergic receptors) as its primary effectors. There is emerging evidence that PNS activity modulates immune responses at the local level to prevent excessive inflammation through both the efferent and afferent fibers of the vagus nerve (Borovikova et al. 2000; Sternberg 2006; Tracey 2009).

The cholinergic anti-inflammatory pathway is the efferent arc of the inflammatory reflex, meaning that its purpose is to send signals down to the periphery to change the response and progression of inflammation. This neural mechanism inhibits macrophage activation through parasympathetic outflow (Borovikova et al. 2000; Tracey

2002, 2009). Specifically, messages sent via action potentials are transmitted by efferent vagus nerve activity to the periphery, including the liver, heart, spleen, and gastrointestinal track, which leads to acetylcholine release. Acetylcholine then interacts with α-bungarotoxin-sensitive nicotinic receptors (ACh receptors) on tissue macrophages and effectively downregulates inflammation by inhibiting the release of TNF, IL-1, and other cytokines (Fig. 5.1f) (Tracey 2002, 2009).

Although the inflammatory reflex is typically described as rapid response to localized inflammation, it may also induce systemic humoral anti-inflammatory response; vagus nerve activity can be relayed to the medullary reticular formation, locus coeruleus, and hypothalamus, leading to increased release of acetylcholine from the anterior pituitary and ultimately a systemic effect to downregulate inflammation (Tracey 2002). Interestingly, based on both in vivo and in vitro experiments, the vagus nerve is selective in that it downregulates the production of proinflammatory cytokines, but not anti-inflammatory cytokines (Tracey 2009). In fact, one abundant peptide, vasoactive intestinal polypeptide (VIP), inhibits TNF-α and IL-12 production and stimulates the secretion of the anti-inflammatory cytokine IL-10, primarily through VPAC1 receptors on immune cells (Ganea and Delgado 2001). Because lymphoid organs receive peptidergic/sensory innervation, this could be one method by which there is a systemic anti-inflammatory effect.

Vagal withdrawal in response to stress might therefore promote inflammation, given the evidence that vagal activity inhibits NF-κB activation (and the release of TNF-alpha from macrophages) via cholinergic signaling through the alpha-7 subunit of the nicotinic acetylcholine receptor (Pavlov and Tracey 2005). Indeed, decreased vagal tone, as manifested by reduced heart rate variability, has been associated with increased inflammatory markers in women with coronary-artery disease (Janszky et al. 2004), healthy controls (Thayer and Fischer 2009), and those with cardiovascular diseases (Haensel et al. 2008).

5.4 When the System Goes Awry: Adaptive and Maladaptive Responses to Stress

The stress response can vary and even go awry, depending on the nature of the stress. In the following section, we discuss how the stress response is typically adaptive in acute stress situations but maladaptive when faced with chronic stressors.

5.4.1 Acute Stress

Psychological acute stressors, such as giving a public speech, and physical acute stressors, such as receiving a cut from a sharp knife, employ the same pathway to activate the stress response system (Maier and Watkins 1998). In both of these hypothetical acute stress situations, the stress response (including activation of the sympathetic nervous system and effects on the immune system) is typically healthy and adaptive for survival.

Acute or short-term stress induces a large-scale redistribution of immune cells in the body. Typically, immune cells stay in certain compartments of the body, including the marginated pool, spleen, bone marrow, and lymph nodes; when acute stressors occur, stress hormones initiate a cascade of events and induce the trafficking of immune cells (e.g., lymphocytes) out of these compartments and into the blood, to ultimately reside in target organs where an individual is most likely to be injured (e.g., skin, gastrointestinal track, urogenital tract, lungs) (Dhabhar et al. 2012). In doing so, the body has increased the immune cell's ability to do defensive maneuvers.

In sum, the body's activation of the sympathetic nervous system and the immune system and concomitant reduction in PNS activity is its way of appropriately responding to the stressor and preparing the body for survival. In the example of public speaking, the brain perceives a stressor, which then warns the body of danger. To promote survival, the body then mounts an immune response to "prepare" for anticipated activation of the immune system (wounding or infection). Although in reality, we do not expect to be physically wounded, for example, when giving a presentation, across evolutionary time stress was a reliable enough signal of impending physical danger that it was adaptive to respond to all fight-or-flight situations (both psychological and physical stressors) by mounting an appropriate biological response, to ultimately ensure survival. Similarly, in the example of receiving a cut, especially prior to modern medicine and hygiene, the immune-enhancement effect of activating the stress response system is advantageous to mount a response against any pathogens that may have entered the wound, as well as to begin the recovery process.

All stress is not necessarily harmful, and all stress is not immunosuppressive. One caveat to this adaptive response to acute stress is that a stress-induced enhancement of the immune system could be harmful if it exacerbates existing inflammatory or autoimmune diseases (Dhabhar and McEwen 2007), possibly due to chronic stress, which we turn to next.

5.4.2 Chronic Stress

Chronic stressors, such as caring for a spouse with dementia, concealing a sexual identity, or coping with childhood physical, or sexual abuse, tell a different story for the stress response system. These types of stressors are considered to be the most toxic because they so often result in long-term changes in the emotional, behavioral, and physiological responses that lead to the risk, development, or progression of diseases (Cohen et al. 2007). In addition to emotional and behavioral changes due to the stressor, such as difficulty in coping with the stressor or changes in health behaviors such as sleeping, physiological changes also occur.

Prolonged or repeated activation of the HPA and SAM axes can disrupt the regulation of other body processes, including the immune system. Individuals experiencing chronic stressors have less effective immune functioning, as demonstrated by their increased susceptibility to the common cold (Cohen et al. 1998), impaired immune response to vaccination, and delayed healing after standardized wound inductions (Glaser et al. 1998, 1999). Additionally, they also experience low-grade, nonspecific inflammation (Segerstrom and Miller 2004). This increase in inflammation is likely due to decreased anti-inflammatory feedback. As previously mentioned, the HPA axis plays an important negative feedback role in suppressing the immune response when it is no longer needed. However, in chronic stress situations, glucocorticoid resistance or insufficient glucocorticoid signaling may occur, which lead to HPA axis-related increases (as opposed to decreases) in inflammation. Possible effects include (1) the adrenal gland can get exhausted and therefore produce less cortisol, which corresponds to decreased anti-inflammatory feedback, or (2) the HPA axis is hyperactivated and the adrenal gland pumps out so much cortisol that the cells' receptors, which typically recognize the cortisol and shut down, become resistant and do not "hear" the cortisol as well (i.e., they are less sensitive) (Hänsel et al. 2010).

In sum, autonomic and neuroendocrine activation in response to stressors is beneficial up to a point, but excessive activation may also have long-term costs. The metabolic requirements posed by psychological stressors to which people are typically exposed in contemporary society are often minimal (Cacioppo 1998). As such, strong autonomic and neuroendocrine activation to psychological stressors is often not needed for effective coping and instead may contribute to chronic diseases over time (Miller et al. 2009b; Robles et al. 2005).

5.5 Intrapersonal Processes and Immune Functioning

How individuals view the world and appraise their own situations and stressors can influence their physiological response to stress. Certain patterns of thought or appraisal of emotions are intrapersonal (i.e., within person) processes that can generate a chronic perception that the world is dangerous, which can create an immune response that "runs hot" and is extra vigilant. Other intrapersonal processes may buffer the effects of stress on immune functioning. In the following section, we discuss various intrapersonal factors that may be associated with, or moderate, the effects of stress on immunity.

5.5.1 Rumination

Rumination is defined as conscious, spontaneous, and recurrent thoughts or images or both about past negative information. For those who ruminate, stress responses may last longer (i.e., slower recovery), based on the Perseverative Cognition Hypothesis (Brosschot et al. 2006). In one study that experimentally induced either rumination or distraction after a public speaking task, the participants in thc rumination condition demonstrated sustained increases in inflammation (measured in plasma CRP) that did not return to prestressor levels by the end of the visit. Conversely, participants' CRP in the distraction group increased post-stressor and then returned to prestressor levels by the end of the visit (Zoccola et al. 2014). Additionally, in a cross-sectional study, older adults who reported being highly ruminative also had greater numbers of leukocytes, lymphocytes, and B cells than those who reported lower rumination (Thomsen et al. 2004), suggesting that rumination may be related to an activation of the acquired immune system and thus may be associated with a more prolonged immune response to stress.

5.5.2 Emotion Regulation

Emotion regulation refers to the processes by which individuals influence which emotions to have, when to have them, and how to experience and express them (Gross 1998). Two common emotion regulation strategies include cognitive reappraisal and expressive suppression. *Cognitive reappraisal*, considered an adaptive strategy, involves altering how to think about an emotion-eliciting situation in order to change its emotional impact. In contrast, *expressive suppression*, generally considered a maladaptive emotion regulatory strategy, involves inhibiting emotional expression in response to an emotion-eliciting event (Gross and John 2003).

To date two studies have examined these emotion regulation strategies in relation to immune functioning and cardiovascular disease (CVD). In one study, CRP levels (a known marker of CVD risk) and reappraisal and suppression were assessed in 379 adults. Reappraisal was associated with lower CRP, and suppression was associated with higher CRP after controlling for demographics, suggesting that adaptive emotion regulation strategies may promote healthy outcomes by lowering inflammatory mechanisms (Appleton et al. 2013). In the second study, IL-6 mediated the association between reappraisal-related engagement of the dorsal anterior cingulate cortex (dACC, a brain region involved in governing the release of neurohormones and neurotransmitters of the HPA axis, SNS, and PNS) and atherosclerosis (Gianaros et al. 2014). One interpretation offered was that elevated inflammation, as reflected by increased IL-6, might have upregulated negative affect or arousal processes that consequently increased the cognitive demands required for the regulation of emotion at the neural level (e.g., there might have been more negative affect to regulate). However, this was a cross-sectional design and so additional research is needed to examine the directionality and pathways linking emotion regulatory processes and immune functioning in clinically relevant populations.

Another emotion regulation strategy, emotional approach coping (EAC), has also been examined in relation to inflammation. EAC is comprised of emotional processing (purposive attempts to acknowledge, explore, and understand one's emotions) and emotional expression (active verbal and nonverbal efforts to communicate emotional experiences) (Stanton et al. 1994). In a sample of 41 men who had undergone prostatectomy or radiation therapy for localized prostate cancer, emotional processing at baseline predicted lower IL-6, sTNF-RII, and CRP 4 months later, whereas emotional expression was associated with higher levels of sTNF-RII (Hoyt et al. 2013). Interestingly, the interaction of emotional processing and expression suggested that expression of emotion is associated with higher inflammation (CRP and sTNF-RII) only in the context of low emotional processing. The expression of emotions without efforts to understand them might promote emotion dysregulation and higher inflammation.

Master et al. (2009) examined emotional approach coping and inflammation before and after a standardized laboratory stressor, the Trier Social Stress Test (TSST). Participants in the TSST paradigm were asked to prepare and give an impromptu public speech and to perform difficult mental arithmetic to a nonresponsive, socially rejecting panel of raters. Findings revealed that, in response to the stressor, higher levels of emotional approach coping, particularly emotional processing, were associated with a less pronounced increases in soluble tumor necrosis factor receptor type-II (sTNF-RII) in oral mucosal transudate. These findings suggest that people who are more likely to cognitively reappraise and cope with stressors by approaching their emotions, particularly through emotional processing and related emotional expression, may demonstrate lower inflammatory outcomes, which could promote more optimal health.

5.5.3 Alexithymia

Alexithymia is a personality trait characterized by impairments in cognitive processing and regulation of emotions that is typically measured using the 20-item Toronto Alexithymia Scale (TAS-20). It is hypothesized that this deficit in affective and cognitive-emotional processing leads to prolonged and amplified physiological arousal to stress thus disturbing the autonomic system and HPA axis and ultimately the immune system (Guilbaud et al. 2003). Rather than there being a clear shift towards either pro- or anti-inflammatory mediators in alexithymic individuals, circulating cytokine profiles and Th1/Th2 responses may be affected (Guilbaud et al. 2003; Mandarelli et al. 2011).

Alexithymia has been linked to lower circulating levels of IL-2R and IL-4α in somatoform disorders (Pedrosa Gil et al. 2007) and IL-6 in healthy participants (Mandarelli et al. 2011). Others have found significant positive correlations between alexithymia and serum levels of TNF-α in patients suffering from rheumatoid arthritis (Bruni et al. 2006), as well as serum levels of IL-4 (a type 2 cytokine) in healthy female participants (Corcos et al. 2004). Alexithymia has also been linked to decreased in vitro production of IL-1β, IL-2, and IL-4, and a skewed Th1/Th2 (IL-2/IL-10) response towards Th2 response (Guilbaud et al. 2009). In addition, various lymphocytes have been found in very low levels in alexithymic men (for the natural killer subset: CD57 − CD16+ and killer effective T-cell CD8+CD11a+cells) (Dewaraja et al. 1997) and women (CD2, CD3, CD4, and CD19) (Todarello et al. 1994, 1997). Lastly, alexithymic patients with HIV exhibited increased norepinephrine-to-cortisol ratio and viral load (McIntosh et al. 2014), suggesting a greater vulnerability to disease progression in these patients. Taken together, these findings suggest that alexithymia may be associated with lower cell-mediated immunity and a skewed Th1/Th2 ratio towards Th2 response. Thus, it has been suggested that the neuroendocrine and immune responses of alexithymics may follow a similar pattern as in persons with chronic stress (Guilbaud et al. 2003).

5.5.4 Psychological Stress

Across a number of studies over the years, psychological stress has been found to be associated with changes in physiological functioning, including changes in immunity. Measures of psychological stress include the Perceived Stress Scale (PSS) (Cohen et al. 1983) and the Life Events and Difficulties Schedule (LEDS) (Brown and Harris 1978). Higher levels of perceived psychological stress have been associated with reduced control of latent herpesviruses, blunted humoral responses to immunization, greater susceptibility to infectious disease, and poorer wound healing (Cohen et al. 2001; Dyck et al. 1999; Glaser and Kiecolt-Glaser 2005; Glaser et al. 1998, 1999; Kiecolt-Glaser et al. 1996a). Interestingly, measures of objective stress do not always yield the same health findings. For example, Cohen and colleagues reported that both perceived stress and stressful live events (objective measure of stress) predicted greater risk for developing the common cold. However, these two measures produced different associations with illness and were mediated by different biological processes (Cohen et al. 1993). Thus, measures of stress based upon the objective environment versus those based upon subjective appraisal relate to different biological mechanisms, predict different aspects of illness, and may ultimately be associated with different disorders and disease (Monroe 2008).

5.5.5 Positive Psychological Well-Being: Optimism and Positive Affect

Positive psychological well-being, including dispositional optimism and positive affect, has also been associated with immune functioning. Evidence suggests that higher levels of optimism and positive affect are generally associated with better immune functioning and may ultimately be protective for health during times of heighted stress, whereas lower levels of these are generally associated with poorer immune functioning.

5.5.5.1 Dispositional Optimism

Dispositional optimism reflects the extent to which individuals hold generalized favorable expectancies for their future and is most often assessed by the Life Orientation Test (LOT (Scheier and Carver 1985)). Greater optimism has been associated with lower levels of IL-6 cross-sectionally (Roy et al. 2010) and IL-6 and soluble intercellular adherence molecule pooled across multiple time points (Ikeda et al. 2011). In a double-blind placebo-controlled study in which men underwent either a placebo or real vaccine and then completed two mental stress tasks, those who reported high levels of dispositional optimism had smaller IL-6 responses to the stress task (independent of age, BMI, trait depression and baseline IL-6) (Brydon et al. 2009). Additionally, in the vaccine/stress group, there was a strong positive association between optimism and antibody responses, indicating that stress accentuated the antibody response to vaccine in optimists (Brydon et al. 2009). Another interesting study examined immune functioning and telomeres, a biological marker of immunosenescence, as related to optimism in men. Men who had shorter telomeres with high telomerase activity (indicative of active cell stress) were less optimistic and showed blunted post-stress recovery in autonomic measures as well as monocyte chemoattractant protein-1 in comparison to men with longer telomeres or men with shorter telomeres and low telomerase activity (Zalli et al. 2014). Together these findings provide support for the stress-buffering hypothesis: optimism may help to buffer the negative effects of stress on immune functioning.

A growing number of studies, however, have demonstrated that difficult stressors have more potentially detrimental effects on the immune systems of more optimistic people (Cohen et al. 1999; Segerstrom 2001, 2005, 2006; Segerstrom et al. 2003). For example, during stressors that are complex, persistent, and uncontrollable, more optimistic people had smaller delayed-type hypersensitivity responses, indicative of less robust in vivo cellular immunity (Segerstrom 2006). This effect may be due to optimists' greater engagement, fatigue, and ultimately physiological stress during difficult stressors. The relation between optimism and immunity is complex and dependent on the duration and type of stressor involved, as well as the individual dealing with the stressor.

5.5.5.2 Positive Affect

There is contention in the field of emotion about how to precisely define positive affect. However, positive affect is broadly defined as reflecting pleasant engagement with the environment (Pressman and Black 2012). Overall, results from investigations into naturally occurring positive affect indicate an association between positive affect and immune function, where higher levels of positive affect are generally associated with enhanced immune function (Pressman and Black 2012). In cross-sectional studies, greater trait positive affect was related to lower levels of circulating IL-6 in the Whitehall study, a large-scale population based study on health (Steptoe et al. 2008), as well as lower levels of stimulated IL-6 in adults after accounting for age, gender, race, BMI, and white blood cell count (Prather et al. 2007). Additionally, the presence of low-grade inflammation (as measured by higher levels of IL-6) and the absence of positive affect were independently predictive of worse subjective health in 347 women of the general population aged 45–90 years (Andreasson et al. 2013).

In other studies, researchers have examined positive affect in the context of an immune challenge and laboratory stressor. In individuals who were experimentally infected with rhino virus, those who had a higher positive emotional style (assessed before infection) demonstrated less symptoms and signs of rhinovirus infection (Doyle et al. 2006); specifically, higher positive emotional style was associated with lower IL-6 levels and lesser symptom and sign responses. Another study examined the maintenance of a positive outlook in the midst of an acute laboratory stress (the Trier Social Stress Test; TSST) in 35 postmenopausal women. Greater acute stress-induced declines in positive outlook were significantly associated with increased IL-1β reactivity, which significantly predicted increases in depressive symptoms over the following year, controlling for age, body mass index, chronic stress, anti-depressant use, and baseline depressive symptoms (Aschbacher et al. 2012). In sum, difficulty maintaining positivity under stress and heightened proinflammatory reactivity may be markers and/or mechanisms of risk for future increases in physical and mental disorders.

5.5.6 Summary

In this section, we identified key intrapersonal factors that are directly or indirectly associated with stress effects on individuals' immune functioning. See Table 5.1 for a summary of intrapersonal factors and their effects on immune parameters. Rumination, the emotion regulation technique of suppression, alexithymia, and perceived psychological stress are generally associated with poorer immune functioning. Interestingly, objective psychological stress may not be associated with the same immune outcomes, or immune pathways. Conversely, the emotion regulation technique of reappraisal and emotional approach coping (particularly emotional processing), optimism, and positive affect are generally associated with better immune functioning and may ultimately be protective for health during times of heightened stress.

5.6 Interpersonal Processes and Immune Functioning

An important extension to the study of relationships between intrapersonal processes and stress and immune functioning acknowledges that individuals live in a social environment and continually interact with others. Individuals' health and emotions influence, and are influenced by, significant others, including, for example, spouses, partners, parents, and children, as well as the broader social ecological contexts in which they live. The following section focuses on how interpersonal processes relate to stress and immune regulation and functioning.

5.6.1 Close Relationships

Broadly, the strength and quality of a person's social connection to other people can predict risk for mortality: stronger social bonds (i.e., better social integration and/or social support) decrease risk for morality by up to 50 % (Holt-Lunstad et al. 2010). Results from another meta-analytic review of 126 published empirical articles over the past 50 years indicate that greater marital quality is related to better health, including lower risk of mortality and lower cardiovascular reactivity during marital

Table 5.1 Summary of intrapersonal factors and effects on immune parameters

Intrapersonal factors	Associated effects on immune parameters and functioning
Rumination	↑ CRP, leukocytes, lymphocytes, and B cells (Thomsen et al. 2004; Zoccola et al. 2014)
Emotion regulation	
Reappraisal	↓ CRP (Appleton et al. 2013)
Suppression	↑ CRP (Appleton et al. 2013)
Emotional approach coping	
Emotional processing	↓ IL-6, sTNF-RII, CRP (Hoyt et al. 2013; Master et al. 2009)
Emotional expression	↑ sTNF-RII (Hoyt et al. 2013)
Alexithymia	↓ IL-2R, IL-4α, IL-6, in vitro production of IL-1β, IL-2, and IL-4, lymphocytes (Dewaraja et al. 1997; Guilbaud et al. 2009; Mandarelli et al. 2011; Pedrosa Gil et al. 2007; Todarello et al. 1994, 1997) ↑ TNF-α, IL-4, in vitro production of IL-1β, IL-2, and IL-4, and Th2 response (Bruni et al. 2006; Corcos et al. 2004; Guilbaud et al. 2009)
Perceived psychological stress	↓ control of latent herpesviruses, blunted humoral responses to immunization, poorer wound healing ↑ susceptibility to infectious disease (Cohen et al. 2001; Dyck et al. 1999; Glaser and Kiecolt-Glaser 2005; Glaser et al. 1998, 1999; Kiecolt-Glaser et al. 1996a)
Dispositional optimism	↓ IL-6, soluble intercellular adherence molecule (Brydon et al. 2009; Ikeda et al. 2011; Roy et al. 2010) ↑ antibody response to vaccine (Brydon et al. 2009)
Positive affect	↓ circulating and stimulated IL-6, ↓ symptoms and signs of rhinovirus infection (Steptoe et al. 2008; Doyle et al. 2006) Declines in positive affect: ↑ IL-1β (Aschbacher et al. 2012)

CRP C-reactive protein, *IL* interleukin, *TNF* tumor necrosis factor, *sTNF-RII* soluble tumor necrosis factor (receptor II), *Th2* T-helper cell type 2

conflict (Robles et al. 2014). Findings from Whisman and Sbarra (2012) suggest that lower marital satisfaction is related to elevated inflammation. It is apparent that close relationships influence health outcomes, and recent growing evidence suggests that immune functioning may be one potential pathway linking close relationships and health (Robles and Kiecolt-Glaser 2003).

5.6.2 Negative Relationship Processes: Anger, Hostility, and Conflict

Negative close relationship processes involving stressful encounters, such as marital strain, conflict, or abuse, can affect immune functioning (Robles and Kiecolt-Glaser 2003). Specifically, how we interact with our close relationship partners (e.g., showing anger, hostility, conflict, blaming or interrupting our partner) may be particularly detrimental and increase both circulating proinflammatory cytokines and stimulated immune inflammatory responses.

For example, couples who displayed higher levels of hostile behaviors during marital conflict showed larger increases in circulating markers of inflammation, including IL-6 and TNF-α, and slower wound healing at 60 % the rate of low-hostile couples (Kiecolt-Glaser et al. 2005). Additionally, high-hostile partners had greater decrements in 24 h immune cell functioning than participants who exhibited fewer negative behaviors (Kiecolt-Glaser et al. 1993). These effects are beginning to be examined longitudinally: Couples who were in more distressful marriages at baseline had larger declines in cellular immune function (proliferative responses to two mitogens, concanavalin A and phytohemagglutinin) 2 years later when compared to spouses in less distressful marriages (Jaremka et al. 2013b). Looking specifically at adaptive immunity, low marital satisfaction and greater hostility during marital conflict were associated with higher Epstein-Barr virus (EBV) antibody titers, indicating poorer ability to control this latent herpesvirus that infects most adults (Kiecolt-Glaser et al. 1987, 1988, 1993, 1997).

There are interesting gender effects in this literature: When comparing a functional measure of the immune system (proliferative response to mitogen) of men and women over the course of a conflict induction, one study found that men's immune functioning increased and women's immune functioning decreased from pre- compared to post-conflict induction (Mayne et al. 1997). These results are corroborated by other findings with similar gender effects, especially for women (Kiecolt-Glaser et al. 1996b, 1998; Malarkey et al. 1994). Taken together, these findings suggest that women may be more sensitive to negative marital interactions than men.

5.6.3 Supportive Relationship Processes

Just as distressing relationships can dysregulate immune function, supportive relationship processes may be immunoprotective. For example, increased positive behaviors exhibited by couples during a social support interaction task predicted faster wound repair from suction blisters (Gouin et al. 2010); positive behaviors were behaviorally indexed by aggregating measures of acceptance, relationship-enhancing attribution, self-disclosure, and humor exhibited during the interaction task. Other behaviors, including warm physical contact, may also be immune enhancing; circulating levels of interferon (IFN)-γ decreased significantly in couples after an hour-long experimental induction of warm physical contact (hugging and kissing), whereas levels did not change in the control condition (couples who read books in separate rooms) (Matsunaga et al. 2009). In other work on HPA and ANS responses to stress, which have important implications on immune functioning, women who received positive physical partner contact (standardized neck and shoulder massage) before undergoing a TSST exhibited significantly reduced subsequent cortisol responses to stress, as well as reduced heart rate increase in response to the stressor (Ditzen et al. 2007).

Supportive communication patterns also promote healthy immune functioning and may be one way to mitigate the stressful effects of martial conflict—or other every-day stressors—on inflammation. Couples who displayed more cognitive engagement, assessed by the number of cognitive processing words used, during a marital conflict discussion had lower systemic IL-6 responses 24 h after the discussion than did those displaying less cognitive engagement (Graham et al. 2009). In another study, researchers examined the effect of communal orientation, which is marked by first-person pronoun use (*we* talk)—as opposed to singular first-person pronoun (*I* talk)—on the trajectory of congestive heart failure, an immune-related disease. Specifically, in couples in which one partner had congestive heart failure, *we* talk by the spouse, but not by the patient (with congestive heart failure), independently predicted positive change in the patient's heart failure symptoms and general health over the next 6 months (Rohrbaugh et al. 2008). Thus, supportive and positive relationship processes, including warm contact and supportive communication patterns within couples, may prove to be a significant area of research for generating interventions to improve partners' health by mitigating inflammatory responses.

5.6.4 Ambivalence

Much research focuses on how the positive or negative aspects of relationships influence health. However, a small but growing area of research examines the effect of ambivalence on health outcomes. Ambivalence is described as simultaneously feeling positive and negative emotions towards a close relationship partner (Uchino et al. 2001). Perceiving ambivalence towards one's spouse in a support context was linked to greater inflammation (higher IL-6 and fibrinogen and marginally higher CRP levels) even when considering health behaviors, relationship-specific romantic attachment style, spouse negativity/positivity ratings, and overall marital satisfaction (Uchino et al. 2013). Perceptions of ambivalence during support may be a particularly important relational context in which close relationship ties influence health. Further work is needed to examine ambivalence in other contexts (e.g., in response to daily stressors and longitudinal assessments of ambivalence ratings, which may change over time), as well its relation to other indicators of immune functioning.

In related work, coronary-artery calcification scores were highest for individuals who both viewed and were viewed by their spouse in an ambivalent manner (Uchino et al. 2014). Importantly, coronary-artery calcification is correlated with the extent of plaque buildup in the coronary arteries and is a robust predictor of cardiovascular disease and stroke—both of which are associated with chronic inflammation (Danesh et al. 2004). Future work that examines inflammatory mediators associated with ambivalence and other clinically relevant diseases may provide insight on possible interventions to improve health outcomes by fostering interpersonal relationship functioning.

5.6.5 Social Rejection and Social Isolation/Loneliness

Social rejection is a major life event that is related to immune functioning. This interpersonal process is often studied in the context of depression due to the sustained inflammatory process that may elicit sickness behaviors and precipitate depression for vulnerable individuals (Slavich et al. 2010). In a longitudinal study of 147 adolescent girls at elevated risk for depression, participants had significantly higher levels of mRNA for both proinflammatory transcription factor NF-κB and inhibitor of κB (I-κB), which regulates the effects of NF-κB, at visits when they had experienced a recent targeted rejection life event compared to visits when no such event had occurred (Murphy et al. 2013). A growing body of research suggests that stressors involving social rejection and exclusion activate neural regions involved in processing negative affect, including the dorsal anterior cingulate cortex (dACC) and anterior insula (Slavich et al. 2010). These neural regions activate multiple biological systems, including, in particular, the HPA axis and sympathetic-adrenal-medullary axis, which produce cortisol and catecholamines that can bind to receptors on immune cells, which then modulate the release of proinflammatory cytokines. Thus, social stress-related implications on the neurocognitive pathway involving the dACC and anterior insula may be one mechanism linking social threat and rejection with elevated inflammation and risk for depression (Slavich et al. 2010).

Social isolation, or loneliness, is another interpersonally distressing state that dysregulates immune function (Jaremka et al. 2013c). Interestingly, social isolation is not broadly immunosuppressive but instead selectively suppresses some groups of immune-response genes (e.g., type I interferons and specific immunoglobulin genes) while simultaneously activating others (e.g., proinflammatory cytokines); social isolation has been related to a downregulation of genes involved in antibody production and an upregulation of expressed genes involved in proinflammatory immune response (Cole et al. 2011). Indeed, lonelier people had smaller antibody responses to an influenza virus vaccine than those who were less lonely (Pressman et al. 2005). Additionally, among healthy adults and posttreatment breast cancer survivors, stimulated TNF-α, IL-6, and IL-1β were higher after laboratory stress tasks (including the TSST and Stroop task) among those experiencing greater loneliness compared with those who were less lonely (Hackett et al. 2012; Jaremka et al. 2013a). Thus, social isolation/loneliness has immune consequences that may be especially relevant to clinical populations, such as women undergoing breast cancer treatment.

5.6.6 Early Life Environment and Adversity

A very well-developed area of research has focused on inflammation and early life adversity. Early life adversity is a term ranging in meaning from poverty and abuse to parental loss and is characterized by unpredictability and interpersonal stress (Slavich and Irwin 2014). Childhood adversity can cause long-term alterations in HPA axis functioning that ultimately affects the immune system. In this context, early life adversity programs the brain and body to run inflammation "hot," likely as a result of evolutionary pressure linking stress to danger of wounding and tissue damage (Raison and Miller 2013). This results in chronically elevated inflammation that, although modest, contributes to shaping the brains and bodies of these individuals to be especially vulnerable to mental and physical health problems, including major depressive disorder, cardiovascular disease, and dementia.

Results from research have demonstrated that adults who experienced early life adversity show exaggerated inflammatory responses to stress (Carpenter et al. 2010; Danese et al. 2007, 2008; Pace et al. 2006). Maltreated children develop higher levels of IL-6 in response to a standardized social stressor (TSST) when tested as adults in comparison to a non-maltreated control group (Carpenter et al. 2010; Pace et al. 2006). Additionally, findings from longitudinal studies showed that greater cumulative stress exposure before age 8 predicted higher levels of IL-6 and CRP at age 10 and higher levels of CRP at age 15 in a sample of 4600 children (Slopen et al. 2013). Additionally, maltreated children tended to have higher levels of CRP 20 years later (Danese et al. 2007).

Interestingly, exposure to coevolved microorganisms in childhood may play an important role in how early life adversity affects immune functioning. Recently, researchers in the Philippines have found that even a childhood trauma as severe as maternal deprivation can fail to result in a raised background CRP in adulthood in those individuals who were heavily exposed to a microbe-rich environment and animal feces in childhood (McDade et al. 2013), whereas individuals raised in clean childhood environments in the Philippines showed strong correlations between early life adversity and elevated CRP in adulthood. In the USA, such adverse childhood events tend to have serious consequences for later health, as previously discussed. These findings suggest that exposure to animal-derived microbes might improve regulation of inflammation and so increase stress resilience, though this observation needs to be confirmed in other populations (Rook et al. 2014). We return to this issue in the Future Considerations below.

5.6.7 Summary

In this section, we have attempted to elucidate interpersonal and interdependent factors that may influence health and immune functioning. See Table 5.2 for a summary of interpersonal factors and their effects on immune parameters. Negative relationship processes, including behaviors such as anger, hostility, and conflict, ambivalence, social rejection, social isolation/loneliness, and early life adversity, are generally associated with poorer immune functioning. Importantly, microbial exposure in childhood may play an important role in moderating the effects of early life adversity on inflammation such that it minimizes the effects of adversity and lessens inflammation. On the other hand, supportive relationship processes, including positive behaviors, supportive communication patterns, and warm touch, are generally associated with better immune functioning and may provide insight on possible interventions to improve health outcomes.

5.7 Conclusions and Future Directions

Immune functioning is essential to health and well-being. Understanding how stress influences the immune system requires knowledge of not only the biological pathways and mechanisms by which stress can "get under the skin" but also the multiple

Table 5.2 Summary of interpersonal factors and effects on immune parameters

Interpersonal factors	Associated effects on immune parameters and functioning
Negative relationship processes	↑ IL-6, TNF-α, EBV-titers (Kiecolt-Glaser et al. 1987, 1988, 1993, 1997, 2005) ↓ wound healing, cellular immune functioning (Kiecolt-Glaser et al. 1993, 2005; Jaremka et al. 2013b) Cellular immune functioning: men >females (Mayne et al. 1997)
Supportive relationship processes	↑ wound repair (Gouin et al. 2010) ↓ IFN-γ, IL-6 (Graham et al. 2009; Matsunaga et al. 2009)
Ambivalence	↑ IL-6, fibrinogen, CRP (Uchino et al. 2013)
Social rejection	↑ mRNA for NF-κB and I-κB (Murphy et al. 2013)
Social Isolation/loneliness	↓ antibody response to influenza; downregulation of genes involved in antibody response (Cole et al. 2011; Pressman et al. 2005); ↑ stimulated TNF-α, IL-6, and IL-1β; upregulation of genes involved in proinflammatory immune response (Cole et al. 2011; Hackett et al. 2012; Jaremka et al. 2013a)
Early life environment and adversity	↑ IL-6, CRP (Carpenter et al. 2010; Danese et al. 2007, 2008; Pace et al. 2006)

IL interleukin, *TNF* tumor necrosis factor, *EBV* Epstein-Barr virus, *IFN* interferon, *CRP* C-reactive protein, *mRNA* messenger ribonucleic acid, *NF* nuclear factor, *I* inhibitor

intrapersonal and interpersonal factors that may exacerbate or buffer the effects of stress on immune functioning. Certain factors may prolong or exacerbate the effects of stress, including rumination, emotional suppression, alexithymia, psychological distress, negative relationships processes, ambivalence, social rejection, social isolation/loneliness, and early life adversity. Other factors may mitigate the effects of stress, including emotional reappraisal, emotional approach coping, dispositional optimism, positive affect, and supportive relationship processes. More research is needed in these areas to further uncover the biological and behavioral mechanisms by which these intrapersonal and interpersonal factors exert their effects on the immune system and ultimately overall health and well-being. In the following sections, we highlight some new substantive and methodological considerations relevant to future research on the effects of stress on immunity.

Research findings on stress and immunity may benefit from being understood and approached from an evolutionary perspective. Evolution can provide a guiding framework to help answer *why* individuals have certain behavioral and immunological responses. For example, Raison and colleagues have put forward the Pathogen Host Defense (PATHOS-D) hypothesis, which suggests that the constellation of behaviors observed in major depression (i.e., symptoms associated with elevated levels of inflammatory cytokines or sickness behavior) should be viewed as having been adaptive, rather than socially maladaptive, across human evolution because they allowed the organism to utilize limited metabolic resources for immune activation and recovery (Raison and Miller 2013). The lens of human evolution can be applied to related work on stress and immune functioning, particularly in the context of interpersonal and group processes. Using this framework, we can begin to address various questions, including, for example, what is the evolved immunology of group processes? In other words, what is advantageous about how couples respond to a conflict? Why might it be advantageous (in an evolutionary sense) that women are seemingly more sensitive to negative social interactions than men? Much of the research reviewed here addresses the "what" (e.g., what factors are associated with altered immune functioning) and "how" (e.g., how does positive social interactions moderate the effects of stress on immune function). The theory of evolution can clarify the underlying logic connecting these issues by beginning to address *why* variables are associated with each other the way that they are. Furthermore, evolution allows us to see stress in a new light—that is, the coordinated stress response is not only a risk factor and source of physiological and behavioral dysregulation but also serves an adaptive and evolutionary function to aid survival. Using this perspective will allow us to continue to move the field forward and examine how and why stress affects immune functioning in individuals, couples, families, and communities.

Future research on stress and immunity may also benefit from accounting for ecological factors, such as exposure, or lack of exposure, to an array of microbes and helminths with which we coevolved and which—while lacking to a large degree in the industrialized world—are still relevant to immune/stress interactions in other geographical and cultural contexts. Much of the present research on stress and inflammation has been conducted exclusively in higher-income, industrialized populations with regimes of sanitation and hygiene that have reduced the frequency and diversity of microbial exposures and burdens of infectious disease (McDade et al. 2013). In other words, we have been studying humans in environments quite different from that in which humans evolved, due to our reconfigured relationship with the microbial and parasitic world. Exposure to these "Old Friend" immunoregulatory organisms may play a paramount role in optimal immune function and should therefore be studied from developmental and life-span perspectives to further advance our knowledge of stress and the immune system.

These "Old Friends" include elements of the gut microbiota, as well as certain pseudocommensal environmental bacterial and helminthes. Exposure to certain ancient viruses at appropriate stages of development also likely programmed appropriate immune function (Rook et al. 2013). Individuals from high-income countries, including the USA, may receive inadequate exposure to immunoregulation-inducting Old Friends. Importantly, infectious exposures in infancy may have lasting effects on the regulation on inflammation in adulthood; to the extent that these pathways become established and carried forward, inflammatory stressors in adulthood may be handled in a similar manner (McDade 2012). Lack of exposure to Old Friends may increase the likelihood of immunoregulatory deficits and uncontrolled inflammation, which, in concert with psychosocial stressors, could contribute to chronic inflammatory and psychiatric diseases (Raison et al. 2010; Rook et al. 2013). As previously discussed, empirical evidence suggested that recent psychosocial stress did not cause detectable increases in CRP in adults who received heavy microbial exposures as infants (McDade et al. 2013). Future research is needed that explores how other ecological factors (including exposure to Old Friends) may help buffer the negative health effects of stress, with possible implications for intervention and prevention in the US and other Westernized cultures.

Lastly, there are a variety of key methodological considerations that can be incorporated into future research on stress and immune functioning. In the current chapter, we focused mostly on one piece of the puzzle—mainly stress effects on *immune functioning*. However, as shown in some of the research findings reviewed here, changes in immune functioning are mediated by bidirectional and interacting effects of the central nervous system (CNS) and endocrine system. Thus, to further tease apart the mechanisms of action and gain a more complete understanding of the physiological and health impacts of stress, advanced computational modeling that accounts for stress-related changes in multiple, dynamic systems (CNS, immune, and endocrine systems) within and between individuals and ecological contexts, over time, should be employed (Reed et al. 2013; Sturmberg and Martin 2013). Ultimately, these methodological advances may allow us to better understand the mechanisms by which intrapersonal and interpersonal factors may moderate and mediate the regulatory effects of stress on immune functioning. Application of these statistical and design methods can help inform future research and practice regarding optimal, targeted ways to improve immune functioning and treat immune-related conditions to improve health.

References

Andreasson AN, Szulkin R, Undén A-L, von Essen J, Nilsson L-G, Lekander M (2013) Inflammation and positive affect are associated with subjective health in women of the general population. J Health Psychol 18(3):311–320

Antoni MH, Lutgendorf SK, Cole SW, Dhabhar FS, Sephton SE, McDonald PG, Stefanek M, Sood AK (2006) The influence of bio-behavioural factors on tumour biology: pathways and mechanisms. Nat Rev Cancer 6(3):240–248

Appleton AA, Buka SL, Loucks EB, Gilman SE, Kubzansky LD (2013) Divergent associations of adaptive and maladaptive emotion regulation strategies with inflammation. Health Psychol 32(7):748–756

Aschbacher K, Epel E, Wolkowitz O, Prather A, Puterman E, Dhabhar F (2012) Maintenance of a positive outlook during acute stress protects against pro-inflammatory reactivity and future depressive symptoms. Brain Behav Immun 26(2):346–352

Barton GM (2008) A calculated response: control of inflammation by the innate immune system. J Clin Invest 118(2):413

Black PH (2002) Stress and the inflammatory response: a review of neurogenic inflammation. Brain Behav Immun 16(6):622–653

Borovikova LV, Ivanova S, Zhang M, Yang H, Botchkina GI, Watkins LR, Wang H, Abumrad N, Eaton JW, Tracey KJ (2000) Vagus nerve stimulation attenuates the systemic inflammatory response to endotoxin. Nature 405(6785):458–462

Brosschot JF, Gerin W, Thayer JF (2006) The perseverative cognition hypothesis: a review of worry, prolonged stress-related physiological activation, and health. J Psychosom Res 60(2):113–124

Brown GW, Harris T (1978) Social origins of depression: a study of psychiatric disorder in women. Free Press, New York

Bruni R, Serino F, Galluzzo S, Coppolino G, Cacciapaglia F, Vadacca M, Nilo S, Terminio N, Afeltra A (2006) Alexithymia and neuroendocrine-immune response in patients with autoimmune diseases. Ann N Y Acad Sci 1069(1):208–211

Brydon L, Walker C, Wawrzyniak AJ, Chart H, Steptoe A (2009) Dispositional optimism and stress-induced changes in immunity and negative mood. Brain Behav Immun 23(6):810–816

Cacioppo JT (1998) Somatic responses to psychological stress: the reactivity hypothesis. In: Sabourin M, Craik F, Robert M (eds) Advances in psychological science, vol 2. Psychology Press, New York, pp 87–112

Carpenter LL, Gawuga CE, Tyrka AR, Lee JK, Anderson GM, Price LH (2010) Association between plasma IL-6 response to acute stress and early-life adversity in healthy adults. Neuropsychopharmacology 35(13):2617–2623

Cohen S, Kamarck T, Mermelstein R (1983) A global measure of perceived stress. J Health Soc Behav 24(4):385–396

Cohen S, Tyrrell DA, Smith AP (1993) Negative life events, perceived stress, negative affect, and susceptibility to the common cold. J Pers Soc Psychol 64(1):131–140

Cohen S, Frank E, Doyle WJ, Skoner DP, Rabin BS, Gwaltney JM Jr (1998) Types of stressors that increase susceptibility to the common cold in healthy adults. Health Psychol 17(3):214–223

Cohen F, Kearney KA, Zegans LS, Kemeny ME, Neuhaus JM, Stites DP (1999) Differential immune system changes with acute and persistent stress for optimists vs pessimists. Brain Behav Immun 13(2):155–174

Cohen S, Miller GE, Rabin BS (2001) Psychological stress and antibody response to immunization: a critical review of the human literature. Psychosom Med 63(1):7–18

Cohen S, Janicki-Deverts D, Miller GE (2007) Psychological stress and disease. JAMA 298(14):1685–1687

Cole SW, Hawkley LC, Arevalo JM, Cacioppo JT (2011) Transcript origin analysis identifies antigen-presenting cells as primary targets of socially regulated gene expression in leukocytes. Proc Natl Acad Sci 108(7):3080–3085

Corcos M, Guilbaud O, Paterniti S, Curt F, Hjalmarsson L, Moussa M, Chambry J, Loas G, Chaouat G, Jeammet P (2004) Correlation between serum levels of interleukin-4 and alexithymia scores in healthy female subjects: preliminary findings. Psychoneuroendocrinology 29(5):686–691

Danese A, Pariante CM, Caspi A, Taylor A, Poulton R (2007) Childhood maltreatment predicts adult inflammation in a life-course study. Proc Natl Acad Sci 104(4):1319–1324

Danese A, Moffitt TE, Pariante CM, Ambler A, Poulton R, Caspi A (2008) Elevated inflammation levels in depressed adults with a history of childhood maltreatment. Arch Gen Psychiatry 65(4):409–415

Danesh J, Wheeler JG, Hirschfield GM, Eda S, Eiriksdottir G, Rumley A, Lowe GD, Pepys MB, Gudnason V (2004) C-reactive protein and other circulating markers of inflammation in the prediction of coronary heart disease. N Engl J Med 350(14):1387–1397

Dantzer R, O'Connor JC, Freund GG, Johnson RW, Kelley KW (2008) From inflammation to sickness and depression: when the immune system subjugates the brain. Nat Rev Neurosci 9(1):46–56

Dewaraja R, Tanigawa T, Araki S, Nakata A, Kawamura N, Ago Y, Sasaki Y (1997) Decreased cytotoxic lymphocyte counts in alexithymia. Psychother Psychosom 66(2):83–86

Dhabhar FS, McEwen BS (1997) Acute stress enhances while chronic stress suppresses cell-mediated immunity in vivo: a potential role for leukocyte trafficking. Brain Behav Immun 11(4):286–306

Dhabhar FS, McEwen BS (2007) Bidirectional effects of stress on immune function: possible explanations for salubrious as well as harmful effects. In: Ader R (ed) Psychoneuroimmunology, vol 2, 4th edn. Elsevier, New York, pp 723–760

Dhabhar FS, Malarkey WB, Neri E, McEwen BS (2012) Stress-induced redistribution of immune cells—from barracks to boulevards to battlefields: a tale of three hormones–Curt Richter Award Winner. Psychoneuroendocrinology 37(9):1345–1368

Ditzen B, Neumann ID, Bodenmann G, von Dawans B, Turner RA, Ehlert U, Heinrichs M (2007) Effects of different kinds of couple interaction on cortisol and heart rate responses to stress in women. Psychoneuroendocrinology 32(5):565–574
Doyle WJ, Gentile DA, Cohen S (2006) Emotional style, nasal cytokines, and illness expression after experimental rhinovirus exposure. Brain Behav Immun 20(2):175–181
Dyck DG, Short R, Vitaliano PP (1999) Predictors of burden and infectious illness in schizophrenia caregivers. Psychosom Med 61(4):411–419
Elenkov IJ (2008) Neurohormonal-cytokine interactions: implications for inflammation, common human diseases and well-being. Neurochem Int 52(1):40–51
Elenkov IJ, Wilder RL, Chrousos GP, Vizi ES (2000) The sympathetic nerve—an integrative interface between two supersystems: the brain and the immune system. Pharmacol Rev 52(4):595–638
Ershler WB, Keller ET (2000) Age-associated increased interleukin-6 gene expression, late-life diseases, and frailty. Annu Rev Med 51(1):245–270
Ganea D, Delgado M (2001) Inhibitory neuropeptide receptors on macrophages. Microbes Infect 3(2):141–147
Gianaros PJ, Marsland AL, Kuan DC-H, Schirda BL, Jennings JR, Sheu LK, Hariri AR, Gross JJ, Manuck SB (2014) An inflammatory pathway links atherosclerotic cardiovascular disease risk to neural activity evoked by the cognitive regulation of emotion. Biol Psychiatry 75:738–745
Glaser R, Kiecolt-Glaser JK (2005) Stress-induced immune dysfunction: implications for health. Nat Rev Immunol 5(3):243–251
Glaser R, Kiecolt-Glaser JK, Malarkey WB, Sheridan JF (1998) The influence of psychological stress on the immune response to vaccines. Ann N Y Acad Sci 840(1):649–655
Glaser R, Kiecolt-Glaser JK, Marucha PT, MacCallum RC, Laskowski BF, Malarkey WB (1999) Stress-related changes in proinflammatory cytokine production in wounds. Arch Gen Psychiatry 56(5):450–456
Gouin J-P, Carter CS, Pournajafi-Nazarloo H, Glaser R, Malarkey WB, Loving TJ, Stowell JJ, Kiecolt-Glaser JK (2010) Marital behavior, oxytocin, vasopressin, and wound healing. Psychoneuroendocrinology 35(7):1082–1090
Graham JE, Glaser R, Loving TJ, Malarkey WB, Stowell JR, Kiecolt-Glaser JK (2009) Cognitive word use during marital conflict and increases in proinflammatory cytokines. Health Psychol 28(5):621–630
Gross JJ (1998) Antecedent-and response-focused emotion regulation: divergent consequences for experience, expression, and physiology. J Pers Soc Psychol 74(1):224–237
Gross JJ, John OP (2003) Individual differences in two emotion regulation processes: implications for affect, relationships, and well-being. J Pers Soc Psychol 85(2):348
Guilbaud O, Corcos M, Hjalmarsson L, Loas G, Jeammet P (2003) Is there a psychoneuroimmunological pathway between alexithymia and immunity? Immune and physiological correlates of alexithymia. Biomed Pharmacother 57(7):292–295
Guilbaud O, Curt F, Perrin C, Chaouat G, Berthoz S, Dugré-Le Bigre C, Wallier J, Strebler M, Touitou C, Jeammet P (2009) Decreased immune response in alexithymic women: a cross-sectional study. Biomed Pharmacother 63(4):297–304
Hackett RA, Hamer M, Endrighi R, Brydon L, Steptoe A (2012) Loneliness and stress-related inflammatory and neuroendocrine responses in older men and women. Psychoneuroendocrinology 37(11):1801–1809
Haensel A, Mills PJ, Nelesen RA, Ziegler MG, Dimsdale JE (2008) The relationship between heart rate variability and inflammatory markers in cardiovascular diseases. Psychoneuroendocrinology 33(10):1305–1312
Hänsel A, Hong S, Cámara RJ, Von Kaenel R (2010) Inflammation as a psychophysiological biomarker in chronic psychosocial stress. Neurosci Biobehav Rev 35(1):115–121
Hansson GK (2005) Inflammation, atherosclerosis, and coronary artery disease. N Engl J Med 352(16):1685–1695
Holt-Lunstad J, Smith TB, Layton JB (2010) Social relationships and mortality risk: a meta-analytic review. PLoS Med 7(7):1–20

Hotamisligil GS (2006) Inflammation and metabolic disorders. Nature 444(7121):860–867
Hoyt MA, Stanton AL, Bower JE, Thomas KS, Litwin MS, Breen EC, Irwin MR (2013) Inflammatory biomarkers and emotional approach coping in men with prostate cancer. Brain Behav Immun 32:173–179
Ikeda A, Schwartz J, Peters JL, Fang S, Spiro A, Sparrow D, Vokonas P, Kubzansky LD (2011) Optimism in relation to inflammation and endothelial dysfunction in older men: the VA Normative Aging Study. Psychosom Med 73(8):664–671
Irwin MR (2008) Human psychoneuroimmunology: 20 years of discovery. Brain Behav Immun 22(2):129–139
Irwin MR, Cole SW (2011) Reciprocal regulation of the neural and innate immune systems. Nat Rev Immunol 11(9):625–632
Janszky I, Ericson M, Lekander M, Blom M, Buhlin K, Georgiades A, Ahnve S (2004) Inflammatory markers and heart rate variability in women with coronary heart disease. J Intern Med 256(5):421–428
Jaremka LM, Fagundes CP, Peng J, Bennett JM, Glaser R, Malarkey WB, Kiecolt-Glaser JK (2013a) Loneliness promotes inflammation during acute stress. Psychol Sci 24(7):1089–1097
Jaremka LM, Glaser R, Malarkey WB, Kiecolt-Glaser JK (2013b) Marital distress prospectively predicts poorer cellular immune function. Psychoneuroendocrinology 38(11):2713–2719
Jaremka LM, Lindgren ME, Kiecolt-Glaser JK (2013c) Synergistic relationships among stress, depression, and troubled relationships: insights from psychoneuroimmunology. Depress Anxiety 30(4):288–296
Kemeny ME, Schedlowski M (2007) Understanding the interaction between psychosocial stress and immune-related diseases: a stepwise progression. Brain Behav Immun 21(8):1009–1018
Kiecolt-Glaser JK, Fisher LD, Ogrocki P, Stout JC, Speicher CE, Glaser R (1987) Marital quality, marital disruption, and immune function. Psychosom Med 49(1):13–34
Kiecolt-Glaser JK, Kennedy S, Malkoff S, Fisher L, Speicher CE, Glaser R (1988) Marital discord and immunity in males. Psychosom Med 50(3):213–229
Kiecolt-Glaser JK, Malarkey WB, Chee MA, Newton T, Cacioppo JT, Mao HY, Glaser R (1993) Negative behavior during marital conflict is associated with immunological down-regulation. Psychosom Med 55(5):395–409
Kiecolt-Glaser JK, Glaser R, Gravenstein S, Malarkey WB, Sheridan J (1996a) Chronic stress alters the immune response to influenza virus vaccine in older adults. Proc Natl Acad Sci 93(7):3043–3047
Kiecolt-Glaser JK, Newton T, Cacioppo JT, MacCallum RC, Glaser R, Malarkey WB (1996b) Marital conflict and endocrine function: are men really more physiologically affected than women? J Consult Clin Psychol 64(2):324–332
Kiecolt-Glaser JK, Glaser R, Cacioppo JT, MacCallum RC, Snydersmith M, Kim C, Malarkey WB (1997) Marital conflict in older adults: endocrinological and immunological correlates. Psychosom Med 59(4):339–349
Kiecolt-Glaser JK, Glaser R, Cacioppo JT, Malarkey WB (1998) Marital stress: immunologic, neuroendocrine, and autonomic correlates. Ann N Y Acad Sci 840:656–663
Kiecolt-Glaser JK, McGuire L, Robles TF, Glaser R (2002) Psychoneuroimmunology: psychological influences on immune function and health. J Consult Clin Psychol 70(3):537
Kiecolt-Glaser JK, Loving TJ, Stowell JR, Malarkey WB, Lemeshow S, Dickinson SL, Glaser R (2005) Hostile marital interactions, proinflammatory cytokine production, and wound healing. Arch Gen Psychiatry 62(12):1377–1384
Lazarus RS, Folkman S (1984) Stress, appraisal, and coping. Springer Publishing Company LLC., New York
Maier SF, Watkins LR (1998) Cytokines for psychologists: implications of bidirectional immune-to-brain communication for understanding behavior, mood, and cognition. Psychol Rev 105(1):83–107
Malarkey WB, Kiecolt-Glaser JK, Pearl D, Glaser R (1994) Hostile behavior during marital conflict alters pituitary and adrenal hormones. Psychosom Med 56(1):41–51

Mandarelli G, Tarsitani L, Ippoliti F, Covotta F, Zerella MP, Mirigliani A, Biondi M (2011) The relationship between alexithymia and circulating cytokine levels in subjects undergoing upper endoscopy. Neuroimmunomodulation 18(1):37–44

Master SL, Amodio DM, Stanton AL, Yee CM, Hilmert CJ, Taylor SE (2009) Neurobiological correlates of coping through emotional approach. Brain Behav Immun 23(1):27–35

Matsunaga M, Sato S, Isowa T, Tsuboi H, Konagaya T, Kaneko H, Ohira H (2009) Profiling of serum proteins influenced by warm partner contact in healthy couples. Neuro Endocrinol Lett 30(2):227–236

Mayne TJ, O'leary A, McCrady B, Contrada R, Labouvie E (1997) The differential effects of acute marital distress on emotional, physiological and immune functions in maritally distressed men and women. Psychol Health 12(2):277–288

McDade TW (2012) Early environments and the ecology of inflammation. Proc Natl Acad Sci 109:17281–17288

McDade TW, Hoke M, Borja JB, Adair LS, Kuzawa C (2013) Do environments in infancy moderate the association between stress and inflammation in adulthood? Initial evidence from a birth cohort in the Philippines. Brain Behav Immun 31:23–30

McIntosh, Roger C, Ironson, Gail, Antoni, Michael, Kumar, Mahendra, Fletcher, Mary Ann, Schneiderman, Neil (2014) Alexithymia is linked to neurocognitive, psychological, neuroendocrine, and immune dysfunction in persons living with HIV. Brain, Behavior, and Immunity, 36:165–175

Medzhitov R (2007) Recognition of microorganisms and activation of the immune response. Nature 449(7164):819–826

Miller AH, Maletic V, Raison CL (2009a) Inflammation and its discontents: the role of cytokines in the pathophysiology of major depression. Biol Psychiatry 65(9):732–741

Miller G, Chen E, Cole SW (2009b) Health psychology: developing biologically plausible models linking the social world and physical health. Annu Rev Psychol 60:501–524

Monroe SM (2008) Modern approaches to conceptualizing and measuring human life stress. Annu Rev Clin Psychol 4(4):33–52

Murphy ML, Slavich GM, Rohleder N, Miller GE (2013) Targeted rejection triggers differential pro-and anti-inflammatory gene expression in adolescents as a function of social status. Clin Psychol Sci 1(1):30–40

Nabipour I, Vahdat K, Jafari SM, Pazoki R, Sanjdideh Z (2006) The association of metabolic syndrome and Chlamydia pneumoniae, Helicobacter pylori, cytomegalovirus, and herpes simplex virus type 1: the Persian Gulf Healthy Heart Study. Cardiovasc Diabetol 5(25):1–6

Nance DM, Sanders VM (2007) Autonomic innervation and regulation of the immune system (1987–2007). Brain Behav Immun 21(6):736–745

Pace T, Mletzko T, Alagbe O, Musselman D, Nemeroff C, Miller A, Heim C (2006) Increased stress-induced inflammatory responses in male patients with major depression and increased early life stress. Am J Psychiatry 163(9):1630–1633

Parkin DM (2006) The global health burden of infection-associated cancers in the year 2002. Int J Cancer 118(12):3030–3044

Pavlov VA, Tracey KJ (2005) The cholinergic anti-inflammatory pathway. Brain Behav Immun 19(6):493–499

Pedrosa Gil F, Nickel M, Ridout N, Schwarz MJ, Schoechlin C, Schmidmaier R (2007) Alexithymia and interleukin variations in somatoform disorder. Neuroimmunomodulation 14(5):235–242

Prather AA, Marsland AL, Muldoon MF, Manuck SB (2007) Positive affective style covaries with stimulated IL-6 and IL-10 production in a middle-aged community sample. Brain Behav Immun 21(8):1033–1037

Pressman, Sarah D, & Black, Lora L. (2012). Positive emotions and immunity. In S.C. Segerstrom (Ed.), The Oxford Handbook of Psychoneuroimmunology. Oxford Press, New York, pp 92–104

Pressman SD, Cohen S, Miller GE, Barkin A, Rabin BS, Treanor JJ (2005) Loneliness, social network size, and immune response to influenza vaccination in college freshmen. Health Psychol 24(3):297–306

Raison C, Miller A (2013) The evolutionary significance of depression in Pathogen Host Defense (PATHOS-D). Mol Psychiatry 18(1):15–37
Raison CL, Capuron L, Miller AH (2006) Cytokines sing the blues: inflammation and the pathogenesis of depression. Trends Immunol 27(1):24–31
Raison CL, Lowry CA, Rook GA (2010) Inflammation, sanitation, and consternation: loss of contact with coevolved, tolerogenic microorganisms and the pathophysiology and treatment of major depression. Arch Gen Psychiatry 67(12):1211–1224
Reed RG, Butler EA, Kenny DA (2013) Dyadic models for the study of health. Soc Personal Psychol Compass 7(4):228–245
Robles TF, Kiecolt-Glaser JK (2003) The physiology of marriage: pathways to health. Physiol Behav 79(3):409–416
Robles TF, Glaser R, Kiecolt-Glaser JK (2005) Out of balance a new look at chronic stress, depression, and immunity. Curr Dir Psychol Sci 14(2):111–115
Robles TF, Slatcher RB, Trombello JM, McGinn MM (2014) Marital quality and health: a meta-analytic review. Psychol Bull 140(1):140–187
Rohrbaugh MJ, Mehl MR, Shoham V, Reilly ES, Ewy GA (2008) Prognostic significance of spouse we talk in couples coping with heart failure. J Consult Clin Psychol 76(5):781
Rook GA, Lowry CA, Raison CL (2013) Microbial 'Old Friends', immunoregulation and stress resilience. Evol Med Public Health 2013(1):46–64
Rook GA, Raison CL, Lowry CA (2014) Microbial "Old Friends", immunoregulation and socioeconomic status. Clin Exp Immunol 177:1–12
Roy B, Diez-Roux AV, Seeman T, Ranjit N, Shea S, Cushman M (2010) Association of optimism and pessimism with inflammation and hemostasis in the Multi-Ethnic Study of Atherosclerosis (MESA). Psychosom Med 72(2):134–140
Sanders VM, Kavelaars A (2007) Adrenergic regulation of immunity. In: Ader R (ed) Psychoneuroimmunology, vol 1, 4th edn. Elsevier/Academic, New York, pp 63–83
Scheier MF, Carver CS (1985) Optimism, coping, and health: assessment and implications of generalized outcome expectancies. Health Psychol 4(3):219–247
Segerstrom SC (2001) Optimism, goal conflict, and stressor-related immune change. J Behav Med 24(5):441–467
Segerstrom SC (2005) Optimism and immunity: do positive thoughts always lead to positive effects? Brain Behav Immun 19(3):195–200
Segerstrom SC (2006) How does optimism suppress immunity? Evaluation of three affective pathways. Health Psychol 25(5):653–657
Segerstrom SC, Miller GE (2004) Psychological stress and the human immune system: a meta-analytic study of 30 years of inquiry. Psychol Bull 130(4):601–630
Segerstrom SC, Castañeda JO, Spencer TE (2003) Optimism effects on cellular immunity: testing the affective and persistence models. Personal Individ Differ 35(7):1615–1624
Slavich GM, Irwin MR (2014) From stress to inflammation and major depressive disorder: a social signal transduction theory of depression. Psychol Bull 140:774–815
Slavich GM, O'Donovan A, Epel ES, Kemeny ME (2010) Black sheep get the blues: a psychobiological model of social rejection and depression. Neurosci Biobehav Rev 35(1):39–45
Slopen N, Kubzansky LD, McLaughlin KA, Koenen KC (2013) Childhood adversity and inflammatory processes in youth: a prospective study. Psychoneuroendocrinology 38(2): 188–200
Stanton AL, Danoff-Burg S, Cameron CL, Ellis AP (1994) Coping through emotional approach: problems of conceptualizaton and confounding. J Pers Soc Psychol 66(2):350–362
Steptoe A, Hamer M, Chida Y (2007) The effects of acute psychological stress on circulating inflammatory factors in humans: a review and meta-analysis. Brain Behav Immun 21(7):901–912
Steptoe A, O'Donnell K, Badrick E, Kumari M, Marmot M (2008) Neuroendocrine and inflammatory factors associated with positive affect in healthy men and women: the Whitehall II Study. Am J Epidemiol 167(1):96–102

Sternberg EM (2006) Neural regulation of innate immunity: a coordinated nonspecific host response to pathogens. Nat Rev Immunol 6(4):318–328
Sturmberg JP, Martin CM (2013) Handbook of systems and complexity in health. Springer, New York
Thayer J, Fischer J (2009) Heart rate variability, overnight urinary norepinephrine and C-reactive protein: evidence for the cholinergic anti-inflammatory pathway in healthy human adults. J Intern Med 265(4):439–447
Thomsen DK, Mehlsen MY, Hokland M, Viidik A, Olesen F, Avlund K, Munk K, Zachariae R (2004) Negative thoughts and health: associations among rumination, immunity, and health care utilization in a young and elderly sample. Psychosom Med 66(3):363–371
Todarello O, Casamassima A, Marinaccio M, La Pesa M, Caradonna L, Valentino L, Marinaccio L (1994) Alexithymia, immunity and cervical intraepithelial neoplasia: a pilot study. Psychother Psychosom 61(3–4):199–204
Todarello O, Casamassima A, Daniele S, Marinaccio M, Fanciullo F, Valentino L, Tedesco N, Wiesel S, Marinaccio L (1997) Alexithymia, immunity and cervical intraepithelial neoplasia: replication. Psychother Psychosom 66(4):208–213
Tracey KJ (2002) The inflammatory reflex. Nature 420:853–859
Tracey KJ (2009) Reflex control of immunity. Nat Rev Immunol 9(6):418–428
Uchino BN, Holt-Lunstad J, Uno D, Flinders JB (2001) Heterogeneity in the social networks of young and older adults: prediction of mental health and cardiovascular reactivity during acute stress. J Behav Med 24(4):361–382
Uchino BN, Bosch JA, Smith TW, Carlisle M, Birmingham W, Bowen KS, Light KC, Heaney J, O'Hartaigh B (2013) Relationships and cardiovascular risk: perceived spousal ambivalence in specific relationship contexts and its links to inflammation. Health Psychol 32(10):1067–1075
Uchino BN, Smith TW, Berg CA (2014) Spousal relationship quality and cardiovascular risk: dyadic perceptions of relationship ambivalence are associated with coronary-artery calcification. Psychol Sci 25(4):1037–1042
Webster JI, Tonelli L, Sternberg EM (2002) Neuroendocrine regulation of immunity. Annu Rev Immunol 20(1):125–163
Whisman MA, Sbarra DA (2012) Marital adjustment and interleukin-6 (IL-6). J Fam Psychol 26(2):290–295
Zalli A, Carvalho LA, Lin J, Hamer M, Erusalimsky JD, Blackburn EH, Steptoe A (2014) Shorter telomeres with high telomerase activity are associated with raised allostatic load and impoverished psychosocial resources. Proc Natl Acad Sci 111(12):4519–4524
Zoccola PM, Figueroa WS, Rabideau EM, Woody A, Benencia F (2014) Differential effects of poststressor rumination and distraction on cortisol and C-reactive protein. Health Psychol 33:1606–1609

Exercise and the Immune System 6

Elisa Couto Gomes and Geraint Florida-James

Contents

6.1 Introduction

It is widely accepted that the stress caused by acute and chronic exercise can alter our immune system. These responses can be observed by alteration in the numbers and temporal pattern of circulating immune mediators, which include the measurement of total white blood cell counts, and concentration and percentage of white cell subsets (Chinda et al. 2003; Kakanis et al. 2010; Scharhag et al. 2005), immunoglobulins (IgA, IgD, IgE, IgG, IgM) and cytokines (Allgrove et al. 2008; Reid

E.C. Gomes • G. Florida-James (✉)
School of Life, Sport and Social Sciences, Edinburgh Napier University, Edinburgh, UK
e-mail: elisacoutogomes@gmail.com; g.florida-james@napier.ac.uk

C. Esser (ed.), *Environmental Influences on the Immune System*,
DOI 10.1007/978-3-7091-1890-0_6

et al. 2004; Tongtako et al. 2012). In addition, measurements of changes in functional parameters of the immune system – leucocytes' oxidative burst and phagocytic functions and lymphocyte phenotype, activity and proliferation – are also assessed to evaluate changes in the immune system (Gleeson and Bishop 2005; Kakanis et al. 2010; Sureda et al. 2009; Syu et al. 2012; Tvede et al. 1994; Woods et al. 1998).

6.2 Part 1: Exercise and Respiratory Infection Risk

Upper respiratory tract infections (URTI), which includes throat infections, coughs and colds, are of concern to the athletic and general population. URTI can negatively affect athletes' well-being, increasing fatigue and decreasing performance during training and competitions or can even cause inability to perform (Pyne et al. 1999, 2000; Reid et al. 2004). Due to considerable personal and economic investment when preparing athletes for competition, it is important to identify factors that make athletes susceptible to URTI. In fact, these types of infections are the most common illnesses reported by athletes at medical centres (Engebretsen et al. 2010; Ruedl et al. 2012). The relationship between risk of URTI and exercise has been suggested as following a J-shaped curve model (Nieman 1994). This model suggests that athletes engaged in prolonged, heavy-intensity exercise or strenuous exercise training periods are more prone to developing URTI compared to their moderately active counterparts. On the other end of the curve, sedentary individuals are shown to also have higher risk of developing URTI compared to their moderately active counterparts. Since this J-shaped model was proposed, there have been various researches that have provided further evidence of its validity.

A number of these studies have had as foundation surveyed-based epidemiological data of signs and symptoms reported by a large cohort of participants (Kostka et al. 2000; Matthews et al. 2002; Nieman et al. 1990). Extensive clinical investigations of URTI with samples collected and pathogens (virus and/or bacteria) being identified are rare, mostly due to the high financial and time costs associated with this kind of investigation (Reid et al. 2004; Spence et al. 2007; Cox et al. 2008). Changes in some immune parameters as a result of exercise do add to the body of evidence that support the J-shaped curve model (discussed below). Nonetheless, the debate on whether URTI are indeed infections caused by a pathogen or a result of upper respiratory inflammation associated with exercise still remains. It is clear though that exercise-induced immune suppression does lead to performance decrements due to an increase in URTI-symptom susceptibility. It is also important to point out that other factors may have influence and should also be considered when investigating possible explanations for the relationship between exercise intensity/volume and URTI risk. These factors include psychological stress, nutritional status and exposure to pathogens. Possible reasons are expanded on below.

6.2.1 Altered Mucosal Immunity

The mucosal immune system is arguably the largest immune component of the body and can be viewed as our body's first line of defence. The innate mucosal defences – immunoglobulins (Ig), lysozyme, lactoferrin and α-amylase, amongst other immune mediators – interact with a variety of potentially infectious antigens, defending not only the respiratory system but also the reproductive system, mouth, gastrointestinal tract and eyes against infections (Gleeson and Pyne 2000; Gleeson 2006). This immune response is mediated mainly by the secretory immunoglobulin A (SIgA) present in the saliva and the predominant antibody in seromucous secretions (Roitt and Delves 2001). IgA appears to function as a multilayered mucosal host defence by three different processes: (1) *Immune exclusion:* IgA inhibits the adherence and penetration of antigens to body tissues. (2) *Intracellular neutralization:* IgA interrupts the replication of intracellular pathogens during transcytosis through epithelial cells. (3) *Immune excretion:* IgA binds antigens to the lamina propria facilitating their excretion through the epithelium back into the lumen of the gut (Lamm 1998). The mucosal immune system has been extensively investigated to determine its link to exercise and URTI. In fact, there is evidence to imply a negative relationship between saliva SIgA and the risk of developing URTI. That is to say that individuals who present low levels or a transient decrease in SIgA have been shown to present a higher risk of developing URTI and vice versa (Carins and Booth 2002; Fahlman and Engels 2005; Gleeson et al. 1995, 1999; Neville et al. 2008).

Salivary SIgA response to exercise is variable and has been studied in different contexts be it through training sessions (Fahlman and Engels 2005; Gleeson et al. 1999; Gleeson and Pyne 2000; Neville et al. 2008), competitive events (Nieman et al. 2002; Libicz et al. 2006; Palmer et al. 2003) or laboratory-based exercise testing (Reid et al. 2004; Li and Gleeson 2004; Allgrove et al. 2008). The consensus in the literature is that an intense and prolonged bout (1.5 h) of exercise decreases saliva SIgA, but a short-duration high- intensity exercise bout increases the secretion rate of salivary SIgA (Allgrove et al. 2008; Blannin et al. 1998; Nieman et al. 2002; Palmer et al. 2003; Steerenberg et al. 1997; Walsh et al. 1999). Moderate bouts of exercise will have little effect on mucosal immunity (Allgrove et al. 2008; Blannin et al. 1998; Sari-Sarraf et al. 2006), but exercise training of moderate intensity has been shown to increase saliva SIgA in previously sedentary individuals (Akimoto et al. 2003; Klentrou et al. 2002). Contrary to that, periods of intense training will impair mucosal immunity which could increase the risk of URTI as suggested in the J-shaped model (Carins and Booth 2002; Fahlman and Engels 2005; Gleeson et al. 1995, 1999, 2002; Neville et al. 2008). Resistance exercise does not seem to affect salivary SIgA levels (Koch 2010), although more studies investigating this type of exercise are necessary. The discrepancies that exist in the literature in relation to these changes in mucosal immunity following exercise can be due to the following factors: training status and physical activity history of the participants, exercise stimuli used in the research, psychological stress, methodological differences between assays, source of saliva or the collection method used, hydration and nutritional status of the participants (Gleeson et al. 2004; Bishop and Gleeson 2009).

The mechanism by which exercise influences mucosal immunity still needs further clarification. Saliva secretion is under neural control, an important factor that affects the saliva flow rate during exercise which is regulated by sympathetic activity: when there is sympathetic stimulation, the blood vessels which supply the salivary glands vasoconstrict, reducing thus saliva secretion. In addition, other variables that might influence the saliva flow rate during exercise are dehydration and evaporative loss of saliva through hyperventilation (Chicharro et al. 1998). Changes in saliva SIgA, following a bout of exercise, appears to be linked to an increase in the transepithelial transport mechanism rather than a regulation at the level of IgA production by plasma cells in the submucosa (Bosch et al. 2002; Proctor et al. 2003). The latter is a process that requires days to develop and, therefore, could explain the SIgA alteration in a training situation (Bishop and Gleeson 2009; Bosch et al. 2002). Proctor et al. (2003) have shown that short-term regulation of saliva IgA secretion is stimulated by sympathetic nerves in a process-denominated transcytosis. Furthermore, they showed that IgA secretion increases in response to acute stimulation of β-adrenoreceptors, above a certain threshold, in a dose-independent manner. However, prolonged stimulation appeared to reduce the secretion of IgA. Consequently, taking into consideration the duration of the exercise protocols in this study, it is interesting to note that the sympathetic stimulation can be enough to increase the IgA in the glandular pool, but not to decrease saliva flow rate. Indeed, the different durations, intensities and types of adrenergic stimulation and the interaction with their receptors could explain the inconsistencies in the literature previously cited.

Other innate mucosal defences present in the saliva have also been shown to be affected by exercise. Salivary α-amylase selectively binds to oral bacteria, interrupting its adherence and growth (Scannapieco et al. 1993; West et al. 2006). This antimicrobial protein, produced mainly by the parotid gland, is sensitive to changes in the adrenergic activity showing responsiveness to both psychological (Kivlighan and Granger 2006; Nater et al. 2006) and physiological stressors (Chatterton et al. 1996; Walsh et al. 1999). In fact, salivary α-amylase activity increases with an increase in exercise intensity (Allgrove et al. 2008; Bishop et al. 2000; Li and Gleeson 2004; West et al. 2006). Similarly, salivary lysozyme and lactoferrin concentrations seem to rise with an increase in exercise intensity of both long and short duration (Allgrove et al. 2008; West et al. 2006, 2010). Lysozyme and lactoferrin are also antimicrobial enzymes – present not only in saliva but also in phagocytic cell granules and tears – they digest peptidoglycans of bacterial cell walls and inhibit bacterial growth and metabolism (Roitt and Delves 2001; Schenkels et al. 1995). The increased concentration of antimicrobial defences in saliva might provide enhanced protection and could be a consequence of a compensatory mechanism within the mucosal immune system. It should be pointed out that if someone has a period of immune deficiency, they will not necessarily develop an infection. This would occur if they enter in contact with a microorganism capable of surviving and reproducing in their body. Therefore, simple measures, such as washing hands, not sharing water bottles and avoiding crowded or enclosed places, can decrease the likelihood of acquiring a pathogen.

6.2.2 Viral Illnesses

Frequent episodes of URTI and/or long-term fatigue in athletes might not be due to an altered mucosal immune system. In fact, studies have shown that the risk of developing URTI symptoms could be due to the presence of herpes group viruses, e.g. cytomegalovirus (CMV) or Epstein-Barr Virus (EBV) (Yoda et al 2000; Gleeson et al. 2002; Reid et al. 2004). CMV is a highly transmittable and prevalent virus, having been shown to affect 30–70 % of the population (Bate et al. 2010). Once infected, this virus is never completely cleared from the organism and, although it usually remains latent, it affects the whole immune system (Britt 2008; Jarvis and Nelson 2002). Recent studies have linked CMV prevalence to a variety of chronic inflammatory diseases such as cardiovascular-related diseases, cancer and increased mortality (Gkrania-Klotsas et al. 2013; Roberts et al. 2010; Simanek et al. 2011).

Similarly to CMV, EBV is very predominant being present in over 90 % of the population worldwide (Chang et al. 2009). Even though the levels of this virus are tightly regulated by our immune system, it persists in our organism as a lifelong low-level infection in lymphocytes, more specifically in the memory B cells (Hochberg et al. 2004; Thorley-Lawson et al. 2013). This herpesvirus has a significant pathological role because it is associated with several benign and malignant conditions: mononucleosis (glandular fever), multiple sclerosis, certain autoimmune diseases and neoplasias such as Hodgkin lymphoma and nasopharyngeal carcinoma (Chang et al. 2009; Thorley-Lawson and Gross 2004; Toussirot and Roudier 2008).

As with the general population, a large number of athletes are infected by CMV and EBV. Nevertheless, viral illness as a cause of URTI symptoms or long-term fatigue often goes undetected, unless specific investigations are requested (Reid et al. 2004). It is thought that intensive training periods can lead to the reactivation of these viruses by alterations in immune mechanisms responsible for their control (Gleeson et al. 2002). The presence of CMV and EBV are not always associated with symptoms of URTI, as elucidated in the study by Cox and colleagues (2004). These researchers administrated an antiviral drug (Valtrex) used in the management of most species of the herpesvirus family – including EBV and CMV – to a group of elite distance runners. The authors reported that even though the treatment was successful in reducing the specific viral load in participants that were EBV positive, it did not affect their upper respiratory symptoms. Therefore, it is important to keep in mind that URTI symptoms are multifactorial and vary between individuals, and hence this should be taken into account when conducting a medical assessment of an athlete and when designing management strategies for URTI.

6.2.3 Allergic Rhinitis

Allergic rhinitis is a global health problem and it is estimated that about 10–25 % of the population is affected by it (Dykewicz and Hamilos 2010). Patients with allergic rhinitis present an exacerbated response of the immune and nervous system to an inhaled allergen. Specific IgE antibodies are released and stimulate mast cells, eosinophils and basophils to produce and release histamines and other allergy mediators into the blood and surrounding tissue. This allergic cascade response is also mediated by neural activity (D'Alonzo 2002; Sarin et al. 2006).

Symptoms of allergic rhinitis, which include sneezing, rhinorrhoea, congestion of the nasal passages, increase in mucus production and watery eyes, can be mistaken for a URTI episode (D'Alonzo 2002). In-depth investigation of athletes with persistent fatigue and recurrent infections showed that 15 % of the participants investigated actually had an episode of allergic rhinitis and not URTI (Reid et al. 2004). The former has shown to be successfully treated with the use of intranasal steroid spray (Reid et al. 2004; Nair 2012) and also with moderate-intensity exercise (Tongtako et al. 2012). The latter has shown to have beneficial effects on the concentration of cytokines in nasal secretions.

6.2.4 Airway Inflammation

Athletes, who are engaged in an endurance training regime, have, at baseline, a higher neutrophil number in the airways compared with the general population. This sign of airway inflammation has been shown to occur with athletes from different sports such as running, cross-country skiing swimming and rowing (Bonsignore et al. 2001; Morice et al. 2004; Belda et al. 2008; Bougault et al. 2009). The number of inflammatory cells in the airways is also higher after competition or intense training, indicating that lung inflammation can be stimulated by both acute and chronic exercise (Denguezli et al. 2008). This occurs because athletes engaged in high-intensity long-duration exercise present a high ventilatory rate. This intense airflow might cause epithelial damage through shear stress on the epithelial wall which triggers an inflammatory process (Bonsignore et al. 2001; Morice et al. 2004). If the air inhaled is of a low temperature, it can cause – in addition to the problems mentioned above – the drying of the airways, consequently leading to repeated thermal or osmotic airway trauma, stimulating a local inflammatory process (Helenius et al. 2005).

The inhalation of air pollution during exercise can also cause airway inflammation. This can happen because some air pollutants, such as ozone and particulate matter, have oxidative properties. This means they react with cells and mediators in the lungs, a process which can lead to local inflammation and oxidative stress (Blomberg 2000; Gomes et al. 2010, 2011). When individuals exercise, they elicit high ventilatory rates, which means that the amount and depth of inspired air is increased, and in an environment with air pollution, this implies that high doses of pollution reach deeper into the lungs, exacerbating the detrimental health effects of air pollution (Bates 2005; Brunekreef and Holgatek 2002).

Airway hyperresponsiveness is a prevalent type of airway inflammation that occurs when there is an exacerbated response to large amounts of inhaled irritants during periods of training. This can lead to airway obstruction and remodelling due to chronic inflammation. Individuals with asthma present similar features, although in athletes it can be asymptomatic (Cockcroft and Davis 2006; Bougault et al. 2010). Some elite athletes have airways hyperresponsiveness, with competitive swimmers being the most affected: about 70 % of them present airway hyperresponsiveness

(Langdeau et al. 2000; Bougault et al. 2009). These individuals, due to long periods of exposure to high volumes of inhaled chlorine, present exacerbated airway inflammation during high-intensity training periods (Boulet et al. 2005a, b; Pedersen 2009). Fortunately, the airway inflammation is reversible in many athletes after, at least, 2 weeks of training cessation (Bougault et al. 2011).

6.2.5 Conclusion to Part 1

Moderate exercise has been shown to exert positive changes in the immune system, and this includes mucosal immunity. Such exercise leads to a decreased risk of URTI and other conditions that present similar symptoms. However, athletes, during intense training periods, may experience multiple stressors – physical, physiological and psychological – that are likely to induce immune, endocrine and neurological alterations that may result in URTI or other conditions with similar symptoms, all of which have a detrimental effect on the athlete's wellbeing and performance. It is important to keep in mind, however, that not all upper respiratory distress is caused by a pathogen. Finally, it should be emphasized that by decreasing sedentary behaviour, the likelihood of experiencing upper respiratory distress symptoms is also decreased.

6.3 Part 2: Effect of Acute Exercise on Innate and Acquired Immune Function

If a pathogen is able to overcome the physical barriers of our body, it soon encounters our first line of defence: the innate immune system. This subsystem, also known as nonspecific immunity, provides immediate protection. Soluble factors and the following immune cells are part of the innate immune system: macrophages, monocytes, neutrophils, eosinophils, basophils, dendritic cells, natural killer cells and mast cells. If these cells are unsuccessful in destroying the invading pathogen, the acquired immune system is activated. This subsystem is also known as adaptive or specific immunity and is composed of highly specialized processes and cells – T- and B-lymphocytes. Contrary to the innate response, the acquired immunity strengthens upon repeated exposure. These two immune subsystems have been studied in the context of exercise; nevertheless, given the short history of exercise immunology research, there is still a great deal to be learned about these responses. This chapter gives an overview on the response of some immune cells to exercise with a focus on neutrophils and lymphocytes.

6.3.1 Innate Immune Function

6.3.1.1 Neutrophils' Response to Acute Exercise

Exercise causes an initial rapid increase in blood neutrophil counts (neutrophilia). This rise in numbers continues, though with a smaller magnitude, for a few hours after exercise cessation. The amplitude and temporal pattern of this neutrophilia depend both on the intensity and duration of the exercise (Blannin et al. 1996; Peake 2002; Sand et al. 2013). Even a very short exercise burst lasting just 60 s, but at maximal intensity, results in an increase in the circulating numbers of neutrophils peaking at around 15 min postexercise (Gabriel et al. 1992).

Various mechanisms are responsible for exercise neutrophilia. The initial rapid increase is explained by a rise in the concentration of circulating catecholamines (adrenaline and noradrenaline). As soon as an exercise bout initiates, there is an increase in catecholamines that quickly returns to basal level when exercise ceases, so its effect on circulating cells also occurs rapidly. Catecholamines elevate cardiac output, stimulates the opening up of capillaries – for higher blood flow to muscles and skin – and releases neutrophils from sites such as spleen, lungs and lymphatic tissue (Benshop et al. 1996; Tvede et al. 1994). The elevated cardiac output and the opening up of capillaries during exercise means that the shear stress within blood vessels mechanically removes adhered neutrophils increasing their numbers in the central circulation (Foster et al. 1986; McCarthy et al. 1992; Pyne 1994). In contrast to catecholamines, an increase in the stress hormone cortisol occurs some time after the exercise has initiated but remains above pre-exercise levels postexercise. Therefore, cortisol and its action on immune cells is more prolonged, with it being the hormone responsible for the continued neutrophilia observed several hours after the termination of an endurance exercise. This hormone mobilizes neutrophils from the bone marrow into the circulation (McCarthy and Dale 1998; McCarthy et al. 1992; Steensberg et al. 2002) (Fig. 6.1).

It has been established that circulating blood neutrophil counts increase after exercise. Yet, what about their function? Neutrophil function can be assessed by its chemotaxis and adherence capacity, by its phagocytosis function and also by its degranulation and oxidative burst activity. Most studies have shown that a moderate or exhaustive exercise bout does not affect neutrophil capacity to adhere to the endothelium – an essential step in the process of transendothelium migration to the inflamed tissue (Lewicki et al. 1987; Ortega et al. 1993). Both moderate and intense exercise bouts have been shown to exert a positive effect on neutrophil chemotaxis, though moderate exercise seems to result in greater benefit (Giraldo et al 2009; Ortega et al. 1993). In fact, exercise to exhaustion seems not to affect the chemotaxis capacity (Rodriguez et al. 1991; Syu et al. 2012).

The benefits of moderate exercise on neutrophil function includes a higher phagocytic capacity, degranulation and oxidative burst activity that can last for up to 24 h (Robson et al. 1999; Peake and Suzuki 2004; Ortega et al. 2005). A higher phagocytic capacity has also been observed as a response to intensive and prolonged exercise (over 2 h). Most studies, however, show a decreased capacity to destroy bacteria – seen by the reduced lipopolysaccharide-stimulated elastase release – and also a decreased phagocytic function commencing 2 h after this type of endurance exercise (Nieman et al. 1998; Bishop et al 2003; Kakanis et al. 2010). Exercise increases the number of circulating neutrophils, and it seems that – despite the phagocytosis being increased – when this capacity is analysed per number of neutrophil, their individual phagocytic capacity is reduced (Blannin 2006).

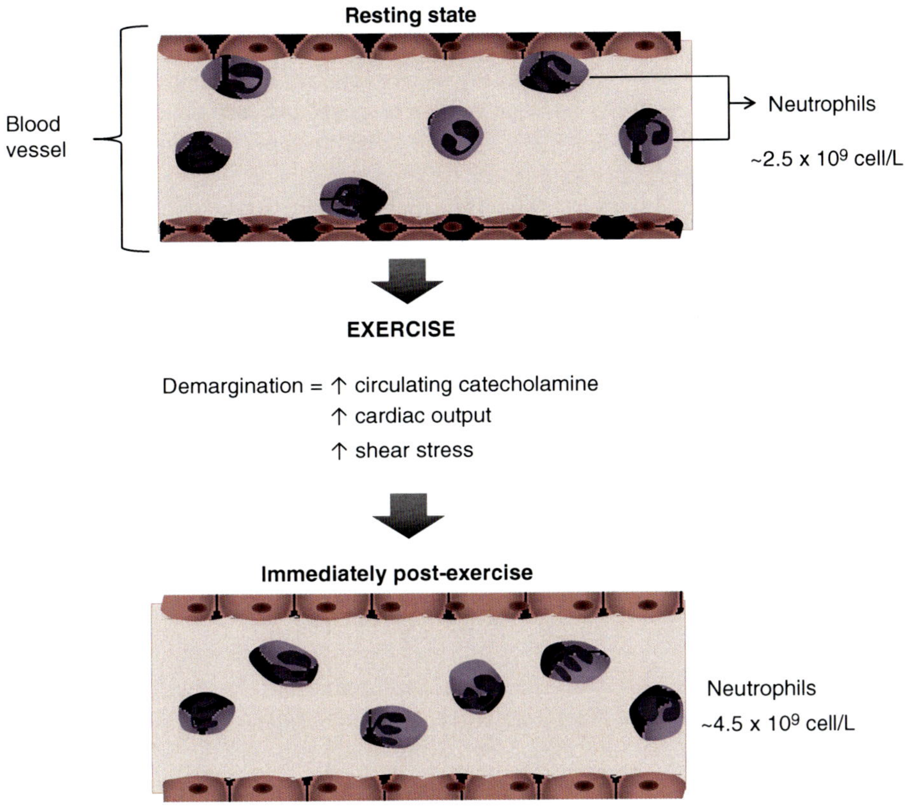

Fig. 6.1 Effect of exercise on circulating neutrophil numbers

Different mechanisms are involved in shaping the neutrophil function after acute exercise. Although they need to be further elucidated, it is known that catecholamines, cortisol and IL-6, released during exercise, are important regulators, and so is the activation of complement (Ortega et al. 2005; Blannin 2006; Camous et al. 2011). The various aspects of neutrophil function are influenced, independently of each other, by exercise intensity, duration and mode; thus, it is tricky to analyse the pertaining literature. All things considered, moderate-intensity exercise seems to exert a positive effect on neutrophil function.

6.3.1.2 Natural Killer (NK) Cells' Response to Acute Exercise

Similar to neutrophils, NK cells are very responsive to a bout of exercise, presenting a sharp increase in circulation. The magnitude of NK cell mobilization seems to be positively related to exercise intensity, with it having been reported to increase by around 480 % (Gabriel et al. 1991; Hoffman-Goetz et al. 1990; Kakanis et al. 2010; Nieman et al. 1993). The downregulation of the expression of adhesion molecules on NK cells, which occurs with rising catecholamine levels, is one of the factors resulting in their increased mobilizing. Another factor would be shear stress of blood to the vessels' walls (Kappel et al. 1991; Timmons and Cieslak 2008; Tvede et al. 1994).

The function of NK cells, assessed by its cytolytic activity, appears to follow a biphasic pattern in response to exercise: there is an enhancement immediately postexercise; however, as short as 1 h after exercise cessation, suppression in its activity has been reported (Woods et al. 1998; Kappel et al. 1991; Nieman et al. 1993). This initial rise in cytolytic activity is largely due to higher NK cell numbers in the circulation and occurs following moderate and intense exercise. Factors that have been proposed as influencing NK cell cytolytic activity include circulation levels of prostaglandins, catecholamine and IL-2 (Kappel et al. 1991; McFarlin et al. 2004a; Walsh et al. 2011).

6.3.1.3 Eosinophil Response to Acute Exercise

Eosinophils are involved in the control mechanisms of asthma and allergies. Its increase with exercise has been shown predominantly in the context of airway inflammation. Eosinophil numbers seem to start rising in the circulation some hours after exercise cessation. Kakanis et al. (2010) reported an increase in eosinophil levels 4 h after a prolonged (2 h) intense cycling exercise, remaining elevated for 4 h, up to the 8 h time point. This is thought to occur due to the release of specific cytokines (Il-3 and IL-5).

6.3.1.4 Monocytes Response to Acute Exercise

Monocytes are one of the main antigen-presenting cells found in the circulatory system, and like neutrophils, they provide important phagocytic function. These cells differentiate into macrophages once they enter into a tissue. Monocytes present on their surface a transmembrane protein that recognizes pathogens and are capable of activating both innate and acquired immune responses (Akira and Hemmi 2003; Brown et al 2011). These proteins are known as toll-like receptors (TLRs). Both monocytes and its TLRs have been studied in the context of exercise. Macrophages have also been investigated, but most studies are with animal model due to the invasive assessment methods (Ortega et al. 1997; Woods et al. 1997; Murphy et al. 2004). Yet, to which extent the results can be extended to the human population remains unknown.

Similar to some of the studies previously mentioned, there is a rise in the circulation counts of monocytes following a bout of exercise, the magnitude of which has not been linked to exercise intensity (Gabriel et al. 1992; Okutsu et al. 2008). Nevertheless, this monocytosis is thought to occur as a result of circulating catecholamines, cortisol and an increase in blood flow in the vessels (Okutsu et al. 2008; Tvede et al. 1994). Although monocytes' phagocytic capacity has been shown to be enhanced following exercise, the same cannot be said about its oxidative burst activity, which seems to remain unaltered (Nieman et al. 1998). As for macrophages, they seem to increase their function, which includes antitumour activity and phagocytosis, as a response to exercise (Murphy et al. 2004; Ortega et al. 1997).

In addition to the role in recognizing and responding to a variety of pathogen-associated molecular patterns, TLRs also help in the control – duration and magnitude – of inflammatory responses (Brown et al. 2011). Studies have shown a decrease in TLR expression in response to an acute bout of exercise (Lancaster et al. 2005a; Oliveira and Gleeson 2010; Simpson et al. 2009). This lower expression is thought to be associated with reductions in accessory signal molecule expression and antigen-presenting cell activation that occur after a bout of exercise. Such alterations may contribute to impaired postexercise immune surveillance (Gleeson et al. 2006). However, Booth et al. (2010) reported an increased expression of TLRs, and consequently further research is needed to fully understand the response of TLRs to exercise based upon the type, duration and intensity of exercise.

6.3.2 Acquired Immune Function

6.3.2.1 Lymphocytes

Lymphocytes have a very distinct biphasic response to exercise. During and immediately after exercise, these immune cells are mobilized from peripheral lymphoid organs. This results in an increase in the number of lymphocytes in the circulation. The magnitude of this lymphocytosis will depend on the type, duration and intensity of the exercise performed (Mooren et al. 2002; Kakanis et al. 2010; Simpson et al. 2006; Steensberg et al. 2002). Nevertheless, the extent of this lymphocytosis is lower than the typical three to fourfold increase in neutrophil numbers following intense exercise. Similarly, an increase in circulating catecholamines is responsible for this immediate rise in the circulating numbers of lymphocytes. These hormones have an effect on the shear stress in the blood vessels and also on lymphocytes' adhesion molecules – present on the surface of the cells and through which they attach to the vascular endothelium (Pedersen et al. 1997; Shephard 2003).

The second phase of the lymphocyte response occurs around 1–3 h after exercise cessation. The lymphocytosis is substituted by lymphopenia, resulting in lymphocyte levels below pre-exercise values for several hours (Simpson et al. 2006; Kakanis et al. 2010). The severity and duration of this decrease will again depend on type, duration and intensity of the exercise performed, as well as the training status of the individual (Steensberg et al. 2002; Mooren et al. 2002; Simpson et al. 2006). This decline in lymphocyte count creates an "open window" of decreased host protection which is speculated to increase the susceptibility to URTI and other infections. The stress hormone cortisol has been shown to contribute to this postexercise lymphopenia, possibly by a redistribution of lymphocyte to other regions of the body (Nieman 1994; Pedersen et al. 1997). The levels of lymphocytes return to baseline 24 h postexercise (Kakanis et al. 2010).

To assess the exercise-induced changes in lymphocyte function, the majority of studies have conducted *in vitro* analysis. The functions assessed include investigating the activation of T cells, by looking at the expression of protein markers on its surface; investigating T-cell cytokine release; and assessing proliferation capacity. However, what must be taken into account is that other factors and changes that occur in the body might affect the cells' function differently than what is observed in a petri dish. In addition, most studies assess the lymphocytes taken from the peripheral circulation, so changes that occur with these cells might not reflect changes of the same cell type in other tissues and organs (Bishop 2006).

There is evidence that an acute exercise bout leads to an increase in T-cell activation (Fry et al 1992; Walsh et al. 2011). This might occur by an increase in the number of already activated T cells present in the circulation (during the lymphocytosis phase) or by an activation of the T cells due to changes in hormone concentration as a result of the exercise. Probably, it is a combination of both processes. The production of cytokines (IL-2 and IFN-y) per T cell has been shown to decrease after exercise; however, because of the higher T-cell numbers in the circulation, this function might not be impaired (Starkie et al. 2001a; Lancaster et al. 2005b). Finally, a decrease in T-cell proliferation (mitogen and antigen stimulated) has been reported following acute exercise, including resistance exercise (Fry et al. 1992; Miles et al. 2003). Nonetheless, caution should be taken when analysing the literature pertaining to T-cell proliferation because of confounding factors, such as the presence of other cells (NK cells and B cells) that do not respond to the proliferation stimulus applied for the T cells. All of the changes in lymphocyte functions mentioned are subject to the intensity and duration of the exercise bout, as with all the other mentioned immune responses.

6.3.3 Conclusion to Part 2

Exercise duration and intensity has a direct effect on circulating numbers and function of immune cells. Immediately after exercise, there is an increase in circulating cell numbers as a result of a demargination process mainly caused by a direct and indirect effect of increased circulating levels of catecholamines. Around 1–3 h after exercise cessation, lymphocyte blood counts decrease below pre-exercise levels, whilst neutrophil numbers remain on the rise, albeit at a slower rate compared to immediately postexercise. These responses are mediated by cortisol. This stress hormone is responsible for the further recruitment of neutrophils from the bone marrow whilst stimulating circulating lymphocytes to enter into tissues and prevent other cells from entering the circulation. Although there are more neutrophils in the circulation, their individual phagocytic capacity appears to be impaired; however, it is increased when all neutrophils are taken into account. A similar pattern is observed for lymphocyte activation. As for neutrophil respiratory burst activity, moderate exercise has a positive effect, whilst exhausting exercise seems to have a negative effect. Finally, lymphocyte production of IL-2 and INF-y is reported to decrease following intensive exercise.

This decline in lymphocyte count and other alterations in the immune system, observed during early recovery from certain types and high-intensity exercise, create an "open window" of decreased host protection. This open-window period of weakened immunity, which can last between 3 and 72 h, represents a vulnerable time period for the individual, during which there might be an increased susceptibility to contracting and developing an infection if there is contact with any pathogen (Nieman and Bishop 2006).

6.4 Part 3: Immunological Effects of Chronic Exercise

Elite athletes, at resting state, have similar counts and functions of immune cells and mediators as nonathletes (Gleeson and Bishop 2005; Nieman 2000). Nevertheless chronic exercise of moderate intensity and periods of heavy training have been shown to elicit changes in the immune system of the general population and of the athletic population, respectfully. In the third part of this chapter, we initially discuss the positive anti-inflammatory changes that occur with people who regularly take part in physical activity; whilst in the second part of the chapter, we outline immune function suppression that can be associated with periods of heavy training.

6.4.1 The Anti-inflammatory Effects of Chronic Exercise

Physical inactivity has been linked to various deleterious health effects. Over the last decades, the relationship between physical inactivity, obesity, metabolic dysfunction and chronic diseases has been widely investigated (Gleeson et al. 2011; Leggate et al. 2010; Pedersen et al. 2001; Peel et al. 2009a; Gratas-Delamarche et al. 2014). Although a great deal of research is still required in this area, it is clear that regular exercise decreases the incidence of chronic diseases by exerting a positive effect on the body's immune system. Chronic diseases, which include cardiovascular diseases, cancers, chronic respiratory diseases, depression and dementia, are responsible for over 80 % of global deaths from noncommunicable diseases (Heymann and Goldsmith 2012; World Health Organization 2012). Interestingly, all have the same predictor of risk: a chronic state of low-grade inflammation. In addition, in 2012, the World Health Organization (2012) released data stating that, considering all deaths around the globe, hypertension was accountable for 13 %, hyperglycaemia for 6 %, physical inactivity for 6 % and obesity for 5 %.

The practice of exercising regularly can be as effective as medicines prescribed to treat the above mentioned conditions (Heymann and Goldsmith 2012; Naci and Ioannidis 2013; Nguyen et al. 2014). In fact, in a recent meta-analysis study (Naci and Ioannidis 2013), data from 305 trials were analysed, involving almost 340,000 patients, for the outcome of drug or exercise treatments in mortality risk. The authors' conclusion was that both exercise and drug treatments have similar outcomes in prevention of coronary heart disease and diabetes, but exercise is more effective for preventing deaths in individuals who suffered a stroke compared with patients using medications. Moreover, scientific evidence shows that the development of endometrial, colon, prostate, lung and breast cancer have an inverse relationship with physical activity, i.e. the more frequent and intense the exercise, the lower the risk of developing these types of cancers (IARC 2002; Slattery 2004; Lee and Oguma 2006; Discacciati and Wolk 2014). Another important factor is how aerobically fit a person is, which is reflected by their cardiorespiratory capacity. When eliminating confounding factors, men with higher cardiorespiratory fitness have a lower probability of death from cancers of the digestive tract, liver and lungs (Peel et al. 2009b; Sui et al 2010).

Regular physical activity is also associated with an improvement in the quality of life reported by cancer patients (Wiggins and Simonavice 2009).

The recognition of skeletal muscle as an endocrine organ capable of producing a variety of metabolic factors when stimulated has increased the knowledge base within this topic. The potential protective mechanisms of regular physical exercise against chronic diseases include reductions in body mass (adipose tissue), increases in endogenous antioxidants concentrations and alterations in functions of the immune system (Walsh et al. 2011). All these contribute to a shift from an inflammatory state to an anti-inflammatory state in the organism as a result of exercise training (Beavers et al. 2010). These mechanisms are further detailed below.

6.4.2 Changes in Immune Functions

6.4.2.1 Reduction in Inflammatory Biomarkers

An increase in circulating levels of inflammatory biomarkers can be used to assess an individual's inflammatory state. These biomarkers include C-reactive protein (CRP) and inflammatory cytokines, such as tumour necrosis factor alpha (TNF-α), IL-1β, IL-8 and IL-6 (Beavers et al. 2010). Observational studies have shown that individuals performing more intense and more frequent bouts of exercise (and this includes leisure-time physical activity) and individuals presenting higher cardiorespiratory fitness present a lower concentration of inflammatory biomarkers (Beavers et al. 2010). In addition, data from interventional studies reinforce the anti-inflammatory effects of exercise, and the inconsistencies between studies can be related to differences in study methodology including, type, duration, intensity and frequency of the exercise training. Although further well-controlled intervention studies are necessary; overall, the studies indicate that regular aerobic exercise enhances the positive decrease in the inflammatory state of individuals who present higher inflammation at baseline. This occurs even without changes in fat mass. In healthy individuals, within the normal inflammatory range, exercise does not seem to have any effect on inflammatory biomarkers, though it could work by preventing increases in these harmful pro-inflammatory biomarkers (Beavers et al. 2010).

The concentration of the IL-6 cytokine is the one marker that is most affected by exercise. Studies show that it can have over a 100-fold increase in the postexercise period (Pedersen et al. 2001). This cytokine is the first to respond to exercise, in a consistent manner depending on exercise mode, duration and intensity (Ostrowski et al. 1998, 2001; Starkie et al. 2001b; Febbraio and Pedersen 2002). Initially it was thought that immune cells – monocytes – were the ones responsible for releasing IL-6 in response to exercise-induced muscle injury. Nevertheless, research studies have shown that this is not the case and that the contracting muscles are responsible for this cytokine's release (monocytes release IL-6 during sepsis) (Febbraio and Pedersen 2002; Pedersen et al. 2001; Starkie et al. 2001b). Therefore, IL-6 is also considered as being a myokine in exercising situations (Pedersen and Fischer 2007). This myokine has an important metabolic role in maintaining homeostasis during episodes of altered metabolic demand. During exercise, IL-6 increases the hepatic glucose production and stimulates lipolysis acting in a neuroendocrine hormone-like manner. In addition, this myokine also has an important anti-inflammatory effect due to its role in the suppression of pro-inflammatory cytokines – TNF-α and IL-1 – and in the increase of circulating levels of anti-inflammatory cytokines, such as IL-10 and IL-1ra and soluble TNF receptors (Pedersen et al. 2001; Tilg et al. 1997). IL-6 also stimulates the production of (CRP), and even though this biomarker is linked with inflammation, it affects immune cells by suppressing the synthesis of pro-inflammatory cytokines by macrophages and inducing anti-inflammatory cytokine release by monocytes (Pue et al. 1996). Whereas immune-cell-derived IL-6 presents pro-inflammatory characteristics, muscle-derived-IL-6 can have positive immunological anti-inflammatory effects. This is achieved by an acute bout of exercise; however, when exercising chronically, individuals maintain a pronounced anti-inflammatory milieu. This in turn protects them against low-grade inflammatory processes, decreasing the risk of chronic diseases. Other immunomodulatory factors that might be released during exercise, such as hormones and heat-shock proteins, also contribute towards an anti-inflammatory state.

6.4.2.2 Reduction of Toll-Like Receptor (TLR) Expression on Monocytes and Macrophages

As mentioned in Sect. 6.2, TLRs have an important role in modulating inflammatory processes. Over the last decade, it has been shown in the majority of studies that regular exercise can lead to a downregulation of TLRs expression on monocytes and macrophages cell surfaces. In fact, active individuals have a reduced TLR 4 expression compared to their less active counterparts. This decrease in TLR, due to chronic exercise, has been observed in both the young and the elderly (Flynn et al. 2003; McFarlin et al. 2004b; Stewart et al. 2005). In addition, a shift in macrophage phenotype from M1-type, which produces inflammatory cytokines, to M2-type, which produces anti-inflammatory cytokines, has also been shown to occur with chronic exercise, consequently leading to an anti-inflammatory milieu (McFarlin et al. 2004b; Gleeson et al. 2006; Timmerman et al. 2008; Coen et al. 2010). These adaptations may occur with a more prolonged training period, as a 2-week study with a high-intensity interval training regimen, in previously sedentary individuals, had the opposite effect – potentially representing a transitory adaptation (Oliveira-Child et al 2013).

6.4.2.3 Reduction of Adipose Tissue

Individuals with a high percentage of body fat will possibly present low-grade inflammation due to their inflamed adipose tissue. This happens because this tissue can produce over 75 inflammatory proteins, amongst which are pro-inflammatory cytokines (adipokines) such as TNF-α, leptin, IL-6 and IL-18 (Wood et al. 2009; Nimmo et al. 2013). Adipocyte hypertrophy, which occurs with obesity, has a key role in initiating a cascade of inflammatory events: adipocytes' volume expansion promotes tissue hypoxia, which leads to cellular stress, resulting in necrosis. This process stimulates T-cell and pro-inflammatory activated macrophage (M1) recruitment into this tissue (Gleeson et al. 2011). Depending on the intensity of this process, the local inflammatory mediators enter into the blood circulation causing

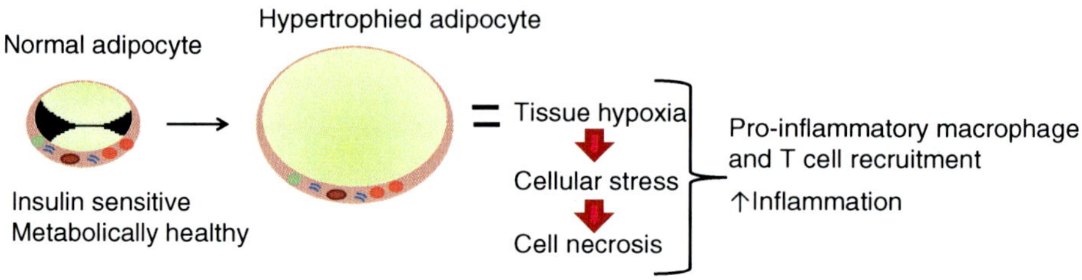

Fig. 6.2 Adipocyte volume change triggering inflammation

systemic inflammation. Fortunately, exercise is a potent stimulus for visceral adipose tissue lipolysis, and regular exercise has been shown to decrease adiposity, even without overall weight change (Lee et al. 2005). The same is true for a reduction in inflammation observed with exercise training, independent of changes in fat mass, because of alterations in the previously mentioned inflammatory pathways. It is important to point out that small adipocytes are insulin sensitive and metabolically healthy, not stimulating health-related issues (Nimmo et al. 2013) (Fig. 6.2).

6.4.3 Increases in Endogenous Antioxidants Concentrations

Pertaining literature provides evidence that performing an exercise bout can lead to oxidative stress. This occurs because, during muscle contraction, there is an increased production in reactive oxygen species and free radicals which might not be buffered by the body's antioxidant defences (Fisher-Wellman and Bloomer 2009; Gomes et al. 2012). The mechanisms of increased reactive species during exercise include (1) ischaemia-reperfusion and activation of endothelial xanthine oxidase, (2) electron leak at the mitochondrial electron transport chain, (3) activation of neutrophils and other phagocytic cell due to muscle injury, (4) activation of NADPH oxidase and (5) auto-oxidation of catecholamines.

Oxidative stress can lead to DNA modification, inhibition of NK cells and of neutrophils' bactericidal activity and locomotion, reduction of T-lymphocytes and B-lymphocytes proliferation capacity and injury of cellular membranes and compounds (Sen and Roy 2001; Niess and Simon 2007). In addition, oxidative stress is known to trigger an inflammatory process, which is why chronic diseases also have as an underlying cause chronic oxidative stress (Chow et al. 2003; Chung et al. 2011; Halliwell and Gutteridge 2007). Although one bout of intense exercise can promote oxidative stress, regular exercise provides a favourable adaptation that protects our body against oxidative stress. This occurs via the reactive species, produced during exercise, acting as signalling molecules. This leads to an up-regulation of the body's antioxidant network system by the enhancement of antioxidants' gene expression and the modulation of other oxidative stress pathways. Both aerobic and anaerobic training have been shown to cause an increase in the antioxidant enzyme activity not only in the muscles but also systemically; hence, this beneficial adjustment also protects vital organs such as the liver and the brain (Gomez-Cabrera et al. 2008; Radák et al. 2003; 2008; Wilson and Johnson 2000).

6.4.4 Heavy Exercise Training

A great number of athletes go through intensified periods of training, sometimes with more than 1 session per day and consequently with insufficient recovery periods. When this happens, the period of suppressed immunity, which occurs after an intense exercise bout (mentioned in Sects. 6.2 and 6.3 of this chapter), is amplified by the subsequent exercise bout. This means that the athletes might remain in a constant "open-window" period, making them more susceptible to infections. Both innate and acquired immune functions have been shown to alter when athletes engage in long periods of intensified training:

- *Innate immune function:* Athletes usually present, at rest, a clinically normal level of blood-circulating neutrophil number, but some studies have reported lower neutrophil counts in trained distance cyclists (Blannin et al. 1996) and elite distance runners (Hack et al. 1994). Elite runners presented a lower neutrophil count at rest during a period of intense training (102 km·wk^{-1}) compared with nonathletes. This baseline difference, however, was not apparent when the training was of moderate intensity (89 km·wk^{-1}) (Hack et al. 1994).
- Neutrophil respiratory burst activity can be impaired by heavy training and can last for several days once training is ceased (Hack et al. 1994; Pyne et al. 1995). Intense training has also been shown to reduce NK cells cytotoxicity (Gleeson et al. 1995; Suzui et al. 2004). Nonetheless, no changes and no increase in NK cells cytotoxicity have been reported in a moderate-intensity training regime in previously sedentary individuals (Nieman et al. 1990; Shephard and Shek 1999; Woods et al. 1999).
- *Acquired immune function:* At rest, lymphocyte count and functions are not different between athletes and nonathletes (Nieman 2000). Studies have shown that when athletes engage in a period of intensified training, they can present a decrease in circulating T cells and a reduction in T-cell proliferative responses and cytokine production. This kind of training also affects the B-cell subsets, by impairment in the synthesis of immunoglobulins when these cells are stimulated (Lancaster et al. 2004). Factors such as circulating levels of stress hormones and an increase in the inflammatory milieu seem to be the triggers to this decrease in acquired immune function.
- *Mucosal immunity:* salivary IgA and other markers of the mucosal immune system can go through a phase of suppression when athletes are engaged in a heavy training period (Gleeson and Robson-Ansley 2006).

Another health-related problem that can happen, during periods of intense training without adequate recovery, is termed *overtraining syndrome.* Although this condition is not fully understood, it is known that when it occurs, athletes are in a constant fatigue state, with decrease in performance and increase in injury rates and infections. The overtraining syndrome is thought to be mediated by low-grade systemic inflammation, characterized by increased circulation levels of pro-inflammatory cytokines and depressed function of the immune system. Other symptoms that can occur are depression, anorexia and sleep disturbance that are thought to be caused by the effect of high pro-inflammatory cytokines acting on the brain (Smith 2000).

6.4.5 Conclusion to Part 3

Moderate exercise training has been shown to have a variety of health-related benefits including a reduction in chronic diseases. Within the mechanisms that lead to this positive outcome, the following are prevalent: a reduction in adipose tissue, an increase in antioxidant defences and immunological alterations that result in attenuation of the inflammatory milieu. A heavy period of exercise training, on the other hand, can exert a detrimental effect on athletes' health if an adequate recovery period is not present. This occurs because an intense exercise bout reduces different aspects of immune functions, and if the body is not given enough time to return to normal levels before another intense exercise bout, the immune functions are further impaired. This state of decreased immunosurveillance can increase the susceptibility to infections. Nevertheless, outside a heavy training period, an athlete's resting immune system appears be similar to nonathletes.

6.5 Summary and Conclusion

- Moderate levels of regular exercise have a positive effect on the working of the immune system in humans. The stress that results from this exercise requires the body to adapt, and as a result, individuals are better placed to withstand other stressors that may come their way. There is an issue with some individuals who stress the body too much through intensive exercise bouts, and they can open themselves up to breakdown as opposed to adaptation – which can manifest initially in the form of upper respiratory tract infections and similar symptoms.
- The response of the immune system to the exercise stress is dependent on the duration, type and intensity of the exercise. On cessation of an exercise, bout circulating cell numbers of lymphocytes and neutrophils are increased, but around 1–3 hr postexercise, the lymphocyte numbers decrease as they enter the tissues, whilst neutrophil numbers continue to rise. These responses appear to be mediated by circulating levels of catecholamines and cortisol.
- The alterations of the immune system postexercise create a period of time, known as an open window were the individual is more prone to contracting infections if exposed to pathogens. Repeated bouts of exercise can stress the immune system, further weakening the individual's ability to adapt to the exercise stress and hence increasing the number of open windows. If appropriate rest and recovery between training bouts is not scheduled to allow the body to adapt, then a negative state of under-performance or overtraining can result.

References

Akimoto T, Kumai Y, Akama T et al (2003) Effects of 12 months of exercise training on salivary secretory IgA levels in elderly subjects. Br J Sports Med 37:76–79

Akira S, Hemmi H (2003) Recognition of pathogen-associated molecular patterns by TLR family. Immunol Lett 85(2):85–95

Allgrove JE, Gomes E, Hough J, Gleeson M (2008) Effects of exercise intensity on salivary antimicrobial proteins and markers of stress in active men. J Sports Sci 26:653–661

Bate SL, Dollard SC, Cannon MJ (2010) Cytomegalovirus seroprevalence in the United States: the national health and nutrition examination surveys, 1988–2004. Clin Infect Dis 50(11):1439–1447

Bates DV (2005) Ambient ozone and mortality. Epidemiology 16(4):427–429

Beavers KM, Brinkley TE, Nicklas BJ (2010) Effect of exercise training on chronic inflammation. Clin Chim Acta 411(11–12):785–793

Belda J, Ricart S, Casan P et al (2008) Airway inflammation in the elite athlete and type of sport. Br J Sports Med 42(4):244–248

Benshop RJ, Rodriguez-Feuerhahn M, Schedlowski M (1996) Catecholamine-induced leukocytosis: early observations, current research, and future directions. Brain Behav Immun 10:77–91

Bishop NC (2006) Acute exercise and acquired immune function. In: Gleeson M (ed) Immune function in sport and exercise, 1st edn. Churchill Livingstone/Elsevier, Edinburgh, pp 91–113

Bishop NC, Gleeson M (2009) Acute and chronic effects of exercise on markers of mucosal immunity. Front Biosci 14:4444–4456

Bishop NC, Blannin AK, Armstrong E et al (2000) Carbohydrate and fluid intake affect the saliva flow rate and IgA response to cycling. Med Sci Sports Exerc 32:2046–2051

Bishop NC, Walsh N, Scanlon GA (2003) Effect of prolonged exercise and carbohydrate on total neutrophil elastase content. Med Sci Sports Exerc 35(8):1326–1332

Blannin AK (2006) Acute exercise and innate immune function. In: Gleeson M (ed) Immune function in sport and exercise, 1st edn. Churchill Livingstone/Elsevier, Edinburgh, pp 67–89

Blannin AK, Robson PJ, Walsh NP et al (1998) The effect of exercising to exhaustion at different intensities on saliva immunoglobulin A, protein and electrolyte secretion. Int J Sports Med 19:547–552

Blomberg A (2000) Airway inflammatory and antioxidant responses to oxidative and particulate air pollutants – experimental exposure studies in humans. Clin Exp Allergy 30:310–317

Bonsignore MR, Morici G, Riccobono L et al (2001) Airway inflammation in nonasthmatic amateur runners. Am J Physiol Lung Cell Mol Physiol 281:668–676

Booth S, Florida-James G, McFarlin B (2010) The impact of acute strenuous exercise on TLR2, TLR4 and HLA.DR expression on human blood monocytes induced by autologous serum. Eur J Appl Physiol 110(6):1259–1268

Bosch JA, Ring C, de Geus EJC, Veerman ECI, Nieuw Amerogen AV (2002) Stress and secretory immunity. Int Rev Neurobiol 52:213–253

Bougault V, Turmel J, St-Laurent J et al (2009) Asthma, airway inflammation and epithelial damage in swimmers and cold-air athletes. Eur Respir J 33(4):740–746

Bougault V, Turmel J, Boulet LP (2010) Bronchial challenges and respiratory symptoms in elite swimmers and winter sport athletes: airway hyperresponsiveness in asthma: its measurement and clinical significance. Chest 138:31S–37S

Boulet LP, Turcotte H, Langdeau JB, Bernier MC (2005a) Lower airway inflammatory responses to high-intensity training in athletes. Clin Invest Med 28(1):15–22

Boulet LP, Prince P, Turcotte H (2005b) Clinical features and airway inflammation in mild asthma versus asymptomatic airway hyperresponsiveness. Respir Med 100(2):292–299

Britt W (2008) Manifestations of human cytomegalovirus infection: proposed mechanisms of acute and chronic disease. Curr Top Microbiol Immunol 325:417–470

Brown J, Wang H, Hajishengallis GN et al (2011) TLR-signaling Networks. J Dent Res 90(4):417–427

Brunekreef B, Holgatek ST (2002) Air pollution and health. Lancet 360:1233–1242
Camous L, Roumenina L, Bigot S et al (2011) Complement alternative pathway acts as a positive feedback amplification of neutrophil activation. Blood 117(4):1340–1349
Carins J, Booth C (2002) Salivary immunoglobulin-A as a marker of stress during strenuous physical training. Aviat Space Environ Med 73:1203–1207
Chang CM, Yu KJ, Mbulaiteye SM et al (2009) The extent of genetic diversity of Epstein-Barr virus and its geographic and disease patterns: a need for reappraisal. Virus Res 143:209–221
Chatterton RT, Vogelsong KM, Lu YC et al (1996) Salivary alpha-amylase as a measure of endogenous adrenergic activity. Clin Physiol 16:433–448
Chicharro JL, Lucía A, Peréz M et al (1998) Saliva composition and exercise. Sports Med 26:17–27
Chinda D, Nakaji S, Umeda T et al (2003) A competitive marathon race decreases neutrophil functions in athletes. Luminescence 18(6):324–329
Chow CW, Abreu MTH, Suzuki T, Downey GP (2003) Oxidative stress and acute lung injury. Am J Respir Cell Mol Biol 29(4):427–431
Chung HY, Lee EK, Choi YJ et al (2011) Molecular inflammation as an underlying mechanism of the aging process and age-related diseases. J Dent Res 90(7):830–840
Cockcroft DW, Davis BE (2006) Mechanisms of airway hyperresponsiveness. J Allergy Clin Immunol 118:551–559
Coen PM, Flynn MG, Markofski MM et al (2010) Adding exercise to rosuvastatin treatment: influence on C-reactive protein, monocyte toll-like receptor 4 expression, and inflammatory monocyte (CD14+CD16+) population. Metabolism 59:1775–1783
Cox AJ, Gleeson M, Pyne DB et al (2004) Valtrex therapy for Epstein-Barr virus reactivation and upper respiratory symptoms in elite runners. Med Sci Sports Exerc 36:1104–1110
Cox AJ, Gleeson M, Pyne DB et al (2008) Clinical and laboratory evaluation of upper respiratory symptoms in elite athletes. Clin J Sport Med 18:438–445
D'Alonzo GE Jr (2002) Scope and impact of allergic rhinitis. J Am Osteopath Assoc 102(6 Suppl 2):S2–S6
Denguezli M, Chiekh IB, Saad HB (2008) One-year endurance training: effects on lung function and airway inflammation. J Sports Sci 26(12):1351–1359
Discacciati A, Wolk A (2014) Lifestyle and dietary factors in prostate cancer prevention. Recent Results Cancer Res 202:27–37
Dykewicz MS, Hamilos DL (2010) Rhinitis and sinusitis. J Allergy Clin Immunol 125(2 Suppl 2):S103–S115
Engebretsen L, Steffen K, Alonso JM et al (2010) Sports injuries and illnesses during the Winter Olympic Games 2010. Br J Sports Med 44(11):772–780
Fahlman MM, Engels HJ (2005) Mucosal IgA and URTI in American college football players: a year longitudinal study. Med Sci Sports Exerc 37:374–380
Febbraio MA, Pedersen BK (2002) Muscle-derived interleukin-6: mechanisms for activation and possible biological roles. FASEB J 16(11):1335–1347
Fisher-Wellman K, Bloomer RJ (2009) Acute exercise and oxidative stress: a 30 year history. Dyn Med 8(1):1–25
Flynn MG, McFarlin BK, Phillips MD et al (2003) Toll-like receptor 4 and CD14 mRNA expression are lower in resistive exercise-trained elderly women. J Appl Physiol 95:1833–1842
Foster NK, Martyn JB, Rangno RE et al (1986) Leukocytosis of exercise: role of cardiac output and catecholamines. J Appl Physiol 61(6):2218–2223
Fry RW, Morton AR, Crawford GP, Keast D (1992) Cell numbers and in vitro responses of leucocytes and lymphocyte subpopulations following maximal exercise and interval training sessions of different intensities. Eur J Appl Physiol Occup Physiol 64:218–227
Gabriel H, Urhausen A, Kindermann W (1991) Circulating leucocyte and lymphocyte subpopulations before and after intensive endurance exercise to exhaustion. Eur J Appl Physiol Occup Physiol 63(6):449–457
Gabriel H, Urhausen A, Kindermann W (1992) Mobilization of circulating leucocyte and lymphocyte subpopulations during and after short, anaerobic exercise. Eur J Appl Physiol Occup Physiol 65(2):164–170

Giraldo E, Garcia JJ, Hinchado MD, Ortega E (2009) Exercise intensity-dependent changes in the inflammatory response in sedentary women: role of neuroendocrine parameters in the neutrophil phagocytic process and the pro-/anti-inflammatory cytokine balance. Neuroimmunomodulation 16(4):237–244

Gkrania-Klotsas E, Langenberg C, Sharp SJ et al (2013) Seropositivity and higher immunoglobulin g antibody levels against cytomegalovirus are associated with mortality in the population-based European prospective investigation of Cancer-Norfolk cohort. Clin Infect Dis 56(10):1421–1427

Gleeson M (2006) Introduction to the immune system. In: Gleeson M (ed) Immune function in sport and exercise, 1st edn. Churchill Livingstone/Elsevier, Edinburgh, pp 15–44

Gleeson M, Bishop NC (2005) The T cell and NK cell immune response to exercise. Ann Transplant 10(4):43–48

Gleeson M, Pyne DB (2000) Special feature for the Olympics: effects of exercise on the immune system: exercise effects on mucosal immunity. Immunol Cell Biol 78:536–544

Gleeson M, Robson-Ansley P (2006) Immune responses to intensified training and overtraining. In: Gleeson M (ed) Immune function in sport and exercise, 1st edn. Churchill Livingstone/Elsevier, Edinburgh, pp 91–113

Gleeson M, McDonald WA, Cripps AW et al (1995) The effect on immunity of long-term intensive training in elite swimmers. Clin Exp Immunol 102:210–216

Gleeson M, McDonald WA, Pyne DB et al (1999) Salivary IgA levels and infection risk in elite swimmers. Med Sci Sports Exerc 31:67–73

Gleeson M, Pyne DB, Austin JP et al (2002) Epstein-Barr virus reactivation and upper-respiratory illness in elite swimmers. Med Sci Sports Exerc 34:411–417

Gleeson M, Pyne DB, Callister R (2004) The missing links in exercise effects on mucosal immunity. Exerc Immunol Rev 4:107–128

Gleeson M, McFarlin B, Flynn M (2006) Exercise and Toll-like receptors. Exerc Immunol Rev 12:34–53

Gleeson M, Bishop NC, Stensel DJ et al (2011) The anti-inflammatory effects of exercise: mechanisms and implications for the prevention and treatment of disease. Nat Rev Immunol 11:607–615

Gomes EC, Stone V, Florida-James G (2010) Investigating performance and lung function in a hot, humid and ozone-polluted environment. Eur J Appl Physiol 110(1):199–205

Gomes EC, Stone V, Florida-James G (2011) Impact of heat and pollution on oxidative stress and CC16 secretion after 8 km run. Eur J Appl Physiol 111(9):2089–2097

Gomes EC, Silva AN, de Oliveira MR (2012) Oxidants, antioxidants, and the beneficial roles of exercise-induced production of reactive species. Oxid Med Cell Longev 2012:756132

Gomez-Cabrera MC, Domenech E, Viña J (2008) Moderate exercise is an antioxidant: upregulation of antioxidant genes by training. Free Radic Biol Med 44(2):126–131

Gratas-Delamarche A, Derbré F, Vincent S, Cillard J (2014) Physical inactivity, insulin resistance, and the oxidative-inflammatory loop. Free Radic Res 48(1):93–108

Hack V, Strobel G, Weiss M, Weicker H (1994) PMN cell counts on phagocytic activity of highly trained athletes depend on training period. J Appl Physiol 77(4):1731–1735

Halliwell B, Gutteridge JMC (2007) Free radicals in biology and medicine. Oxford University Press, New York

Helenius I, Lumme A, Haahtela T (2005) Asthma, airway inflammation and treatment in elite athletes. Sports Med 35(7):565–574

Heymann EP, Goldsmith D (2012) Best approaches in the battle against Globesity? Learning lessons from our experience tackling HIV-AIDS and tobacco smoking. JRSM Short Rep 3(7):45

Hochberg D, Souza T, Catalina M et al (2004) Acute infection with Epstein-Barr virus targets and overwhelms the peripheral memory B-cell compartment with resting, latently infected cells. J Virol 78(10):5194–5204

Hoffman-Goetz L, Simpson JR, Cipp N et al (1990) Lymphocyte subset responses to repeated submaximal exercise in men. J Appl Physiol 68(3):1069–1074

IARC (2002) Handbooks of cancer prevention. In: Weight control and physical activity, vol 6. IARC press, Lyon

Jarvis MA, Nelson JA (2002) Mechanisms of human cytomegalovirus persistence and latency. Front Biosci 7:d1575–d1582

Kakanis MW, Peake J, Brenu EW et al (2010) The open window of susceptibility to infection after acute exercise in healthy young male elite athletes. Exerc Immunol Rev 16:119–137

Kappel M, Tvede N, Galbo H et al (1991) Evidence that the effect of physical exercise on NK cell activity is mediated by epinephrine. J Appl Physiol 70:2530–2534

Kivlighan KT, Granger DA (2006) Salivary alpha-amylase response to competition: Relation to gender, previous experience, and attitudes. Psychoneuroendocrinology 31:703–714

Klentrou P, Cieslak T, MacNeil M et al (2002) Effect of moderate exercise on salivary immunoglobulin A and infection risk in humans. Eur J Appl Physiol 87:153–158

Koch A (2010) Immune response to resistance exercise. Am J Lifestyle Med 4:244–252

Kostka T, Berthouze SE, Lacour J, Bonnefoy M (2000) The symptomatology of upper respiratory tract infections and exercise in elderly people. Med Sci Sports Exerc 32(1):46–51

Lamm ME (1998) Current concepts in mucosal immunity IV. How epithelial transport of IgA antibodies relates to host defense. Am J Physiol 274:G614–G617

Lancaster GI, Halson SL, Khan Q et al (2004) Effects of acute exhaustive exercise and chronic exercise training on type 1 and type 2 T lymphocytes. Exerc Immunol Rev 10:91–106

Lancaster GI, Khan Q, Drysdale P et al (2005a) The physiological regulation of toll-like receptor expression and function in humans. J Physiol 563(3):945–955

Lancaster GI, Khan Q, Drysdale PT et al (2005b) Effect of prolonged exercise and carbohydrate ingestion on type 1 and type 2 T lymphocyte distribution and intracellular cytokine production in humans. J Appl Physiol 98(2):565–571

Langdeau JB, Turcotte H, Bowie DM et al (2000) Airway hyperresponsiveness in elite athletes. Am J Respir Crit Care Med 161(5):1479–1484

Lee I, Oguma Y (2006) Physical activity. In: Schottenfeld D, Fraumeni JF (eds) Cancer epidemiology and prevention, 3rd edn. Oxford University Press, New York

Lee S, Kuk JL, Davidson LE et al (2005) Exercise without weight loss is an effective strategy for obesity reduction in obese individuals with and without Type 2 diabetes. J Appl Physiol 99:1220–1225

Leggate M, Nowell MA, Jones SA, Nimmo MA (2010) The response of interleukin-6 and soluble interleukin-6 receptor isoforms following intermittent high intensity and continuous moderate intensity cycling. Cell Stress Chaperones 15(6):827–833

Lewicki R, Tchórzewski H, Denys A et al (1987) Effect of physical exercise on some parameters of immunity in conditioned sportsmen. Int J Sports Med 8(5):309–314

Li TL, Gleeson M (2004) The effect of single and repeated bouts of prolonged cycling and circadian variation on saliva flow rate, immunoglobulin A and alpha-amylase responses. J Sports Sci 22:1015–1024

Libicz S, Mercier B, Bigou N, Le Gallais D, Castex F (2006) Salivary IgA response of triathletes participating in the French Iron Tour. Int J Sports Med 27(5):389–394

Matthews CE, Ockene IS, Freedson PS et al (2002) Moderate to vigorous physical activity and risk of upper-respiratory tract infection. Med Sci Sports Exerc 34(8):1242–1248

McCarthy DA, Dale MM (1998) The leucocytosis of exercise. A review and model. Sports Med 6(6):333–363

McCarthy DA, Macdonald I, Grant M et al (1992) Studies on the immediate and delayed leucocytosis elicited by brief (30-min) strenuous exercise. Eur J Appl Physiol Occup Physiol 64:513–517

McFarlin BK, Flynn MG, Stewart LK, Timmerman KL (2004a) Carbohydrate intake during endurance exercise increases natural killer cell responsiveness to IL-2. J Appl Physiol 96(1):271–275

McFarlin BK, Flynn MG, Campbell WW et al (2004b) TLR4 is lower in resistance-trained older women and related to inflammatory cytokines. Med Sci Sports Exerc 36(11):1876–1883

Miles MP, Kraemer WJ, Nindl BC et al (2003) Strength, workload, anaerobic intensity and the immune response to resistance exercise in women. Acta Physiol Scand 178(2):155–163

Mooren FC, Blöming D, Lechtermann A et al (2002) Lymphocyte apoptosis after exhaustive and moderate exercise. J Appl Physiol 93(1):147–153

Morice G, Bonsignore MR, Zangala D et al (2004) Airway cell composition at rest and after an all-out test in competitive runners. Med Sci Sports Exerc 36(10):1723–1729
Murphy EA, Davis JM, Brown AS et al (2004) Effects of moderate exercise and oat beta-glucan on lung tumor metastases and macrophage antitumor cytotoxicity. J Appl Physiol 97(3):955–959
Naci H, Ioannidis JP (2013) Comparative effectiveness of exercise and drug interventions on mortality outcomes: metaepidemiological study. BMJ 347:f5577
Nair S (2012) Nasal breathing exercise and its effect on symptoms of allergic rhinitis. Indian J Otolaryngol Head Neck Surg 64(2):172–176
Nater UM, La Marca R, Florin L et al (2006) Stress-induced changes in human salivary alpha-amylase activity - associations with adrenergic activity. Psychoneuroendocrinology 31:49–58
Neville V, Gleeson M, Folland JP (2008) Salivary IgA as a risk factor for upper respiratory infections in elite professional athletes. Med Sci Sports Exerc 40:1228–1236
Nguyen HQ, Chu L, Liu IL et al (2014) Associations between physical activity and 30-Day readmission risk in chronic obstructive pulmonary disease. Ann Am Thorac Soc 11(5):695–705
Nieman DC (1994) Exercise, infection and immunity. Int J Sports Med 15:S131–S141
Nieman DC (2000) Is infection risk linked to exercise workload? Med Sci Sports Exerc 32:S406–S411
Nieman DC, Bishop NC (2006) Nutritional strategies to counter stress to the immune system in athletes, with special reference to football. J Sports Sci 24(7):763–772
Nieman DC, Nehlsen-Cannarella SL, Markoff PA et al (1990) The effects of moderate exercise training on natural killer cells and acute upper respiratory tract infections. Int J Sports Med 11(6):467–473
Nieman DC, Miller AR, Henson DA et al (1993) Effects of high- vs moderate-intensity exercise on natural killer cell activity. Med Sci Sports Exerc 25(10):1126–1134
Nieman DC, Nehlsen-Cannarella SL, Fagoaga OR et al (1998) Effects of mode and carbohydrate on the granulocyte and monocyte response to intensive, prolonged exercise. J Appl Physiol 84(4):1252–1259
Nieman DC, Henson DA, Fagoaga OR et al (2002) Change in salivary IgA following a competitive marathon race. Int J Sports Med 23:69–75
Niess AM, Simon P (2007) Response and adaptation of skeletal muscle to exercise—the role of reactive oxygen species. Front Biosci 12:4826–4838
Nimmo MA, Leggate M, Viana JL, King JA (2013) The effect of physical activity on mediators of inflammation. Diabetes Obes Metab 15(Suppl 3):51–60
Okutsu M, Suzuki K, Ishijima T et al (2008) The effects of acute exercise-induced cortisol on CCR2 expression on human monocytes. Brain Behav Immun 22(7):1066–1071
Oliveira-Child M, Leggate M, Gleeson M (2013) Effects of two weeks of high-intensity interval training (HIIT) on monocyte TLR2 and TLR4 expression in high BMI sedentary Men. Int J Exerc Sci 6:81–90
Ortega E, Collazos ME, Maynar M et al (1993) Stimulation of the phagocytic function of neutrophils in sedentary men after acute moderate exercise. Eur J Appl Physiol Occup Physiol 66(1):60–64
Ortega E, Forner MA, Barriga C (1997) Exercise-induced stimulation of murine macrophage chemotaxis: role of corticosterone and prolactin as mediators. J Physiol 498(Pt 3):729–734
Ortega E, Marchena JM, García JJ et al (2005) Norepinephrine as mediator in the stimulation of phagocytosis induced by moderate exercise. Eur J Appl Physiol 93(5–6):714–718
Ostrowski K, Rohde T, Zacho M et al (1998) Evidence that IL-6 is produced in skeletal muscle during intense long-term muscle activity. J Physiol 508:949–953
Ostrowski K, Rohde T, Asp S et al (2001) Chemokines are elevated in plasma after strenuous exercise. Eur J Cell Physiol 84:244–245
Palmer FM, Nieman DC, Henson DA et al (2003) Influence of vitamin C supplementation on oxidative and salivary IgA changes following an ultramarathon. Eur J Appl Physiol 89:100–107
Peake JM (2002) Exercise-induced alterations in neutrophil degranulation and respiratory burst activity: possible mechanisms of action. Exerc Immunol Rev 8:49–100

Peake J, Suzuki K (2004) Neutrophil activation, antioxidant supplements and exercise induced oxidative stress. Exerc Immunol Rev 10:129–141

Pedersen L (2009) Airway hyperresponsiveness and airway inflammation in elite swimmers. Clin Respir J 3(1):62

Pedersen BK, Fischer CP (2007) Beneficial health effects of exercise–the role of IL-6 as a myokine. Trends Pharmacol Sci 28(4):152–156

Pedersen BK, Bruunsgaard H, Klokker M et al (1997) Exercise-induced immunomodulation – possible roles of neuroendocrine and metabolic factors. Int J Sports Med 18(Suppl 1):S2–S7

Pedersen BK, Steensberg A, Schjerling P (2001) Muscle-derived interleukin-6: possible biological effects. J Physiol 536(Pt 2):329–337

Peel JB, Sui X, Adams SA et al (2009a) A prospective study of cardiorespiratory fitness and breast cancer mortality. Med Sci Sports Exerc 41(4):742–748

Peel JB, Sui X, Matthews CE et al (2009b) Cardiorespiratory fitness and digestive cancer mortality: findings from the aerobics center longitudinal study. Cancer Epidemiol Biomarkers Prev 18(4):1111–1117

Proctor GB, Garrett JR, Carpenter GH, Ebersole LE (2003) Salivary secretion of immunoglobulin A by submandibular glands in response to autonomimetic infusions in anaesthetised rats. J Neuroimmunol 136:17–24

Pue CA, Mortensen RF, Marsh CB et al (1996) Acute phase levels of C-reactive protein enhance IL-1 beta and IL-1ra production by human blood monocytes but inhibit IL-1 beta and IL-1ra production by alveolar macrophages. J Immunol 156:1594–1600

Pyne DB (1994) Regulation of neutrophil function during exercise. Sports Med 17(4):245–258

Pyne DB, Baker MS, Fricker PA et al (1995) Effects of an intensive 12-wk training program by elite swimmers on neutrophil oxidative activity. Med Sci Sports Exerc 27:536–542

Pyne D, McDonald W, Gleeson M et al (1999) Mucosal immunity, respiratory illness, and competitive performance in elite swimmers. Med Sci Sports Exerc 33:348–353

Pyne DB, Gleeson M, McDonald WA et al (2000) Training strategies to maintain immunocompetence in athletes. Int J Sports Med 21(suppl 1):S51–S60

Radák Z, Apor P, Pucsok J et al (2003) Marathon running alters the DNA base excision repair in human skeletal muscle. Life Sci 72(14):1627–1633

Radak Z, Chung HY, Goto S (2008) Systemic adaptation to oxidative challenge induced by regular exercise. Free Radic Biol Med 44(2):153–159

Reid VL, Gleeson M, Williams N, Clancy RL (2004) Clinical investigation of athletes with persistent fatigue and/or recurrent infections. Br J Sports Med 38:42–45

Roberts ET, Haan MN, Dowd JB, Aiello AE (2010) Cytomegalovirus antibody levels, inflammation, and mortality among elderly Latinos over 9 years of follow-up. Am J Epidemiol 172(4):363–371

Robson PJ, Blannin AK, Walsh NP et al (1999) Effects of exercise intensity, duration and recovery on in vitro neutrophil function in male athletes. Int J Sports Med 20(2):128–135

Rodriguez AB, Barriga C, De la Fuente M (1991) Phagocytic function of blood neutrophils in sedentary Young people after physical exercise. Int J Sports Med 12(3):276–280

Roitt IM, Delves PJ (2001) Roitt's essential immunology, 10th edn. Blackwell Science, London

Ruedl G, Schobersberger W, Pocecco E et al (2012) Sport injuries and illnesses during the first Winter Youth Olympic Games 2012 in Innsbruck, Austria. Br J Sports Med 46(15):1030–1037

Sand KL, Flatebo T, Andersen MB et al (2013) Effects of exercise on leukocytosis and blood hemostasis in 800 healthy young females and males. World J Exp Med 3(1):11–20

Sarin S, Undem B, Sanico A et al (2006) The role of the nervous system in rhinitis. J Allergy Clin Immunol 118(5):999–1016

Sari-Sarraf V, Reilly T, Doran DA (2006) Salivary IgA response to intermittent and continuous exercise. Int J Sports Med 27:849–855

Scannapieco FA, Torres G, Levine MJ (1993) Salivary alpha-amylase: role in dental plaque and caries formation. Crit Rev Oral Biol Med 4:301–307

Scharhag J, Meyer T, Gabriel HH et al (2005) Does prolonged cycling of moderate intensity affect immune cell function? Br J Sports Med 39(3):171–177
Schenkels LC, Veerman EC, Nieuw Amerongen AV (1995) Biochemical composition of human saliva in relation to other mucosal fluids. Crit Rev Oral Biol Med 6:161–175
Sen CK, Roy S (2001) Antioxidant regulation of cell adhesion. Med Sci Sports Exerc 33(3):377–381
Shephard RJ (2003) Adhesion molecules, catecholamines and leucocyte redistribution during and following exercise. Sports Med 33(4):261–284
Shephard RJ, Shek PN (1999) Effects of exercise and training on natural killer cell counts and cytolytic activity: a meta-analysis. Sports Med 28:177–195
Simanek AM, Dowd JB, Pawelec G et al (2011) Seropositivity to cytomegalovirus, inflammation, all-cause and cardiovascular disease-related mortality in the United States. PLoS One 6(2), e16103
Simpson RJ, Florida-James GD, Whyte PG et al (2006) The effects of intensive, moderate and downhill treadmill running on human blood lymphocytes expressing the adhesion/activation molecules CD54 (ICAM-1), CD18 (B2 integrin) and CD53. Eur J Appl Physiol 97:109–121
Simpson RJ, McFarlin BK, McSporran C et al (2009) Toll-like receptor expression on classic and pro-inflammatory blood monocytes after acute exercise in humans. Brain Behav Immun 23(2):232–239
Slattery ML (2004) Physical activity and colorectal cancer. Sports Med 34(4):239–252
Smith LL (2000) Cytokine hypothesis of overtraining: a physiological adaptation to excessive stress? Med Sci Sports Exerc 32:317–331
Spence L, Brown WJ, Pyne DB et al (2007) Incidence, etiology, and symptomatology of upper respiratory illness in elite athletes. Med Sci Sports Exerc 39(4):577–586
Starkie RL, Rolland J, Febbraio MA (2001a) Effect of adrenergic blockade on lymphocyte cytokine production at rest and during exercise. Am J Physiol Cell Physiol 281(4):C1233–C1240
Starkie RL, Rolland J, Angus DJ et al (2001b) Circulating monocytes are not the source of elevations in plasma IL-6 and TNF-alpha levels. Am J Physiol Cell Physiol 280(4):C769–C774
Steensberg A, Morrow J, Toft AD et al (2002) Prolonged exercise, lymphocyte apoptosis and F_2-isoprostanes. Eur J Appl Physiol 87:38–42
Steerenberg PA, van Asperen IA, van Nieuw AA et al (1997) Salivary levels of immunoglobulin A in triathletes. Eur J Oral Sci 105:305–309
Stewart LK, Flynn MG, Campbell WW et al (2005) Influence of exercise training and age on CD14+ cell-surface expression of toll-like receptor 2 and 4. Brain Behav Immun 19:389–397
Sui X, Lee DC, Matthews CE et al (2010) Influence of cardiorespiratory fitness on lung cancer mortality. Med Sci Sports Exerc 42(5):872–878
Sureda A, Cordova A, Ferrer MD et al (2009) Effects of L-citrulline oral supplementation on polymorphonuclear neutrophils oxidative burst and nitric oxide production after exercise. Free Radic Res 43(9):828–835
Suzui M, Kawai T, Kimura H, Takeda K et al (2004) Natural killer cell lytic activity and CD56(dim) and CD56(bright) cell distributions during and after intensive training. J Appl Physiol 96:2167–2173
Syu GD, Chen HI, Jen CJ (2012) Differential effects of acute and chronic exercise on human neutrophil functions. Med Sci Sports Exerc 44(6):1021–1027
Thorley-Lawson DA, Gross A (2004) Persistence of the Epstein-Barr virus and the origins of associated lymphomas. N Engl J Med 350:1328–1337
Thorley-Lawson DA, Hawkins JB, Tracy SI et al (2013) The pathogenesis of Epstein-Barr vírus persistent infection. Curr Opin Virol 3(3):227–232
Tilg H, Dinarello CA, Mier JW (1997) IL-6 and APPs: anti-inflammatory and immunosuppressive mediators. Immunol Today 18:428–432
Timmerman KL, Flynn MG, Coen PM et al (2008) Exercise training-induced lowering of inflammatory (CD14+CD16+) monocytes: a role in the anti-inflammatory influence of exercise? J Leukoc Biol 84:1271–1278

Timmons BW, Cieslak T (2008) Human natural killer cell subsets and acute exercise: a brief review. Exerc Immunol Rev 14:8–23

Tongtako W, Klaewsongkram J, Jaronsukwimal N et al (2012) The effect of acute exhaustive and moderate intensity exercises on nasal cytokine secretion and clinical symptoms in allergic rhinitis patients. Asian Pac J Allergy Immunol 30(3):185–192

Toussirot E, Roudier J (2008) Epstein–Barr virus in autoimmune diseases. Best Pract Res Clin Rheumatol 22(5):883–896

Tvede N, Kappel M, Klarlund K et al (1994) Evidence that the effect of bicycle exercise on blood mononuclear cell proliferative responses and subsets is mediated by epinephrine. Int J Sports Med 15(2):100–104

Walsh NP, Blannin AK, Clark AM et al (1999) The effects of high-intensity intermittent exercise on saliva IgA, total protein and alpha-amylase. J Sports Sci 17:129–134

Walsh NP, Gleeson M, Shephard RJ et al (2011) Position statement. Part one: immune function and exercise. Exerc Immunol Rev 17:6–63

West NP, Pyne DB, Renshaw G et al (2006) Antimicrobial peptides and proteins, exercise and innate mucosal immunity. FEMS Immunol Med Microbiol 48:293–304

West NP, Pyne DB, Kyd JM et al (2010) The effect of exercise on innate mucosal immunity. Br J Sports Med 44:227–231

Wiggins MS, Simonavice EM (2009) Quality of life benefits in cancer survivorship with supervised exercise. Psychol Rep 104(2):421–424

Wilson DO, Johnson P (2000) Exercise modulates antioxidant enzyme gene expression in rat myocardium and liver. J Appl Physiol 88(5):1791–1796

Wood IS, de Heredia FP, Wang B et al (2009) Cellular hypoxia and adipose tissue dysfunction in obesity. Proc Nutr Soc 68:370–377

Woods JA, Ceddia MA, Kozak C et al (1997) Effects of exercise on the macrophage MHC II response to inflammation. Int J Sports Med 18:483–488

Woods JA, Evans JK, Wolters BW et al (1998) Effects of maximal exercise on natural killer (NK) cell cytotoxicity and responsiveness to interferon-alpha in the young and old. J Gerontol A Biol Sci Med Sci 53:B430–B437

Woods JA, Ceddia MA, Wolters BW et al (1999) Effects of 6 months of moderate aerobic exercise training on immune function in the elderly. Mech Ageing Dev 109:1–19

World Health Organization (2012) World Health Statistics 2012. http://apps.who.int/iris/bitstream/10665/44844/1/9789241564441_eng.pdf?ua=1&ua=1Assessed. Accessed 7 Jan 2012

Yoda K, Sata T, Kurata T et al (2000) Oropharyngotonsillitis associated with nonprimary Epstein-Barr virus infection. Arch Otolaryngol Head Neck Surg 126:185–193

Part II

Chemicals and Pollutants

Mechanisms by Which UV Radiation, a Natural Component of Sunlight, Suppresses the Immune Response

7

Stephen E. Ullrich

Contents

S.E. Ullrich, PhD
Department of Immunology, The University of Texas MD Anderson Cancer Center, Houston, TX, USA
e-mail: sullrich@mdanderson.org

C. Esser (ed.), *Environmental Influences on the Immune System*,
DOI 10.1007/978-3-7091-1890-0_7

7.1 What Wavelengths of UV Radiation Provide the Toxic Effect?

The UV radiation present in sunlight is divided into three bands: UVC (100–290 nm), UVB (290–320 nm), and UVA (320–400 nm). Because UVC is absorbed by the ozone layer, and its toxic effect on humans is limited to those who are exposed to artificial sources of sunlight (mercury arc lamps and germicidal lamps), it will not be discussed here. Most of the UV radiation reaching the earth's surface is UVA (≈95 %), and approximately 5 % is in the UVB range. There is a wealth of data in the scientific literature indicating that UVB radiation is responsible for carcinogenesis and immunosuppression in experimental animals and in humans (Halliday 2005; Ullrich 2005). The immunosuppressive role of UVA has perplexed photobiologists and dermatologists over the years, but it is safe to say that the most recent findings indicate that UVA is immunosuppressive and carcinogenic (for a more complete review of this subject, see Halliday et al. 2011). Some basic differences between the immunosuppressive properties of UVB and UVA may partially explain some of the past controversy. Unlike UVB, which suppresses primary immune reactions, UVA suppresses the recall response (Nghiem et al. 2001; Moyal and Fourtanier 2001). Also, unlike UVB, in which the UV dose-response curve is linear (Kim et al. 1998; Matthews et al. 2010), UVA-induced immunosuppression displays a bell-shaped dose-response curve (Byrne et al. 2002). Moreover, studies with human volunteers suggest that doses of UVA that are not suppressive by themselves (i.e., the ends of the bell-shaped curve) can cooperate with UVB to enhance immunosuppression (Poon et al. 2005). Regardless of the exact mechanisms involved, most photoimmunologists readily accept that wavelengths in both the UVB and UVA regions of the solar spectrum can cause immunosuppression.

7.2 Interaction with Photoreceptors in the Skin, First Step in the Induction of Immunosuppression

The first step in the process is the activation of photoreceptors in the skin that convert the electromagnetic energy into a biological signal. For many years only two photoreceptors were recognized, urocanic acid and DNA. However, over the past few years, a number of other UV photoreceptors have been identified. These include membrane phospholipids, tryptophan and the aryl hydrocarbon receptor, 7-dehydrocholesterol, and the elements in the complement pathway (Table 7.1).

7.2.1 Urocanic Acid

In 1983 based on the analysis of an action spectrum for UV-induced suppression of contact hypersensitivity (CHS) in mice, De Fabo and Noonan proposed that *trans-urocanic* acid, a compound that is abundant in the outermost layers of the skin, is a

Table 7.1 UV photoreceptors in the skin that activate immunosuppression

Photoreceptor	Mode of action	Downstream target	Reference
Urocanic acid	APC function Cytokine secretion	Dendritic cells/mast cells	Gibbs and Norval (2013), Gibbs et al. (2008)
Nucleic acids			
DNA	APC function	Dendritic cells/mast cells	Vink et al. (1996a, b)
	Cytokine secretion	Keratinocytes	Nishigori et al. (1996), Stege et al. (2000), Wolf et al. (2000)
RNA	Cytokine secretion		Bernard et al. (2012)
Membrane lipid oxidation	PAF agonists	Cytokine secretion and mast cell migration	Walterscheid et al. (2002), Chacón-Salinas et al. (2014)
Tryptophan	Metabolites bind Aryl-hydrocarbon receptor	Cytokine secretion and Treg induction	Fritsche et al. (2007), Navid et al. (2013), Bruhs et al. (2015), Esser et al. (2013)
7-dehydrocholestrol	Vitamin D production	APC function/Treg activation	Kurtitzky et al. (2007), Hart et al. (2011)
Complement	Cytokine secretion	Keratinocytes	Esser et al. (2013), Yoshida et al. (1998)

UVB photoreceptor (De Fabo and Noonan 1983). Upon UV exposure *trans*-urocanic acid is converted to the *cis*-isomer, and subsequent studies by De Fabo and Noonan and many others confirmed that *cis*-urocanic acid is immunosuppressive (Ullrich 2005; Gibbs and Norval 2013; Gibbs et al. 2008; Walterscheid et al. 2006). More recent studies have indicated that the receptor for *cis*-urocanic acid is the serotonin receptor (Walterscheid et al. 2006; Correale and Farez 2013). Serotonin receptors are expressed on a variety of skin cells, including Langerhans cells, mast cells, and keratinocytes (Nordlind et al. 2008). What is not entirely clear is the cellular target of *cis*-urocanic acid. Studies have indicated that *cis*-urocanic acid can trigger mast cell activation (Wille et al. 1999) and can interfere with antigen-presenting cell function (Noonan et al. 1988). It also triggers cytokine release by primary human keratinocytes, but in this case, the serotonin receptor does not appear to be involved (Kaneko et al. 2009).

7.2.2 DNA and UV-Modified Nucleic Acids

A series of papers from Margaret Kripke's laboratory demonstrated that that UV-damaged DNA serves as a photoreceptor for UV-induced immunosuppression. In these studies liposomes containing DNA repair enzymes were applied to the skin after UV exposure, and the reversal of immunosuppression, abrogation of

T regulatory (Treg) cell induction, and the reversal of cytokine release provided evidence supporting a role for DNA as a photoreceptor for immunosuppression (Kripke et al. 1992; Vink et al. 1996a, 1997, 1998; Nishigori et al. 1996). Similarly, applying liposomes containing HindIII to the skin, which causes double-stranded breaks in the DNA, resulted in cytokine release and the induction of immunosuppression (O'Connor et al. 1996). A number of studies done with human volunteers confirmed that UV-induced DNA damage triggers cytokine secretion and immunosuppression in humans (Stege et al. 2000; Wolf et al. 2000; Kuchel et al. 2005).

A more recent finding has provided evidence that UV-damaged RNA also plays a role in the cellular response to UV radiation (Bernard et al. 2012). Normal human epidermal keratinocytes were exposed to UV radiation, and lysates from these cells were added to nonirradiated keratinocytes or peripheral blood mononuclear cells (PBMC). Lysates from the irradiated cells induced the production of proinflammatory cytokines such as tumor necrosis factor-alpha (TNF-α) and interleukin (IL)-6 by the nonirradiated keratinocytes. When the lysates were treated with RNase, proinflammatory cytokine release was ablated. In addition, treating the nonirradiated keratinocytes with RNA from the UV-treated cells induces the upregulation of toll-like receptor 3 (TLR3) on the surface of the treated cells. RNA sequencing studies indicated that UV-induced modifications in the U1 stem-loop structure of noncoding RNA generated ligands for TLR3. Subsequent studies with synthetic RNA confirmed these results and also indicated that the synthetic U1 RNA activated inflammation in vivo. These findings provide yet another example of the role that UV-modified nucleic acids play in UV-induced inflammation and potentially immunosuppression.

7.2.3 Membrane Lipid Oxidation

UV exposure modulates the function of a variety of molecules. The transformation of *trans*-urocanic acid to the immunosuppressive *cis*-isomer (De Fabo and Noonan 1983) and UV-induced DNA damage, which triggers the induction of immunosuppression, are two prominent examples (Kripke et al. 1992). Early studies using cell-free cytosolic extracts (Devary et al. 1993), or agents that modulate membrane redox potential (Simon et al. 1994), suggested that UV-induced oxidative stress and membrane lipid oxidation could contribute UV modulation of cellular function. One prime example is platelet-activating factor (PAF, 1-alkyl-2-acetyl-*sn*-glycero-3-phosphocholine). Keratinocytes release *bona fide* PAF and PAF agonists (i.e., oxidized phosphocholine) following UV exposure (Marathe et al. 2005; Travers et al. 2010), which is inhibited by antioxidant treatment in vivo (Sahu et al. 2012). UV-induced osmotic stress also results in the release of PAF by keratinocytes (Rockel et al. 2007). A number of reports have indicated that PAF plays a critical role in UV-induced immunosuppression (Rockel et al. 2007; Walterscheid et al. 2002; Wolf et al. 2006; Zhang et al. 2008). The mechanisms by which PAF activates immunosuppression will be described below.

7.2.4 7-Dehydrocholesterol and Vitamin D

The photosynthesis of previtamin D3 in UV-irradiated skin is well known (MacLaughlin et al. 1982). Vitamin D can affect immune cells and modulate immune reactions, supporting a role for 7-dehydrocholesterol as a photoreceptor for immunosuppression. However, the literature regarding the effects of vitamin D on the immune system is somewhat controversial and somewhat reminiscent of the early reports describing UVA-induced immunosuppression. Some reports indicate that vitamin D can suppress immune reactions, and others indicate it can enhance immune reactions (Kurtitzky et al. 2007). Also it is very clear that vitamin D is not the only factor involved in UV-induced immunosuppression, in that immunosuppression is found in UV-irradiated vitamin D receptor knockout mice (Schwarz et al. 2012). Vitamin D can also suppress photocarcinogenesis (Dixon et al. 2011), in part by enhancing cellular defense mechanisms in the skin (Gordon-Thomson et al. 2012). Vitamin D insufficiency has also been associated with an increased risk of nonmelanoma skin cancer (Eide et al. 2011) (for more detail on this subject, see Hart et al. 2011).

7.2.5 Tryptophan and the Aryl Hydrocarbon Receptor (AhR)

Exposing aqueous solutions of tryptophan to UV light results in the formation of 6-diformylindolo[2,3-*b*]carbazole (FICZ), a tryptophan dimer that binds to the AhR (Rannug et al. 1987). Studies by Fritsche and colleagues demonstrated that UV-induced FICZ binds to the AhR receptor, which then translocates from the cytosol to the nucleus and induces Cyp1A1 and cyclooxygenase-2 (COX-2) gene expression. Moreover, no activation of Cyp1A1 or Cox-2 mRNA expression was found in the skin of AhR-deficient mice (Fritsche et al. 2007). These findings identify AhR as a cytoplasmic target for UV radiation. A subsequent study from this group confirmed a role for AhR in UV-induced immunosuppression. In these experiments, AhR antagonists blocked immunosuppression and the induction of Tregs in UV-irradiated mice. Conversely, the introduction of AhR agonists activated suppressive activity, and UV-induced immunosuppression was significantly reduced in AhR knockout mice (Navid et al. 2013).

The mechanism involves a modulation of dendritic cell function. Dendritic cells were activated with an AhR receptor agonist (4-n-nonylphenol: NP) and then coupled with a hapten. When the NP-treated antigen-presenting cells were injected into the mice, rather than antigen sensitization, immunosuppression was noted. The suppression was associated with the induction of negative costimulatory molecules (B7-H4) on the surface of the dendritic cells and the activation of T regulatory cells (Bruhs et al. 2015). AhR also regulates a number of functions in UV-irradiated skin, including melanogenesis, tanning, and cytokine production (Esser et al. 2013).

7.2.6 Complement

Two interesting papers published a number of years ago implicated a role for activated complement in UV-induced immunosuppression. No immunosuppression was found in complement (C3)-deficient UV-irradiated mice (Hammerberg et al. 1998). The activated C3 fragment, iC3b, is found in UV-irradiated human skin, and binding of iC3b to CD11b on skin monocytes activates increased IL-10 secretion and decreased IL-12 secretion (Yoshida et al. 1998). Stapelberg and colleagues (Stapelberg et al. 2009) provided additional data supporting complement as a UV photoreceptor. Mice were exposed to an immunosuppressive dose of UVA radiation, and 24 h later, the skin was removed and RNA isolated for microarray analysis. The authors report activation of genes involved in the alternative complement pathway (C3, complement factor B, and properdin). These findings, and the data indicating that no immunosuppression is found in UV-irradiated C3-deficient mice, indicate that components of the complement pathway serve as environmental sensors for UV radiation.

7.3 How Is the Immunosuppressive Signal Transmitted from the Skin to the Immune System?

Once the photoreceptors are activated, and an immunosuppressive signal is generated in the skin, that signal must be transmitted to the immune system. To date, most findings support a role for migrating immunocytes and immune modulatory soluble factors as prominent mediators of immunosuppression (Table 7.2).

7.3.1 Langerhans Cells

Once considered to be the primary antigen-presenting cell of the skin, recent findings have led immunologists to question their understanding of the role of Langerhans cells in initiating immune reactions. It now appears that the primary function of Langerhans cells is immune regulation (Kaplan et al. 2005; Bennett et al. 2005; Ritter et al. 2004; Fukunaga et al. 2008; Shklovskaya et al. 2011).

Table 7.2 Transmitting the suppressive signal from the skin to the immune system

Cell	Activating signal	Mechanism	Reference
Langerhans cell	RANKL	Treg ↑	Loser et al. (2007)
Mast cell	PAF	Mast cell derived IL-10 ↑	Byrne et al. (2008, 2011), Sarchio et al. (2014), Chacón-Salinas et al. (2014)
Bone marrow derived CD11c+ dendritic cell	PGE_2	APC function ↓ DNA methylation ↑	Ng et al. (2010, 2013a, b), Scott et al. (2014), Prasad and Katiyar (2013)

Fundamental studies in photoimmunology clearly demonstrated many years ago that Langerhans cells play an important role the induction of immunosuppression (reviewed in Ullrich 2005). Data from more recent studies, using selective depletion of Langerhans cells, confirm that Langerhans cells are absolutely required for UV-induced immunosuppression (Yoshiki et al. 2010; Schwarz et al. 2010; Fukunaga et al. 2010). Keratinocyte-derived gene products activate epidermal Langerhans cells to migrate from the skin to draining lymph nodes to induce immunosuppression, in part by activating Tregs. Loser and colleagues (Loser et al. 2007) noted that the expression of *r*eceptor *a*ctivator of *N*F-κβ (RANK) and its ligand (RANKL) was upregulated in psoriatic skin and following UV exposure of keratinocytes in vitro. Vitamin D may serve as the photoreceptor for this reaction because others have shown that vitamin D upregulates RANKL expression (Kitazawa et al. 2008). The role of RANKL in CHS was addressed by generating a transgenic mouse in which the keratin-14 promoter was employed to drive the expression of RANKL in keratinocytes. They noted a meager CHS response in mice where RANKL was overexpressed. They also observed a dramatic increase in CD4+ CD25+ Tregs in the lymph modes of RANKL+ mice. Because Langerhans cell trafficking from the skin to the lymph nodes is a well-known phenomenon, Loser and colleagues surmised that epidermal Langerhans cells were migrating from the skin of the RANKL+ mice to the lymph nodes and activating the Tregs. To directly address this hypothesis, two experiments were performed. In the first, dendritic cells from the RANKL+ mice were cultured with wild-type naive CD4+ CD25+ T cells. Proliferation of the CD4+ CD25+ T cells indicated that the dendritic cells from the RANKL+ mice could activate T-cell proliferation. This study was confirmed by an in vivo experiment in which depleting Langerhans cells from the skin of RANKL+ mice with mometasone furoate blocked the upregulation of CD4+ CD25+ cells in the lymph nodes of these mice.

To confirm that this mechanism was involved in UV-induced immunosuppression, two further experiments were performed. The first skin from RANKL-deficient mice was grafted onto the backs of wild-type mice and exposed to UV radiation. No immunosuppression was found in this situation. In the second, wild-type UV-irradiated mice were injected with RANK-Fc, which blocks the binding of RANK to its ligand. Because this procedure blocked the induction of immunosuppression, Loser and colleagues concluded that upregulation of RANKL on UV-irradiated keratinocytes modulated the function of epidermal Langerhans cells, which suppressed CHS in vivo, resulting in the activation of Tregs in the draining lymph node. They propose that "RANKL expressed on inflamed or activated keratinocytes seems to rewire local Langerhans cells that then have the capacity to regulate the number of peripheral regulatory T cells" (Loser et al. 2007).

7.3.2 Mast Cells

Our understanding of mast cells in the immune response has undergone a minor revolution in the past few years. The conventional view of mast cells was that they were limited to immune reactions associated with allergy and protection against parasites, but more recent studies have clearly demonstrated that these cells have potent immunoregulatory function (Kalesnikoff and Galli 2008; Galli et al. 2005). One of the very first indications that mast cells can regulate adaptive immune reactions came from a study published by Hart and colleagues demonstrating that mast cells were essential for UV-induced suppression in vivo. In this seminal study, Hart and colleagues observed no immunosuppression in mast cell-deficient UV-irradiated mice. Moreover, when the mast cell-deficient mice were reconstituted with normal bone marrow-derived mast cells, UV-induced immunosuppression was restored (Hart et al. 1998). Others have confirmed these observations over the years (Alard et al. 2001), and mast cells have also been shown to play an essential role in UVA-induced suppression of secondary immune reactions (Ullrich et al. 2007). Mast cell density in the skin correlates with the incidence of basal cell carcinoma and melanoma, suggesting that the immunoregulatory function of mast cells may contribute to skin carcinogenesis in humans (Grimbaldeston et al. 2000, 2004), similar to what has been demonstrated in mice (Sarchio et al. 2012).

One of the most intriguing questions regarding UV-induced immunosuppression is how can a physical substance like UV radiation, that barely penetrates past the dermal-epidermal junction of the skin, induce system-wide immunosuppression? Studying the role of mast cells in UV-induced immunosuppression provides some unique insights into the mechanisms involved. Like Langerhans cells, mast cells migrate from the dermis to the draining lymph nodes, and blocking mast cell migration in vivo will block UV-induced immunosuppression. Evidence to support the migratory role of mast cells in UV-induced immunosuppression was first provided by Byrne and colleagues (Byrne et al. 2008). They observed that mast cells quickly accumulate in the skin following UV exposure, a phenomenon latter shown to be dependent upon UV-induced IL-33 (Byrne et al. 2011). This was expected because UV irradiation induces skin inflammation, and mast cells migrate to areas of inflammation. What was surprising was the finding that within 24 h, the dermal mast cell density returned to normal, with a concomitant increase in mast cell density in skin draining lymph nodes. This suggested that mast cells were migrating away from the inflamed skin to the draining lymph nodes. To prove this was the case, skin from green fluorescent protein-positive mice was grafted onto the backs of syngeneic mast cell-deficient mice and the animals were then exposed to UV radiation. The appearance of green fluorescent protein-positive mast cells in the lymph nodes of the UV-irradiated mice, and not in the nodes of skin-grafted, nonirradiated control mice, demonstrated that the mast cells were indeed migrating from the skin to the lymph nodes. Because the CXCR4-CXCL12 axis is important for mast cell migration, an attempt was made to use a selective CXCR4 antagonist, AMD3100, to block in vivo mast cell migration. Abrogation of mast cell migration as well as immunosuppression in UV-irradiated, AMD3100-treated mice confirmed that mast cells carry the immunosuppressive signal from the skin to the immune system. Recently, Sarchio and colleagues confirmed that blocking mast cell migration with AMD3100 prevents immunosuppression in UV-irradiated mice and further found that blocking mast cell migration suppresses tumor development (Sarchio et al.

2014). These observations once again confirm the association between UV-induced immunosuppression and UV-induced skin carcinogenesis.

The signal that triggers mast cell migration from the skin to the draining lymph nodes is PAF. This was shown in a series of experiments recently published by Chacón-Salinas and colleagues (Chacón-Salinas et al. 2014). As mentioned above, mast cell-deficient mice are resistant to the immunosuppressive effects of UV radiation, and the immunosuppression can be restored by reconstituting mast cell-deficient mice with bone marrow-derived mast cells (Hart et al. 1998; Byrne et al. 2008). However, when mast cell-deficient mice were reconstituted with bone marrow-derived mast cells taken from PAF receptor knockout mice, no immunosuppression was observed. Further, when mast cell-deficient mice were reconstituted with bone marrow-derived mast cells isolated from PAF receptor knockout mice, and then exposed to UV radiation, no mast cell migration from the skin to the lymph node was noted. Similarly, no mast cell migration was observed in UV-irradiated PAF receptor-deficient mice. Alternatively, injecting wild-type mice with PAF activates mast cell migration, and this migration was suppressed by the use of selective PAF receptor antagonists. Chacón-Salinas and colleagues also found that injecting wild-type mice with PAF causes the upregulation of CXCR4 on mast cells and the upregulation of its ligand CXCL12 on lymph node cells (Chacón-Salinas et al. 2014).

Once mast cells arrive at the draining lymph node, they suppress by releasing interleukin (IL)-10 (Chacón-Salinas et al. 2011). Many of the papers in the literature documenting UV-induced immunosuppression employ in vivo measures of cell-mediated immunity (Ullrich 2005; Halliday et al. 2011). UV exposure also suppresses antibody formation in vivo, particularly T-dependent antibody reactions (Ullrich 1987; Brown et al. 1995). Chacón-Salinas et al. took advantage of this phenomenon when studying the effects of UV exposure on germinal center formation (Chacón-Salinas et al. 2011). They found that UV exposure suppresses antibody formation and germinal center formation in wild-type UV-irradiated mice. The mechanism involves suppressing the function of T follicular helper cells, which are the unique T helper cells found in the lymph nodes that are essential for germinal center formation. No suppression of antibody formation, T follicular helper cell function, or germinal center formation was found in mast cell-deficient mice. Reminiscent of the situation found in UV-induced suppression of CHS, reconstituting mast cell-deficient mice with wild-type bone marrow-derived mast cells reconstituted UV-induced suppression of T follicular helper cell function, germinal center formation, and antibody formation in vivo. However, when the mast cell-deficient mice were reconstituted with mast cells derived from IL-10-deficient animals, UV exposure did not suppress T follicular helper cell function, germinal center formation, and antibody formation, indicating that mast cell-derived IL-10 is essential for immunosuppression in vivo.

The interaction between PAF and mast cells provides an excellent example of the process that mediates the induction of systemic photo-immunosuppression. UV exposure causes the production of PAF by keratinocytes. Presumably the photoreceptor is membrane lipid oxidation or UV-induced osmotic stress. PAF then activates mast cell migration from the dermis to the B-cell region of the draining lymph node (Byrne et al. 2008). Once there the PAF-activated mast cells secrete IL-10, which depresses the function of T follicular helper cells, thus suppressing germinal center formation and antibody production in vivo. Moreover, since UV-induced IL-10 has been shown to suppress cell-mediated immune reactions, such as delayed type hypersensitivity (Rivas and Ullrich 1992), it is reasonable to assume that mast cell-derived IL-10 is playing a role here as well Fig. 7.1.

A similar story can be constructed when considering the mechanisms governing Langerhans cell migration and the induction of systemic immunosuppression. For many years photoimmunologists have recognized that UV irradiation of the skin alters the morphology and function of epidermal Langerhans cells (Toews et al. 1980). Under steady-state conditions, Langerhans cells normally migrate from the skin to the lymph nodes (Ginhoux et al. 2007), but UV exposure drives increased migration of Langerhans cells into the draining lymph nodes (illustrated in Fukunaga et al. 2010). Studies by Kripke and colleagues clearly show that UV-induced DNA damage is a trigger for altering Langerhans cell function (Vink et al. 1996a, 1997). Similarly, UV-induced DNA damage signals increased cytokine production by keratinocytes (Nishigori et al. 1996), and whether DNA damage plays a role in upregulating other factors that alter Langerhans cells function, such as RANKL, remains to be seen. It is also interesting to note that one report in the literature indicates that PAF regulates in vivo Langerhans cell migration (Fukunaga et al. 2008). Immunohistochemical analysis of lymph nodes from UV-irradiated mice indicates that Langerhans cells migrate to the T-cell area of the node (Fukunaga et al. 2010), where they activate T cells that have suppressive activity, namely, Tregs (Schwarz 2008) and natural killer T (NKT) cells (Moodycliffe et al. 2000) (Fig. 7.1).

7.3.3 UV-Induced Alteration of Bone Marrow-Derived Dendritic Cells by a Prostaglandin-Dependent Mechanism

Both Langerhans cells and mast cells reside in the skin (epidermis and dermis, respectively) and migrate to the lymph nodes to mediate their suppressive effects. However, a recent series of reports from Hart and her colleagues have shed light on another mechanism by which UV exposure induces systemic immunosuppression (Fig. 7.1). The data indicates that prostaglandin produced in response to epidermal UV exposure alters the antigen-presenting function of bone marrow-derived dendritic cells to induce immunosuppression. In the first series of experiments, mice were exposed to UV radiation, and 3 days later their bone marrow cells were isolated. CD11c+ cells were propagated by culturing the bone marrow cells in IL-4 and granulocyte macrophage colony-stimulating factor (GM-CSF), a standard technique for generating dendritic cells. The dendritic cells were pulsed with hapten (dinitrobenzene sulfonic acid) and injected into the ears of nonirradiated mice. After 7 days a cross-reacting contact allergen (dinitrofluorobenzene) was applied to the ears, and the elicitation of CHS was measured 24 h later. Bone marrow-derived CD11c+ cells were isolated from nonirradiated controls and treated in an identical

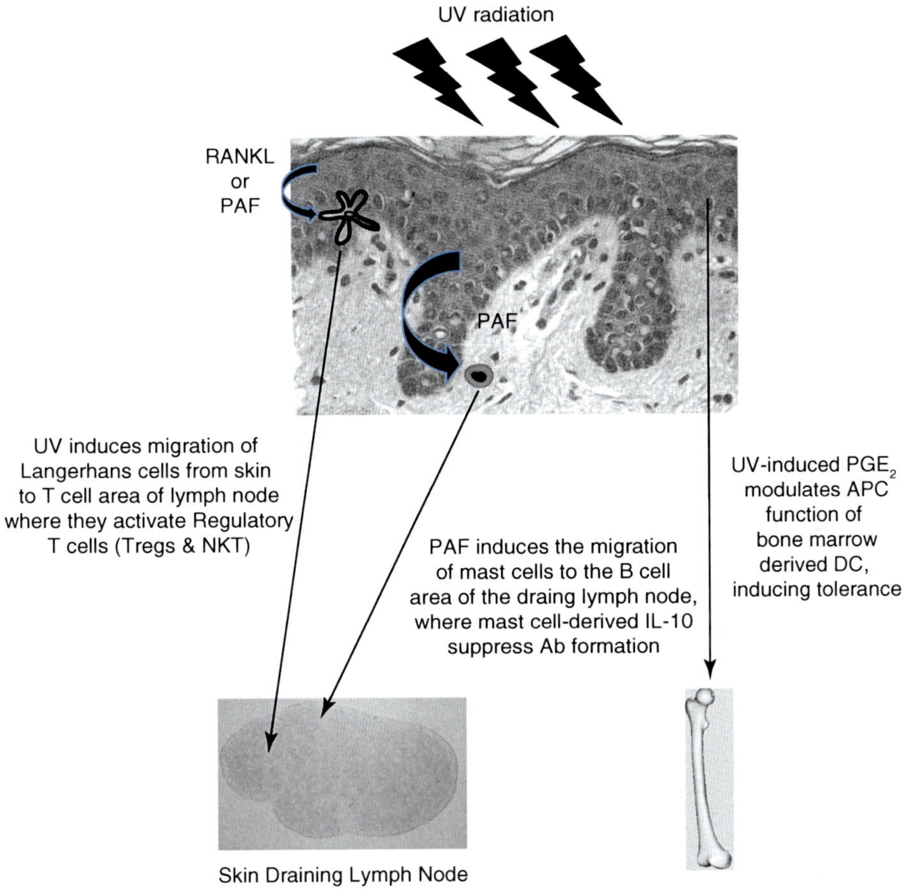

Fig. 7.1 Mechanisms by which UV exposure induces systemic immunosuppression. For details see references and relevant text

fashion. As expected, CD11c+ derived from nonirradiated mice presented the hapten to the immune system and generated a vigorous CHS response. On the other hand, when antigen-presenting CD11c+ cells were propagated from the bone marrow of UV-irradiated mice, their ability to present antigen was significantly suppressed (Ng et al. 2010). Both primary and secondary immune reactions were suppressed, and the effect was long lasting (Ng et al. 2013a, b). Characterization of the CD11c+ cells isolated from UV-irradiated mice failed to find any obvious defect (i.e., decreased maturation, altered antigen uptake, differential expression of costimulatory or regulatory molecules) to explain the decreased antigen-presenting cell function (Ng et al. 2010, 2013b). The authors did note that the dendritic cells secreted IL-10 and PGE_2; however, subsequent studies showing decreased antigen presentation when the CD11c+ cells were isolated from IL-10-deficient mice suggest that IL-10 secretion by the CD11c+ cells is not responsible (Ng et al. 2013a).

Production of prostaglandin, presumably by UV-irradiated skin cells (keratinocytes and/or mast cells), appears to be the signal that activates deficient antigen presentation by bone marrow-derived dendritic cells isolated from UV-irradiated mice. The suppressive effect was reversed when the UV-irradiated mice were treated with indomethacin, and the effect could be mimicked when the UV irradiation was replaced by PGE_2 (Ng et al. 2010; Scott et al. 2014). Moreover, inhibiting the effect by treating the UV-irradiated mice with 5-aza-2′-deoxycytidine suggests that an epigenetic mechanism may be involved (Ng et al. 2013a). One other recently published paper supports a role for epigenetic mechanisms in UV-induced immunosuppression. Prasad and Katiyar report that UV-induced PGE_2 promotes immunosuppression by altering DNA methylation, in particular by increasing the expression of DNA methyltransferases in the skin. Here also, treating UV-irradiated mice with 5-aza-2′-deoxycytidine reversed the upregulation of the DNA methyltransferases in the skin and reversed the induction of immunosuppression (Prasad and Katiyar 2013). Hart and colleagues suggest that upon UV irradiation of the skin, dermal dendritic cells and Langerhans cells migrate from the skin to the draining lymph node. These cells are then replaced by bone marrow-derived dendritic cells whose ability to present antigen is impaired. This novel mechanism may explain longlasting immune tolerance following in vivo UV exposure. It is interesting to note that inflammatory insults other than UV radiation that also activate PGE_2 production promote the development of immunoregulatory CD11c+ cells in the bone marrow, suggesting this maybe a widespread homeostatic mechanism to check inflammation (Scott et al. 2012).

7.4 Advantageous Effects of UV-Induced Immunosuppression

The term that is most often used regarding UV-induced modulation of immunity is immunosuppression, primarily because UV exposure suppresses inflammatory T helper-1 (Th1)-mediated immune reactions, which are protective in skin cancer and drive cell-mediated immune reactions. I have used the conventional nomenclature throughout this chapter, but the more appropriate term is UV-induced immune deviation. Not all immune reactions are suppressed by UV irradiation. The best example of this is the immune response to infectious agents. When immune reactions that are mediated by Th1 cells are examined, such as the immune response to *Mycobacterium bovis* (Jeevan et al. 1996; Cestari et al. 1995), *Candida albicans* (Denkins and Kripke 1993; Denkins et al. 1989), or Herpes simplex virus (Ross et al. 1986), to name a few, UV is immunosuppressive. However, when the immune response to parasitic infections is examined, such as the response to *Schistosoma mansoni* (Jeevan et al. 1992) or *Leishmania* (Giannini 1992), UV exposure of infected individuals does not reduce infectivity, nor does it reduce pathogenesis. Similarly, UV-induced suppressor T cells (now known as Tregs) suppress antibody production to T-dependent, but not T-independent, antigens (Ullrich 1987). This may reflect the fact that UV irradiation suppresses Th-1-mediated reactions and shifts the immune response toward a T-helper 2 (Th2)-mediated immune reaction, in part through a differential effect on cytokine production (Ullrich 1996). In this section I will discuss some examples where UV exposure provides a beneficial effect, primarily by suppressing Th1 and T helper-17 (Th17)-mediated immune reactions.

7.4.1 Phototherapy

Phototherapy, either narrow band UVB or psoralen plus UVA (PUVA), is the first-line treatment for psoriasis and is considered to be the most safe and most effective treatment regimen for this disease (Lapolla et al. 2011), even when compared to newer systemic biological treatments (Inzinger et al. 2011). The potential mechanisms involved have been nicely reviewed in the past (Weichenthal and Schwarz 2005), so a detained discussion will not be provided here. However, recent reports using a transgenic animal model that spontaneously develops psoriasis have provided some new information regarding the mechanisms of immunosuppression. Singh et al. find that PUVA treatment suppresses the production of inflammatory cytokines including IL-1α and IL1β and IL-6, IL-12, IL-17, and IL-23 by suppressing the expression of two relevant transcription factors: STAT3 and the orphan nuclear receptor RORγt. At the same time PUVA treatment activates IL-10-secreting Tregs that suppress the disease (Singh et al. 2010). Injecting PAF accelerates psoriasis induction in these mice, and injecting PAF receptor antagonists blocked psoriasis, by suppressing inflammatory cytokine (IL-6, IL-12, IL-17, IL-23) production in vivo (Singh et al. 2011). These data shed new details on how phototherapy suppresses psoriasis and may provide new insights into novel therapeutic targets.

7.4.2 Multiple Sclerosis (MS)

MS is a debilitating inflammatory autoimmune disease characterized by demyelination and destruction of axons in the central nervous system. It has been recognized for years that genetic and environmental risk factors contribute to MS, including sunlight exposure. A recent meta-analysis using data generated in nations populated by people of European descent confirmed a positive association between increased prevalence of MS and higher latitudes (Simpson et al. 2011). Because MS is mediated by Th1 and Th17 cells (Petermann and Korn 2011), higher MS prevalence at higher latitudes suggest UV-induced immunosuppression, or lack thereof, may play a role.

One of the first questions addressed was the role of UV-induced vitamin D. Studies with large cohorts of volunteers (Nurses' Health Study I and II) indicated a protective effect of vitamin D on the incidence of MS and suggested that a low serum level of vitamin D is a risk factor (Munger et al. 2004). This would suggest that UV-induced vitamin D and the immune regulation that results from increased vitamin D serum levels may limit autoimmune inflammation, thereby playing a protective role. As discussed above, vitamin D has been shown to suppress immune reactions (Hart et al. 2011). What is not clear is whether high-dose vitamin D supplementation affects MS. A phase I/II dose escalation trial with high-dose vitamin D (10,000–40,000 international units/day for 28 days) indicated that high-dose supplementation was safe, serum levels of 413 nmol/l were achieved, and no significant adverse effects were noted. The authors report that "clinical outcomes appeared to favor the treatment group," but because the trial was small and insufficiently powered, no conclusions as to the efficiency of dietary vitamin D supplementation on MS risk could be made (Burton et al. 2010). A much larger placebo-controlled randomized clinical trial is needed to adequately address this question.

What other mechanisms may be involved? The immunological profile of MS patients is characterized by low serum IL-10 (Salmaggi et al. 1996), increased production of inflammatory cytokines such as IL-12 (Salmaggi et al. 1996) and TNF-α (Sharief and Hentges 1991), and loss of Treg function (Viglietta et al. 2004). A recent study suggests that sunlight-induced *cis*-urocanic acid may modulate the profile of MS patients in a positive way (Correale and Farez 2013). Correale and colleagues report that plasma levels of *cis*-urocanic acid were significantly lower in MS patients compared to healthy controls. When myelin basic protein and/or myelin oligodendrocyte glycoprotein-specific T cell lines from MS patients were stimulated in vitro with *cis*-urocanic acid, they secreted more IL-10 and IFN-γ production was suppressed. When PBMC were cultured with *cis*-urocanic acid, the number of CD4+ CD25+ FoxP3+ Tregs in the culture increased, and the treated cells secreted more IL-10 than cells treated with *trans*-urocanic acid. This resulted regardless of whether the PBMC were isolated from MS patients or healthy controls and confirmed an early report using murine T cells (Holan et al. 1998). The effects of *cis-urocanic* acid on cytokine production (increased IL-10 with a concomitant decrease in IFN-γ production) were suppressed by adding either anti-serotonin (5HT) monoclonal antibody or serotonin $5HT_{2a}$ receptor antagonists to the cultures, confirming an earlier report that *cis*-urocanic acid mediates its effects through the $5HT_{2a}$ receptor (Walterscheid et al. 2006). These findings support the hypothesis that sunlight-induced immunosuppression in humans may be suppressing the induction of MS.

7.4.3 UV Irradiation and Asthma

Asthma is as a chronic lung disease that inflames and constricts the airways, resulting in shortness of breath, wheezing, and coughing. Asthma is triggered by many factors, including allergens, irritants, cigarette smoking, air pollution, chemicals in our environment, and viral upper respiratory tract infections, just to name a few. Data supporting a latitude gradient for asthma prevalence is conflicting, some support a higher prevalence of asthma close to the equator (Staples et al. 2003), and others support a higher prevalence of asthma as one moves away from the equator (Krstic 2011). In addition, findings from yet another study suggest that once the data is corrected for average daily temperature, the latitude gradient disappears (Hughes et al. 2011). Others have suggested that asthma prevalence correlates negatively with the time spent in the sun (Hughes et al. 2011; Loh et al. 2011; Chen et al. 2006).

Although not yet studied in man, data from a series of animal experiments suggest that UV irradiation can suppress asthma incidence, albeit by using a mechanism that is somewhat unique. In the initial report demonstrating that UV exposure suppresses allergic lung disease, BALB/c mice were exposed to an erythemal dose of UV prior to immunization by aspiration of a fungal antigen. UV exposure prior to immunization suppressed immune-mediated inflammation as measured by eosinophil influx and total IgE secretion. No suppression of either measure of asthma was noted if the UV was applied after immunization (Ward et al. 2004). A subsequent study by McGlade and colleagues confirmed the UV effect using an ovalbumin (OVA) model of asthma. Here again, the mice were exposed to an erythemal dose of UV radiation (8 kJ/m^2) 3 days prior to intraperitoneal immunization with OVA. The mice were boosted 14 days later and challenged on day 21 by aerosol administration of OVA. A single UV exposure suppressed airway hyperresponsiveness; suppressed the in vitro proliferation of lymph node cells to OVA; suppressed the production of IL-4, IL-5, and IL-10 by lymph node cells; and blocked the secretion of IL-5 into the lung lavage fluid. Also, adoptive transfer of lymph node cells from the UV-irradiated mice could transfer suppression to naive unirradiated-recipient mice, although the identity of the suppressor cells was not determined (McGlade et al. 2007a). Similar results were observed when C57BL/6 mice were used, and the failure to see UV-induced immunosuppression in histamine-2 receptor-deficient mice, but not in histamine-1 receptor-deficient mice, indicates a role for the histamine-2 receptor in immunosuppression (McGlade et al. 2007b). As mentioned above, McGlade and colleagues found evidence to support the existence of antigen-specific suppressor cells (McGlade et al. 2007a), but in a subsequent report they failed to find any evidence for classic Tregs (CD4+ CD25+, IL-10+) in the suppression of airway hyperresponsiveness. Nor could they find evidence for the existence of other UV-induced "suppressor cells" such as B regulatory cells (Matsumura et al. 2006; Byrne et al. 2005) or NKT cells (Moodycliffe et al. 2000). Rather they found fewer activated CD4+ T effector cells in the lymph nodes of UV-irradiated mice, and the few CD4+ cells present had a defect in cytokine production and a lower rate of proliferation (McGlade et al. 2010). The Treg-independent downregulation of immunity by UV radiation in this model is unique. The reason for this unique method of immune regulation may be derived from the fact that asthma is a complicated immune reaction in which both Th1 and Th2 cells contribute to its induction (Randolph et al. 1999a, b).

7.5 Is a Common Pathway to Immunosuppression Utilized by Other Agents That Traumatize the Skin?

In the 1980s, the US Military, lead by the US Air Force, started converting to a new universal fuel, jet propulsion (JP)-8. JP-8 was formulated to be a safer fuel than JP-4, the fuel it replaced, in that its higher vapor pressure and higher flash point rendered it less flammable and less likely to explode upon accidents. Also,

JP-8 had a much lower concentration of carcinogenic benzene and neurotoxic hexane than JP-4. The higher vapor pressure had the added advantage of reducing evaporative loss during storage. Unfortunately, the higher vapor pressure and higher flash point of JP-8 meant less evaporation when the fuel was released into the environment, resulting in increased human exposure and increased risk of toxicity. During the conversion to JP-8, individual complaints of health problems prompted the US Air Force Office of Scientific Research to form a task force and fund research to study the potential toxic effects of JP-8 (reviewed by Witten et al. 2011). My laboratory was funded to examine the immunotoxicity of dermal JP-8 exposure, and what was surprising to me was the similarity of the immune mechanisms that were activated after exposure to UV radiation and JP-8. In the initial experiments, we found that dermal exposure to JP-8, either one single large exposure or multiple smaller exposures, over a short period of time resulted in immunosuppression. Both delayed type hypersensitivity to a bacterial antigen and CHS to a contact allergen were suppressed. T-cell proliferation was also suppressed, but antibody production was not. IL-10 was found in the serum of JP-8-treated mice, and antibodies to IL-10 blocked immunosuppression. Injecting IL-12, which counteracts the activity of IL-10 into JP-8-treated mice, reversed immunosuppression, similar to its effects in UV-irradiated mice (Schmitt et al. 1995, 2000; Müller et al. 1995; Schwarz et al. 1996; Schmitt and Ullrich 2000). In addition, when JP-8-treated mice received a selective COX-2 inhibitor, the induction of immunosuppression was blocked (Ullrich 1999; Ullrich and Lyons 2000). Applying JP-8 to the skin of mice also suppressed secondary immune reactions, and here again, PGE_2 was involved since the use of a selective COX-2 inhibitor overcame immunosuppression (Ramos et al. 2002). Because we knew that other toxins besides UV radiation upregulate dermal PAF production, and because PAF upregulates PGE_2 production by keratinocytes, we asked if PAF was involved. Injecting a series of PAF receptor antagonists into JP-8-treated mice totally reversed the induction of immunosuppression (Ramos et al. 2004). Like UV radiation, JP-8 treatment activates oxidative stress, which activates NF-κβ and contributes to COX-2 expression and the production of PGE_2. Treating JP-8-treated mice with agents that scavenge reactive oxygen species or reverse NF-κβ activation block COX-2 upregulation and block immunosuppression (Ramos et al. 2009). Finally, applying JP-8 to the skin of mast cell-deficient mice failed to induce immunosuppression, and reconstituting these mice with normal bone marrow-derived mast cells restored the suppressive effect. We also found that JP-8 activated mast cell migration from the skin to the draining lymph nodes and blocking this unconventional mast cell migration with AMD3100 blocked the induction of immunosuppression (Limón-Flores et al. 2009). So even though the two environmental immunotoxins are very different in composition, both have amazingly similar consequences and modes of action. Do other environmental agents that interact with the skin and induce immune regulation employ similar mechanisms to limit immune reactivity? Do the mechanisms outlined above reflect common homeostatic mechanisms to check inflammation?

7.6 Summary

Skin cancer is the most prevalent cancer found in the industrialized world (Siegel et al. 2012), and in the United States the cost of treating all skin cancers is estimated to be in excess of 30 billion dollars a year (Housman et al. 2003; Bickers et al. 2006). The driving force behind studies designed to understand the mechanisms underlying UV-induced immunosuppression is the association between skin cancer induction and immunosuppression by sunlight (Kripke 1974; Yoshikawa et al. 1990; Penn 2000). What I have attempted to do in this chapter is to follow a photon of UV light after it activates a photoreceptor in the skin and then initiates a cascade of events that lead to UV-induced systemic immunosuppression. Because UV-induced systemic immunosuppression is a major risk factor for skin cancer induction, this topic has been an active area of research for many years, and many outstanding reviews have been published describing the mechanisms involved. Here I have reviewed some of the more recent data concerning the mechanisms by which this common environmental agent induces immunosuppression. I have also attempted to outline some examples where UV-induced suppression of inflammatory immune reactions may provide a beneficial effect on human health. Of course balancing the harmful effects of UV-induced immunosuppression (skin cancer induction) versus its beneficial effects (suppressing autoimmunity) will be the subject of many serious debates in the future.

The lessons learned from studying the immune regulation induced by UV exposure are relevant to anyone interested in environmental influences on the immune system. First of all, the UV radiation in sunlight is the most common immunotoxin in our biosphere. Second, although UV radiation is absorbed in the upper layers of the skin, UV exposure causes systemic immunosuppression. We now know that UV irradiation of the skin activates a cascade of effects that results in immunosuppression. What about other dermal immunotoxins? I find it curious that the mechanism of immunosuppression activated by UV exposure is remarkably similar to that seen following dermal application of jet fuel, a complex chemical mixture of aliphatic and aromatic hydrocarbons. Do other agents that traumatize the skin and induce skin inflammation activate the same immunosuppressive pathways? We can see jet fuel-induced immunosuppression using repeated application of small volumes of fuel that probably do not penetrate to the underlying immune tissues (Ullrich 1999; McDougal et al. 2000). Does this imply that the toxin does not have to directly interact with immune cells, but indirect mechanisms, as described here are involved? Third, although most of the studies reviewed above used experimental animals, there is a wealth of information on UV-induced immunosuppression in humans, and for the most part, the mechanisms are identical (Norval and Halliday 2011; Norval and Woods 2011). Finally, a better understanding of the mechanisms involved has led to strategies to overcome the immunosuppressive effects. Besides the obvious, and best strategy of avoiding the toxin by staying out of the sun, wearing protective clothing or using a good broad band sunscreen, a number of therapies have been designed to overcome the immunosuppressive effects of UV radiation and in some cases suppress skin cancer induction. For example, liposomes containing DNA repair enzymes block immunosuppression and progression to skin cancer in humans (Stege et al. 2000; Yarosh et al. 2001). Nutritional supplements that modulate cytokine production in the skin have been shown to block immunosuppression and photocarcinogenesis (Katiyar 2007). Topical application of nicotinamide (vitamin B3) has also been shown to reverse UV-induced DNA damage and reverse immunosuppression (Surjana et al. 2013; Damian et al. 2007). Applying PAF and/or serotonin receptor antagonists blocks immunosuppression and photocarcinogenesis in mice (Walterscheid et al. 2006; Sreevidya et al. 2008, 2010). On the other hand, UV-induced vitamin D production may limit MS, so perhaps vitamin D supplementation is a way to segregate the desired immunoregulatory effects from the carcinogenic potential of total body UV exposure. These are just a few examples how understanding the basic immunological mechanisms leading to immunosuppression allows for the design of therapies that can be used before and/or after UV exposure to prevent its toxic effects. Moreover, the information gained by an in-depth understanding of the mechanisms underlying UV-induced immunosuppression may also allow us to use the immunoregulatory effects of UV in an advantageous way to suppress unwanted immune reactions such as autoimmunity.

References

Alard P, Kurimoto I, Niizeki H, Doherty JM, Streilein JW (2001) Hapten-specific tolerance induced by acute, low-dose ultraviolet B radiation of skin requires mast cell degranulation. Eur J Immunol 31:1736–1746

Bennett CL, van Rijn E, Jung S, Inaba K, Steinman RM, Kapsenberg ML et al (2005) Inducible ablation of mouse Langerhans cells diminishes but fails to abrogate contact hypersensitivity. J Cell Biol 169:569–576

Bernard JJ, Cowing-Zitron C, Nakatsuji T, Muehleisen B, Muto J, Borkowski AW et al (2012) Ultraviolet radiation damages self noncoding RNA and is detected by TLR3. Nat Med 18:1286–1290

Bickers DR, Lim HW, Margolis D, Weinstock MA, Goodman C, Faulkner E et al (2006) The burden of skin diseases: 2004 a joint project of the American Academy of Dermatology Association and the Society for Investigative Dermatology. J Am Acad Dermatol 55:490–500

Brown EL, Rivas JM, Ullrich SE, Young CR, Norris SJ, Kripke ML (1995) Modulation of immunity to Borrelia burgdorferi by ultraviolet irradiation: differential effect on Th1 and Th2 immune responses. Eur J Immunol 25:3017–3022

Bruhs A, Haarmann-Stemmann T, Frauenstein K, Krutmann J, Schwarz T, Schwarz A (2015) Activation of the arylhydrocarbon receptor causes immunosuppression primarily by modulating dendritic cells. J Invest Dermatol 135:435–444

Burton JM, Kimball S, Vieth R, Bar-Or A, Dosch HM, Cheung R et al (2010) A phase I/II dose-escalation trial of vitamin D3 and calcium in multiple sclerosis. Neurology 74: 1852–1859

Byrne SN, Spinks N, Halliday GM (2002) Ultraviolet a irradiation of C57BL/6 mice suppresses systemic contact hypersensitivity or enhances secondary immunity depending on dose. J Invest Dermatol 119:858–864

Byrne SN, Ahmed J, Halliday GM (2005) Ultraviolet B but not A radiation activates suppressor B cells in draining lymph nodes. Photochem Photobiol 81:1366–1370

Byrne SN, Limón-Flores AY, Ullrich SE (2008) Mast cell migration from the skin to the draining lymph nodes upon ultraviolet irradiation represents a key step in the induction of immune suppression. J Immunol 180:4648–4655

Byrne SN, Beaugie C, O'Sullivan C, Leighton S, Halliday GM (2011) The immune-modulating cytokine and endogenous Alarmin interleukin-33 is upregulated in skin exposed to inflammatory UVB radiation. Am J Pathol 179:211–222

Cestari TF, Kripke ML, Baptista PL, Bakos L, Bucana CD (1995) Ultraviolet radiation decreases the granulomatous response to lepromin in humans. J Invest Dermatol 105:8–13

Chacón-Salinas R, Limón-Flores AY, Chávez-Blanco AD, Gonzalez-Estrada A, Ullrich SE (2011) Mast cell-derived IL-10 suppresses germinal center formation by affecting T follicular helper cell function. J Immunol 186:25–31

Chacón-Salinas R, Chen L, Chávez-Blanco AD, Limón-Flores AY, Ma Y, Ullrich SE (2014) An essential role for platelet-activating factor in activating mast cell migration following ultraviolet irradiation. J Leukoc Biol 95:139–148

Chen CH, Xirasagar S, Lin HC (2006) Seasonality in adult asthma admissions, air pollutant levels, and climate: a population-based study. J Asthma 43:287–292

Correale J, Farez MF (2013) Modulation of multiple sclerosis by sunlight exposure: role of cis-urocanic acid. J Neuroimmunol 261:134–140

Damian DL, Patterson CR, Stapelberg M, Park J, Barnetson RS, Halliday GM (2007) UV radiation-induced immunosuppression is greater in men and prevented by topical nicotinamide. J Invest Dermatol 128:447–454

De Fabo EC, Noonan FP (1983) Mechanism of immune suppression by ultraviolet irradiation in vivo. I. Evidence for the existence of a unique photoreceptor in skin and its role in photoimmunology. J Exp Med 158:84–98

Denkins YM, Kripke ML (1993) Effect of UV irradiation on lethal infection of mice with Candida albicans. Photochem Photobiol 57:266–271

Denkins Y, Fidler IJ, Kripke ML (1989) Exposure of mice to UV-B radiation suppresses delayed hypersensitivity to Candida albicans. Photochem Photobiol 49:615–619

Devary Y, Rosette C, DiDonato JA, Karin M (1993) NF-κB activation by ultraviolet light is not dependent on a nuclear signal. Science 261:1442–1445

Dixon KM, Norman AW, Sequeira VB, Mohan R, Rybchyn MS, Reeve VE et al (2011) 1{alpha},25(OH)2-vitamin D and a non-genomic vitamin D analog inhibit ultraviolet radiation-induced skin carcinogenesis. Cancer Prev Res (Phila) 4:1485–1494

Eide MJ, Johnson DA, Jacobsen GR, Krajenta RJ, Rao DS, Lim HW et al (2011) Vitamin D and nonmelanoma skin cancer in a health maintenance organization cohort. Arch Dermatol 147:1379–1384

Esser C, Bargen I, Weighardt H, Haarmann-Stemmann T, Krutmann J (2013) Functions of the aryl hydrocarbon receptor in the skin. Semin Immunopathol 35:677–691

Fritsche E, Schafer C, Calles C, Bernsmann T, Bernshausen T, Wurm M et al (2007) Lightening up the UV response by identification of the arylhydrocarbon receptor as a cytoplasmatic target for ultraviolet B radiation. Proc Natl Acad Sci U S A 104:8851–8856

Fukunaga A, Khaskhely NM, Sreevidya CS, Byrne SN, Ullrich SE (2008) Dermal dendritic cells, and not Langerhans cells, play an essential role in inducing an immune response. J Immunol 180:3057–3064

Fukunaga A, Khaskhely NM, Ma Y, Sreevidya CS, Taguchi K, Nishigori C et al (2010) Langerhans cells serve as immunoregulatory cells by activating NKT cells. J Immunol 185:4633–4640

Galli SJ, Kalesnikoff J, Grimbaldeston MA, Piliponsky AM, Williams CM, Tsai M (2005) Mast cells as "tunable" effector and immunoregulatory cells: recent advances. Annu Rev Immunol 23:749–786

Giannini SH (1992) Effects of ultraviolet B irradiation on cutaneous leishmaniasis. Parasitol Today 8:44–48

Gibbs NK, Norval M (2013) Photoimmunosuppression: a brief overview. Photodermatol Photoimmunol Photomed 29:57–64

Gibbs NK, Tye J, Norval M (2008) Recent advances in urocanic acid photochemistry, photobiology and photoimmunology. Photochem Photobiol Sci 7:655–667

Ginhoux F, Collin MP, Bogunovic M, Abel M, Leboeuf M, Helft J et al (2007) Blood-derived dermal langerin+dendritic cells survey the skin in the steady state. J Exp Med 204:3133–3146

Gordon-Thomson C, Gupta R, Tongkao-on W, Ryan A, Halliday GM, Mason RS (2012) 1alpha, 25 dihydroxyvitamin D3 enhances cellular defences against UV-induced oxidative and other forms of DNA damage in skin. Photochem Photobiol Sci 11:1837–1847

Grimbaldeston MA, Skov L, Baadsgaard O, Skov BG, Marshman G, Finlay-Jones JJ et al (2000) Communications: high dermal mast cell prevalence is a predisposing factor for basal cell carcinoma in humans. J Invest Dermatol 115:317–320

Grimbaldeston MA, Pearce AL, Robertson BO, Coventry BJ, Marshman G, Finlay-Jones JJ et al (2004) Association between melanoma and dermal mast cell prevalence in sun-unexposed skin. Br J Dermatol 150:895–903

Halliday GM (2005) Inflammation, gene mutation and photoimmunosuppression in response to UVR-induced oxidative damage contributes to photocarcinogenesis. Mutat Res 571:107–120

Halliday GM, Byrne SN, Damian DL (2011) Ultraviolet a radiation: its role in immunosuppression and carcinogenesis. Semin Cutan Med Surg 30:214–221

Hammerberg C, Katiyar SK, Carroll MC, Cooper KD (1998) Activated complement component 3 (C3) is required for ultraviolet induction of immunosuppression and antigenic tolerance. J Exp Med 187:1133–1138

Hart PH, Grimbaldeston MA, Swift GJ, Jaksic A, Noonan FP, Finlay-Jones JJ (1998) Dermal mast cells determine susceptibility to Ultraviolet B-induced systemic suppression of contact hypersensitivity responses in mice. J Exp Med 187:2045–2053

Hart PH, Gorman S, Finlay-Jones JJ (2011) Modulation of the immune system by UV radiation: more than just the effects of vitamin D? Nat Rev Immunol 11:584–596

Holan V, Kuffová L, Zajícová A, Krulová M, Filipec M, Holler P et al (1998) Urocanic acid enhances IL-10 production in activated CD4+ T cells. J Immunol 161:3237–3241

Housman TS, Feldman SR, Williford PM, Fleischer AB Jr, Goldman ND, Acostamadiedo JM et al (2003) Skin cancer is among the most costly of all cancers to treat for the medicare population. J Am Acad Dermatol 48:425–429

Hughes AM, Lucas RM, Ponsonby AL, Chapman C, Coulthard A, Dear K et al (2011) The role of latitude, ultraviolet radiation exposure and vitamin D in childhood asthma and hayfever: an Australian multicenter study. Pediatr Allergy Immunol 22:327–333

Inzinger M, Heschl B, Weger W, Hofer A, Legat FJ, Gruber-Wackernagel A et al (2011) Efficacy of psoralen plus ultraviolet A therapy vs. biologics in moderate to severe chronic plaque psoriasis: retrospective data analysis of a patient registry. Br J Dermatol 165:640–645

Jeevan A, Evans R, Brown EL, Kripke ML (1992) Effect of local ultraviolet irradiation on infections of mice with Candida albicans, Mycobacterium bovis BCG, and Schistosoma mansoni. J Invest Dermatol 99:59–64

Jeevan A, Ullrich SE, De Gracia M, Shah R, Sun Y (1996) Mechanism of UVB-induced suppression of the immune response to Mycobacterium bovis bacillus Calmette-Guerin: role of cytokines on macrophage function. Photochem Photobiol 64:259–266

Kalesnikoff J, Galli SJ (2008) New developments in mast cell biology. Nat Immunol 9:1215–1223

Kaneko K, Travers JB, Matsui MS, Young AR, Norval M, Walker SL (2009) cis-Urocanic acid stimulates primary human keratinocytes independently of serotonin or platelet-activating factor receptors. J Invest Dermatol 129:2567–2573

Kaplan DH, Jenison MC, Saeland S, Shlomchik WD, Shlomchik MJ (2005) Epidermal langerhans cell-deficient mice develop enhanced contact hypersensitivity. Immunity 23:611–620

Katiyar SK (2007) UV-induced immune suppression and photocarcinogenesis: chemoprevention by dietary botanical agents. Cancer Lett 255:1–11

Kim TH, Ullrich SE, Ananthaswamy HN, Zimmerman S, Kripke ML (1998) Suppression of delayed and contact hypersensitivity responses in mice have different UV dose responses. Photochem Photobiol 68:738–744

Kitazawa R, Mori K, Yamaguchi A, Kondo T, Kitazawa S (2008) Modulation of mouse RANKL gene expression by Runx2 and vitamin D3. J Cell Biochem 105:1289–1297

Kripke ML (1974) Antigenicity of murine skin tumors induced by ultraviolet light. J Natl Cancer Inst 53:1333–1336

Kripke ML, Cox PA, Alas LG, Yarosh DB (1992) Pyrimidine dimers in DNA initiate systemic immunosuppression in UV-irradiated mice. Proc Natl Acad Sci U S A 89:7516–7520

Krstic G (2011) Asthma prevalence associated with geographical latitude and regional insolation in the United States of America and Australia. PLoS One 6, e18492

Kuchel JM, Barnetson RS, Halliday GM (2005) Cyclobutane pyrimidine dimer formation is a molecular trigger for solar-simulated ultraviolet radiation-induced suppression of memory immunity in humans. Photochem Photobiol Sci 4:577–582

Kurtitzky LA, Finlay-Jones JJ, Hart PH (2007) The controversial role of vitamin D in the skin: immunosuppression vs. photoprotection. Clin Exp Dermatol 33:167–170

Lapolla W, Yentzer BA, Bagel J, Halvorson CR, Feldman SR (2011) A review of phototherapy protocols for psoriasis treatment. J Am Acad Dermatol 64:936–949

Limón-Flores AY, Chacón-Salinas R, Ramos G, Ullrich SE (2009) Mast cells mediate the immune suppression induced by dermal exposure to JP-8 jet fuel. Toxicol Sci 112:144–152

Loh TP, Lai FY, Tan ES, Thoon KC, Tee NW, Cutter J et al (2011) Correlations between clinical illness, respiratory virus infections and climate factors in a tropical paediatric population. Epidemiol Infect 139:1884–1894

Loser K, Mehling A, Loeser S, Apelt J, Kuhn A, Grabbe S et al (2007) Epidermal RANKL controls regulatory T-cell numbers via activation of dendritic cells. Nat Med 12:1372–1379

MacLaughlin JA, Anderson RR, Holick MF (1982) Spectral character of sunlight modulates photosynthesis of previtamin D3 and its photoisomers in human skin. Science 216: 1001–1003

Marathe GK, Johnson C, Billings SD, Southall MD, Pei Y, Spandau D et al (2005) Ultraviolet B radiation generates platelet-activating factor-like phospholipids underlying cutaneous damage. J Biol Chem 280:35448–35457

Matsumura Y, Byrne SN, Nghiem DX, Miyahara Y, Ullrich SE (2006) A role for inflammatory mediators in the induction of immunoregulatory B cells. J Immunol 177:4810–4817

Matthews YJ, Halliday GM, Phan TA, Damian DL (2010) Wavelength dependency for UVA-induced suppression of recall immunity in humans. J Dermatol Sci 59:192–197

McDougal JN, Pollard DL, Weisman W, Garrett CM, Miller TE (2000) Assessment of skin absorption and penetration of JP-8 jet fuel and its components. Toxicol Sci 55:247–255

McGlade JP, Gorman S, Zosky GR, Larcombe AN, Sly PD, Finlay-Jones JJ et al (2007a) Suppression of the asthmatic phenotype by ultraviolet B-induced, antigen-specific regulatory cells. Clin Exp Allergy 37:1267–1276

McGlade JP, Gorman S, Lenzo JC, Tan JW, Watanabe T, Finlay-Jones JJ et al (2007b) Effect of both ultraviolet B irradiation and histamine receptor function on allergic responses to an inhaled antigen. J Immunol 178:2794–2802

McGlade JP, Strickland DH, Lambert MJ, Gorman S, Thomas JA, Judge MA et al (2010) UV inhibits allergic airways disease in mice by reducing effector CD4 T cells. Clin Exp Allergy 40:772–785

Moodycliffe AM, Nghiem D, Clydesdale G, Ullrich SE (2000) Immune suppression and skin cancer development: regulation by NKT cells. Nat Immunol 1:521–525

Moyal DD, Fourtanier AM (2001) Broad-spectrum sunscreens provide better protection from the suppression of the elicitation phase of delayed-type hypersensitivity response in humans. J Invest Dermatol 117:1186–1192

Müller G, Salonga J, Germann T, Schuler G, Knopp J, Enk AH (1995) IL-12 as mediator and adjuvant for the induction of contact sensitivity in vivo. J Immunol 155:4661–4668

Munger KL, Zhang SM, O'Reilly E, Hernan MA, Olek MJ, Willett WC et al (2004) Vitamin D intake and incidence of multiple sclerosis. Neurology 62:60–65

Navid F, Bruhs A, Schuller W, Fritsche E, Krutmann J, Schwarz T et al (2013) The aryl hydrocarbon receptor is involved in UVR-induced immunosuppression. J Invest Dermatol 133:2763–2770

Ng RL, Bisley JL, Gorman S, Norval M, Hart PH (2010) Ultraviolet irradiation of mice reduces the competency of bone marrow-derived CD11c+cells via an indomethacin-inhibitable pathway. J Immunol 185:7207–7215

Ng RL, Scott NM, Strickland DH, Gorman S, Grimbaldeston MA, Norval M et al (2013a) Altered immunity and dendritic cell activity in the periphery of mice after long-term engraftment with bone marrow from ultraviolet-irradiated mice. J Immunol 190:5471–5484

Ng RL, Scott NM, Bisley JL, Lambert MJ, Gorman S, Norval M et al (2013b) Characterisation of regulatory dendritic cells differentiated from the bone marrow of UV-irradiated mice. Immunology 140:399–412

Nghiem DX, Kazimi N, Clydesdale G, Ananthaswamy HN, Kripke ML, Ullrich SE (2001) Ultraviolet a radiation suppresses an established immune response: implications for sunscreen design. J Invest Dermatol 117:1193–1199

Nishigori C, Yarosh DB, Ullrich SE, Vink AA, Bucana CD, Roza L et al (1996) Evidence that DNA damage triggers interleukin 10 cytokine production in UV-irradiated murine keratinocytes. Proc Natl Acad Sci U S A 93:10354–10359

Noonan FP, De Fabo EC, Morrison H (1988) Cis-urocanic acid, a product formed by UVB irradiation of the skin, initiates an antigen presentation defect in splenic cells in vivo. J Invest Dermatol 90:92–99

Nordlind K, Azmitia EC, Slominski A (2008) The skin as a mirror of the soul: exploring the possible roles of serotonin. Exp Dermatol 17:301–311

Norval M, Halliday GM (2011) The consequences of UV-induced immunosuppression for human health. Photochem Photobiol 87:965–977

Norval M, Woods GM (2011) UV-induced immunosuppression and the efficacy of vaccination. Photochem Photobiol Sci 10:1267–1274

O'Connor A, Nishigori C, Yarosh D, Alas L, Kibitel J, Burley L et al (1996) DNA double strand breaks in epidermal cells cause immune suppression in vivo and cytokine production in vitro. J Immunol 157:271–278

Penn I (2000) Post-transplant malignancy: the role of immunosuppression. Drug Saf 23:101–113

Petermann F, Korn T (2011) Cytokines and effector T cell subsets causing autoimmune CNS disease. FEBS Lett 585:3747–3757

Poon TS, Barnetson RS, Halliday GM (2005) Sunlight-induced immunosuppression in humans is initially because of UVB, then UVA, followed by interactive effects. J Invest Dermatol 125:840–846

Prasad R, Katiyar SK (2013) Prostaglandin E2 promotes UV radiation-induced immune suppression through DNA hypermethylation. Neoplasia 15:795–804

Ramos G, Nghiem DX, Walterscheid JP, Ullrich SE (2002) Dermal application of jet fuel suppresses secondary immune reactions. Toxicol Appl Pharmacol 180:136–144

Ramos G, Kazimi N, Nghiem DX, Walterscheid JP, Ullrich SE (2004) Platelet activating factor receptor binding plays a critical role in jet fuel-induced immune suppression. Toxicol Appl Pharmacol 195:331–338

Ramos G, Limón-Flores AY, Ullrich SE (2009) JP-8 induces immune suppression via a reactive oxygen species NF-kappa beta-dependent mechanism. Toxicol Sci 108:100–109

Randolph DA, Stephens R, Carruthers CJ, Chaplin DD (1999a) Cooperation between Th1 and Th2 cells in a murine model of eosinophilic airway inflammation. J Clin Invest 104:1021–1029

Randolph DA, Carruthers CJ, Szabo SJ, Murphy KM, Chaplin DD (1999b) Modulation of airway inflammation by passive transfer of allergen-specific Th1 and Th2 cells in a mouse model of asthma. J Immunol 162:2375–2383

Rannug A, Rannug U, Rosenkranz HS, Winqvist L, Westerholm R, Agurell E et al (1987) Certain photooxidized derivatives of tryptophan bind with very high affinity to the Ah receptor and are likely to be endogenous signal substances. J Biol Chem 262:15422–15427

Ritter U, Meissner A, Scheidig C, Korner H (2004) CD8 alpha- and Langerin-negative dendritic cells, but not Langerhans cells, act as principal antigen-presenting cells in leishmaniasis. Eur J Immunol 34:1542–1550

Rivas JM, Ullrich SE (1992) Systemic suppression of delayed-type hypersensitivity by supernatants from UV-irradiated keratinocytes. An essential role for keratinocyte-derived IL-10. J Immunol 149:3865–3871

Rockel N, Esser C, Grether-Beck S, Warskulat U, Flogel U, Schwarz A et al (2007) The osmolyte taurine protects against ultraviolet B radiation-induced immunosuppression. J Immunol 179:3604–3612

Ross JA, Howie SEM, Norval M, Maingay J, Simpson TJ (1986) Ultraviolet-irradiated urocanic acid suppresses delayed type hypersensitivity to herpes simplex virus in mice. J Invest Dermatol 87:630–633

Sahu RP, Turner MJ, DaSilva SC, Rashid BM, Ocana JA, Perkins SM et al (2012) The environmental stressor ultraviolet B radiation inhibits murine antitumor immunity through its ability to generate platelet-activating factor agonists. Carcinogenesis 33:1360–1367

Salmaggi A, Dufour A, Eoli M, Corsini E, La Mantia L, Massa G et al (1996) Low serum interleukin-10 levels in multiple sclerosis: further evidence for decreased systemic immunosuppression? J Neurol 243:13–17

Sarchio SN, Kok LF, O'Sullivan C, Halliday GM, Byrne SN (2012) Dermal mast cells affect the development of sunlight-induced skin tumours. Exp Dermatol 21:241–248

Sarchio SN, Scolyer RA, Beaugie C, McDounald D, Marsh-Wakefield F, Halliday GM, Byrne SN (2014) Pharmacologically antagonizing the CXCR4-CXCL12 chemokine pathway with AMD3100 inhibits sunlight-induced skin cancer. J Invest Dermatol 134:1091–1100

Schmitt DA, Ullrich SE (2000) Exposure to ultraviolet radiation causes dendritic cells/macrophages to secrete immune suppressive IL-12p40 homodimers. J Immunol 165:3162–3167

Schmitt DA, Owen-Schaub L, Ullrich SE (1995) Effect of IL-12 on immune suppression and suppressor cell induction by ultraviolet radiation. J Immunol 154:5114–5120

Schmitt DA, Walterscheid JP, Ullrich SE (2000) Reversal of ultraviolet radiation-induced immune suppression by recombinant interleukin-12: suppression of cytokine production. Immunology 101:90–96

Schwarz T (2008) 25 years of UV-induced immunosuppression mediated by T cells-from disregarded T suppressor cells to highly respected regulatory T cells. Photochem Photobiol 84:10–18

Schwarz A, Grabbe S, Aragane Y, Sandkuhl K, Riemann H, Luger TA et al (1996) Interleukin-12 prevents ultraviolet B-induced local immunosuppression and overcomes UVB-induced tolerance. J Invest Dermatol 106:1187–1191

Schwarz A, Noordegraaf M, Maeda A, Torii K, Clausen BE, Schwarz T (2010) Langerhans cells are required for UVR-induced immunosuppression. J Invest Dermatol 130:1419–1427

Schwarz A, Navid F, Sparwasser T, Clausen BE, Schwarz T (2012) 1,25-dihydroxyvitamin D exerts similar immunosuppressive effects as UVR but is dispensable for local UVR-induced immunosuppression. J Invest Dermatol 132:2762–2769

Scott NM, Ng RL, Strickland DH, Bisley JL, Bazely SA, Gorman S et al (2012) Toward homeostasis: regulatory dendritic cells from the bone marrow of mice with inflammation of the airways and peritoneal cavity. Am J Pathol 181:535–547

Scott NM, Ng RL, Gorman S, Norval M, Waithman J, Hart PH (2014) Prostaglandin E2 imprints a long-lasting effect on dendritic cell progenitors in the bone marrow. J Leukoc Biol 95:225–232

Sharief MK, Hentges R (1991) Association between tumor necrosis factor-alpha and disease progression in patients with multiple sclerosis. N Engl J Med 325:467–472

Shklovskaya E, O'Sullivan BJ, Ng LG, Roediger B, Thomas R, Weninger W et al (2011) Langerhans cells are precommitted to immune tolerance induction. Proc Natl Acad Sci U S A 108:18049–18054

Siegel R, Naishadham D, Jemal A (2012) Cancer statistics, 2012. CA Cancer J Clin 62:10–29

Simon MM, Aragane Y, Schwarz A, Luger TA, Schwarz T (1994) UVB light induces a nuclear factor κB (NFκB) activity independently from chromosomal DNA damage in cell-free cytosolic extracts. J Invest Dermatol 102:422–427

Simpson S Jr, Blizzard L, Otahal P, Van der Mei I, Taylor B (2011) Latitude is significantly associated with the prevalence of multiple sclerosis: a meta-analysis. J Neurol Neurosurg Psychiatry 82:1132–1141

Singh TP, Schon MP, Wallbrecht K, Michaelis K, Rinner B, Mayer G et al (2010) 8-methoxypsoralen plus ultraviolet A therapy acts via inhibition of the IL-23/Th17 axis and induction of Foxp3+ regulatory T cells involving CTLA4 signaling in a psoriasis-like skin disorder. J Immunol 184:7257–7267

Singh TP, Huettner B, Koefeler H, Mayer G, Bambach I, Wallbrecht K et al (2011) Platelet-activating factor blockade inhibits the T-helper type 17 cell pathway and suppresses psoriasis-like skin disease in K5.hTGF-beta1 transgenic mice. Am J Pathol 178:699–708

Sreevidya CS, Khaskhely NM, Fukunaga A, Khaskina P, Ullrich SE (2008) Inhibition of photocarcinogenesis by platelet-activating factor or serotonin receptor antagonists. Cancer Res 68:3978–3984

Sreevidya CS, Fukunaga A, Khaskhely NM, Masaki T, Ono R, Nishigori C et al (2010) Agents that reverse UV-induced immune suppression and photocarcinogenesis affect DNA repair. J Invest Dermatol 130:1428–1437

Stapelberg MP, Williams RB, Byrne SN, Halliday GM (2009) The alternative complement pathway seems to be a UVA sensor that leads to systemic immunosuppression. J Invest Dermatol 129:2694–2701

Staples JA, Ponsonby AL, Lim LL, McMichael AJ (2003) Ecologic analysis of some immune-related disorders, including type 1 diabetes, in Australia: latitude, regional ultraviolet radiation, and disease prevalence. Environ Health Perspect 111:518–523

Stege H, Roza L, Vink AA, Grewe M, Ruzicka T, Grether-Beck S et al (2000) Enzyme plus light therapy to repair DNA damage in ultraviolet-B-irradiated human skin. Proc Natl Acad Sci U S A 97:1790–1795

Surjana D, Halliday GM, Damian DL (2013) Nicotinamide enhances repair of ultraviolet radiation-induced DNA damage in human keratinocytes and ex vivo skin. Carcinogenesis 34:1144–1149

Toews GB, Bergstresser PR, Streilein JW (1980) Epidermal langerhans cell density determines whether contact hypersensitivity or unresponsiveness follows skin painting with DNFB. J Immunol 124:445–449

Travers JB, Berry D, Yao Y, Yi Q, Konger RL (2010) Ultraviolet B radiation of human skin generates platelet-activating factor receptor agonists. Photochem Photobiol 86:949–954

Ullrich SE (1987) The effect of ultraviolet radiation-induced suppressor cells on T cell activity. Immunology 60:353–360

Ullrich SE (1996) Does exposure to UV radiation induce a shift to a Th-2-like immune reaction? Photochem Photobiol 64:254–258

Ullrich SE (1999) Dermal application of JP-8 jet fuel induces immune suppression. Toxicol Sci 52:61–67

Ullrich SE (2005) Mechanisms underlying UV-induced immune suppression. Mutat Res 571:185–205

Ullrich SE, Lyons HJ (2000) Mechanisms involved in the immunotoxicity induced by dermal application of JP-8 jet fuel. Toxicol Sci 58:290–298

Ullrich SE, Nghiem DX, Khaskina P (2007) Suppression of an established immune response by UVA-a critical role for mast cells. Photochem Photobiol 83:1095–1100

Viglietta V, Baecher-Allan C, Weiner HL, Hafler DA (2004) Loss of functional suppression by CD4+ CD25+ regulatory T cells in patients with multiple sclerosis. J Exp Med 199:971–979

Vink AA, Strickland FM, Bucana C, Cox PA, Roza L, Yarosh DB et al (1996a) Localization of DNA damage and its role in altered antigen-presenting cell function in ultraviolet-irradiated mice. J Exp Med 183:1491–1500

Vink AA, Yarosh DB, Kripke ML (1996b) Chromophore for UV-induced immunosuppression: DNA. Photochem Photobiol 63:383–386

Vink AA, Moodycliffe AM, Shreedhar V, Ullrich SE, Roza L, Yarosh DB et al (1997) The inhibition of antigen-presenting activity of dendritic cells resulting from UV irradiation of murine skin is restored by in vitro photorepair of cyclobutane pyrimidine dimers. Proc Natl Acad Sci U S A 94:5255–5260

Vink AA, Shreedhar V, Roza L, Krutmann J, Kripke ML (1998) Cellular target of UVB-induced DNA damage resulting in local suppression of contact hypersensitivity. J Photochem Photobiol B 44:107–111

Walterscheid JP, Ullrich SE, Nghiem DX (2002) Platelet-activating factor, a molecular sensor for cellular damage, activates systemic immune suppression. J Exp Med 195:171–179

Walterscheid JP, Nghiem DX, Kazimi N, Nutt LK, McConkey DJ, Norval M et al (2006) *Cis-urocanic* acid, a sunlight-induced immunosuppressive factor, activates immune suppression via the 5-HT_{2A} receptor. Proc Natl Acad Sci U S A 103:17420–17425

Ward MD, Sailstad DM, Andrews DL, Boykin EH, Selgrade MK (2004) Ultraviolet radiation downregulates allergy in BALB/c mice. J Toxicol Environ Health A 67:73–85

Weichenthal M, Schwarz T (2005) Phototherapy: how does UV work? Photodermatol Photoimmunol Photomed 21:260–266

Wille JJ, Kydonieus AF, Murphy GF (1999) cis-urocanic acid induces mast cell degranulation and release of preformed TNF-alpha: a possible mechanism linking UVB and cis-urocanic acid to immunosuppression of contact hypersensitivity. Skin Pharmacol Appl Skin Physiol 12:18–27

Witten ML, Zeiger E, Richie GD (eds) (2011) Jet fuel toxicology. CRC Press, Boca Raton

Wolf P, Maier H, Mullegger RR, Chadwick CA, Hofmann-Wellenhof R, Soyer HP et al (2000) Topical treatment with liposomes containing T4 endonuclease V protects human skin in vivo from ultraviolet-induced upregulation of interleukin-10 and tumor necrosis factor-alpha. J Invest Dermatol 114:149–156

Wolf P, Nghiem DX, Walterscheid JP, Byrne S, Matsumura Y, Matsumura Y et al (2006) Platelet-activating factor is crucial in psoralen and ultraviolet A-induced immune suppression, inflammation, and apoptosis. Am J Pathol 169:795–805

Yarosh D, Klein J, O'Connor A, Hawk J, Rafal E, Wolf P (2001) Effect of topically applied T4 endonuclease V in liposomes on skin cancer in xeroderma pigmentosum: a randomised study. Xeroderma Pigmentosum Study Group. Lancet 357:926–929

Yoshida Y, Kang K, Berger M, Chen G, Gilliam AC, Moser A et al (1998) Monocyte induction of IL-10 and down-regulation of IL-12 by iC3b deposited in ultraviolet-exposed human skin. J Immunol 161:5873–5879

Yoshikawa T, Rae V, Bruins-Slot W, Van den Berg JW, Taylor JR, Streilein JW (1990) Susceptibility to effects of UVB radiation on induction of contact hypersensitivity as a risk factor for skin cancer in humans. J Invest Dermatol 95:530–536

Yoshiki R, Kabashima K, Sakabe J, Sugita K, Bito T, Nakamura M et al (2010) The mandatory role of IL-10-producing and OX40 ligand-expressing mature Langerhans cells in local UVB-induced immunosuppression. J Immunol 184:5670–5677

Zhang Q, Yao Y, Konger RL, Sinn AL, Cai S, Pollok KE et al (2008) UVB radiation-mediated inhibition of contact hypersensitivity reactions is dependent on the platelet-activating factor system. J Invest Dermatol 128:1780–1787

Vaccination Efficacy and Environmental Pollution

8

Katrine Kielsen, Zaiba Shamim, Lars P. Ryder, Philippe Grandjean, and Carsten Heilmann

Contents

K. Kielsen, MD (✉) • Z. Shamim, MSc, PhD
Institute for Inflammation Research, Department of Paediatric and Adolescent Medicine, Copenhagen University Hospital Rigshospitalet, Copenhagen, Denmark
e-mail: katrine.kielsen@regionh.dk

L.P. Ryder, MSc
Department of Clinical Immunology, Copenhagen University Hospital Rigshospitalet, Copenhagen, Denmark

P. Grandjean, MD, DMSc
Department of Environmental Medicine, University of Southern Denmark, Odense, Denmark

Department of Environmental Health, Harvard School of Public Health, Boston, MA, USA

C. Heilmann
Department of Paediatric and Adolescent Medicine, Copenhagen University Hospital Rigshospitalet, Copenhagen, Denmark
e-mail: carsten.heilmann@regionh.dk

C. Esser (ed.), *Environmental Influences on the Immune System*,
DOI 10.1007/978-3-7091-1890-0_8

8.1 Vaccinations

The immune system has evolved to combat the environmental influences to which humans are constantly exposed, such as bacteria, viruses, fungi, and parasites. Furthermore, immunization programs have *helped* the immune system to reach protection through vaccination against life-threatening diseases.

For generations, it has been known that there was no second occurrence after suffering from certain illnesses. The modern history of vaccination began in England in 1796, when Edward Jenner discovered that inoculation of children with cowpox virus prevented smallpox disease (André 2003; Hilleman 2000; Plotkin and Plotkin 2011). Subsequent development of vaccines against various bacteria and viruses led to substantial progress in prevention of infectious diseases, and today vaccinations are widely used and considered one of the greatest successes of medical science. Vaccination has led to the total extinction of smallpox and the nearly elimination of polio, which now only exists in a few remote areas in Asia.

In developed countries, the structured programs of vaccination of small children against severe childhood diseases such as diphtheria, tetanus, polio, pertussis, measles, rubella, and mumps have significantly reduced the mortality of infants as well as the infection-associated morbidity caused by these diseases. Furthermore, vaccinations against viruses such as hepatitis, yellow fever, tick-borne encephalitis, etc. have reduced the risk of traveling in endemic areas (André 2003).

In recent years, vaccines against microorganisms causing cancer, such as the human papillomavirus, have been evolved, and a lot of effort is directed against the development of vaccines against severe infections such as HIV and malaria (Plotkin and Plotkin 2011).

The common goal of vaccinations is to reduce the high risk of mortality or chronic disabilities caused by the natural diseases, which overcomes the side effects of vaccination. Furthermore, establishment of herd immunity, with most individuals in a population vaccinated against diseases, reduces the risk of epidemic outbreaks and thereby also indirectly protects unvaccinated individuals such as small infants and individuals with immunodeficiencies from transmission of the disease.

8.1.1 Immunological Control of Infections

Humans are constantly in contact with microorganisms, but the vast majority is prevented from causing infections due to intact epithelial barriers on the body's surface. Microorganisms penetrating these barriers meet fixed mechanisms of the innate immune system, such as the complement system, phagocytosis from macrophages and neutrophils, and inflammatory cytokines, which will overcome most infections before they have caused symptoms. Medical practice is mostly concerned with diseases that results from the small proportion of infections that the innate immune system fails to terminate and for which the spread of the pathogen to secondary lymphoid tissues stimulates an adaptive immune response.

Upon the first encounter with a pathogen, a primary adaptive immune response is raised. This involves an extensive process in which pathogen-specific naïve T- and B-lymphocyte clones are selected, expanded, and differentiated into effector cells within the lymphoid tissue. Effector CD4+ Th1 cells and cytotoxic CD8+ T cells travel from the lymphoid tissue to the infected site and activate macrophages to destroy extracellular pathogens or kill infected human cells. Within the secondary lymphoid tissue, T cells activate pathogen-specific B cells and drive isotype switching and somatic hypermutation to produce high-affinity plasma cells, from whence secreted antibodies then travel to the site of infection.

In addition to successful clearance of the specific pathogen, an adaptive immune response also establishes a long-term protective immunity. The next encounter with a specific pathogen will provoke a faster and stronger response against the pathogen. This is produced by circulating antibodies and clones of long-lived memory B and T cells formed by the primary response and reactivated when confronted with the antigen. As a consequence, the infection is cleared quickly by the secondary response, with few or no symptoms of disease and little mortality from even life-threatening pathogens.

8.1.2 Immunological Stimulation by Immunizations and Vaccination Efficacy

Vaccinations are given to prime the human body against a specific pathogen, before the pathogen itself is encountered. Successful vaccination involves a range of key immunological processes involved in antibody formation. Establishment of long-lasting protective immunology depends on isotype switch and affinity maturation of B cells in the germinal center and on production of memory T and B lymphocytes, which are maintained throughout life and will be recruited upon renewed exposure to the antigen. Antibody titers can be used to measure any given vaccination's efficacy (Fig. 8.1). An antibody concentration >0.1 IU/mL is considered essential to achieve full long-term protection against diphtheria and tetanus in accordance with the public health purpose of routine vaccinations. However, the immunological response to routine prophylactic vaccinations varies substantially, and the reasons for this wide variation are poorly understood apart from well-known immune deficiencies.

8.1.3 Vaccine Subtypes

Vaccines can be divided into live (attenuated) or inactivated vaccines. Attenuated vaccines cause a nonpathogenic infection (e.g., measles, rubella, mumps, and tuberculosis), mimicking a natural infection through replication of the microorganism within the host and presentation of antigens on MHC class I and II to the adaptive immune system, causing diverse and long-lasting immunity.

Inactivated vaccines consist of either dead microorganisms (killed bacteria or inactivated viruses, e.g., polio, cholera, and rabies) or components of microorganisms.

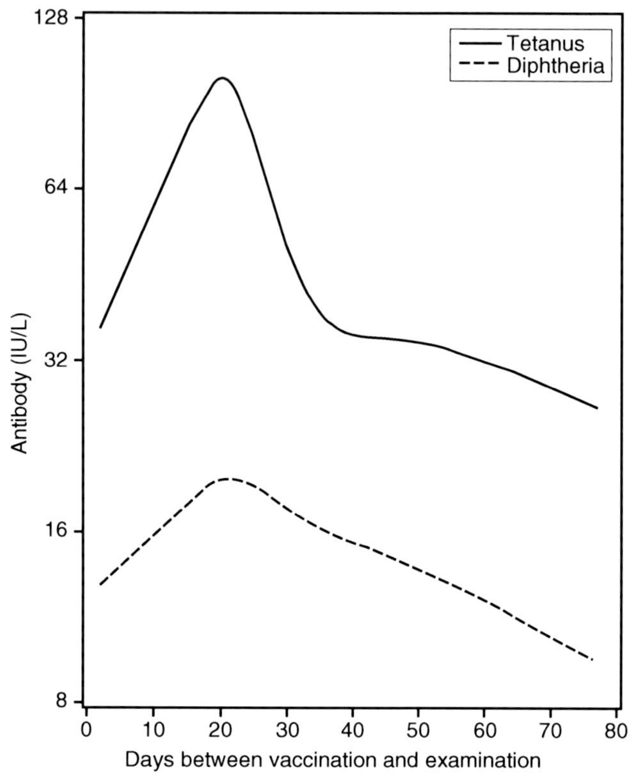

Fig. 8.1 Antibody response against tetanus and diphtheria. Antibody response to vaccinations with tetanus and diphtheria. The antibody titers rise during the first weeks after vaccination and then gradually decline to a steady level

Component vaccines contain purified components of inactivated microorganisms (e.g., influenza), recombinant antigens (e.g., hepatitis B), purified exotoxins processed with formaldehyde to make them nonpoisonous (e.g., tetanus, diphtheria), or conjugated vaccines of carbohydrates linked to a toxoid (e.g., pneumococcus). Component vaccines activate immunity against the specific antigen without copying the challenge of a natural pathogen. Even though adjuvant is added to increase the immunogenicity by stimulation of an inflammatory response and transportation of the antigen to the regional lymph nodes, nonliving vaccines are less effective than living vaccines, and repeated inoculations and boosters are often required to achieve the needed level of protection against microorganisms (Hilleman 2000; Plotkin 2014).

8.1.4 Use of Vaccination Efficacy for Epidemiological Studies of Immune Function

Findings identifying dose–response relationships, biological plausibility, and mode of action from experimental models of immune dysfunction caused by environmental pollutants may be telling, but they need to be verified in humans to assess the public health risk. Thus, epidemiological studies of immune dysfunction in exposed individuals are needed, as randomized studies of immunotoxicity are ethically unjustifiable. As such, longitudinal studies are preferable as they follow subjects for a sufficient time period to assess the health outcome of environmental pollutants. International guidelines recommend that the evaluation of immune-related health outcomes should be based on validated direct biological measures of exposure and effect and to a smaller extent on self-reported qualities, such as obtained by questionnaires and diaries (van Loveren et al. 1999; WHO 2012). In contrast to the findings from patients with severe defects in the immune system such as inborn errors or HIV infection, common clinical immune test can hardly detect potential mild-to-moderate states of immunodeficiency caused by immunotoxic agents. However, small changes in the immune system can be detected at a population level under the appropriate conditions. Antibody responses to vaccinations with standardized doses of foreign antigens are recommended as the best tool for evaluation of immune suppression, since antibody responses are measurable indicators of major immune system functions. Therefore, data from vaccination responses reflect the overall efficacy of the immune system to protect individuals from infections with pathogens beyond the specific vaccinations and represent the strongest evidence of immunosuppression (van Loveren et al. 1999; WHO 2012).

Most relevant for assessment of immunotoxic effects are primary immune responses to antigens to which the immunized persons have not previously encountered. This includes standardized pediatric immunizations for assessment of developmental immunotoxicity and vaccines against infections without prior exposure in adults (such as hepatitis B in non-epidemic areas) (Dietert 2008). Furthermore, well-known vaccines with availability to standardized methods for antibody assessment and extensive experience on vaccine efficacy are most suitable (Dietert 2008; Luster et al. 2005; van Loveren et al. 1999; WHO 2012).

8.2 Persistent Organic Pollutants

Toxic chemicals like groups of polychlorinated biphenyls (PCBs) and perfluorinated alkylate substances (PFAS) are considered persistent organic pollutant as noted under the Stockholm Convention as they are highly persistent and bioaccumulative in the environment. Many are now phased out from production due to their toxic effects, but are still widespread and found in humans, animals, and the environment. Besides being linked to human hepatotoxicity, developmental toxicity, hormonal effects, and a carcinogenic potency, some compounds have been found to cause immune dysfunction.

Immunotoxicology is defined as the study of adverse effects on the immune system resulting from occupational, environmental, or therapeutic exposure to agents suspected of toxicity toward the immune system. Due to its systemic distribution, the immune system is a moving toxicological target for interactions with organic pollutants (Dietert 2008). As human studies regarding quantitative exposure data and dose–response relationships are very limited, to expand our knowledge base, large epidemiological studies addressing the importance of environmental pollutants for human health have been carried out.

Immunological toxicity includes immune dysfunction and activation causing asthma, allergies, and autoimmune diseases as well as immune suppression with increased incidence or severity of infections and cancer. The immunosuppressive capacity of PCBs and PFAS has been investigated using cohort studies of vaccination efficacy conducted in children vaccinated according to their national vaccination program. These studies have addressed both the prenatal and postnatal exposure to potential immunotoxicants to establish their effect on both developmental and mature immunity.

When investigating childhood immunotoxicity, it is important to notice that the human immune system is developing throughout the fetal stage as well as postnatal. In the early gestational stage, hematopoietic stem cell formation, tissue migration, and progenitor cell expansion take place. The bone marrow and thymus are colonized late in prenatal life, while the development of the thymus includes both the late gestational period as well as early postnatal life. At birth, the child is still dependent on transferred immunity from the mother, as critical maturation of child's capacity for immune responses and immunological memory is ongoing throughout the first year of life. Of special interest to vaccination studies, an infant can achieve adult levels of antibodies by approximately the end of the first year and can maintain a concentrated level of antibodies for several years (Dietert 2008; Holladay and Smialowicz 2000; Holsapple et al. 2004; Luster et al. 2005; WHO 2012).

Due to the timing of the development of immunity, the susceptibility of the immune system to environment pollutants may be greater during fetal and early postnatal development than later in life. As some persistent pollutants are transferred to the fetus during pregnancy, it is important to examine any long-lasting effects these chemicals may cause during this period of life as they could lead to irreversible dysfunctions. Furthermore, young children are most important for assessment of immunotoxicity in humans, as young children, apart from their increased dose sensitivity of immune-modulating drugs, have a short history of environmental exposures and lack the personal confounders found in adults (WHO 2012).

8.2.1 Polychlorinated Biphenyls

Polychlorinated biphenyls (PCBs) are organic compounds consisting of 209 different isomeric congeners with variable chlorination of a biphenyl group, which displays individual physicochemical properties depending on their particular chlorine substitution pattern (Safe 1994).

The production of PCBs, lasted until the late 1970s when it was banned in most Western countries, but many PCBs are resistant to degradation processes due to their inflammability, chemical stability, and lipophilic nature. They have been used as insulators in electrical equipment and fire retardants and persist widespread in global ecosystems. They are still found in air, water, and soil, as well as in fish, birds, animals, and humans, with a tendency to bioaccumulate in food chains and cause high tissue concentrations in top predator species (Atlas and Giam 1981; Bacon et al. 1992; Kannan et al. 1989; Safe 1994; Skaare et al. 2000).

Their long experimental half-lives and tendency to bioaccumulate in adipose tissues have raised concern of the possible hazards of PCBs to humans. One structural aspect considered to be very important in both biostability and toxicological properties of PCBs is the planarity of the congener. PCBs are divided into coplanar non-ortho PCBs and noncoplanar mono-, di-, tetra-, or hydra-ortho-substituted PCBs according to their chemical structure. The recent concern about immunotoxicity of PCBs is largely based on the dioxin-like coplanar congeners, which share a structural similarity with 2,3,7,8-tetrachlorodibenzo-*p*-dioxin (TCDD) and might produce similar health effects, including metabolic, reproductive, developmental, and immunotoxic responses (Van den Berg et al. 2006).

Most biological and toxic effects of the dioxin-like PCB congeners are mediated through the aryl hydrocarbon receptor (AhR), a cytosolic receptor protein that regulates the induction of cytochrome P450 enzymes (Safe 1994). The AhR is found in many cells of the immune system and shows high affinity for coplanar and mono-ortho-substituted PCB congeners as well as for TCDD (Safe 1994). In contrast, the non-dioxin-like ortho-substituted PCB congeners exhibit low affinity for the AhR and mediate toxicity via a less clearly described mechanism that involves altered signal transduction pathways and interruption of intracellular Ca^{2+} homeostasis (Fischer et al. 1998).

Dietary intake of especially fish and animal fats is assumed to be the main source of human exposure to PCB, and humans retain dozens of different PCB congeners in their adipose tissues, blood, and milk (Domingo and Bocio 2007; Liem et al. 2000; Schecter et al. 1994). Populations whose diet includes consumption of PCB-contaminated fish or seafood are still particularly affected, although concentrations of these substances in humans in general are declining (Dallaire et al. 2004; Fängström et al. 2005; Schecter et al. 2005).

Studies have shown that PCBs and dioxins cross the placenta and reach the fetus (Covaci et al. 2002). Additionally, children are exposed to PCBs by the mother through breastfeeding and from dietary intake when older (Brouwer et al. 1998).

8.2.1.1 Vaccination Efficacy

Due to their wide distribution and long half-lives of several years, chronic low-level PCB exposures remain a significant public health concern. Several studies have investigated the importance of PCB for humoral immune response to routine childhood immunizations in birth cohorts (summarized in Table 8.1). Vaccinations included attenuated virus vaccines of mumps, measles, and rubella, toxoid vaccines of tetanus and diphtheria, and the bacterial polysaccharide–protein conjugated vaccines of *Haemophilus influenzae* type b (Hib).

Weisglas-Kuperus et al. and Støvlevik et al. found negative effects of in utero PCB exposure on antibody levels against mumps, measles, and rubella at 3–3.5 years of age (Stølevik et al. 2013; Weisglas-Kuperus et al. 2000). In line with this, studies by Heilmann et al. in two cohorts showed reduced antibody responses against

Table 8.1 Studies of PCB exposure and childhood immunizations

Authors	Weisglas-Kuperus et al (2000)	Heilmann et al. (2006)	Heilmann et al. (2010)	Jusko et al. (2010)	Stølevik et al. (2013)
Geographical location	The Netherlands	The Faroe Islands	The Faroe Islands	Slovakia	Norway
Population size	193 children	Two cohorts of 124 and 116 children	587 children	384 children	111 children
Inclusion and exclusion criteria	Healthy first- or second-born infants at term	Consecutive spontaneous singleton births at term	Consecutive spontaneous singleton births at term	Healthy mothers above 18 years with 0–4 previous births, who had lived in the specific area for at least 5 years, and infants without severe birth defects	Singleton births at term with mothers without autoimmune diseases or use of steroids and anti-inflammatory or epileptic drugs during pregnancy
PCB exposure	Prenatal: Maternal serum during pregnancy, cord blood, and milk	Pre- and postnatal: Maternal pregnancy serum, cord blood, and transition milk and serum at age 4.5 and 7.5 years (cohort A) and 18 months (cohort B)	Pre- and postnatal: Maternal pregnancy serum, cord blood, and transition milk Serum at age 12 months, 18 months, 5 years, and 7 years	Pre- and postnatal: Maternal serum at birth, cord blood, and serum at age 6 months	Prenatal: Calculated maternal PCB intake during pregnancy

Vaccinations	Mumps, measles, and rubella at age 14 months	Tetanus and diphtheria at age 5, 6, and 15 months and booster at age 5 years (cohort A) or at age 3, 5, and 12 months (cohort B)	Tetanus and diphtheria at age 3, 5, and 12 months. Booster at age 5 years	Tetanus, diphtheria, and *Haemophilus influenzae* type b at age 3–4 months and 5–6 months	Tetanus and *Haemophilus influenzae* type b at age 3, 5, and 12 months Measles and rubella at age 15 months
Antibody response	Age 3.5 years	Age 7.5 years and 18 months, respectively	Age 5 years and 7	Age 6 months	Age 3 years
Statistical model	Pearson correlation	Structural equation models adjusted for age and sex	Multiple regression analyses adjusted for age, sex, and type of booster vaccination	Linear regression model adjusted for ethnicity, maternal smoking, and maternal age	Multiple regression analyses with manual backward deletion of potential confounding variables
Findings	Associations between high levels of PCB in maternal serum and cord blood and reduced antibody response to mumps and rubella	Associations between high levels of PCB prenatal and early postnatal and reduced antibody response to diphtheria and tetanus	Associations between high prenatal and postnatal PCB exposure and the risk of an antibody concentration below a clinically protective level of 0.1 UI/ml	No associations between PCB exposure and antibody concentrations	Associations between high maternal PCB exposure and reduced antibody response to measles

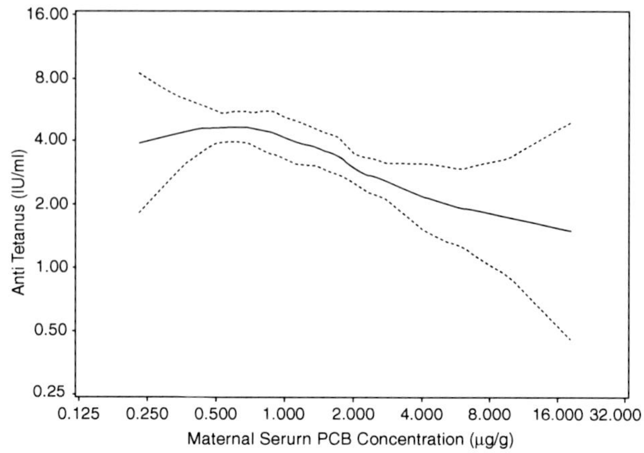

Fig. 8.2 Dose–effect PCB level in mother and antibody response in child. Dose–effect response between total PCB concentration in maternal serum and antibody responses to childhood vaccinations at age 18 months. *Broken lines* indicate the 95 % CI (Heilmann et al. 2006, reprinted with permission)

tetanus and diphtheria at age 5 years and 7.5 years in children with high PCB levels prenatally and early postnatal (Fig. 8.2). The serum PCB concentration at 18 months was especially significant in regard to the decrease in antibody concentrations, with an estimated 18 % lower diphtheria toxoid antibody level and estimated 22 % lower tetanus toxoid antibody level at age 7 years for each doubling of the PCB exposure at 18 months of age (Heilmann et al. 2006, 2010). In contrast to these findings, Jusko et al. did not show any associations of prenatal PCB exposure and antibody responses to immunizations with tetanus, diphtheria, and Hib at 6 months of age (Jusko et al. 2010).

The levels of antibodies varied substantially, but most of the children had antibody concentrations well within the range assumed to provide long-term protection against the infectious diseases vaccinated against. However, in the largest Faroese study conducted by Heilmann et al., 26–37 % of children had diphtheria and tetanus toxoid antibody concentrations below the limit for clinically protection of 0.1 IU/ml before the booster vaccination at 5 years of age. The odds of an anti-diphtheria antibody concentration below a clinically protective level of 0.1 IU/L at age 5 years increased by about 30 % for a doubling of PCB concentrations in milk and 18-month serum (Heilmann et al. 2010). These findings were not reported elsewhere; however, the time from vaccination to antibody measurement differed, and the PCB levels in the Dutch, Slovakian, or Norwegian cohorts were lower than among the Faroese mothers and children due to the traditional habit of eating pilot whale blubber in the Faroese Islands (Grandjean et al. 1995; Weihe et al. 1996). Although the Faroese population has a substantially increased PCB exposure, these data suggest that possible adverse influences on the immune function may well occur also at lower exposure ranges prevalent worldwide.

Even though the published studies are not consistent regarding the effect of PCB to immune responses after specific vaccines, data from these studies taken together points to an adverse effect of PCB exposure on antibody responses to routine childhood vaccinations. Overall, these effects are consistent for both attenuated virus vaccinations and bacterial toxoid vaccinations, suggesting a depression of several components of the immune system.

Discrepancies between studies might be caused by differences in sample sizes and vaccines used (including adjuvants) and dissimilarities in determination of PCB exposure. The time of assessment of antibody response also have an impact. Jusko et al. explain their lack of associations between prenatal PCB exposure and antibody responses at 6 months to the immaturity of the immune system in young infants. However this study has general flaws in the assessment of the vaccination responses. The number of vaccinations differed from 1 to 2 in this study, and the timing of both vaccinations and collection of blood samples for antibody responses varied within the study population. As the vaccination responses were measured very shortly after vaccination, the resulting antibody levels might reflect the difference in individual immunization responses as well as the interindividual variation in timing of the blood sample collection.

Furthermore, the differences in findings from the vaccination studies suggest that the estimated effect on antibody concentrations depends on the time window that analysis represents. Data regarding prenatal PCB exposure have shown associations with some vaccinations, but not others. In the prospective follow-up studies, the highest levels of immunosuppressive associations were caused by serum levels of PCB at 18 months in the Faroese study. This suggests that PCB levels in the fetal stage and early childhood coincide with a highly vulnerable stage of immune system development and that the antibody response may be influenced by accumulated PCB burden from placental transfer and lactation during the early postnatal period.

8.2.1.2 Immunological Parameters in Humans

In line with the findings regarding childhood vaccinations, exposure to PCBs during pregnancy is found to be associated with immune-related health outcomes and immune functionality parameters in small children. Several studies report that PCB-exposed children have more childhood infections during infancy, preschool, and school age, including respiratory tract infections, gastroenteritis, exanthema subitum, recurrent middle ear infections, and chicken pox (Dallaire et al. 2004; Stølevik et al. 2011, 2013; Weisglas-Kuperus et al. 2000, 2004).

High prenatal PCB exposure in utero and early postnatal has been associated with a decreased thymus size among neonates born in an area with high environmental load of PCBs in Eastern Slovakia, suggesting that infant immune development might be altered by in utero exposure to PCB (Jusko et al. 2012). Mostly nonplanar PCBs were investigated, but concentrations of nonplanar PCBs are moderately correlated with dioxin-like activity in several populations. Therefore, the results could be supported by experimental studies in pregnant mice exposed for TCDD, in which offspring revealed decreases in thymic weight and cellularity (Fine et al. 1989). In addition, in vitro studies of TCDD have suggested a change in the kinetics of thymocyte maturation with skewing of the thymocyte differentiation toward the CD8+ phenotypically more mature TcR αβ+ T cells following exposure to PCBs and dioxins (Lai et al. 1998).

In addition to toxic effects on the thymus, high maternal and cord blood PCB levels have been associated with increased numbers of leukocytes, lymphocytes, and T cells, including cytotoxic T cells, memory T cells, TcR αβ+ T cells, and activated T cells, in infants and small children (Belles-Isles et al. 2002; Weisglas-Kuperus et al. 2000), although this was not confirmed by others (Glynn et al. 2008; Stølevik et al. 2013).

Additionally, a recent substudy within the Norwegian vaccination cohort linked high maternal PCB exposure during pregnancy to representation of immune-related genes at the transcriptomic levels in cord blood, which were also correlated to measles vaccination responses at 3 years of age. The involved genes include regulation of intracellular signaling cascades, lymphocyte activation and T-cell proliferation, cytokine production, and antigen presentation (Hochstenbach et al. 2012), which may further indicate immunosuppressive effects of prenatal exposure to PCBs.

The present findings associating early exposure to PCBs and reduced antibody response to immunizations together with increased risk of infections, decreased thymic volume, and altered composition of T-lymphocyte subsets are all in accordance with the proposed immunosuppressive properties of dioxins and dioxin-like PCBs. The findings of PCB-related impaired resistance to infections and impairment of lymphocyte function and antibody responses have also been demonstrated in animals (Ross et al. 1996). These results suggest that further efforts are needed to minimize the hazard from PCB exposure.

8.2.2 Perfluorinated Alkylate Substances

Perfluorinated alkylate substances (PFAS) are a group of chemicals with many important industrial and manufacturing applications, which are used widely in surfactants and repellents in food packaging and textile impregnation (Vestergren and Cousins 2009). PFAS comprise a heterogeneous class of chemicals consisting of an alkyl chain (4–14 carbons), which is partially or fully fluorinated and have different functional groups attached. As a group, they display unique characteristics such as chemical and thermal stability, low surface free energy, and surface-active properties.

The most studied compounds to date are perfluoroalkyl sulfonic acids and perfluoroalkyl carboxylic acids. Among these, perfluorooctane sulfonic acid (PFOS) and perfluorooctanoic acid (PFOA) are of greatest concern due to their high level of persistence in the environment.

Perfluorooctane sulfonyl fluorides and perfluoroalkyl carboxylic acids have been produced with increasing intensities since the 1950s. The production of PFOS and related compounds were phased out by the major global manufacturer in 2002, resulting in a rapid decrease in release of these substances; the use of PFOS also is now restricted in the USA and the European Union due to their persistency and toxicity. The production of perfluoroalkyl carboxylic acids is however still ongoing, although several major PFOA-producing companies have committed to reduce their emissions. The historical releases of PFOS, PFOA, and related agents to land, air, and ground and surface water during production and use have caused dispersal to the global environment, especially in the oceans. Additionally, both PFOS and PFOA can be generated through environmental degradation processes of commercially synthesized precursors (Paul et al. 2009; Wang et al. 2014).

As several PFAS are highly persistent substances due to a strong carbon–fluorine bond, they remain a major environmental concern. PFAS are commonly detected in wildlife and have been demonstrated to bioaccumulate in fish all over the globe (Suja et al. 2009). It has been shown that the longer the perfluorinated carbon chains (>C5), the stronger the persistence and strength of binding to proteins is and the more bioaccumulative they become (Martin et al. 2003).

Analyses of occupational and nonoccupational human blood samples from all over the world contain PFAS at detectable levels, in which PFOS is the most dominant compound followed by PFOA and perfluorohexane sulfonic acid (PFHxS) (Calafat et al. 2007; Kannan et al. 2004; Toms et al. 2014; Weihe et al. 2008). Analyses conducted by the Centers for Disease Control and Prevention in the USA show that PFOS and PFOA are detectable in virtually all Americans (Calafat et al. 2007), with children often showing higher serum concentrations than adults (Kato et al. 2009). Studies of retired workers from PFAS production facilities show a long elimination half-life of 5.4 years for PFOS, 3.8 years for PFOA, and 8.5 years for PFHxS (Olsen et al. 2007).

Dietary intake is suspected to be the dominant cause of human PFAS exposure, and elevated concentrations of PFAS are especially associated with consumption of seafood and meat as PFAS pile up in proteins (Berger et al. 2009, 2014; Fromme et al. 2009; Haug et al. 2010). However, PFAS have also been detected in other food items, where they could be transferred from packaging or preparation with PFAS-contaminated applications in addition to bioaccumulation. Dust in the indoor environment may also be an important contributor to PFAS exposure (Fromme et al. 2009). Early-life exposure occurs through placental transfer from the mother during pregnancy as well as through lactation postnatal (Liu et al. 2011; Needham et al. 2011; Vestergren and Cousins 2009). Therefore, fetuses and infants are exposed to these compounds at critical developmental stages of the immune system.

8.2.2.1 Vaccination Studies

A study by Grandjean et al. has investigated the influence of PFAS on antibody responses after childhood immunizations (Grandjean et al. 2012). The study focused on the homogenous fishing community of the Faroe Islands in the North Atlantic Sea, where frequent intake of marine food is associated with increased exposures to PFAS (Weihe et al. 2008). A birth cohort of 587 consecutive singleton births was included and followed for 7 years. All children were vaccinated according to the official Faroese vaccination program, which includes vaccinations against diphtheria and tetanus at age 3, 5, and 12 months and a booster vaccination at age 5 years. Exposure to PFAS was assessed prenatally from analyses of serum from the mother during pregnancy and from the child at age 5 years before the booster vaccination, and antibody responses were measured at age 5 years prebooster, approximately 4 weeks after the booster, and at age 7 years. Similar to findings from other locations,

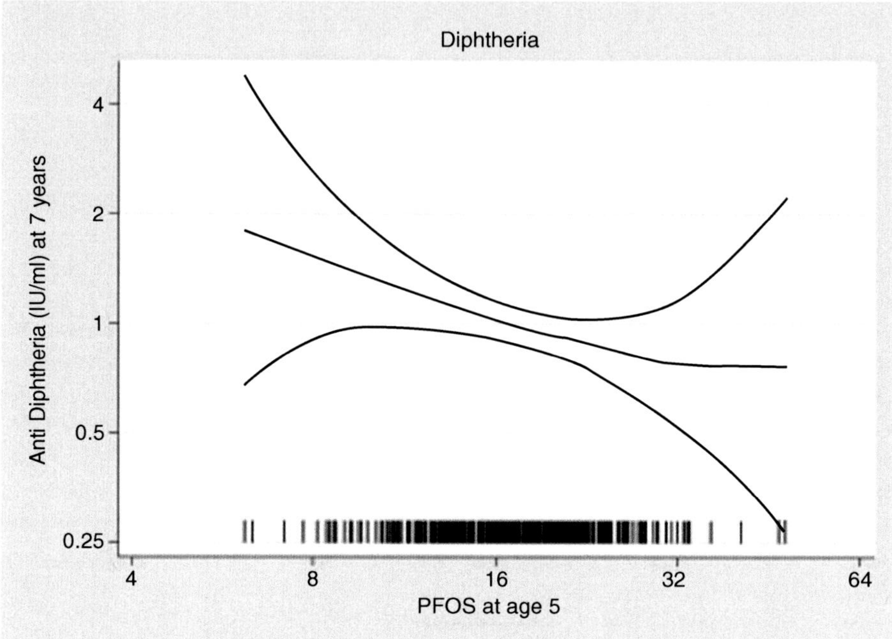

Fig. 8.3 Dose–effect between PFOS exposure and antibody response. Dose–effect response between PFOS exposure at age 5 years and antibody responses to childhood vaccinations with diphtheria at 7 years. *Lines* indicate the 95 % CI. The *vertical bars* on the *horizontal scale* indicate individual observations (Grandjean et al. 2012, with permission)

the highest serum concentrations of PFAS were PFOS, PFOA, and PFHxS, which all were interrelated within the sample, but showed only weak associations between prenatal and postnatal exposures.

Antibody responses against diphtheria and tetanus were conducted at age 5 years before the booster, approximately 4 weeks after the booster, and at age 7 years. Among PFAS in maternal pregnancy serum, PFOS showed the strongest negative correlations with antibody concentrations at age 5 years. In a structural equation model adjusting for age, sex, and type of booster vaccination, a twofold greater concentration of the three major PFAS in child serum at age 5 years was associated with a halving of the antibody concentration (Fig. 8.3). Furthermore, a twofold increase in PFOS and PFOA concentrations at age 5 years was associated with odds ratios 2.38 and 4.20 for having antibody concentrations below the clinically protective level of 0.1 IU/mL for tetanus and diphtheria at age 7 years, respectively. Data presented in this study suggest a decreased effect of routine childhood vaccines in children aged 5 and 7 years with elevated exposures to PFAS. This may reflect a more general immune system deficit caused by exposure to PFAS (Grandjean et al. 2012).

Based on the human immunotoxicity data from the Faroese cohort, calculations of benchmark doses have been conducted. It relies on the serum PFAS measurements at age 5 and the serum concentrations of specific antibodies 2 years later.

Under different linear assumptions regarding dose dependence of the effects, benchmark dose levels were about 0.13 ng/mL serum for PFOS and 0.03 ng/mL serum for PFOA (Grandjean and Budtz-Jørgensen 2013). These doses are below most serum concentrations reported in recent population studies (Calafat et al. 2007; Kannan et al. 2004; Toms et al. 2014; Weihe et al. 2008). When converted to approximate exposure limits for drinking water, current limits appear to be several hundredfold too high in light of the observed immunotoxicity associated with PFAS exposure (Grandjean and Budtz-Jørgensen 2013). Furthermore, the human exposure to PFAS exceeds in several locations.

The Faroese results have been supported by the only other study regarding PFAS exposure and childhood vaccinations conducted in Norway by Granum et al. Prenatal PFAS exposure and antibody responses to childhood vaccinations at age 3 years were analyzed in 56 children. Vaccination included tetanus and *Haemophilus influenzae* type b given at ages 3, 5, and 12 months and measles and rubella given at 15 months of age. Increased concentrations of PFOA, PFHxS, perfluorononanoate (PFNA), and PFOS were significantly associated with reduced levels of anti-rubella antibodies in children at 3 years of age, with the highest estimated strength of association for PFNA and the lowest for PFOS. No significant associations were found between the concentrations of PFAS and antibody response to the other vaccines (Granum et al. 2013). Although a discrepancy of results regarding the influence of PFAS and tetanus response exists between the two studies, this study in a very limited number of children in general supports that prenatal exposure to PFAS may suppress responses to some childhood vaccines.

Furthermore, a study of humoral immune response to the seasonal influenza vaccination in 403 adults exposed to PFOA through contaminated drinking water in the USA has recently been published by Looker et al. The authors found evidence of a reduced antibody response 3 weeks after immunization with the influenza strain A/H3N2 in adults with higher PFOA concentrations. This response was reflected in titer rise, titer ratio, and long-term seroprotection after vaccination, although not seroconversion. No associations were found for PFOS exposure or for the response to the influenza serotypes A/H1N1 (swine flu) and influenza B. Even though these results are very supportive of the immunosuppressive effect of PFAS, especially since the antibody responses involved a mixture of primary and secondary reactions as all participants had titers for some of these common viruses prior to the immunization (Looker et al. 2014).

8.2.2.2 Immunological Parameters in Humans

Epidemiologic data related to PFAS exposure and immunotoxicity in humans are limited. The effect of prenatal PFAS exposure has been investigated with regard to specific childhood infections. In a Danish study of 14,000 children, prenatal exposure to PFOS and PFOA at approximately 8 weeks of pregnancy was not associated with increased risk of infectious diseases leading to hospitalization in the first 8 years of childhood (Fei et al. 2010). Similar to this, no associations were found between PFAS exposure during pregnancy and occurrence of middle ear infection in 18-month-old Japanese children (Okada et al. 2012). However, a positive association between prenatal exposure to PFAS and the self-reported number of episodes of common cold and gastroenteritis, but not middle ear infection, during the first 3 years of life was found in a small Norwegian study (Granum et al. 2013). In adults, no associations were found between PFAS levels and recent self-reported cold or influenza episodes (Looker et al. 2014).

However, occurrence of specific common infections may be a less sensitive or appropriate test of the presence of immune system dysfunction than immunization responses, as these infections are very prevalent and multiple social and demographic factors could affect the results.

The C8 Health Project investigated immune markers in blood from healthy individuals exposed to PFOA-contaminated drinking water for at least 1 year due to residency in the vicinity of a PFOA manufacturing site in the USA. High serum levels of PFOA were significantly associated with lower serum concentrations of C-reactive protein, total IgA, and total IgE (in females only), though not with IgG levels. Furthermore, increasing concentrations of PFOA showed a positive relationship with total antinuclear antibodies, which may indicate an increase in the risk of autoimmune diseases (Fletcher et al. 2009).

Corresponding to this, small changes in serum immunoglobulin levels have been reported in workers with occupational exposure to PFOA (Costa et al. 2009). Although supportive of immunosuppression, these basic tests cannot be considered as either sensitive or specific for immunotoxicity or predictive enough of an adverse effect of the immune system (van Loveren et al. 1999).

8.2.2.3 Animal Models of Immunotoxicity

Although confirmation of the findings from immunization studies in human studies is sparse, immunotoxicity of PFAS has been demonstrated in rodent, avian, and reptilian models as well as in mammalian and nonmammalian wildlife, affecting both cellular and humoral immunity (DeWitt et al. 2012).

High-dose dietary treatment of mice with PFOS or PFOA has been shown to cause atrophy of the thymus and spleen, decreased thymocyte and splenocyte counts, as well as significantly reduced numbers of bone marrow cells and myeloid cells, pro-/pre-B cells, immature B cells, and early mature B cells. These adverse effects were reversed partially or completely after withdrawal of these compounds (Qazi et al. 2012; Yang et al. 2000).

Furthermore, oral PFOS and PFOA exposure in mice has been found to cause severe suppression of the adaptive immune system in several studies. Findings include decreased numbers of immunoglobulin-producing splenic cells and a dose-dependent reduction in plasma levels of specific IgM and IgG upon immunization with horse or sheep red blood cells in PFOA-exposed mice. The immunosuppressive effect of PFOA appeared to recover following the administration of normal feeding (Dewitt et al. 2008; Yang et al. 2002).

Additionally, immune suppression has been found in mice exposed to PFOS in doses equivalent to the high end of nonoccupational exposure levels in humans. Alteration included especially suppression of specific IgM production by the plaque-forming cell response, a functional assay revealing the attack, and destruction of an antigen by antibodies. Production of both T-cell-dependent and T-cell-independent antibodies was suppressed, suggesting that B cells or antigen-presenting cells might be the specific target of PFOS (Peden-Adams et al. 2008). PFOS exposure in mice has also been associated with reduced host defense to influenza A virus infection, resulting in increased weight loss and reduced survival (Guruge et al. 2009). Additionally, cytokine expression and signaling related to inflammation and T-cell responses are altered by exposure to PFAS (DeWitt et al. 2012).

The importance of in utero exposure to PFAS and immunotoxicity has been investigated in two mice studies with different conclusions. Prenatal PFOS exposure was shown to suppress T-cell-dependent IgM antibody responses upon immunization in male offspring (Keil et al. 2008), although this was not confirmed for PFOA by others (Hu et al. 2010).

Taken together, experimental studies support an increased susceptibility to pathogens due to exposure to PFAS.

8.2.2.4 Molecular Mechanisms Involved in Immunosuppression by PFAS

The molecular mode of action of PFAS-induced immune suppression has been investigated by examining essential cell populations as well as cell signaling and activation. IgM secretion by B lymphocytes is controlled through production of IL-6. The onset of inflammatory gene expression is driven by the nuclear transcription factor NF-κB, whose transcriptional activity is regulated at multiple levels. In B cells, engagement of CD40 by its ligand leads to signaling through c-Jun and NF-κB, with subsequent IL-6 production (Baccam et al. 2003). This process is critical for humoral responses to T-dependent antigens and results in B-cell proliferation, differentiation, and IL-6 stimulation of immunoglobulin production (Pérez-Melgosa et al. 1999).

Current evidence suggests that the immunotoxic mechanisms of PFAS involve interaction with peroxisome proliferator-activated receptors (PPARs) as well as an PPAR-independent pathways (DeWitt et al. 2009, 2012). In vitro stimulation of human peripheral blood leukocytes with lipopolysaccharide (LPS) in the presence of PFAS for 24–72 hours showed a dose-dependent suppression of TNF-α and IL-6 production. Furthermore, in vitro exposure to PFAS inhibited production of the regulatory cytokines IL-4 and IL-10. The immunotoxic effect of PFOS was PPAR independent through interference with LPS-induced I-κB degradation, NF-κB transactivation, and DNA binding. In contrast, PFOA acted through PPAR to prevent p65 phosphorylation and NF-κB-mediated transcription (Corsini et al. 2011). Both mechanisms of action could theoretically lead to reduced antibody secretion by B lymphocytes, although further evidence for this is needed.

8.3 Perspectives

Vaccination is a mainstay in modern prevention of infectious diseases. While antibody responses of the average child to routine prophylactic immunizations can vary substantially, the reasons for poor responses are poorly understood, though exposure to environmental pollutants may account for at least some of the variation.

The presented studies provide epidemiological evidence for an association between exposure to persistent organic pollutants and a reduced humoral immune response to routine childhood vaccinations. Specifically, PCB reduces the immune response perinatally and early postnatal and is suspected to be toxic to the development of thymic function. PFAS, whereas, reduce the antibody levels in a more direct manner. The results are supported by animals and cellular studies, although a clear mechanism of action has not yet been demonstrated for all compounds.

The children examined came from population-based birth cohorts and were overall in good health. Although the changes in antibody response are subtle, they could become clinically important when the immune function is challenged by other risk factors such as preterm birth, chronic infections, increasing age, or competing diseases. Furthermore, high exposure levels in populations at risk may hinder a long-term protective immune response and increase the risk of a child not being protected against severe infections, despite a full schedule of vaccinations. In addition, even slight impairments could be important at a population level, e.g., if the herd immunity fails and epidemics of highly infectious and pathogenic infections brake out. In countries without access to routine pediatric immunizations, tetanus, diphtheria, measles, mumps, and rubella still give rise to increased levels of childhood mortality and morbidity.

In addition, one should keep in mind that antibody responses to standardized antigen stimulations reflect the overall efficacy of the immune system in relation to infections, including antigen presentation and T- and B-lymphocyte function. Therefore, the presented immune suppression may also increase children's susceptibility to common infections, as have been suggested for PCB in some human studies. Therefore, even the small changes in immune function, suggested by the presented vaccination data, could be clinically important and might affect both the general health of children and the degree of protection against infectious diseases that vaccination provides.

Some of the immunotoxic effects from the immunization studies are revealed in children exposed to immunotoxicants through perinatal and early postnatal exposure, while others are associated with concentrations at the time of vaccination. The immune function returns to normal following exposure to immunosuppressive drugs in adults, especially if the drugs do not affect precursor or stem cells. However, the special concerns of low-level exposure to immunotoxicants during the development of the immune system are that developmental immunity may be more sensitive to the dose of immunotoxicants and that an early-life insult prenatally or early postnatal might cause immune defects on a permanent basis, leading to reduced host resistance in children and adults.

8.4 Conclusion

In conclusion, vaccination efficacy is found to be the most relevant parameter for assessment of immune suppression by environmental persistent pollutants. Results arising from vaccination studies in children support the hypothesis that some people today could be immunocompromised because of exposure to environmental pollutants such as PCBs and PFAS. In contrast to investigations from the presented study, human populations will often be exposed to a mixture of immunomodulating agents, which may in combination cause even more severe effects on the immune system than seen in these studies. While PCBs have been banned since the 1970s and are now found at reduced levels, the production of PFAS has just recently been limited. The persistence of these compounds means that they will still be widespread in the environment and in humans for a long period of time. However, the presented findings suggest that efforts must be stepped up to reduce exposure levels to protect the human immune systems from these as well as other potent immunotoxicants.

Acknowledgments The research summarized in this chapter has been supported by National Institute of Environmental Health Sciences, NIH (ES012199 and ES021993), the National Science Foundation (OCE-1321612), the US Environmental Protection Agency (R830758), the Danish Council for Strategic Research (09–063094), and the Danish Environmental Protection Agency as part of the environmental support program DANCEA (Danish Cooperation for Environment in the Arctic). The authors are solely responsible for the conclusions expressed in this chapter, which do not necessarily reflect the position of any of the funding agencies. We thank Katie T. Herz for editorial comments.

References

André FE (2003) Vaccinology: past achievements, present roadblocks and future promises. Vaccine 21:593–595

Atlas E, Giam CS (1981) Global transport of organic pollutants: ambient concentrations in the remote marine atmosphere. Science 211:163–165

Baccam M, Woo S-Y, Vinson C, Bishop GA (2003) CD40-mediated transcriptional regulation of the IL-6 gene in B lymphocytes: involvement of NF-kappa B, AP-1, and C/EBP. J Immunol 170:3099–3108

Bacon CE, Jarman WM, Costa DP (1992) Organochlorine and polychlorinated biphenyl levels in pinniped milk from the Arctic, the Antarctic. Chemosphere 24:779–791

Belles-Isles M, Ayotte P, Dewailly E, Weber J-P, Roy R (2002) Cord blood lymphocyte functions in newborns from a remote maritime population exposed to organochlorines and methylmercury. J Toxicol Environ Health A 65:165–182

Berg V, Nøst TH, Huber S et al (2014) Maternal serum concentrations of per- and polyfluoroalkyl substances and their predictors in years with reduced production and use. Environ Int 69: 58–66

Berger U, Glynn A, Holmström KE, Berglund M, Ankarberg EH, Törnkvist A (2009) Fish consumption as a source of human exposure to perfluorinated alkyl substances in Sweden – analysis of edible fish from Lake Vättern and the Baltic Sea. Chemosphere 76:799–804

Brouwer A, Ahlborg UG, van Leeuwen FX, Feeley MM (1998) Report of the WHO working group on the assessment of health risks for human infants from exposure to PCDDs, PCDFs and PCBs. Chemosphere 37:1627–1643

Calafat AM, Kuklenyik Z, Reidy JA, Caudill SP, Tully JS, Needham LL (2007) Serum concentrations of 11 polyfluoroalkyl compounds in the U.S. Population: data from the National Health and Nutrition Examination Survey (NHANES) 1999–2000. Environ Sci Technol 41:2237–2242

Corsini E, Avogadro A, Galbiati V, Dell'Agli M, Marinovich M, Galli CL, Germolec DR (2011) In vitro evaluation of the immunotoxic potential of perfluorinated compounds (PFCs). Toxicol Appl Pharmacol 250:108–116

Costa G, Sartori S, Consonni D (2009) Thirty years of medical surveillance in perfluooctanoic acid production workers. J Occup Environ Med 51:364–372
Covaci A, Jorens P, Jacquemyn Y, Schepens P (2002) Distribution of PCBs and organochlorine pesticides in umbilical cord and maternal serum. Sci Total Environ 298:45–53
Dallaire F, Dewailly E, Muckle G, Vézina C, Jacobson SW, Jacobson JL, Ayotte P (2004) Acute infections and environmental exposure to organochlorines in Inuit infants from Nunavik. Environ Health Perspect 112:1359–1365
Dewitt JC, Copeland CB, Strynar MJ, Luebke RW (2008) Perfluorooctanoic acid-induced immunomodulation in adult C57BL/6J or C57BL/6N female mice. Environ Health Perspect 116:644–650
DeWitt JC, Shnyra A, Badr MZ et al (2009) Immunotoxicity of perfluorooctanoic acid and perfluorooctane sulfonate and the role of peroxisome proliferator-activated receptor alpha. Crit Rev Toxicol 39:76–94
DeWitt JC, Peden-Adams MM, Keller JM, Germolec DR (2012) Immunotoxicity of perfluorinated compounds: recent developments. Toxicol Pathol 40:300–311
Dietert RR (2008) Developmental immunotoxicology (DIT): windows of vulnerability, immune dysfunction and safety assessment. J Immunotoxicol 5:401–412
Domingo JL, Bocio A (2007) Levels of PCDD/PCDFs and PCBs in edible marine species and human intake: a literature review. Environ Int 33:397–405
Fängström B, Strid A, Grandjean P, Weihe P, Bergman A (2005) A retrospective study of PBDEs and PCBs in human milk from the Faroe Islands. Environ Health 4:12
Fei C, McLaughlin JK, Lipworth L, Olsen J (2010) Prenatal exposure to PFOA and PFOS and risk of hospitalization for infectious diseases in early childhood. Environ Res 110:773–777
Fine JS, Gasiewicz TA, Silverstone AE (1989) Lymphocyte stem cell alterations following perinatal exposure to 2,3,7,8-tetrachlorodibenzo-p-dioxin. Mol Pharmacol 35:18–25
Fischer LJ, Seegal RF, Ganey PE, Pessah IN, Kodavanti PR (1998) Symposium overview: toxicity of non-coplanar PCBs. Toxicol Sci 41:49–61
Fletcher T, Steenland K, Savitz D (2009) C8 Science Panel: Status Report: PFOA and immune biomarkers in adults exposed to PFOA in drinking water in the mid Ohio valley. Available: http://www.c8sciencepanel.org/index. html. Accessed on Dec 2014
Fromme H, Tittlemier SA, Völkel W, Wilhelm M, Twardella D (2009) Perfluorinated compounds–exposure assessment for the general population in Western countries. Int J Hyg Environ Health 212:239–270
Glynn A, Thuvander A, Aune M, Johannisson A, Darnerud PO, Ronquist G, Cnattingius S (2008) Immune cell counts and risks of respiratory infections among infants exposed pre- and postnatally to organochlorine compounds: a prospective study. Environ Health 7:62
Grandjean P, Budtz-Jørgensen E (2013) Immunotoxicity of perfluorinated alkylates: calculation of benchmark doses based on serum concentrations in children. Environ Health 12:35
Grandjean P, Weihe P, Needham L, Burse V, Patterson DJ, Sampson E, Jørgensen P, Vahter M (1995) Relation of a seafood diet to mercury, selenium, arsenic, and polychlorinated biphenyl and other organochlorine concentrations in human milk. Environ Res 71:29–38
Grandjean P, Andersen EW, Budtz-Jørgensen E, Nielsen F, Mølbak K, Weihe P, Heilmann C (2012) Serum vaccine antibody concentrations in children exposed to perfluorinated compounds. JAMA 307:391–397
Granum B, Haug LS, Namork E et al (2013) Pre-natal exposure to perfluoroalkyl substances may be associated with altered vaccine antibody levels and immune-related health outcomes in early childhood. J Immunotoxicol 10:373–379
Guruge KS, Hikono H, Shimada N, Murakami K, Hasegawa J, Yeung LWY, Yamanaka N, Yamashita N (2009) Effect of perfluorooctane sulfonate (PFOS) on influenza A virus-induced mortality in female B6C3F1 mice. J Toxicol Sci 34:687–691
Haug LS, Thomsen C, Brantsaeter AL et al (2010) Diet and particularly seafood are major sources of perfluorinated compounds in humans. Environ Int 36:772–778
Heilmann C, Grandjean P, Weihe P, Nielsen F, Budtz-Jørgensen E (2006) Reduced antibody responses to vaccinations in children exposed to polychlorinated biphenyls. PLoS Med 3:e311

Heilmann C, Budtz-Jørgensen E, Nielsen F, Heinzow B, Weihe P, Grandjean P (2010) Serum concentrations of antibodies against vaccine toxoids in children exposed perinatally to immunotoxicants. Environ Health Perspect 118:1434–1438
Hilleman MR (2000) Vaccines in historic evolution and perspective: a narrative of vaccine discoveries. Vaccine 18:1436–1447
Hochstenbach K, van Leeuwen DM, Gmuender H et al (2012) Toxicogenomic profiles in relation to maternal immunotoxic exposure and immune functionality in newborns. Toxicol Sci 129:315–324
Holladay SD, Smialowicz RJ (2000) Development of the murine and human immune system: differential effects of immunotoxicants depend on time of exposure. Environ Health Perspect 108(Suppl):463–473
Holsapple MP, Paustenbach DJ, Charnley G, West LJ, Luster MI, Dietert RR, Burns-Naas LA (2004) Symposium summary: children's health risk–what's so special about the developing immune system? Toxicol Appl Pharmacol 199:61–70
Hu Q, Strynar MJ, DeWitt JC (2010) Are developmentally exposed C57BL/6 mice insensitive to suppression of TDAR by PFOA? J Immunotoxicol 7:344–349
Jusko TA, De Roos AJ, Schwartz SM et al (2010) A cohort study of developmental polychlorinated biphenyl (PCB) exposure in relation to post-vaccination antibody response at 6-months of age. Environ Res 110:388–395
Jusko TA, Sonneborn D, Palkovicova L, Kocan A, Drobna B, Trnovec T, Hertz-Picciotto I (2012) Pre- and postnatal polychlorinated biphenyl concentrations and longitudinal measures of thymus volume in infants. Environ Health Perspect 120:595–600
Kannan N, Tanabe S, Ono M, Tatsukawa R (1989) Critical evaluation of polychlorinated biphenyl toxicity in terrestrial and marine mammals: increasing impact of non-ortho and mono-ortho coplanar polychlorinated biphenyls from land to ocean. Arch Environ Contam Toxicol 18:850–857
Kannan K, Corsolini S, Falandysz J et al (2004) Perfluorooctanesulfonate and related fluorochemicals in human blood from several countries. Environ Sci Technol 38:4489–4495
Kato K, Calafat AM, Wong L-Y, Wanigatunga AA, Caudill SP, Needham LL (2009) Polyfluoroalkyl compounds in pooled sera from children participating in the National Health and Nutrition Examination Survey 2001–2002. Environ Sci Technol 43:2641–2647
Keil DE, Mehlmann T, Butterworth L, Peden-Adams MM (2008) Gestational exposure to perfluorooctane sulfonate suppresses immune function in B6C3F1 mice. Toxicol Sci 103:77–85
Lai ZW, Fiore NC, Gasiewicz TA, Silverstone AE (1998) 2,3,7,8-Tetrachlorodibenzo-p-dioxin and diethylstilbestrol affect thymocytes at different stages of development in fetal thymus organ culture. Toxicol Appl Pharmacol 149:167–177
Liem AK, Fürst P, Rappe C (2000) Exposure of populations to dioxins and related compounds. Food Addit Contam 17:241–259
Liu J, Li J, Liu Y, Chan HM, Zhao Y, Cai Z, Wu Y (2011) Comparison on gestation and lactation exposure of perfluorinated compounds for newborns. Environ Int 37:1206–1212
Looker C, Luster MI, Calafat AM, Johnson VJ, Burleson GR, Burleson FG, Fletcher T (2014) Influenza vaccine response in adults exposed to perfluorooctanoate and perfluorooctanesulfonate. Toxicol Sci 138:76–88
Luster MI, Johnson VJ, Yucesoy B, Simeonova PP (2005) Biomarkers to assess potential developmental immunotoxicity in children. Toxicol Appl Pharmacol 206:229–236
Martin JW, Mabury SA, Solomon KR, Muir DCG (2003) Bioconcentration and tissue distribution of perfluorinated acids in rainbow trout (Oncorhynchus mykiss). Environ Toxicol Chem 22:196–204
Needham LL, Grandjean P, Heinzow B et al (2011) Partition of environmental chemicals between maternal and fetal blood and tissues. Environ Sci Technol 45:1121–1126
Okada E, Sasaki S, Saijo Y et al (2012) Prenatal exposure to perfluorinated chemicals and relationship with allergies and infectious diseases in infants. Environ Res 112:118–125
Olsen GW, Burris JM, Ehresman DJ, Froehlich JW, Seacat AM, Butenhoff JL, Zobel LR (2007) Half-life of serum elimination of perfluorooctanesulfonate, perfluorohexanesulfonate, and perfluorooctanoate in retired fluorochemical production workers. Environ Health Perspect 115:1298–1305

Paul AG, Jones KC, Sweetman AJ (2009) A first global production, emission, and environmental inventory for perfluorooctane sulfonate. Environ Sci Technol 43:386–392

Peden-Adams MM, Keller JM, Eudaly JG, Berger J, Gilkeson GS, Keil DE (2008) Suppression of humoral immunity in mice following exposure to perfluorooctane sulfonate. Toxicol Sci 104:144–154

Pérez-Melgosa M, Hollenbaugh D, Wilson CB (1999) Cutting edge: CD40 ligand is a limiting factor in the humoral response to T cell-dependent antigens. J Immunol 163:1123–1127

Plotkin S (2014) History of vaccination. Proc Natl Acad Sci U S A 111:12283–12287

Plotkin SA, Plotkin SL (2011) The development of vaccines: how the past led to the future. Nat Rev Microbiol 9:889–893

Qazi MR, Nelson BD, DePierre JW, Abedi-Valugerdi M (2012) High-dose dietary exposure of mice to perfluorooctanoate or perfluorooctane sulfonate exerts toxic effects on myeloid and B-lymphoid cells in the bone marrow and these effects are partially dependent on reduced food consumption. Food Chem Toxicol 50:2955–2963

Ross P, De Swart R, Addison R, Van Loveren H, Vos J, Osterhaus A (1996) Contaminant-induced immunotoxicity in harbour seals: wildlife at risk? Toxicology 112:157–169

Safe SH (1994) Polychlorinated biphenyls (PCBs): environmental impact, biochemical and toxic responses, and implications for risk assessment. Crit Rev Toxicol 24:87–149

Schecter A, Stanley J, Boggess K et al (1994) Polychlorinated biphenyl levels in the tissues of exposed and nonexposed humans. Environ Health Perspect 102(Suppl):149–158

Schecter A, Päpke O, Tung KC, Joseph J, Harris TR, Dahlgren J (2005) Polybrominated diphenyl ether flame retardants in the U.S. population: current levels, temporal trends, and comparison with dioxins, dibenzofurans, and polychlorinated biphenyls. J Occup Environ Med 47:199–211

Skaare JU, Bernhoft A, Derocher A et al (2000) Organochlorines in top predators at Svalbard–occurrence, levels and effects. Toxicol Lett 112–113:103–109

Stølevik SB, Nygaard UC, Namork E et al (2011) Prenatal exposure to polychlorinated biphenyls and dioxins is associated with increased risk of wheeze and infections in infants. Food Chem Toxicol 49:1843–1848

Stølevik SB, Nygaard UC, Namork E et al (2013) Prenatal exposure to polychlorinated biphenyls and dioxins from the maternal diet may be associated with immunosuppressive effects that persist into early childhood. Food Chem Toxicol 51:165–172

Suja F, Pramanik BK, Zain SM (2009) Contamination, bioaccumulation and toxic effects of perfluorinated chemicals (PFCs) in the water environment: a review paper. Water Sci Technol 60:1533–1544

Toms L-ML, Thompson J, Rotander A et al (2014) Decline in perfluorooctane sulfonate and perfluorooctanoate serum concentrations in an Australian population from 2002 to 2011. Environ Int 71:74–80

Van den Berg M, Birnbaum LS, Denison M et al (2006) The 2005 World Health Organization reevaluation of human and Mammalian toxic equivalency factors for dioxins and dioxin-like compounds. Toxicol Sci 93:223–241

Van Loveren H, Germolec D, Koren HS et al (1999) Report of the bilthoven symposium: advancement of epidemiological studies in assessing the human health effects of immunotoxic agents in the environment and the workplace. Biomarkers 4:135–157

Vestergren R, Cousins IT (2009) Tracking the pathways of human exposure to perfluorocarboxylates. Environ Sci Technol 43:5565–5575

Wang Z, Cousins IT, Scheringer M, Buck RC, Hungerbühler K (2014) Global emission inventories for C4-C14 perfluoroalkyl carboxylic acid (PFCA) homologues from 1951 to 2030, Part I: production and emissions from quantifiable sources. Environ Int 70:62–75

Weihe P, Grandjean P, Debes F, White R (1996) Health implications for Faroe islanders of heavy metals and PCBs from pilot whales. Sci Total Environ 186:141–148

Weihe P, Kato K, Calafat AM, Nielsen F, Wanigatunga AA, Needham LL, Grandjean P (2008) Serum concentrations of polyfluoroalkyl compounds in Faroese whale meat consumers. Environ Sci Technol 42:6291–6295

Weisglas-Kuperus N, Patandin S, Berbers GA, Sas TC, Mulder PG, Sauer PJ, Hooijkaas H (2000) Immunologic effects of background exposure to polychlorinated biphenyls and dioxins in Dutch preschool children. Environ Health Perspect 108:1203–1207

Weisglas-Kuperus N, Vreugdenhil HJI, Mulder PGH (2004) Immunological effects of environmental exposure to polychlorinated biphenyls and dioxins in Dutch school children. Toxicol Lett 149:281–285

WHO (2012) Guidance for immunotoxicity risk assessment for chemicals, IPCS harmonization project document; No. 10 World Health Organization, Geneva. ISBN 978 92 4150 330 3

Yang Q, Xie Y, Depierre JW (2000) Effects of peroxisome proliferators on the thymus and spleen of mice. Clin Exp Immunol 122:219–226

Yang Q, Abedi-Valugerdi M, Xie Y, Zhao X-Y, Möller G, Dean Nelson B, DePierre JW (2002) Potent suppression of the adaptive immune response in mice upon dietary exposure to the potent peroxisome proliferator, perfluorooctanoic acid. Int Immunopharmacol 2:389–397

Engineered Nanoparticles and the Immune System: Interaction and Consequences

9

Paola Italiani and Diana Boraschi

Contents

9.1 Introduction: The Human Immune System

The immune system is the guardian of the human body and works to protect its integrity and to defend it against external and internal injuries. In physiological conditions, the immune system monitors the body in order to remove dead and damaged cells, thereby maintaining tissue homeostasis. Upon encounter with invading pathogens or environmentally borne dangerous chemicals, the immune system reacts to the foreign agents by launching a destructive attack aiming at eliminating

P. Italiani
Institute of Biomedical Technologies, National Research Council (CNR), Segrate, Italy

D. Boraschi (✉)
Institute of Protein Biochemistry, National Research Council (CNR), Naples, Italy
e-mail: d.boraschi@ibp.cnr.it

C. Esser (ed.), *Environmental Influences on the Immune System*,
DOI 10.1007/978-3-7091-1890-0_9

the danger. Effective protection must be rapid (leaving no time for the dangerous agent to enter the inner parts of the body and cause damage) and specific (targeting the dangerous agent only and sparing the surrounding cells and tissue). To guarantee both characteristics, in mammals, including man, immunity relies on two different sets of effector mechanisms: innate immunity and adaptive immunity. Innate immunity is an ancient defensive system (almost identical in invertebrates) that has the advantage of immediate reaction (rapidity) every time that a challenge is met. Innate immunity is also able to make a first "selection" of foreign agents and mount an inflammatory reaction to those perceived as potentially dangerous. Despite this selection capacity, innate immunity is not truly specific, as it recognises broad arrays of molecular patterns or signatures. Its rapidity blocks most of the dangerous agents, for instance, infectious microorganisms, at the level of the body barriers (skin, mucosal barriers in the lung and gut). For the very limited number of agents that can overcome innate immunity and pass the barrier thereby accessing the inner body, a second immune system comes into play. Adaptive immunity is much slower than innate immunity, needing time for building specific weapons against the invaders, but its high specificity allows a perfect recognition of the invader (dumping the risk of collateral damage to the body's tissues) and a very efficient elimination. Table 9.1 summarises the main characteristics of innate and adaptive immunity (for more details, see Murphy 2011).

It is worth to briefly remind the main components of innate and adaptive immunity. The innate immune response is based on phagocytosis of foreign agents and particles by specialised cells, such as polymorphonuclear leukocytes (PMN or neutrophils), which circulate in the blood, and the mononuclear phagocytes (monocytes/macrophages and dendritic cells (DC)), which are distributed throughout all tissues and body fluids. The short-lived PMN are excellent phagocytes and have

Table 9.1 The human immune system: innate vs. adaptive immunity

	Innate immunity	Adaptive immunity
Receptors	Germ-line-encoded PRR (e.g. TLR, CLR, NLR)	Generated by gene rearrangement (e.g. TCR, BCR, MHC)
Recognition capacity	Broad	Specific
Molecules recognised	Conserved molecular patterns (e.g. LPS, glycans, mannan, glycolipids)	Specific molecular epitopes (e.g. short peptide sequences)
Immune response	Non-specific broad reaction	Antigen-specific antibody or cellular response
Time lag	Immediate response	Delayed response
Memory	No	Yes
Effector cells	PMN, monocytes/macrophages, DC	T and B lymphocytes
Effector molecules	Cytokines, pentraxins, collectins, alarmins, complement system	Antibodies

TCR T-cell receptor, *BCR* B-cell receptor, *MHC* major histocompatibility complex potent destructive activity, through degranulation and release of proteases, oxidising enzymes, and reactive oxygen species (ROS) and by releasing the neutrophil extracellular traps (NET), which are true nets of DNA filaments decorated with granules filled with enzymes and toxic peptides able in trapping and killing pathogens (Papayannopoulos and Zychlinsky 2009). The monocytes/macrophages are potent phagocytes and the main producers of inflammatory cytokines (TNF-α, IL-6, IL-1β, etc.), soluble proteins that signal to other immune cells the presence of a dangerous agent, and anti-inflammatory cytokines (IL-10, IL-1Ra, TGF-β, etc.), which start the resolution phase at the end of the inflammatory reaction once the danger is successfully eliminated. Monocytes/macrophages are key players of all phases of the inflammatory defensive reaction, from blood monocyte recruitment in infected/damaged tissues to differentiation into inflammatory macrophages and activation of defensive activities (phagocytosis, killing of bacteria or infected cells, release of inflammatory mediators, etc.), elimination of the dangerous agents, and eventual resolution of inflammation and re-establishment of tissue integrity. Other innate cells are the DC, similar to macrophages but less efficient in phagocytosis and more efficient to present antigens to T cells, and natural killer cells (NK), able to distinguish between normal and tumour cells and potent producers of IFN-γ, an inflammatory cytokine of major importance in the amplification of innate immunity.

Besides effector cells, innate immunity encompasses a series of soluble factors able to bind foreign agents in order to facilitate their phagocytosis and destruction. These include the proteins of the collectin family (e.g. the surfactant proteins A and B), lipid transport proteins (e.g. apolipoproteins, serum amyloid A), acute-phase proteins such as short pentraxins (C-reactive protein) and the long pentraxin PTX3, and the complement system. The complement system includes more than 40 soluble proteins and enzymes (e.g. Cq1, mannose-binding lectin (MBL), MASP-1 and MASP-2, mediators of inflammation such as C5a, C3a, and C4a), able to sense all kinds of invaders, from pathogens to synthetic chemicals.

Adaptive immunity relies on two types of cells: T and B lymphocytes. The antigen recognition by specific receptors leads to lymphocyte activation and differentiation into effector cells. $CD8^+$ T cells become cytotoxic while $CD4^+$ T cells (helper T cells, Th) are specialised in supporting B-cell differentiation into antibody-secreting cells. Th cells can be differentially polarised to participate to type 1 immune response against bacteria and viruses (Th1) or to type 2 immune response involved in anti-parasite defence and in allergic reactivity (Th2). Eventually, regulatory T cells (Treg) downregulate adaptive immune responses. Table 9.2 summarises the major function of innate and adaptive immune cells.

There are instances in which immune response can become harmful. The lack of specificity of innate immunity (necessary to ensure the broadest possible coverage against possible dangers) may lead in some circumstances to the anomalous recognition (or cross recognition) and consequent presentation to T cells of self-antigens, thereby initiating autoimmune disorders. In other cases persistence of the triggering agent or exaggerated defensive reaction against it can cause chronic inflammatory diseases.

Table 9.2 Major immune effector cells

Immune cells	Functions
Monocytes and macrophages	Powerful phagocytes, scavengers, release ROS and nitric oxide (NO), and inflammatory and anti-inflammatory cytokines, present antigen to T cells
PMN	Phagocytic activity, release ROS, NET, degranulation
DC	Antigen uptake in the tissue and antigen presentation to T cells in the lymph nodes
NK	IFN-γ production, antitumor surveillance, strong cytolytic activity
T lymphocytes	Help in type 1 and type 2 immune response (Th1 and Th2), immunosuppressive (Treg), and cytotoxic (CTL). Can maintain memory
B lymphocytes	Mature into plasma cells that are able to produce antibodies

As mentioned above, if the innate immune system does not sense an incurring molecule/agent as dangerous, no defence response is initiated. Innate effector cells can sense the surrounding molecules by using the so-called pattern recognition receptors (PRR). PRR include several types of receptors expressed on the plasma membrane, such as Toll-like receptors (TLR), the C-type lectin receptors (CLR), the scavenger receptors (SR-A and SR-B), and different classes of intracytoplasmic receptor families, such as the NOD-like receptors (NLR) and the RIG-I-like receptors. PRR recognise molecular signatures, shared by different microbial agents (bacteria, viruses, fungi) and called pathogen-associated molecular patterns (PAMP), and molecules from damaged and necrotic/dead cells known as damage-associated molecular patterns (DAMP). Typical PAMP are bacterial endotoxin (or LPS), ssRNA, unmethylated CpG DNA, and mannan, while DAMP encompass heat shock proteins, high mobility group protein B1, fragments of matrix component (fibrinogen, hyaluronate, heparin sulphate), and uric acid crystals. The recent development of nanotechnologies has brought a new class of agents into the focus of the immune system, the engineered nanoparticles (NP), which may have shape and size similar to those of bacteria or uric acid crystals and have the ability to absorb biomolecules onto their surface, thus either masking themselves or becoming more recognisable. If NP are "opsonised" by complement factors or other innate soluble factors, they can be readily taken up by phagocytes and destroyed into phagolysosomes. Actually it is very unlikely that NP may persist in the organism for long, as those that are not rapidly excreted are taken up by phagocytes. This ability of the immune system to recognise and eliminate engineered NP could become an issue in nanomedicine (e.g. in vaccination, in diagnostic imaging procedures, and as therapeutic drug delivery) (Stern and McNeil 2008).

Thus, we need to know more of the interaction of the immune system with NP for two main reasons. The first is for evaluating the possible immunotoxicity of NP, including the risk of pre-pathological alterations of immune responses. Immunonanosafety is a major issue in the evaluation of engineered NP safety, which needs to be considered in the implementation of nanosafety regulations. The second reason is to understand how nanomedicines could persist in the organism long enough

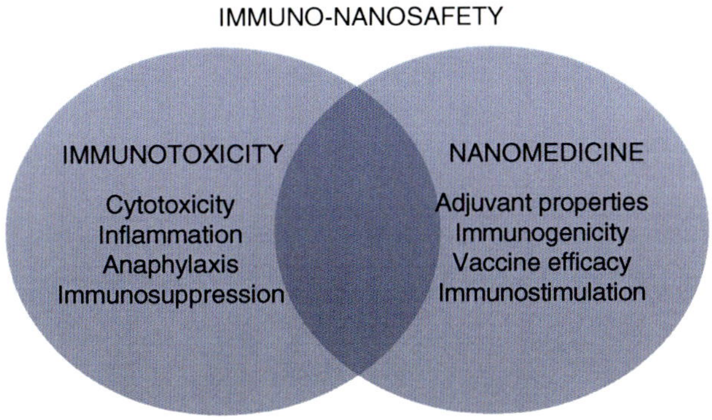

Fig. 9.1 Immuno-nanosafety, immunotoxicity, and nanomedicine. The aims of immuno-nanosafety are the evaluation of immunotoxicity of the new NP in order to avoid undesirable effects (e.g. cytotoxicity, inflammation, anaphylaxis, immunosuppression) and the design of NP with desirable effects for medical application in drug delivery, vaccination carrier, diagnostic imaging agents, and immunostimulation

for exerting their beneficial effects, before being recognised and eliminated by our defensive systems (Fig. 9.1). These questions and the features of the interaction between NP and immune effector molecules and cells will be discussed more in detail in the following sections.

9.2 Nanoparticles: Do They Pose a New Challenge to Our Immune System? Interaction and Consequences

Physico-chemical properties of NP, such as size (ranging 1–500 nm), surface charge (positive, negative, neutral), solubility (colloidal suspensions, polydispersity and formation of agglomerates and aggregates, ion release), hydrophobicity/hydrophilicity, and surface functionality (including the steric effects of particle coatings), influence NP interaction with the immune system (Dobrovolskaia and McNeil 2007; Aggarwal et al. 2009; Dobrovolskaia et al. 2008). In general, larger particles are taken up more efficiently than smaller particles of the same composition and surface properties (Fang et al. 2006). Particles with cationic or anionic surface charge have been shown to be more attractive to phagocytes than neutral particles of the same size (Zahr et al. 2006). Cationic particles are more likely to induce inflammatory reactions than anionic and neutral species (Tan et al. 1999). Hydrophobic NP easily adsorb proteins on their surface, whereas hydrophilic NP are less prone to protein binding (Esmaeili et al. 2008), and positively charged NP adsorb different proteins on their surface as compared to negatively charged particles, thereby inducing distinct types of interaction with cells (Fleischer and Payne 2012). NP can interact with blood or human plasma components (e.g. albumin, apolipoproteins, immunoglobulins, complement components, fibrinogen) and bind these molecules on their surface to form a biomolecular corona (Mahon et al. 2012; Casals et al. 2010; Monopoli et al. 2013). The formation of a bio-corona involves various consequences: (1) NP can be recognised and taken up by innate immune cells much more efficiently when coated with opsonins, immunoglobulins, or complement molecules; (2) the bio-corona may provide a new biological "identity" to NP and consequently affect the type of immune response (of lack of response) stimulated upon NP-immune system interaction; and (3) different NP with a similar bio-corona may nevertheless induce diverse immune reactions (e.g. cytokine release) (Deng et al. 2012), depending on how proteins bind to the surface (e.g. their orientation) and whether and how much binding could cause protein unfolding (Cukalevski et al. 2011).

A large number of studies are focused on the interaction between the immune system and NP that, for medical purposes, have been specifically modified to stimulate immunity or to avoid immune recognition, as in the case of vaccine carriers/adjuvants or drug delivery systems, respectively. In the case of nanomedicine, exposure to NP takes place with a defined administration schedule (route, dose, frequency). Immuno-nanosafety studies, on the other hand, consider accidental exposure to NP that are released in the environment, which may occur by contact (skin), ingestion, or inhalation, at doses and with a frequency that are not known (Table 9.3).

In the case of nanomedicine, several surface modifications are being studied for prolonging the persistence of NP in circulation (thereby allowing them to reach their final target, e.g., brain tumours), by avoiding immune recognition and consequent destruction. Likewise, manipulation of NP size and surface characteristics is also used for improving NP recognition by phagocytic cells in the case of antigen delivery to immune cells for efficient vaccination. For example, surface coating with a polymer such as polyethylene glycol (PEG) is used with some success for shielding NP from recognition by immune cells (Csaba et al. 2006; Goppert and Muller 2005; Muller et al. 1996; Paciotti et al. 2004; Redhead et al. 2001), whereas coating with polyethyleneimine of carbon nanotubes (CNT) favours cellular uptake (Li et al. 2013). Other examples of improved immune targeting for vaccine applications encompass the NP surface functionalisation with polysaccharides such as chitosan or mannose, which mimic a bacterial surface and enhance recognition and uptake by macrophages and DC, and consequently improve immune response to NP-bound antigens (Kim et al. 2006; Cui and Mumper 2002; Cuna et al. 2006).

Table 9.3 Engineered NP and their immune effects

	Intentional NP exposure	Occasional NP exposure
Source	Nanomedicine	Nanotechnologies
Exposure	Medical	Occupational and unintentional
Route	i.v., s.c., i.d., p.o., i.m.	Contact (skin), ingestion, inhalation
Surface coating	With plasma or other body fluids (depending on administration route)	With environment agents (e.g. allergens, LPS) and body surface barriers (e.g. mucus, surfactant)
Immune interaction	Immunostimulation (e.g. adjuvant) Immunosuppression (if loaded with cytotoxic drugs)	Variable (depending on NP type, exposure, interacting factors)

i.v. intravenous, *s.c.* subcutaneous, *i.d.* intradermal, *p.o.* per os, *i.m.* intramuscular

In the case of NP released in the environment, these come generally in contact with the immune cells in the human body's barriers (skin, respiratory mucosa, digestive mucosa) as particles coated with other environmentally borne agents and never as pristine NP. Besides aggregation due to moisture, NP can be coated with a number of other molecules, such as allergens or bacterial products, of which the most common is bacterial endotoxin or lipopolysaccharide (LPS), a membrane component of Gram-negative bacteria that is ubiquitously present in the environment (for reviews, see Lieder et al. 2013; Smulders et al. 2012). Also in the case of engineered NP as synthesised, if special precautions are not taken during the synthetic process, it is highly likely that NP are contaminated with LPS. Since LPS is an excellent stimulator of immune and inflammatory responses, in particular in humans, its presence in NP preparations, if undetected, may significantly interfere with the NP safety profiling during risk assessment testing and induce to erroneously attributing to NP detrimental effects (toxicity, production of reactive oxygen species (ROS), induction of inflammation in vitro and in vivo) caused by LPS-induced macrophage activation (Oostingh et al. 2011; Vetten et al. 2014).

In the following sections, we will review more in depth the interaction of NP with the innate and adaptive immune system, highlighting the cases in which the interaction produces a measurable effect, either immunostimulation or immunosuppression.

9.3 Nanoparticle Interactions with the Innate Immune System

NP can interact with the innate immune system at various levels. NP-induced innate effects that may have detrimental consequences for the host are the activation of the complement system (the major component of the soluble innate immune system) and activation of inflammasomes in macrophages and DC.

When NP reach the bloodstream (e.g. in the case of nanocarriers for anticancer drugs), the first molecules that they meet are the proteins of the complement system. Complement is the most powerful weapon we have against infections and is composed by a number of different proteins and enzymes that activate each other in a cascade in which the various components can identify foreign entities, generate chemotactic and inflammatory factors that trigger a more general reaction, produce opsonins that facilitate phagocytosis and destruction of the foreign element, and physically drill holes in the membrane of the agent (usually a bacterium or an anomalous cell). Depending on the type of recognition, three distinct pathways can be defined (classical, alternative, and lectin) (Murphy 2011). An excessive activation of complement can lead to severe tissue injury, hypersensitivity (allergic) reactions, and anaphylaxis. Therefore, the ability of NP to cause complement activation is a major issue in nanosafety, in particular when these particles are intended for in vivo administration for biomedical applications. NP recognition by complement most likely depends on size, shape, charge, configuration on the NP surface, and accessibility of reactive groups (Moghimi et al. 2011). Thus, the NP surface characteristics must be tuned and modified in order to decrease interaction with and activation of the complement system to an acceptable level, in the case of drug delivery nano-platforms. On the other hand, in the case of vaccination, in which immunostimulation is the goal, local and limited activation of complement by NP may be desirable for enhancing antigen presentation. In a clinical study, it has been shown that nanoliposomes induced hypersensitivity due to complement activation (Chanan-Khan et al. 2003), while another recent study demonstrated that complement activation by NP was responsible for NP uptake by DC and subsequent activation of T cells and generation of an antigen-specific immune response (Reddy et al. 2007). In the case of CNT, conflicting data have been reported with respect to binding and activation of complement (Salvador-Morales et al. 2006; Hamad et al. 2008; Ling et al. 2011).

Innate effector cells encompass phagocytes such as monocytes/macrophages, PMN, and DC. These cells are able to internalise pathogens and any foreign particulate element and to degrade them. In addition, macrophages and DC are also able to "present" fragments of the degraded agent to T lymphocytes and initiate adaptive immunity. These cells are those that first encounter NP when they access the body (either by accidental exposure or upon biomedical application) (Table 9.4). NP recognition by innate immune cells depends on NP surface functionalisation, biocorona, size, and shape. Covering the surface of foreign particles with proteins such as immunoglobulins or complement components is one of the ways for the immune system for improving uptake by phagocytes and elimination of foreign agents. Therefore, the ability of NP to adsorb different types of molecules in a particular tissue microenvironment may make a huge difference in the capacity of the innate immune system to recognise them as foreign entities and to mount an inflammatory response. In general macrophages and DC tend to take up NP covered with altered self-proteins, and they do it by using active endocytic mechanisms depending on the particle size and surface coating (Boraschi et al. 2012; Kettiger et al. 2013). DC, less active than macrophages in particle uptake, can take up NP of 500 nm or smaller, while positive surface charge increases uptake by both DC and macrophages, overall resulting in a better capacity of triggering adaptive immunity (Thiele et al. 2001; Foged et al. 2005; Peer 2012). NP target distinct DC populations in vivo in a size-dependent manner (Manolova et al. 2008). Moreover, NP size can quantitatively affect innate immune responses, and the immune system apparently distinguishes the size of PAMP structures, such as single-stranded RNA, in such a way as to

Table 9.4 NP interaction with innate immune cells

Innate cell	Interaction
Monocytes and macrophages	Phagocytosis, clearance of NP, granuloma formation
PMN	Phagocytosis, digestion of NP by granule enzymes, entrapping NP with NET
DC	Phagocytosis, processing, and presentation of NP-carried antigens

selectively trigger either antiviral (IFN-α) or antibacterial/antifungal (TNF-α) immune responses (Rettig et al. 2010). Particles possessing the longest dimension in the range of 2–3 μm exhibited the highest attachment to macrophages (Champion and Mitragotri 2006), and spherical NP are taken up much faster and more efficiently than rod-shaped NP (Champion et al. 2007; Ferrari 2008). It is evident that a better understanding of the mechanisms by which cells are capable to sensing such shape and size differences in NP could be exploited to achieve more efficient drug delivery (Meng et al. 2011).

NP-loaded macrophages can form granulomas (Bartneck et al. 2010), while PMN can interact with NP also by entrapping them with NET (Bartneck et al. 2010). It has been shown that PMN can degrade NP (both intracellularly and extracellularly) with enzymes and oxidising compounds. In the case of CNT, it is been observed that myeloperoxidase, produced by PMN and other innate leukocytes, can effectively degrade the nanotubes that, once degraded, lose their ability to cause lung inflammation when inhaled by mice (Kagan et al. 2010; Kotchey et al. 2012).

When the cells of innate immune system recognise pathogens or harmful substances or damaged cells, this recognition triggers an innate/inflammatory reaction. The reaction starts with the production of chemokines and chemotactic complement components by resident macrophages in response to the dangerous agent (the most potent inducer being bacterial LPS), which attract more leukocytes to inflamed tissue. The inflammatory reaction then involves phagocytosis and production of toxic factors and degrading enzymes, in order to destroy the invaders. Once the goal is achieved, the reaction stops and the same macrophages take care of repairing the damages suffered by the surrounding tissues so as to restore tissue homeostasis. Testing the ability of NP to induce inflammation, either in vivo or in vitro, is one of the hallmarks of nanohazard research and regulatory assessments. However, this is not an easy task because of the NP capacity to adsorb LPS on the surface (Lieder et al. 2013). In fact, trace amounts of LPS are very frequently present in NP preparations, which may be partly if not completely responsible for the inflammatory effects attributed to NP in many publications. This may possibly explain the controversial results reported in the literature about the toxic and inflammatory effects of several types of NP. For instance, it has been shown that CNT do not induce an inflammatory reaction when purified (Pulskamp et al. 2007). Therefore, an important initial step in the preclinical characterisation of NP side effects, and in nanosafety assessment in general, is testing for LPS contamination. This can be done, after adequate validation, by adapting to NP the Limulus amoebocyte lysate (LAL) assay, approved by regulatory agencies worldwide. The importance of the LAL test in NP characterisation and the challenging aspects of its application are reviewed elsewhere (Smulders et al. 2012; Hall et al. 2007; Dobrovolskaia et al. 2013). It has been reported that cationic particles are more able to induce inflammation than anionic and neutral species (Tan et al. 1999) and that cationic nanoliposomes can induce DC maturation, which is important in the inflammatory response (Cui et al. 2005). Also, induction of inflammatory cytokines (which may be responsible for severe inflammation) has been reported in studies in which in vitro screening of NP-mediated cytokine response correlated well with in vivo cytokine induction

(Moyano et al. 2012). However, the lack of characterisation of the possible LPS contamination of NP in these studies does not allow the straightforward interpretation of the data.

The ability of triggering a localised and controlled innate/inflammatory reaction is at the basis of immunostimulation by vaccine adjuvants. Indeed, micro- and nanoparticles have been successfully used for many years in vaccines for efficient establishment of protective immunity. Nowadays, nanomaterials can be intentionally engineered to optimise their ability to stimulate the innate immune reaction required for successful immunisation (Xiang et al. 2012). The central event in initiation of the innate/inflammatory response is the activation of the inflammasome. The major inflammasome in innate immune cells is the NLRP3 (NLR-related protein 3) inflammasome, a cytoplasmic multi-protein complex that assembles in phagocytic cells in response to inflammatory stimuli (Latz et al. 2013). The NLRP3 inflammasome coordinates the maturation and secretion of the inflammatory cytokines IL-1β and IL-18 through caspase-1 activation. Activation of NLRP3 and subsequent inflammasome assembly occur in response to foreign molecules and agents that gain access to the cell cytoplasm, including bacterial components, viruses, and micro- and nanocrystals (e.g. uric acid crystals) (Martinon et al. 2006). Particles such as crystalline silica, asbestos fibres, and aluminium salts were also found able to induce NLRP3 inflammasome activation (Hornung et al. 2008; Dostert et al. 2008; Cassel et al. 2008). The NLRP3 inflammasome is the choice molecular target of vaccine adjuvants (Eisenbarth et al. 2008); therefore, NP are being exploited in optimising vaccine design by taking advantage, in addition to their capacity to carry antigens, of their ability to target and activate the inflammasome (Demento et al. 2009). For example, the exposure of macrophages to TiO_2 and SiO_2 NP activates the NLRP3 inflammasome leading to IL-1β release (Yazdi et al. 2010). Long needle-like multiwalled CNT (MWCNT) can activate the NLRP3 inflammasome upon uptake by LPS-primed human macrophages (Palomäki et al. 2011), but also metal contamination in MWCNT preparations can have the same effect of NLRP3 inflammasome activation (Hamilton et al. 2012).

There is a subtle distinction between induction of controlled inflammation (as it occurs naturally during a defensive innate response and as it is designed in vaccination) and triggering of a destructive inflammatory reaction (as it occurs acutely in some infections, meningitis, streptococcal pneumonia, etc., and chronically in diseases such as rheumatoid arthritis), a distinction mainly based on the persistence of the triggering agent and on regulatory mechanisms that control the extent and the duration of the reaction. This suggests us that caution should be taken in distinguishing between NP inflammatory effects that can be good (if properly controlled) or detrimental (when uncontrolled). In nanosafety assessment, the duration of the inflammatory reaction, rather than its occurrence, should be considered as a possible indication of hazard. Likewise, in the design of novel vaccines, the use of short-lived nanoparticulate carriers (such as biodegradable liposomes) may allow us to control the duration of the inflammatory stimulus, so as to obtain the immunostimulatory effect while avoiding chronicisation of the reaction. Table 9.5 summarises undesirable and desirable interactions between innate immune system and NP.

Table 9.5 NP interaction with the innate immune system: undesirable and desirable effects

Interaction	Undesirable effects	Desirable effects
Complement system	Anaphylaxis, allergic reaction	Facilitation uptake by APC, enhanced antigen presentation and immunisation
Monocytes, macrophages, DC	Acute/chronic inflammation	Drug delivery to innate cells, presentation of NP-carried antigens
Inflammasome activation	Excessive inflammatory cytokines (cytokine storm)	Adjuvanticity, immunostimulation

9.4 NP Interactions with Adaptive Immune System

The effector cells of the adaptive immunity are T and B lymphocytes, the former principally involved in receiving innate signals and in helping B-cell activation and the latter able to produce specific antibodies that promote pathogen elimination. Depending on the signals they receive during interaction with antigen-presenting cells, T cells can differentiate into several different functional types that have different roles. $CD8^+$ T cells may become cytotoxic and able to kill virus-infected cells. Th1, Th17, and Th22 cells are associated with type 1 immune response, Th2 and Th9 are responsible for type 2 immune response, and regulatory T cells (Treg) are responsible for suppressing immune responses and inducing tolerance. Type 1 immune response is the classical antibacterial response, while type 2 immune response is involved in anti-parasite defence and allergic diseases (Hirahara et al. 2013; Nakayamada et al. 2012). B cells (that become plasma cells upon activation) are the cells able to produce and secrete specific antibodies during an immune response against invading microorganisms or other dangerous agents.

The effects of NP on adaptive immunity have been only partially studied, with conflicting results. For example, polystyrene NP (<100 nm) promoted CD8 and CD4 T-cell responses and were associated with higher antibody levels than larger particles (>500 nm) (Xiang et al. 2006). Inhalation of CNT suppresses B-cell functions through TGF-β produced by alveolar macrophages (Mitchell et al. 2009) or IL-10 (Mitchell et al. 2007).

An effect of NP on adaptive immunity that is worth investigating is the ability to induce Th1 vs. Th2 responses. This is not easy to investigate, since rat and mouse strains can be differently biased towards Th1 or Th2 responses, and therefore, the results may depend on the specific animal model used, rather than to the specific capacity of NP to promote type 1 or type 2 responses. Small engineered NP were found able to induce Th1 responses (Lutsiak et al. 2006; Chong et al. 2005; Cui and Mumper 2002; Cui et al. 2004), while other NP can induce Th2 cytokine production and enhanced immunoglobulin production (Rajananthanan et al. 1999; van Zijverden and Granum 2000). Particle size has been reported as a factor that can influence the type or T-dependent response to NP-loaded antigens, this being possibly due to the different endocytic pathway by which the antigen is taken up and processed (Mottram et al. 2007).

Little is known on the possible antigenicity (capacity of being recognised by antibodies and T cells) and immunogenicity (capacity of inducing an adaptive immune response) of NP. NP antigenicity has never been formally demonstrated. Two studies have reported the generation of specific antibodies upon immunisation with C_{60} fullerene bound to bovine serum albumin (BSA) (Braden et al. 2000; Chen et al. 1998) or with polyamidoamine dendrimers conjugated to BSA (Lee et al. 2004). This kind of response, however, is only triggered in the case that the NP (very small) are carried by a large protein, the so-called carrier effect, while no response can be raised against the small entities alone, even in the presence of strong adjuvant (Agashe et al. 2006; Andreev et al. 2000; Kreuter 1995; Masalova et al. 1999; Roberts et al. 1996; Tomii and Masugi 1991). However, the carrier effect may work also the other way around, as it has been reported that administration of PEG-coated NP could induce an antibody response against PEG, a fact that may hamper the use of PEG in the design of stealth NP (Judge et al. 2006; Wang et al. 2007; Ishida et al. 2007; Koide et al. 2010). Table 9.6 shows the main consequences of immune reaction to NP.

9.5 NP and Healthy vs. Frail Immunity

When NP come in contact with a healthy body, the interaction with the cells and factors of innate immune system is immediate and can result in two different situations. If the NP are seen as harmless, there will be no reaction, and NP will be rapidly excreted through renal filtration. On the other hand, if NP are considered a danger, an inflammatory defensive reaction will ensue (Murphy 2011). In a healthy subject, an inflammatory reaction towards NP (as for the majority of particles and agents that we come daily in contact with) is not symptomatic and has no pathological consequence. On the contrary, the innate/inflammatory reaction is the normal defensive immune response, required to eliminate possible dangers and to subsequently re-establish tissue integrity. This, however, may not be the case in subjects with frail immunity. Disease and ageing are the most frequent causes of immunological frailty. Immunological immaturity, and consequent inadequate immune

Table 9.6 Immune reactions to NP

Reaction	Consequence
Complement activation	Tissue damage, intravascular coagulation, anaphylaxis
Opsonisation (with complement, immunoglobulins, or collectins)	Enhanced phagocytosis and destruction of NP
Macrophage activation	Inflammation Beneficial (if controlled, e.g., adjuvanticity) Detrimental (cytokine storm, acute/chronic inflammation, autoimmunity)
Antibody response to haptens on NP surface (e.g. PEG)	"Carrier effect", i.e., production of antibodies that neutralise/eliminate the hapten-carrying NP

reactivity, is typical in babies and very young children. Some diseases cause immunosuppression, such as tumours and infections (e.g. HIV-1 infection), while others are characterised by immunostimulation, as in the case of hypersensitivity reactions (e.g. allergies) and chronic and degenerative inflammatory diseases autoimmunity. In the case of ageing, the elderly immune system is generally characterised by a reduced frequency of naïve T cells and by increased memory T cells. The relative inability to recognise novel infections and generally less adequate capacity of reacting to them make the elderly population more susceptible to diseases (in particular respiratory illnesses) (El Solh and Ramadan 2006).

It is therefore possible that foreign agents such as NP, even when unable to trigger a detrimental inflammatory response in immunocompetent adult healthy people, may pose a risk in immunologically frail individuals. Thus, it is of particular importance that nanosafety studies target immunocompromised and immunologically frail people, as this group is more at risk of developing pathological consequences upon exposure to challenges (including NP) that are harmless in immunologically healthy individuals.

Despite this urgency, little is known at present about the interaction of NP with frail immunity. Among the few studies available, it has been shown that inhalation of high doses of SiO_2 NP caused more severe cardiopulmonary disorders in old rats as compared to young animals (Chen et al. 2008), while high doses of metal NP showed more pronounced neurotoxicity in very young or very old rats as compared to adult animals (Sharma et al. 2013). A very nice recent study has shown that SiO_2 NP and CNT increase protein citrullination in vitro and in vivo, suggesting a possible contribution to pathogenesis of rheumatoid arthritis, in which autoantibodies to citrullinated self-protein are a hallmark (Mohamed et al. 2012).

9.6 How to Assess Immuno-Nanosafety: Animal Models vs. In Vitro Models

Understanding the impact of NP on the immune system is rapidly developing as an important area in modern toxicology, considering the high impact of nanotechnology-based formulations for diagnostic and therapeutic applications and the generalised public concern for the possible risks posed by the nanomaterials present in a large number of consumers' products. The goal of immunosafety studies is to identify potential concerns before a new drug or a new product is marketed (Fig. 9.2). In the case of NP, which have a particulate nature that facilitates their recognition by the innate immune system, it is particularly important to make sure that NP (in particular nanomedicines that are administered deliberately and in high doses to immunologically compromised patients) do not induce excessive immune/inflammatory reactions that may damage the body and cause pathologies. Assessing immuno-nanosafety requires reliable in vitro and in vivo tests/models, in which the immune mechanisms under scrutiny are the same as they occur in man, to avoid assessing effects that are not representative of human responses. In vivo models rely on the use of animal species, either rodents (rat and mouse) or non-rodents (rabbit, pig, etc.), and provide a

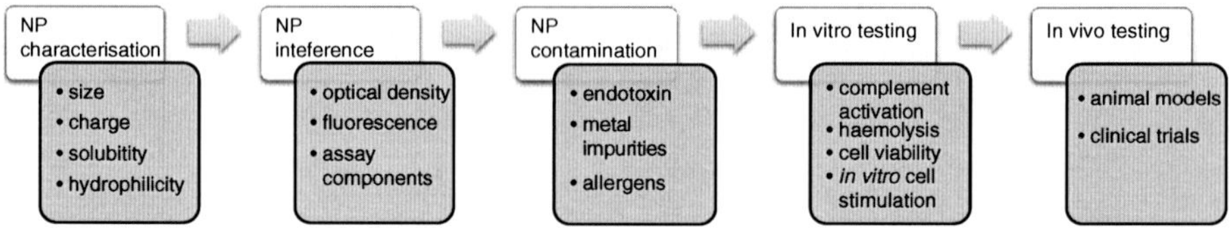

Fig. 9.2 Experimental design flow for immuno-nanotoxicological studies. Several problems are encountered during the evaluation of the effects of NP on the human immune system, from accurate NP characterisation, interference with assay components and readouts, and chemical or endotoxin contaminations to appropriate in vitro and in vivo tests. Considering and resolving these challenges is of great importance for the reliable assessment of NP immunosafety

series of information that cannot be obtained otherwise, in terms of pharmacokinetics, pharmacodynamics, biodistribution, long-term effects, and most of all the development of localised immune responses in the context of a specific tissue microenvironment. However, extrapolation of findings from animal models to humans is sometimes challenging, due to some differences in immune responses and molecular pathway usage between mice and men (Seok et al. 2013). In vitro assays have the advantage of rapidity and robustness when using murine and human continuous cell lines with stable characteristics. The drawback is that these cell lines are mostly tumour derived or transformed; thus they may not reflect the reactivity of normal primary cells. The great variability of published results describing the inflammatory or toxic effects of NP for a variety of cell lines in culture may in part depend on the poorly representative systems used. This issue has been recently reviewed (Oostingh et al. 2011).

Here, we will focus on the correlation between in vitro tests and in vivo experiments. It is important to mention that in vitro assays are prone to significant interference when NP are used. In fact, the majority of standard assays for cytotoxicity, cell proliferation, and cytokine production are based on optical or fluorescent readouts (e.g. MTT-based tests and ELISA assays). The presence of NP in the assay may cause, depending on their specific optical properties, false-positive results. Interference or false-positive results may also arise from surfactants, capping agents, or synthesis by-products present in the NP preparation (e.g. Triton X-100, sodium citrate, cetyltrimethylammonium bromide) or from the NP catalytic properties and intrinsic fluorescence. On the other hand, false-negative results may be caused by the capacity of NP to absorb assay-specific test substance (e.g. cytokines, LAL substrate or enzyme) or to quench luminescence and fluorescence. Thus, application of these assays for assessing NP effects requires a case-to-case validation, to make sure that those NP in that particular assay set-up do not interfere with the test readouts (Dobrovolskaia et al. 2009) or, if NP are eliminated by centrifugation before the test, to make sure that NP did not bind and subtract the factors to be measured.

With the precautions mentioned above (validation for reliability and predictability of human responses), there is no doubt that in vitro screening of NP formulations allows for rapid, low-cost, time-saving, and high-throughput evaluation compared to in vivo models. For NP that are administered intravenously, for drug delivery, or for diagnostic imaging procedures, it is particularly important to examine haematocompatibility and possible immunotoxicity. The international recognised standard ISO 10993 guidelines recommend the following in vitro tests to identify severe acute toxicities: haemolysis (to evaluate the effects on erythrocytes), anaphylaxis (to evaluate the activation of complement system), and thrombogenicity (to evaluate the effects on platelets and the induction blood clotting). Recently, other markers for NP acute toxicities have been included, i.e. phagocytosis, pyrogenicity, cytokine production, and leukocyte/lymphocyte proliferation. For these measures, different murine and human monocytic cell lines are commonly used, such as RAW264.7 (murine macrophages), U937 (human monocyte-macrophage cell line), THP-1 (human monocytic cell line), and human primary cells such as peripheral blood mononuclear cells (PBMC), monocytes, and macrophages. Innate and adaptive immune responses often work together to induce an efficient protection against foreign intrusions, and only in vivo experiments allow considering the complex molecular and cellular network that link the two. Since the aim of in vitro test is to evaluate the NP formulations that can be toxic in vivo, thus one of the major challenges in the in vitro testing of NP immunotoxicity is understanding in vitro assay predictability of corresponding immunotoxicities in vivo. Dobrovolskaia and McNeil reviewed the literature comparing performance of in vitro and in vivo immunotoxicity tests in order to establish assays with "good or fair" in vitro-in vivo correlation (Dobrovolskaia and McNeil 2013). They claim that in vitro assays for haemolysis, complement activation, opsonisation and phagocytosis, and cytokine production are among the best predictive in vitro tests (performed principally on mononuclear phagocytes) of NP in vivo toxicities (performed on rabbits, dogs, and rats). Three points arise from the studies examined by Dobrovolskaia and McNeil to assess immuno-nanotoxicity: (1) the importance of the animal models chosen, (2) the importance of cytokine production test in vitro prior to in vivo testing, and (3) the use of the in vitro phagocytosis test for predicting in vivo capture of NP by the mononuclear phagocyte system (MPS). For example, different animal species have different complement-mediated hypersensitivity reaction to complement-activating substances, with pigs and dogs responding more similarly to man than rats (Szebeni et al. 2007). There are examples in which in vitro screening of NP-induced cytokine response correlates well with in vivo cytokine induction and that emphasised the importance of the cytokine test in vitro prior to testing nanomaterials in vivo (Moyano et al. 2012). Two nanoformulations of metal oxide NP with identical cores and different surface chemistries were tested in vivo in rats and rabbits and in vitro using primary human PBMC. Only one formulation caused cytokine production and inflammation in animals and induced inflammatory cytokines in vitro in human PBMC. It should be emphasised, however, that cytokine production represents a reaction to NP, and it does not necessarily imply a pathological consequence, unless the reaction is exaggerated and persistent. Moreover, it is known that in vitro NP phagocytosis correlates with in vivo particle retention by the MPS (Caron et al. 2012). For example, in vitro NP phagocytosis by RAW264.7 murine macrophages was shown to correlate with the accumulation of polymeric NP in the spleens and livers of rats in an in vivo biodistribution study (Gaucher et al. 2009). Likewise, in vitro phagocytosis of cross-linked albumin nanospheres by human U937 monocyte-like tumour cells and by murine peritoneal macrophages correlated with liver uptake in rats (Roser et al. 1998). NP entrapment in murine spleen was predicted in an in vitro spleen tissue culture model (Demoy et al. 1999). All these studies prove that the in vitro phagocytosis assay is robust (it works with primary cells or cell lines, of human or mouse origin) and predictive of in vivo capture by the MPS. This finding however cannot be generalised. In assessing differential uptake of polystyrene functionalised NP, it has been shown that macrophage-like cell lines do not reproduce the activity of primary macrophages (Lunov et al. 2011). This suggests that each assay should be individually validated, before cell lines or animals can be used for predicting human reactivity.

9.7 The "Ecological Immunity": Nanoparticles as Environmental Stressors (Damage vs. Evolutionary Shaping)

It is worth considering that the immune system of the environmental species is a key target of toxic compounds and that, similarly to what we have described for human health effects, the integrity of environmental immune responses is central to the well-being of environmental species. Thus, environmental nanosafety studies must consider the effects of NP of the immune responses of environmental species as a major element in risk assessment.

The immune system of invertebrates presents two advantages to the researcher: conservation of the fundamental mechanisms of the innate immune response, with high defensive value (Kvell et al. 2007), and adaptation to a wide range of physical, chemical, and biological molecules as they are widespread in all types of environment. Invertebrates are the most abundant animal species, widespread in every type of the environment and exposed to a wide range of chemical and biological attacks. Thus, invertebrate immunity provides an ideal model system for investigating the host survival in different habitats and the evolution of immune defence to cope with both natural and anthropogenic stressors (Cooper 2010; Chang 2009), including environmental contamination with NP. The "ecological immunity" is an expanding field that examines the impact of environmental stressors on the immune reaction and how these stresses act to create and maintain different immune functions in the context of evolution and ecology (Söderäll 2010). NP can be considered as a new kind of stressor molecules present in the environment, which may be able to contribute to the evolutionary shaping of the immune system, besides their immediate capacity to come in contact with the immune system of environmental species, and possibly inducing detrimental responses. Based on their number and species diversity, their role in different habitats, and their potential to transfer NP through food chains, invertebrates are excellent organisms for assessing the environmental impact of NP as well as for directly testing nanotoxicity (Marsh and May 2012). The impact of NP on the invertebrate immune functions is still poorly investigated, and most of the available data have been generated in bivalve molluscs (Mollusca, Lophotrochozoa) and earthworms

(Lumbricidae, Oligochaeta, Annelida). Both are good models that can allow us to investigate NP effects in two different environments, i.e. the marine environment and the earth environment.

Bivalves are widespread in freshwater, estuarine, and coastal environment, where they function as sentinel organisms to evaluate the biological impact of aggregated/ agglomerated NP (as they are expected to be accumulated in these areas). The marine mussel *Mytilus* is the invertebrate species so far most used for assessing the NP effects on immune cells and responses in vitro and in vivo (Barmo et al. 2013; Ciacci et al. 2012; Canesi et al. 2012). While it is impossible to generalise the reaction of *Mytilus* haemocytes to NP, it can be said that stimulation of several immune functions can be observed (e.g. stimulation of oxidative burst and NO production, increased transcription of antimicrobial peptides, changes in total haemocyte counts), but no direct cytotoxicity.

Due to their permanent contact with soil, earthworms are widely used in standard toxicity tests for studies of soil pollution, including nanopollution. Exposure to NP may have different effects on the earthworm immune responses, these being mainly represented by stress reactions (e.g. oxidative stress; Whitfield Aslund et al. 2012) that can bring about reduced growth and reproduction and also mortality (Li et al. 2011; El-Temsah and Joner 2012; Unrine et al. 2010).

The use of invertebrate immunity as a nanocontamination biomonitoring tool is very promising. Given the impressive similarities between the innate immune mechanisms of invertebrates and vertebrates (including man), it is also tempting to speculate that these models may be exploited for predicting effects on human health.

9.8 Conclusions

The immune system, responsible for maintaining the body's physical and functional integrity against detrimental agents, is the first system coming in contact with engineered NP that enter the human body. The immune reaction to NP is dictated by the circumstances in which this interaction occurs, but it generally results in recognition and elimination of those NP that were not already excreted. Nanomedicines that need to persist long in the organism must be engineered in a way that decreases immune recognition of their surface and consequent elimination. In addition, they should avoid triggering innate immunity, in particular the complement system and subsequent inflammation, to avoid causing serious tissue and organ damage. On the other hand, nanovaccines that must specifically target immune cells should have different surface characteristics, since they need to be efficiently taken up by phagocytes and stimulate a local immune/inflammatory reaction.

Assessing the immune response of invertebrates to NP has a double advantage, evaluating the nanorisk for the environment and predicting human health-related effects, since invertebrate and human innate immunity/inflammation share many common features.

Acknowledgements This work was supported by the EU FP7 grant HUMUNITY (PITN-GA-2012-316383) and by the grant 2011–2114 of Fondazione Cariplo, Milano, Italy.

References

Agashe HB, Dutta T, Garg M et al (2006) Investigations on the toxicological profile of functionalized fifth-generation poly (propylene-imine) dendrimer. J Pharm Pharmacol 58:1491–1498

Aggarwal P, Hall JB, McLeland CB et al (2009) Nanoparticle interaction with plasma proteins as it relates to particle biodistribution, biocompatibility and therapeutic efficacy. Adv Drug Deliv Rev 61:428–437

Andreev SM, Babakhin AA, Petrukhina AO et al (2000) Immunogenic and allergenic properties of fullerene conjugates with amino acids and proteins. Dokl Biochem 370:4–7

Barmo C, Ciacci C, Canonico B et al (2013) In vivo effects of n-TiO2 on digestive gland and immune function of the marine bivalve Mytilus galloprovincialis. Aquat Toxicol 9:132–133

Bartneck M, Keul HA, Zwadlo-Klarwasser G et al (2010) Phagocytosis independent extracellular nanoparticle clearance by human immune cells. Nano Lett 10:59–63

Boraschi D, Costantino L, Italiani P (2012) Interaction of nanoparticles with immunocompetent cells: nanosafety considerations. Nanomedicine (Lond) 7:121–131

Braden BC, Goldbaum FA, Chen BX et al (2000) X-ray crystal structure of an anti-Buckminsterfullerene antibody fab fragment: biomolecular recognition of C(60). Proc Natl Acad Sci U S A 97:12193–12197

Canesi L, Ciacci C, Fabbri R et al (2012) Bivalve molluscs as a unique target group for nanoparticle toxicity. Mar Environ Res 76:16–21

Caron W, Rawal S, Song G et al (2012) Bidirectional interaction between nanoparticles and cells of the mononuclear phagocyte system. In: Yarmush ML, Shi D (eds) Frontiers in nanobiomedical research. World Scientific Publishing, Singapore

Casals E, Pfaller T, Duschl A et al (2010) Time evolution of the nanoparticle protein corona. ACS Nano 4:3623–3632

Cassel SL, Eisenbarth SC, Iyer SS et al (2008) The Nalp3 inflammasome is essential for the development of silicosis. Proc Natl Acad Sci U S A 105:9035–9040

Champion JA, Mitragotri S (2006) Role of target geometry in phagocytosis. Proc Natl Acad Sci U S A 103:4930–4934

Champion JA, Katare YK, Mitragotri S (2007) Particle shape: a new design parameter for micro- and nanoscale drug delivery carriers. J Control Release 121:3–9

Chanan-Khan A, Szebeni J, Savay S et al (2003) Complement activation following first exposure to pegylated liposomal doxorubicin (Doxil): possible role in hypersensitivity reactions. Ann Oncol 14:1430–1437

Chang ZL (2009) Recent development of the mononuclear phagocyte system: in memory of Metchnikoff and Ehrlich on the 100th anniversary of the 1908 Nobel Prize in Physiology or Medicine. Biol Cell 101:709–721

Chen BX, Wilson SR, Das M et al (1998) Antigenicity of fullerenes: antibodies specific for fullerenes and their characteristics. Proc Natl Acad Sci U S A 95:10809–10813

Chen Z, Meng H, Xing G et al (2008) Age-related differences in pulmonary and cardiovascular responses to SiO2 nanoparticle inhalation: nanotoxicity has susceptible population. Environ Sci Technol 42:8985–8992

Chong CS, Cao M, Wong WW et al (2005) Enhancement of T helper type 1 immune responses against hepatitis B virus core antigen by PLGA nanoparticle vaccine delivery. J Control Release 102:85–99

Ciacci C, Canonico B, Bilanicova D et al (2012) Immunomodulation by different types of N-oxides in the hemocytes of the marine bivalve Mytilus galloprovincialis. PLoS One 7:e36937

Cooper EL (2010) Evolution of immune systems from self/not self to danger to artificial immune systems (AIS). Phys Life Rev 7:55–78

Csaba N, Sanchez A, Alonso MJ (2006) PLGA:poloxamer and PLGA:poloxamine blend nanostructures as carriers for nasal gene delivery. J Control Release 113:164–172

Cui Z, Mumper RJ (2002) Coating of cationized protein on engineered nanoparticles results in enhanced immune responses. Int J Pharm 238:229–239

Cui Z, Patel J, Tuzova M et al (2004) Strong T cell type-1 immune responses to HIV-1 Tat (1–72) protein-coated nanoparticles. Vaccine 22:2631–2640
Cui Z, Han SJ, Vangasseri DP et al (2005) Immunostimulation mechanism of LPD nanoparticle as a vaccine carrier. Mol Pharm 2:22–28
Cukalevski R, Lundqvist M, Oslakovic C et al (2011) Structural changes in apolipoproteins bound to nanoparticles. Langmuir 27:14360–14369
Cuna M, Alonso-Sandel M, Remunan-Lopez C et al (2006) Development of phosphorylated glucomannan-coated chitosan nanoparticles as nanocarriers for protein delivery. J Nanosci Nanotechnol 6:2887–2895
Demento SL, Eisenbarth SC, Foellmer HG et al (2009) Inflammasome-activating nanoparticles as modular systems for optimizing vaccine efficacy. Vaccine 27:3013–3021
Demoy M, Andreux JP, Weingarten C et al (1999) In vitro evaluation of nanoparticles spleen capture. Life Sci 64:1329–1337
Deng ZJ, Liang M, Toth I et al (2012) Plasma protein binding of positively and negatively charged polymer-coated gold nanoparticles elicits different biological responses. Nanotoxicology 7:314–322
Dobrovolskaia MA, McNeil SE (2007) Immunological properties of engineered nanomaterials. Nat Nanotechnol 2:469–478
Dobrovolskaia MA, McNeil SE (2013) Understanding the correlation between in vitro and in vivo immunotoxicity tests for nanomedicines. J Control Release 172:456–466
Dobrovolskaia MA, Aggarwal P, Hall JB et al (2008) Preclinical studies to understand nanoparticle interaction with the immune system and its potential effects on nanoparticle biodistribution. Mol Pharm 5:487–495
Dobrovolskaia MA, Germolec DR, Weaver J (2009) Evaluation of nanoparticle immunotoxicity. Nat Nanotechnol 4:411–414
Dobrovolskaia MA, Neun BW, Clogston JD et al (2013) Choice of method for endotoxin detection depends on nanoformulation. Nanomedicine (Lond) 9(12):1847–56. [Epub ahead of print]
Dostert C, Pétrilli V, Van Bruggen R et al (2008) Innate immune activation through Nalp3 inflammasome sensing of asbestos and silica. Science 320:674–677
Eisenbarth SC, Colegio OR, O'Connor W et al (2008) Crucial role for the Nalp3 inflammasome in the immunostimulatory properties of aluminium adjuvants. Nature 453:1122–1126
El Solh AA, Ramadan H (2006) Overview of respiratory failure in older adults. J Intensive Care Med 21:345–351
El-Temsah YS, Joner EJ (2012) Ecotoxicological effects on earthworms of fresh and aged nano-sized zero-valent iron (nZVI) in soil. Chemosphere 89:76–82
Esmaeili F, Ghahremani MH, Esmaeili B et al (2008) PLGA nanoparticles of different surface properties: preparation and evaluation of their body distribution. Int J Pharm 349:249–255
Fang C, Shi B, Pei YY et al (2006) In vivo tumor targeting of tumor necrosis factor alpha-loaded stealth nanoparticles: effect of MePEG molecular weight and particle size. Eur J Pharm Sci 27:27–36
Ferrari M (2008) Nanogeometry: beyond drug delivery. Nat Nanotechnol 3:131–132
Fleischer CC, Payne CK (2012) Nanoparticle surface charge mediates the cellular receptors used by protein-nanoparticle complexes. J Phys Chem B 116:8901–8907
Foged C, Brodin B, Frokjaer S et al (2005) Particle size and surface charge affect particle uptake by human dendritic cells in an in vitro model. Int J Pharm 298:315–322
Gaucher G, Asahina K, Wang J et al (2009) Effect of poly(N-vinyl-pyrrolidone)-block-poly(D, L-lactide) as coating agent on the opsonization, phagocytosis, and pharmacokinetics of biodegradable nanoparticles. Biomacromolecules 10:408–416
Goppert TM, Muller RH (2005) Protein adsorption patterns on poloxamer- and poloxamine-stabilized solid lipid nanoparticles (SLN). Eur J Pharm Biopharm 60:361–372
Hall JB, Dobrovolskaia MA, Patri AK et al (2007) Characterization of nanoparticles for therapeutics. Nanomedicine 2:789–803
Hamad I, Hunter AC, Rutt KJ et al (2008) Complement activation by PEGylated single-walled carbon nanotubes is independent of C1q and alternative pathway turnover. Mol Immunol 45:3797–3803

Hamilton RF Jr, Buford M, Xiang C et al (2012) NLRP3 inflammasome activation in murine alveolar macrophages and related lung pathology is associated with MWCNT nickel contamination. Inhal Toxicol 24:995–1008

Hirahara K, Poholek A, Vahedi G et al (2013) Mechanisms underlying helper T cell plasticity: implications for immune-mediated disease. J Allergy Clin Immunol 131:1276–1287

Hornung V, Bauernfeind F, Halle A et al (2008) Silica crystals and aluminum salts activate the NALP3 inflammasome through phagosomal destabilization. Nat Immunol 9:847–856

Ishida T, Wang X, Shimizu T et al (2007) PEGylated liposomes elicit an anti-PEG IgM response in a T cell-independent manner. J Control Release 122:349–355

Judge A, McClintock K, Phelps JR et al (2006) Hypersensitivity and loss of disease site targeting caused by antibody responses to PEGylated liposomes. Mol Ther 13:328–337

Kagan VE, Konduru NV, Feng W et al (2010) Carbon nanotubes degraded by neutrophil myeloperoxidase induce less pulmonary inflammation. Nat Nanotechnol 5:354–359

Kettiger H, Schipanski A, Wick P et al (2013) Engineered nanomaterial uptake and tissue distribution: from cell to organism. Int J Nanomedicine 8:3255–3269

Kim TH, Nah JW, Cho MH et al (2006) Receptor-mediated gene delivery into antigen presenting cells using mannosylated chitosan/DNA nanoparticles. J Nanosci Nanotechnol 6:2796–2803

Koide H, Asai T, Hatanaka K et al (2010) T cell-independent B cell response is responsible for ABC phenomenon induced by repeated injection of PEGylated liposomes. Int J Pharm 392:218–223

Kotchey GP, Hasan SA, Kaparlov AA et al (2012) A natural vanishing act: the enzyme-catalyzed degradation of carbon nanomaterials. Acc Chem Res 45:1770–1781

Kreuter J (1995) Nanoparticles as adjuvants for vaccines. Pharm Biotechnol 6:463–472

Kvell K, Cooper EL, Engelmann P et al (2007) Blurring borders: innate immunity with adaptive features. Clin Dev Immunol 2007:83671

Latz E, Xiao TS, Stutz A (2013) Activation and regulation of the inflammasomes. Nat Rev Immunol 13:397–411

Lee SC, Parthasarathy R, Botwin K et al (2004) Biochemical and immunological properties of cytokines conjugated to dendritic polymers. Biomed Microdevices 6:191–202

Li LZ, Zhou DM, Peijnenburg WJ et al (2011) Toxicity of zinc oxide nanoparticles in the earthworm, Eisenia fetida and subcellular fractionation of Zn. Environ Int 37:1098–1104

Li R, Wang X, Ji Z et al (2013) Surface charge and cellular processing of covalently functionalized multiwall carbon nanotubes determine pulmonary toxicity. ACS Nano 7:2352–2368

Lieder R, Petersen PH, Sigurjónsson OE (2013) Endotoxins—the invisible companion in biomaterials research. Tissue Eng Part B Rev 19:391–402

Ling WL, Biro A, Bally I et al (2011) Proteins of the innate immune system crystallize on carbon nanotubes but are not activated. ACS Nano 5:730–737

Lunov O, Syrovets T, Loos C et al (2011) Differential uptake of functionalized polystyrene nanoparticles by human macrophages and a monocytic cell line. ACS Nano 5:1657–1669

Lutsiak ME, Kwon GS, Samuel J (2006) Biodegradable nanoparticle delivery of a Th2-biased peptide for induction of Th1 immune responses. J Pharm Pharmacol 58:739–747

Mahon E, Salvati A, Baldelli Bombelli F et al (2012) Designing the nanoparticle-biomolecule interface for "targeting and therapeutic delivery". J Control Release 161:164–174

Manolova V, Flace A, Bauer M et al (2008) Nanoparticles target distinct dendritic cell populations according to their size. Eur J Immunol 38:1404–1413

Marsh EK, May RC (2012) Caenorhabditis elegans, a model organism for investigating immunity. Appl Environ Microbiol 78:2075–2081

Martinon F, Pétrilli V, Mayor A et al (2006) Gout-associated uric acid crystals activate the NALP3 inflammasome. Nature 440:237–241

Masalova OV, Shepelev AV, Atanadze SN et al (1999) Immunostimulating effect of water-soluble fullerene derivatives—perspective adjuvants for a new generation of vaccine. Dokl Akad Nauk 369:411–413

Meng H, Yang S, Li Z et al (2011) Aspect ratio determines the quantity of mesoporous silica nanoparticle uptake by a small GTPase-dependent macropinocytosis mechanism. ACS Nano 5:4434–4447

Mitchell LA, Gao J, Wal RV et al (2007) Pulmonary and systemic immune response to inhaled multiwalled carbon nanotubes. Toxicol Sci 100:203–214

Mitchell LA, Lauer FT, Burchiel SW et al (2009) Mechanisms for how inhaled multiwalled carbon nanotubes suppress systemic immune function in mice. Nat Nanotechnol 4:451–456

Moghimi SM, Andersen AJ, Ahmadvand D et al (2011) Material properties in complement activation. Adv Drug Deliv Rev 63:1000–1007

Mohamed BM, Verma NK, Davies AM et al (2012) Citrullination of proteins: a common post-translational modification pathway induced by different nanoparticles in vitro and in vivo. Nanomedicine (Lond) 7:1181–1195

Monopoli MP, Pitek AS, Lynch I et al (2013) Formation and characterization of the nanoparticle-protein corona. Methods Mol Biol 1025:137–155

Mottram PL, Leong D, Crimeen-Irwin B et al (2007) Type 1 and 2 immunity following vaccination is influenced by nanoparticle size: formulation of a model vaccine for respiratory syncytial virus. Mol Pharm 4:73–84

Moyano DF, Goldsmith M, Solfiell DJ et al (2012) Nanoparticle hydrophobicity dictates immune response. J Am Chem Soc 134:3965–3967

Muller RH, Maassen S, Weyhers H et al (1996) Phagocytic uptake and cytotoxicity of solid lipid nanoparticles (SLN) sterically stabilized with poloxamine 908 and poloxamer 407. J Drug Target 4:161–170

Murphy KM (ed) (2011) Janeway's immunobiology, 8th edn. Garland Science, New York

Nakayamada S, Takahashi H, Kanno Y et al (2012) Helper T cell diversity and plasticity. Curr Opin Immunol 24:297–302

Oostingh GJ, Casals E, Italiani P et al (2011) Problems and challenges in the development and validation of human cell-based assays to determine nanoparticle-induced immunomodulatory effects. Part Fibre Toxicol 8:8

Paciotti GF, Myer L, Weinreich D et al (2004) Colloidal gold: a novel nanoparticle vector for tumor directed drug delivery. Drug Deliv 11:169–183

Palomäki J, Välimäki E, Sund J et al (2011) Long, needle-like carbon nanotubes and asbestos activate the NLRP3 inflammasome through a similar mechanism. ACS Nano 5:6861–6870

Papayannopoulos V, Zychlinsky A (2009) NETs: a new strategy for using old weapons. Trends Immunol 30:513–521

Peer D (2012) Immunotoxicity derived from manipulating leukocytes with lipid-based nanoparticles. Adv Drug Deliv Rev 64:1738–1748

Pulskamp K, Diabate S, Krug HF (2007) Carbon nanotubes show no sign of acute toxicity but induce intracellular reactive oxygen species in dependence on contaminants. Toxicol Lett 168:58–74

Rajananthanan P, Attard GS, Sheikh NA et al (1999) Evaluation of novel aggregate structures as adjuvants: composition, toxicity studies and humoral responses. Vaccine 17:715–730

Reddy ST, van der Vlies AJ, Simeoni E et al (2007) Exploiting lymphatic transport and complement activation in nanoparticle vaccines. Nat Biotechnol 25:1159–1164

Redhead HM, Davis SS, Illum L (2001) Drug delivery in poly(lactide-co-glycolide) nanoparticles surface modified with poloxamer 407 and poloxamine 908: in vitro characterisation and in vivo evaluation. J Control Release 70:353–363

Rettig L, Haen SP, Bittermann AG et al (2010) Particle size and activation threshold: a new dimension of danger signaling. Blood 115:4533–4541

Roberts JC, Bhalgat MK, Zera RT (1996) Preliminary biological evaluation of polyamidoamine (PAMAM) Starburst dendrimers. J Biomed Mater Res 30:53–65

Roser M, Fischer D, Kissel T (1998) Surface-modified biodegradable albumin nano and microspheres. II: effect of surface charges on in vitro phagocytosis and biodistribution in rats. Eur J Pharm Biopharm 46:255–263

Salvador-Morales C, Flahaut E, Sim E et al (2006) Complement activation and protein adsorption by carbon nanotubes. Mol Immunol 43:193–201

Seok J, Warren HS, Cuenca AG et al (2013) Genomic responses in mouse models poorly mimic human inflammatory diseases. Proc Natl Acad Sci USA 110:3507

Sharma A, Muresanu DF, Patnail R et al (2013) Size- and age-dependent neurotoxicity of engineered metal nanoparticles in rats. Mol Neurobiol 48:386–396

Smulders S, Kaiser JP, Zuin S et al (2012) Contamination of nanoparticles by endotoxin: evaluation of different test methods. Part Fibre Toxicol 9:41

Söderäll K (2010) Invertebrate immunity. Advances in experimental medicine and biology. Landes Bioscience and Springer Science + Business Media, LCC, New York, p 314

Stern ST, McNeil SE (2008) Nanotechnology safety concerns revisited. Toxicol Sci 101:4–21

Szebeni J, Alving CR, Rosivall L et al (2007) Animal models of complement-mediated hypersensitivity reactions to liposomes and other lipid-based nanoparticles. J Liposome Res 17:107–117

Tan Y, Li S, Pitt BR et al (1999) The inhibitory role of CpG immunostimulatory motifs in cationic lipid vector-mediated transgene expression in vivo. Hum Gene Ther 10:2153–2161

Thiele L, Rothen-Rutishauser B, Jilek S et al (2001) Evaluation of particle uptake in human blood monocyte-derived cells in vitro. Does phagocytosis activity of dendritic cells measure up with macrophages? J Control Release 76:59–71

Tomii A, Masugi F (1991) Production of anti-platelet-activating factor antibodies by the use of colloidal gold as carrier. Jpn J Med Sci Biol 44:75–80

Unrine JM, Hunyadi SE, Tsyusko OV et al (2010) Evidence for bioavailability of Au nanoparticles from soil and biodistribution within earthworms (Eisenia fetida). Environ Sci Technol 44:830813

van Zijverden M, Granum B (2000) Adjuvant activity of particulate pollutants in different mouse models. Toxicology 152:69–77

Vetten MA, Yah CS, Singh T et al (2014) Challenges facing sterilization and depyrogenation of nanoparticles: effects on structural stability and biomedical applications. Nanomedicine 10(7):1391–1399

Wang X, Ishida T, Kiwada H (2007) Anti-PEG IgM elicited by injection of liposomes is involved in the enhanced blood clearance of a subsequent dose of PEGylated liposomes. J Control Release 119:236–244

Whitfield Aslund ML, McShane H, Simpson MJ et al (2012) Earthworm sublethal responses to titanium dioxide nanomaterial in soil detected by (1)H NMR metabolomics. Environ Sci Technol 46:1111–1118

Xiang SD, Scholzen A, Minigo G et al (2006) Pathogen recognition and development of particulate vaccines: does size matter? Methods 40:1–9

Xiang SD, Fuchsberger M, Karlson TDL et al (2012) Nanoparticles, immune modulation and vaccine delivery. In: Yarmush ML, Shi D (eds) Frontiers in nanobiomedical research. World Scientific Publishing, Singapore

Yazdi AS, Guarda G, Riteau N et al (2010) Nanoparticles activate the NLR pyrin domain containing 3 (Nlrp3) inflammasome and cause pulmonary inflammation through release of IL-1α and IL-1β. Proc Natl Acad Sci U S A 107:19449–19454

Zahr AS, Davis CA, Pishko MV (2006) Macrophage uptake of core-shell nanoparticles surface modified with poly(ethylene glycol). Langmuir 22:8178–8185

Air Pollution and Allergy in Germany: Surprising Results of Data Obtained After Reunification

10

Ursula Krämer

Contents

10.1 Introduction

The German reunification gave the unique opportunity to investigate the development of allergies in two population groups with similar genetic background but a different environment. Therefore, shortly after 1989 first epidemiological comparison studies between East and West Germany were planned. At that time most scientists assumed

U. Krämer
IUF – Leibniz Research Institute for Environmental Medicine, Düsseldorf, Germany
e-mail: Ursula.Kraemer@IUF-duesseldorf.de

C. Esser (ed.), *Environmental Influences on the Immune System*,
DOI 10.1007/978-3-7091-1890-0_10

that the risk factors for the development of allergies would be more or less identical to those already known for airway diseases such as bronchitis: Traditional outdoor air pollution, i.e., sulfur dioxide (SO_2) and particles from coal burning, or unfavorable indoor factors like single room heating or crowding all would induce more allergies. Because it was known that the concentration of air pollution was higher in East Germany than in West Germany and also unfavorable indoor conditions were more prevalent in East Germany, it was anticipated that (1) the prevalence of allergies would be higher in East than in West Germany, (2) allergies in East Germany would decrease after the reunification due to decreasing outdoor air pollution, and (3) the change to more favorable indoor conditions in East Germany would lead to fewer allergies. However, surprisingly all three expectations did not come true. In the following chapters we will describe these three surprising results in more detail. The results were driving forces for a change in the paradigms of allergy research.

10.2 The First Surprising Result: Prevalence of Allergies in East and West Germany

The first surprise when comparing allergy data between West and East Germany was that the prevalence of allergies was higher in West Germany than in East Germany. This contradicted the expectations, which had assumed the same pattern of differences for allergies as those observed for "classical" airway diseases like bronchitis. Today (in 2015) most people assume that all allergic manifestations, asthma, hay fever, eczema, and allergic sensitization to pollen or mites, had been more prevalent in West than in East Germany and that these differences are no longer visible. This, however, is not true. Therefore, in this paragraph, we will describe the extent of these differences in allergy prevalence for different allergic manifestations from reunification until today. Allergy differences can be derived from 14 different West/East German comparison studies which have been conducted since 1989 and were recently summarized (Krämer et al. 2015). Most of these studies only compared the result of one place in East Germany with one place in West Germany and cannot be considered representative. Six studies, three in children and adolescents and three in adults, followed a different design, and their design is very shortly summarized below:

SAWO1 and SAWO2 (*S*chul*a*nfängerstudie in *W*est und *O*stdeutschland) Study on school beginners in East and West Germany. Between 1991 and 2000, 6-year-old children from 4 areas in East Germany participated in annual investigations. Every third year, a parallel investigation was performed in 4 areas of West Germany. 31,903 children were included. SAWO1 summarizes the data for children born before the reunification and SAWO2 for those children born after reunification (Krämer et al. 1999, 2010).

KiGGS German Health Interview and Examination Survey for Children and Adolescents. A German representative study including data about allergies in 0–17-year-old children and adolescents was conducted between 2003 and 2006. In East Germany, 5,900 persons participated and 11,741 in West Germany (Schlaud et al. 2007).

OW1 First representative federal health survey in adults including data from 2617 East Germans and 5313 West Germans, which was conducted between 1990 and 1992 (Hermann-Kunz 1999).

BGS98 Representative federal health survey in adults from 1997 to 1999 with 2419 East German and 4705 West German participants (Hermann-Kunz 2000).

DEGS1 The first German Health Interview and Examination Survey, a federal health survey in adults which was conducted between 2008 and 2011 with 2520 participants from East and 5468 from West Germany (Haftenberger et al. 2013; Langen et al. 2013).

Table 10.1 demonstrates the prevalence of different allergic manifestations (doctor-diagnosed asthma, hay fever, and eczema as well as allergic sensitization to birch pollen, grass pollen, and house dust mites) as observed in these six studies.

Table 10.2 depicts the prevalence ratios of West/East for these allergic manifestations. A ratio above one indicates a higher prevalence in West than in East Germany, which is significant if the 95 % confidence interval does not include the one.

As can be seen in Table 10.2, hay fever and birch pollen sensitization most clearly demonstrate higher prevalence in West than in East at the time of the German reunification with fast convergence after that time in children studies. In adult populations however, the West/East difference in prevalence can be seen until today. The differences for asthma and house dust mite sensitization at the time of the German reunification were smaller than those observed for hay fever and birch pollen sensitization. However, the differences prevailed for a longer time. The adult studies also show a steeply increasing trend in the prevalence of doctor-diagnosed asthma in

Table 10.1 Allergic manifestations in Germany: Prevalence from six West/East German studies

			Ever doctor diagnosed			Specific IgE sensitization (>0.35 kU/l) to		
Study	Age of participants	Year of study	Asthma	Hay fever	Eczema	Birch pollen	Grass pollen	House dust mite
SAWO1	6	1991–1995 yearly	2.0	1.6	14.0	2.0[a]	11.6	10.9
SAWO2	6	1996–2000 yearly	2.7	3.2	17.9	3.1[a]	11.4	12.0
KiGGS	0–17	2003–2006	4.7	10.2	12.7	14.0	22.8	21.1
OW1	25–69	1990–1992	2.0	9.2	–	–	–	–
BGS98	18–79	1997–1999	5.6	15.1	3.3	–	–	–
DEGS1	18–79	2008–2011	8.7	15.6	3.7	17.6	18.4	15.9

[a]Specific IgE >3.5 kU/l

Table 10.2 West/East German allergy differences: Prevalence ratios West/East and 95 % confidence intervals

Study	Ever doctor diagnosed			Specific IgE sensitization		
	Asthma	Hay fever	Eczema	Birch pollen	Grass pollen	House dust mite
SAWO1	1.4 (1.1–1.7)	1.9 (1.5–2.4)	0.8 (0.7–0.9)	3.0 (1.9–4.8)	0.9 (0.8–1.1)	1.3 (1.1–1.5)
SAWO2	1.3 (1.0–1.6)	1.0 (0.8–1.2)	0.9 (0.9–1.0)	1.3 (0.8–2.1)	0.9 (0.7–1.1)	1.3 (1.0–1.6)
KiGGS	0.9 (0.8–1.1)	0.9 (0.8–1.0)	1.0 (0.9–1.0)	0.9 (0.9–1.0)	1.0 (1.0–1.1)	1.0 (1.0–1.1)
OW1	1.2 (0.8–1.8)	1.7 (1.4–2.1)	–	–	–	–
BGS98	1.5 (1.2–1.9)	1.5 (1.3–1.7)	1.3 (1.0–1.8)	–	–	–
DEGS1	1.2 (1.0–1.4)	1.2 (1.0–1.3)	1.2 (0.9–1.5)	1.1 (1.0–1.2)	1.1 (1.0–1.2)	1.3 (1.1–1.4)

both parts of Germany. Eczema however showed the opposite pattern of differences and trends in children studies than those observed for hay fever and birch pollen sensitization. At the time of the German reunification, the prevalence of eczema was higher in East than in West, and this difference diminished.

More distinct differences in hay fever and pollen sensitization than those described here were reported between Leipzig (East Germany) and Munich (West Germany) (von Mutius et al. 1994). Munich however is not representative for West Germany and Leipzig not for East Germany. However, Munich might be a better example for an affluent westernized region than the rest of West Germany.

In all West/East German comparison studies, differences in reactivity occurred: Hay fever and birch pollen sensitization were the most sensitive markers associated with "western lifestyle." Such differences have also been reported in other contexts. Studies dealing with allergies in a farm environment (Riedler et al. 2000) or comparing allergies between West and East Karelia (Vartiainen et al. 2002) also found the most marked effects on hay fever and birch pollen sensitization. The effects on eczema and sensitizations to house dust mites were less marked or even the opposite. Thus, the individual allergic manifestations apparently have different sets of environmental triggers.

The given data already demonstrate the importance of influences in early childhood. In the KiGGS study, where children and adolescents were investigated, who all have been born after the German reunification, no West/East German difference emerged anymore. However, in all adult studies, even in the newest ones, all participating individuals were born before the reunification and the allergy differences prevail.

Since hay fever showed the clearest West > East pattern and the clearest trend, Fig. 10.1 summarizes the observations on hay fever made in SAWO in an annual resolution. In the next paragraph we will describe outdoor air pollution in East and West Germany during the first 10 years after reunification. We will see whether the pattern of outdoor air pollution can be responsible for the induction of the pattern of hay fever development as shown in Fig. 10.1.

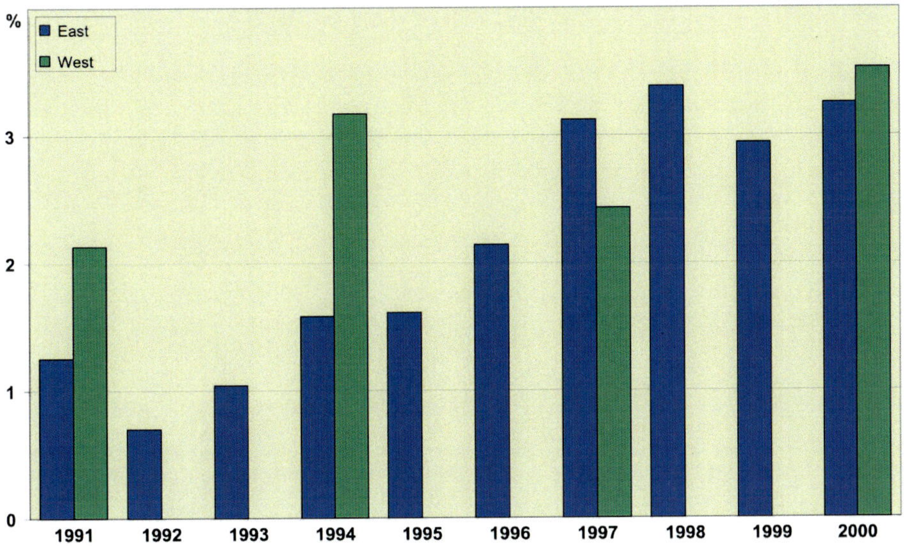

Fig. 10.1 Prevalence of hay fever in 6-year-old children as ever diagnosed by a physician (Results from SAWO 1991–2000)

10.3 The Second Surprising Result: The Role of Outdoor Air Pollution

10.3.1 Outdoor Air Pollution in East and West Germany (1989–2000)

During the time of the German reunification, outdoor air in East Germany was highly polluted with the "classical" type of pollution dominated by particles and sulfur dioxide (SO_2), which was generated by coal combustion from heating of private homes and industry (Fig. 10.2). Concentration of SO_2 in outdoor air in East German industrialized areas (Leipzig/Halle) was 10 times higher than in West German industrialized areas (Ruhr area) and 15 times higher than in nonindustrialized areas. Particle pollution in East was 1.5–2 times higher than in the West. Total suspended particles (TSP) and SO_2 concentrations in outdoor air decreased mainly in East Germany; however, a slight decrease was also found in West Germany. Already in 1994, annual TSP levels were similar in East and West. Differences in SO_2 between East and West Germany and the decrease of SO_2 in East Germany were much stronger than for TSP. Since 1997, annual SO_2 levels were similar in East and West.

The pattern of traffic-related pollution however was different from the just described pattern in the "traditional" pollutants TSP and SO_2. In East Germany, the number of automobiles quickly increased in the first 10 years after reunification: In Saxony (East Germany) from 1.2 to 2.1 million (Statistisches Landesamt des Freistaates Sachsen 2000), in Saxony-Anhalt (East Germany) from 0.8 to 1.2 million (Landesamt für Umweltschutz Sachsen-Anhalt 2004), but in North

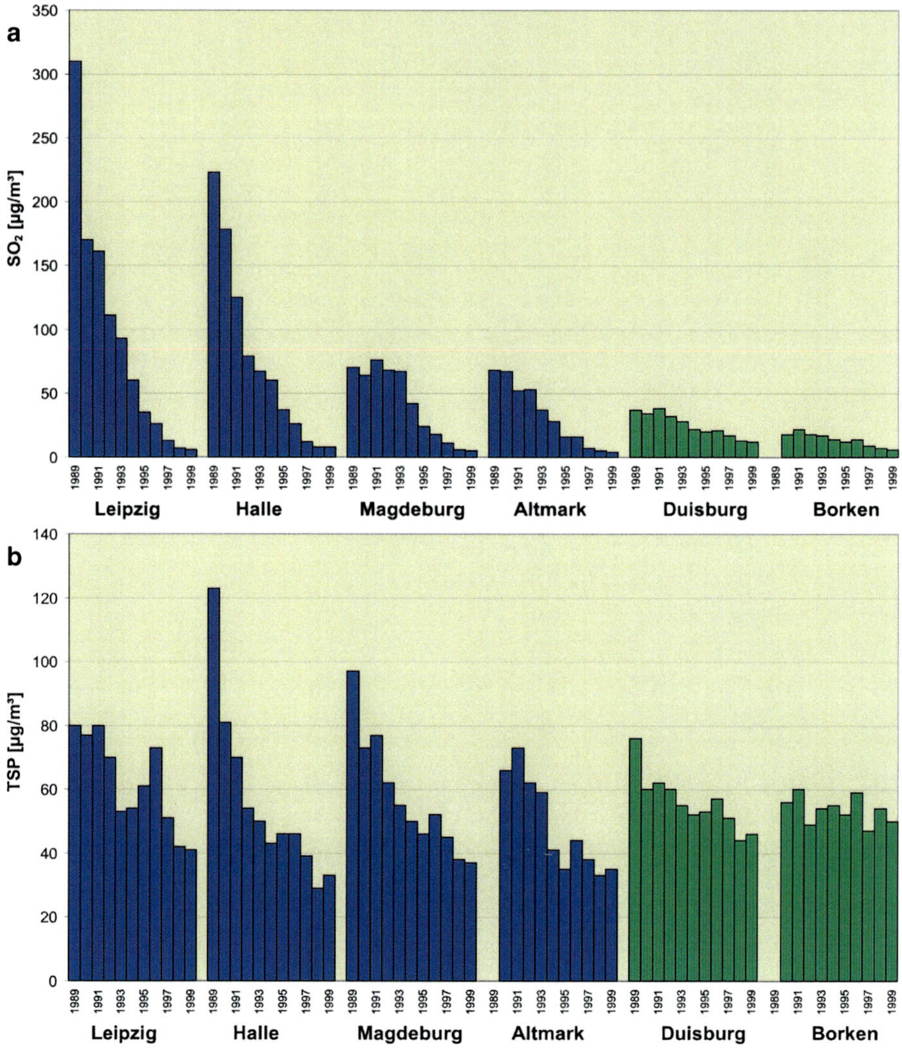

Fig. 10.2 Air pollution in East and West Germany between 1991 and 2000. Air pollution in East (= *blue*) German study areas (Leipzig, Halle, Magdeburg, and Altmark) as well as in West (=*green*) German study areas (Duisburg and Borken) between 1991 and 2000. Air pollution measurements were done by the regional authorities. Total suspended particles (TSP) were sampled with a low volume sampler and concentrations determined by a radiometric technique (β-ray absorption monitor). Sulfur dioxide (SO_2) concentrations were determined by a UV fluorescence method. Arithmetic means of the values gained at the monitoring station(s) (one to three) in the areas during the 2 years preceding the investigation are presented

Rhine-Westphalia (West Germany) only from eight to nine million (Ministerium für Verkehr, Energie und Landesplanung des Landes Nordrhein-Westfalen 2004). The number of cars per 1000 inhabitants changed in East Germany from 245 to 488 between the years 1989 and 1997, resulting in a number nearly equal to that found in West Germany (517). Due to propagation of catalytic converters, particulate matter (PM) and nitrogen oxide (NO_x) emissions from traffic-related sources in East Germany did not increase proportionally to the increase in automobile numbers, but peaked in 1993 (particulate matter, PM) and 1995 (NO_x) (Landesamt für Umwelt und Geologie, Freistaat Sachsen 1997). The relative contribution of traffic-related sources to all emissions increased between 1989 and 1997 for PM from 2 to 22 % and for NO_x from 30 to 48 %. A measurement station with traffic exposure in Leipzig, one of our study areas, showed a steady increase of NO_2 annual means between 39 μg/m³ in 1991 and 53 μg/m³ in 1996 (Landesamt für Umwelt und Geologie 1997). The number concentration of ultrafine particles (0.01–0.02 μm diameter) in East Germany increased after 1991 despite decreasing TSP concentrations and decreasing concentrations of fine particulate matter ($PM_{2.5}$) (Ebelt et al. 2001; Kreyling et al. 2003). Ultrafine particles from automobile emissions vary on a small spatial scale; they disappear exponentially with the distance from a major road and reach background levels at a distance of 300 m (Zhu et al. 2002). Similar effects can be observed for other traffic-related air pollutants. In East Germany, overall background TSP concentrations decreased, whereas concentrations of certain traffic-related substances near roads with heavy traffic actually increased.

In SAWO no direct measurements of traffic-related pollution were available. Instead parents were asked (standardized questionnaire) "How far away is your address (beeline) from a busy street (rush hour traffic/ through traffic)," and answer categories were predefined as "less than fifty meter" and "more than fifty meter." This information was used to define two categories (high and low) of traffic exposure. Parent's judgment about distance to a busy street was shown to highly correlate with objective measure of traffic density in the subsample of children in West Germany 2000, where geo-coded residential addresses and data of traffic density were available (Sugiri et al. 2006).

10.3.2 Effects of Outdoor Air Pollution on Allergic Manifestations in East and West Germany

When initiating the West/East German studies to compare the prevalence of airway diseases and allergies in population groups from East and West German areas, it was anticipated that prevalence would be higher in the Eastern areas. This came true for bronchitis, tonsillitis, number of colds, and parameters of lung function as total lung capacity (Krämer et al. 1999; Sugiri et al. 2006). This however was neither true for bronchial asthma nor hay fever or allergic sensitization. Surprisingly in the West German study areas, the prevalence was higher. Likewise the improvement of air quality in East Germany after the reunification leads to decreases in respiratory diseases and improvement in lung function (Sugiri et al. 2006); however, bronchial asthma, hay fever, and allergic sensitization tended to increase in East Germany (Krämer et al. 2010). When comparing Fig. 10.1 with Fig. 10.2, it is apparent that TSP and SO_2 pollution in outdoor air are no major factors to drive allergy differences between East and West Germany or the increasing trend of hay fever in East

Germany. This was the second surprising result of the data gained after the German reunification.

The possible conclusion, however, that outdoor pollution is not associated with allergies at all is not correct. In spite of generally higher prevalence of asthma in West Germany, a comparison of differently polluted areas in East Germany from 1991 to 1995 (Krämer et al. 1999) showed that bronchial asthma was associated with SO_2 pollution (adjusted odds ratio per 200 (!) μg/m^3 SO_2: 2.46 95 % confidence interval: 1.18–5.15). Likewise another cross-sectional comparison in East German children showed higher lifetime prevalence of asthma in the industrialized Bitterfeld region compared to the not industrialized Zerbst region. These results from East Germany demonstrate that the risk factors for asthma and nonallergic airway diseases partly overlap. An association with hay fever or allergic sensitization could not be demonstrated. The Bitterfeld study showed that allergies and allergic sensitizations were highest in the mining region of Hettstedt, with its high lead and cadmium content in dust (Heinrich et al. 1999). This hints at a special importance of metals for the induction of allergic diseases. However, on an individual scale (cadmium in urine), this type of pollution could not be associated with more sensitization (Ritz et al. 1998). Additionally the internal body burden with lead and cadmium was lower in West than in East German areas (Begerow et al. 1994). Therefore, the role of metals in the induction of allergies could not be solved in the West/East comparison studies.

Contrary to SO_2 and TSP pollution, small-scale traffic-related pollution was higher in West Germany in 1989 and increased in East Germany since that time. Traffic-related pollution might therefore be a candidate to explain allergy differences as well as different allergy trends. In SAWO the effect of living near a major road on hay fever diagnosis and attacks of sneezing in East Germany was 50 % stronger after 1995 than between 1991 and 1995 (Krämer et al. 2010) and explained a small portion of the rising trend observed in East Germany. This effect was stronger for a symptom often related to hay fever: sneezing attacks without a cold. Just after the reunification, this symptom occurred most often in the industrialized towns. Children living near a major road had only slightly higher prevalence. Over time there was a steep increase in prevalence of sneezing attacks in children living near major roads. The increase for children not living near major roads was not significant. Fig. 10.3 summarizes the results gained. A similar effect for allergic sensitizations could not be observed.

Until today (2015) epidemiological studies on the effects of traffic-related pollution on allergic sensitizations show no consistent associations (Gruzieva et al. 2014). Traffic-related pollution might partly explain an increase in hay fever and hay fever symptoms in East Germany but is probably not the causal factor for the increase in allergic sensitization and asthma. Other factors have to be considered which are more important to drive trends in allergies. In the next paragraph, we will shortly summarize factors which have been identified in the context of SAWO to drive trends and—maybe still more important—factors which were thought to be important when the study began but were irrelevant in explaining the rising trends (Krämer et al. 2010), the third surprising result.

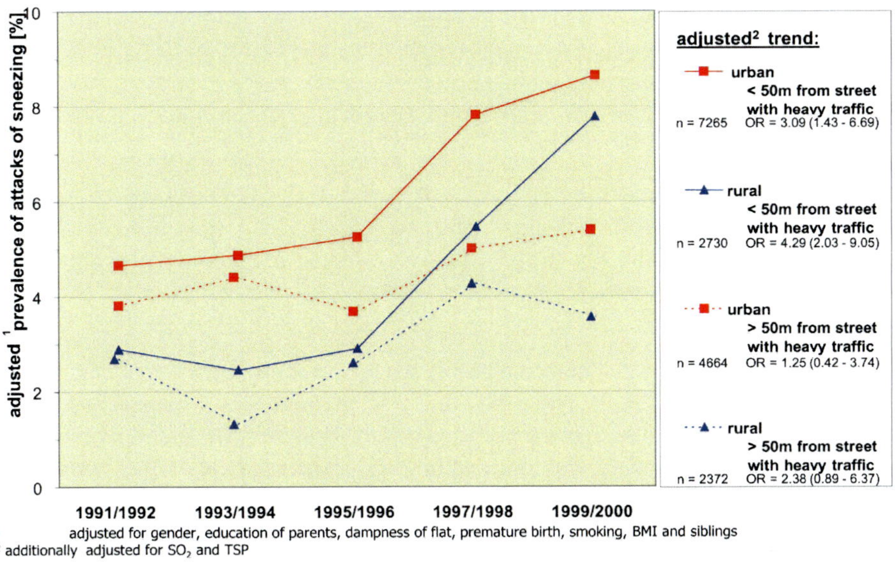

Fig. 10.3 Attacks of sneezing in 6-year-old children from urban and rural areas in East Germany living more or less than 50 m from a major street with heavy traffic (SAWO)

10.4 The Third Surprising Result: Associations with Indoor Factors and Other Potential Risk Factors

10.4.1 Pollen Exposure

Exposure with Pollen Exposure is a necessary determinant for the development of an allergic sensitization against pollen. Due to the climatic conditions in Germany, there is a gradient from southwest (warmer winters) to northeast (colder winters) in the start of pollen seasons and even a temporal trend toward earlier pollination due to climate change (Krämer et al. 2001); however, these trends cannot explain the special patterns observed in Germany after reunification. The only exception is the exposure to mugwort pollen, which is higher in the more continental climate of East Germany and might explain the higher sensitization to mugwort pollen in East Germany, which is visible until today (Krämer et al. 2015).

10.4.2 Indoor Factors

Carpets Rooms with carpets showed higher mite allergen levels (van Strien et al. 1994). Therefore, it was anticipated that the increase in wall to wall carpeting in East Germany after reunification might be the reason for the increase in allergies. In SAWO, however, reverse causation was identified. Families with allergies removed the carpets; thus, a fictitiously protective effect of carpeting was introduced.

Duvets Use of synthetic beddings increased the risk to have perennial rhinitis (Frosh et al. 1999). Therefore, the replacement of feather beddings in East Germany after the reunification was thought to cause the increase in allergies. However, again only reverse causation was identified. Families with allergies primarily removed the feather beddings.

Smoking Environmental tobacco smoke is a risk factor for asthma in children (Strachan and Cook 1998). However, inclusion of passive smoking into the statistical evaluation models did not explain the development of asthma in both parts of Germany. Doctor diagnoses of asthma increased, whereas passive smoking decreased.

Pets There is conflicting evidence whether when and what pet is beneficial or bad (Lødrup Carlsen et al. 2012). After the reunification, there was a strong increase of pets in families in East Germany (exception birds). However, we found a protective influence of dogs; therefore, the increase in dogs could not explain the increase in allergies (Apfelbacher et al. 2010).

Single Room Heating with Fossil Fuels This is a known risk factor for respiratory diseases (Kodgule and Salvi 2012); however, a couple of studies show a beneficial effect of single room heating on allergy development (Nowak et al. 1996; Duhme et al. 1998; von Mutius et al. 1998). Single room heating was less prevalent in West Germany than in East Germany during the time of reunification. It steeply decreased in East Germany after reunification. Inclusion of single room heating in the statistical evaluation model explained a considerable part of the pattern of hay fever and sensitization development in SAWO. The loss of single room heating is associated with an increase in allergies on an individual scale. However, the final cause of this association is still unclear. Single room heating causes much higher temperature gradients between rooms than central heating. Whether that effect causes a positive stimulation of the immune system has not been investigated so far.

Exposure with House Dust Mites Exposure with house dust mites is a necessary determinant for house dust mite sensitization. If investigated at all, a higher exposure in West German households was found than in East German households (Gehring et al. 2001; Oppermann et al. 2001). This is probably due to the climatic differences between East (less humid) and West Germany (humid). Exposure measurements for all investigated children were not available, but these differences in exposure might have caused differences in sensitization.

Endotoxin Endotoxin is a known risk factor for airway diseases, but high exposure was shown to be associated with less allergies (Braun-Fahrländer et al. 2002). This effect is consistent with the hygiene hypotheses. Exposure with endotoxin, however, was not consistently higher in East than in West Germany. In Hamburg (West), for instance, it was higher than in Erfurt (East) (Bischof et al. 2002). Therefore, this exposure is no explanation for the West/East German patterns in allergies.

10.4.3 Early Childhood Influences

Preterm Birth Children born preterm wheeze more often during childhood (Gold et al. 1999) but have less sensitizations (Siltanen et al. 2001). There was a decrease of preterm births in both parts of Germany. This factor was no explanation for the allergy pattern observed.

Birthweight High birthweight was associated with more eczema (Tedner et al. 2012). Birthweight increased in both parts of Germany during the investigation period, but the point prevalence of eczema in the 6 years old decreased.

Age of Mother at Birth It was hypothesized that the increase in allergies might be related to the age of mothers (Ring et al. 2001). There was a steep increase of maternal age at birth in both parts of Germany and steeper so in East Germany. Age of mother, however, did not explain any of the patterns of allergies in East and West Germany, when included in the statistical evaluation model.

Breast Feeding There was a publication apparently demonstrating a positive association between breast feeding and allergies (Sears et al. 2002); however, this is much debated. Breast feeding increased in both parts of Germany in a similar fashion and was no explanation for the increase in allergies. Hints for a reverse causation were also found in SAWO: Mothers with a predisposition for allergies tended to breast fed their child longer.

One-Child Family It is well known that siblings, and especially older ones, protect from allergies (Karmaus and Botezan 2002). We observed a steep increase of one-child families in East Germany but not in West Germany. This was associated with less crowding and partly explained the observed trends, when included in the statistical evaluation.

Day Care Centers Day care was much less prevalent in West Germany than in East Germany throughout the observation period. On an individual scale, day care did not explain the observed pattern of allergies. Day care visits, however, explained higher prevalence of eczema in children (contrary to the expectations of the hygiene hypothesis). When included in the statistical evaluation models, the differences in day care visits totally explained the higher prevalence of eczema in East Germany when compared to West Germany (Cramer et al. 2011).

Worm Infection Parasitic infection protects from allergies (Perzanowski et al. 2002). There were less worm infections in West than in East Germany and a steep decrease in East Germany. On an individual scale, this explains the decrease in total IgE but not the increase in allergies.

Vaccination Vaccination might lead to allergies by preventing infectious diseases (never proven). There was a higher vaccination coverage in East Germany at the time of the German reunification, and only protective effects of vaccination were proven (Grüber et al. 2002; Ring et al. 2004).

10.4.4 Other Factors

Nutrition Nutrition was seldom investigated in the frame of East/West comparison studies. Margarine consumption was associated with more allergies (Sausenthaler et al. 2006; von Mutius et al. 1998). Whether that can explain the observed trends is unclear.

Weight Increase in asthma might be due to an increase in overweight (Wickens et al. 2005). Increase in overweight, however, was equally pronounced in both parts of Germany; the different trends cannot be explained (Apfelbacher et al. 2008a, b).

Genetics Genetic susceptibility clearly differs between individuals. This factor was not investigated in detail; most studies assumed similarity between East and West. Genetic changes, however, in such a short time and restricted to a geographic subarea do not seem plausible.

Awareness Diagnostic habits may change over time. A recent meta-analysis of European birth cohort studies demonstrates that the discrepancy between wheeze in the last year (symptom of asthma) and doctor diagnosis of asthma in the last year is most pronounced in the German birth cohorts. The diagnosis is expressed much less often than would be indicated by the symptom (Mölter et al. 2015). The same concern is also expressed by Ellsäßer (Ellsäßer and Diepgen 2002), when reflecting about the rising trend of asthma diagnoses in Brandenburg (East Germany). The increase in doctor-diagnosed asthma seen in the adult studies might be due to a shift in diagnostic habits of physicians in Germany.

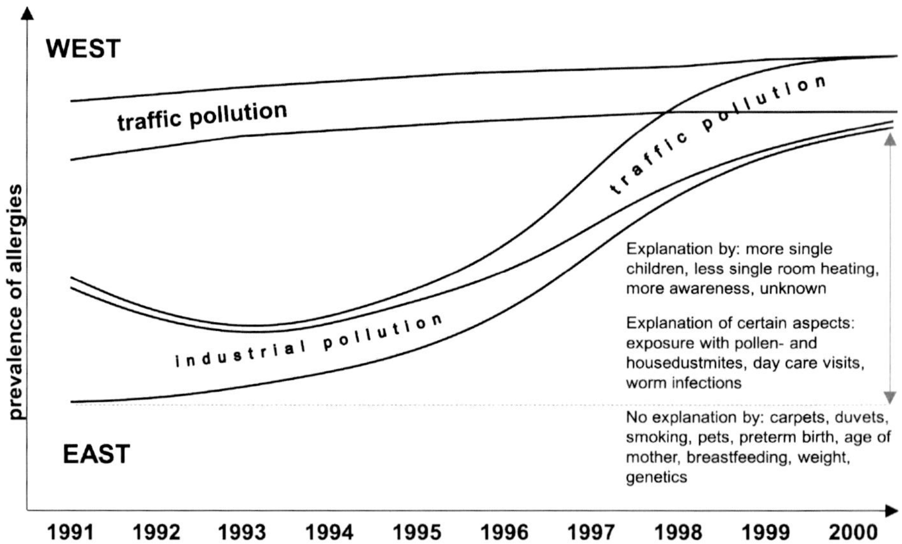

Fig. 10.4 Development of allergies in East and West Germany between 1991 and 2000 and the influence of traffic-related and industrial pollution; explanations of West–East differences and trends by risk factors investigated in SAWO are also indicated

10.5 Summary and Conclusion

Figure 10.4 summarizes the surprising results about the development of allergies in the first 10 years after reunification gained in children studies: The prevalence of allergies, especially of hay fever and birch pollen sensitization, was higher in West than in East Germany shortly after the reunification. Then there was a small influence of industrial pollution on asthma. Traffic pollution gained influence after 1995 especially for hay fever symptoms. Unexpectedly unfavorable indoor conditions like single room heating and crowding were associated with less hay fever and pollen sensitization and partly explained the observed trends in allergy development.

The results gained in the East–West comparison studies were driving forces for a change in the paradigms of allergy research. Up to the beginning of the 1990, focus of that research was the identification of risk factors for the development of allergies. Since then the focus changed to the identification of protective factors, which might prevent the development of allergies. Early microbial stimulation (hygiene hypothesis), in SAWO, for instance, the existence of siblings, is one of the factors identified as protective, and the increasing absence of siblings in East Germany is partly driving the increasing trend of hay fever and allergic sensitization.

References

Apfelbacher CJ, Loerbroks A, Cairns J, Behrendt H, Ring J, Krämer U (2008a) Predictors of overweight and obesity in five to seven-year-old children in Germany: results from cross-sectional studies. BMC Public Health 8:171

Apfelbacher CJ, Cairns J, Bruckner T, Möhrenschlager M, Behrendt H, Ring J et al (2008b) Prevalence of overweight and obesity in East and West German children in the decade after reunification: population-based series of cross-sectional studies. J Epidemiol Community Health 62:125–130

Apfelbacher CJ, Ollert M, Ring J, Behrendt H, Krämer U (2010) Contact to cat or dog, allergies and parental education. Pediatr Allergy Immunol 21:284–291

Begerow J, Freier I, Turfeld M, Krämer U, Dunemann L (1994) Internal lead and cadmium exposure in 6-year-old children from western and eastern Germany. Int Arch Occup Environ Health 66:243–248

Bischof W, Koch A, Gehring U, Fahlbusch B, Wichmann HE, Heinrich J et al (2002) Predictors of high endotoxin concentrations in the settled dust of German homes. Indoor Air 12:2–9

Braun-Fahrländer C, Riedler J, Herz U, Eder W, Waser M, Grize L et al (2002) Environmental exposure to endotoxin and its relation to asthma in school-age children. N Engl J Med 347:869–877

Cramer C, Link E, Bauer CP, Hoffmann U, von Berg A, Lehmann I et al (2011) Association between attendance of day care centres and increased prevalence of eczema in the German birth cohort study LISAplus. Allergy 66:68–75

Duhme H, Weiland SK, Rudolph P, Wienke A, Kramer A, Keil U (1998) Asthma and allergies among children in West and East Germany: a comparison between Münster and Greifswald using the ISAAC phase I protocol. Eur Respir J 11:840–847

Ebelt S, Brauer M, Cyrys J, Tuch T, Kreyling WG, Wichmann H-E et al (2001) Air quality in postunification Erfurt, East Germany: associating changes in pollutant concentrations with changes in emissions. Environ Health Perspect 109:325–333

Ellsäßer G, Diepgen TL (2002) Atopische Erkrankungen und soziale Lage bei Einschulungskindern im Land -Brandburg -Trendanalyse 1994–2000. Monatsschr Kinderheilkd 150:839–847

Frosh AC, Sandhu G, Joyce R, Strachan DP (1999) Prevalence of rhinitis, pillow type and past and present ownership of furred pets. Clin Exp Allergy 29:457–460

Gehring U, Heinrich J, Jacob B, Richter K, Fahlbusch B, Schlenvoigt G et al (2001) Respiratory symptoms in relation to indoor exposure to mite and cat allergens and endotoxins. Eur Respir J 18:555–563

Gold DR, Burge HA, Carey V, Milton DK, Platts-Mills TA, Weiss ST (1999) Predictors of repeated wheeze in the first year of life: the relative roles of cockroach, birth weight, acute lower respiratory illness, and maternal smoking. Am J Respir Crit Care Med 160:227–236

Grüber C, Meinlschmidt G, Bergmann R, Wahn U, Stark K (2002) Is early BCG vaccination associated with less atopic disease? An epidemiological study in German preschool children with different ethnic backgrounds. Pediatr Allergy Immunol 13:177–181

Gruzieva O, Gehring U, Aalberse R, Agius R, Beelen R, Behrendt H et al (2014) Meta-analysis of air pollution exposure association with allergic sensitization in European birth cohorts. J Allergy Clin Immunol 133:767–776.e7

Haftenberger M, Laußmann D, Ellert U, Kalcklösch M, Langen U, Schlaud M et al (2013) Prävalenz von Sensibilisierungen gegen Inhalations- und Nahrungsmittelallergene: Ergebnisse der Studie zur Gesundheit Erwachsener in Deutschland (DEGS1). [Prevalence of sensitisation to aeraoallergens and food allergens: results of the German Health Interview and Examination Survey for Adults (DEGS1)]. Bundesgesundheitsblatt Gesundheitsforschung Gesundheitsschutz 56:687–697

Heinrich J, Hoelscher B, Wjst M, Ritz B, Cyrys J, Wichmann HE (1999) Respiratory diseases and allergies in two polluted areas in East Germany. Environ Health Perspect 107:53–62

Hermann-Kunz E (1999) Häufigkeit allergischer Krankheiten in Ost- und Westdeutschland. [Incidence of allergic diseases in East and West Germany]. Gesundheitswesen 61 Sonderheft 2:S100–S105

Hermann-Kunz E (2000) Allergische Krankheiten in Deutschland. Bundesgesundheitsblatt Gesundheitsforschung Gesundheitsschutz 43:400–406

Karmaus W, Botezan C (2002) Does a higher number of siblings protect against the development of allergy and asthma? A review. J Epidemiol Community Health 56:209–217

Kodgule R, Salvi S (2012) Exposure to biomass smoke as a cause for airway disease in women and children. Curr Opin Allergy Clin Immunol 12:82–90

Krämer U, Behrendt H, Dolgner R, Ranft U, Ring J, Willer J et al (1999) Airway diseases and allergies in East and West German children during the first 5 years after reunification: time trends and the impact of sulphur dioxide and total suspended particles. Int J Epidemiol 28:865–873

Krämer U, Link E, Behrendt H (2001) Geografische und zeitliche Trends der Birken-, Gras- und Beifußpollenbelastung in Deutschland. [Geographic and time trends of pollen count due to beeches, grass and mugwort (Artemisia) in Germany]. Pneumologie 55:229–230

Krämer U, Oppermann H, Ranft U, Schäfer T, Ring J, Behrendt H (2010) Differences in allergy trends between East and West Germany and possible explanations. Clin Exp Allergy 40:289–298

Krämer U, Schmitz R, Ring J, Behrendt H (2015) What can Reunification of East and West Germany tell us about the cause of the allergy epidemic? Clin Exp Allergy 45:94–107

Kreyling WG, Tuch T, Peters A, Pitz M, Heinrich J, Stölzel M et al (2003) Diverging long-term trends in ambient urban particle mass and number concentrations associated with emission changes caused by the German unification. Atmos Environ 37:3841–3848

Langen U, Schmitz R, Steppuhn H (2013) Häufigkeit allergischer Erkrankungen in Deutschland: Ergebnisse der Studie zur Gesundheit Erwachsener in Deutschland (DEGS1). [Prevalence of allergic diseases in Germany: results of the German Health Interview and Examination Survey for Adults (DEGS1)]. Bundesgesundheitsblatt Gesundheitsforschung Gesundheitsschutz 56:698–706

Lødrup Carlsen KC, Roll S, Carlsen KH, Mowinckel P, Wijga AH, Brunekreef B et al (2012) Does pet ownership in infancy lead to asthma or allergy at school age? Pooled analysis of individual participant data from 11 European birth cohorts. PLoS One 7, e43214

Mölter A, Simpson A, Berdel D, Brunekreef B, Custovic A, Cyrys J et al (2015) A multicentre study of air pollution exposure and childhood asthma prevalence: the ESCAPE project. Eur Respir J 45:610–624

Nowak D, Heinrich J, Jörres R, Wassmer G, Berger J, Beck E et al (1996) Prevalence of respiratory symptoms, bronchial hyperresponsiveness and atopy among adults: West and East Germany. Eur Respir J 9:2541–2552

Oppermann H, Doering C, Sobottka A, Krämer U, Thriene B (2001) Belastungssituation ost- und westdeutscher Haushalte mit Hausstaubmilben und Schimmelpilzen. Gesundheitswesen 63:85–89

Perzanowski M, Nganga L, Carter M, Odhiambo J, Ngari P, Vaughan J et al (2002) Atopy, asthma and antibodies to Ascaris among rural and urban children in Kenya. J Pediatr 140:582–588

Riedler J, Eder W, Oberfeld G, Schreuer M (2000) Austrian children living on a farm have less hay fever, asthma and allergic sensitization. Clin Exp Allergy 30:194–200

Ring J, Krämer U, Schäfer T, Behrendt H (2001) Why are allergies increasing? Curr Opin Immunol 13:701–708

Ring J, Krämer U, Oppermann H, Ranft U, Behrendt H (2004) Influence of pertussis / Pertussis vaccination on asthma and allergy, prevalence in East and West Germany. Critical remarks on the Hygiene Hypothesis. Allergy Clin Immunol Int 1:17–24

Ritz B, Heinrich J, Wjst M, Wichmann E, Krause C (1998) Effect of cadmium body burden on immune response of school children. Arch Environ Health 53:272–280

Sausenthaler S, Kompauer I, Borte M, Herbarth O, Schaaf B, Berg A et al (2006) Margarine and butter consumption, eczema and allergic sensitization in children. The LISA birth cohort study. Pediatr Allergy Immunol 17:85–93

Schlaud M, Atzpodien K, Thierfelder W (2007) Allergische Erkrankungen. Ergebnisse aus dem Kinder- und Jugendgesundheitssurvey (KiGGS). [Allergic diseases. Results from the German Health Interview and Examination Survey for Children and Adolescents (KiGGS)]. Bundesgesundheitsblatt Gesundheitsforschung Gesundheitsschutz 50:701–710

Sears MR, Greene JM, Willian AR, Taylor DR, Flannery EM, Cowan JO et al (2002) Long-term relation between breastfeeding and development of atopy and asthma in children and young adults: a longitudinal study. Lancet 360:901–907

Siltanen M, Kajosaari M, Pohjavuori M, Savilahti E (2001) Prematurity at birth reduces the long-term risk of atopy. J Allergy Clin Immunol 107:229–234

Strachan D, Cook DG (1998) Parental smoking and childhood asthma: longitudinal and case control studies. Thorax 53:204–212

Sugiri D, Ranft U, Schikowski T, Krämer U (2006) The influence of large-scale airborne particle decline and traffic-related exposure on children's lung function. Environ Health Perspect 114:282–288

Tedner SG, Örtqvist AK, Almqvist C (2012) Fetal growth and risk of childhood asthma and allergic disease. Clin Exp Allergy 42:1430–1447

van Strien RT, Verhoeff AP, Brunekreef B, van Wijnen JH (1994) Mite antigen in house dust: relationship with different housing characteristics in the Netherlands. Clin Exp Allergy 24:843–853

Vartiainen E, Petays T, Haahtela T, Jousilahti P, Pekkanen J (2002) Allergic diseases, skin prick test responses, and IgE levels in North Karelia, Finland, and the Republic of Karelia, Russia. J Allergy Clin Immunol 109:643–648

von Mutius E, Martinez FD, Fritzsch C, Nicolai T, Roell G, Thiemann H (1994) Prevalence of asthma and atopy in two areas of West and East Germany. Am J Respir Crit Care Med 49:358–364

von Mutius E, Weiland SK, Fritzsch C, Duhme H, Keil U (1998) Increasing prevalence of hay fever and atopy among children in Leipzig, East Germany. Lancet 351:862–866

Wickens K, Barry D, Friezema A, Rhodius R, Bone N, Purdie G et al (2005) Obesity and asthma in 11–12 year old New Zealand children in 1989 and 2000. Thorax 60:7–12

Zhu Y, Hinds WC, Kim S, Sioutas C (2002) Concentration and size distribution of ultrafine particles near a major highway. J Air Waste Manag Assoc 52:1032–1042

Air Pollution, Subclinical Inflammation and the Risk of Type 2 Diabetes

11

Tom Teichert and Christian Herder

Contents

T. Teichert • C. Herder (✉)
Institute for Clinical Diabetology, German Diabetes Center, Leibniz Center for Diabetes Research at Heinrich Heine University Düsseldorf, Düsseldorf, Germany

German Center for Diabetes Research (DZD e.V.), Düsseldorf, Germany
e-mail: christian.herder@ddz.uni-duesseldorf.de

C. Esser (ed.), *Environmental Influences on the Immune System*,
DOI 10.1007/978-3-7091-1890-0_11

11.1 Introduction

Type 2 diabetes accounts for more than 90 % of all diabetes cases worldwide and is characterised by insulin resistance (i.e. reduced action of insulin) and beta-cell dysfunction (i.e. insufficient release of insulin to regulate blood glucose levels) (IDF 2013; Tabák et al. 2012; Kahn et al. 2014). The International Diabetes Federation (IDF) estimated that in 2013, 382 million adults between 20 and 79 years had diabetes. This number has been projected to rise by 55 % to 592 million by 2035 with the highest increases in low- and middle-income countries (IDF 2013). The rise in the cases of diabetes along with its chronic complications (e.g. cardiovascular diseases, kidney disease, retinopathy, polyneuropathy, cognitive decline) will be challenging, not only at the individual level but also because of the economic burden to health services and societies as a whole (Tamayo et al. 2014b). The precise mechanisms why people develop diabetes are not completely understood, but several important factors are consistently related to the risk of type 2 diabetes. Advancing age, ethnicity and genetic predisposition are not modifiable, whereas obesity, hypercaloric diet, physical inactivity and smoking represent modifiable risk factors and thus potential targets for prevention (Chen et al. 2012b; Tamayo et al. 2014b).

It is interesting to note that the burden of individually modifiable risk factors increased in many countries during the last decades. However, currently known risk factors are not able to predict the individual diabetes risk with sufficiently high precision (Herder et al. 2014) and explain only partially the growth in diabetes incidence and prevalence that parallels industrialisation and urbanisation in many countries (Chen et al. 2012b). A range of recent studies provided evidence for the impact of psychosocial factors (Pouwer et al. 2010; Stringhini et al. 2013), exposures to environmental toxins mainly in air and water (Hectors et al. 2011) and infectious diseases (Echouffo-Tcheugui and Dagogo-Jack 2012; Sima and Glogauer 2013). Environmental toxins that have been discussed in the context of type 2 diabetes include air pollution (Rajagopalan and Brook 2012), persistent organic pollutants (Hectors et al. 2011; Taylor et al. 2013) and metals (Hectors et al. 2011). In our article, we will focus on air pollution as a risk factor of disease first of all because of its ubiquitous exposure and also because currently high exposure levels are continuing to rise in many urban areas worldwide (WHO 2014b).

Air pollution has traditionally been regarded as a risk factor for diseases of the respiratory tract (e.g. chronic obstructive pulmonary disease, lung cancer) and cardiovascular diseases (Brook et al. 2010; Gold and Mittleman 2013; Raaschou-Nielsen et al. 2013), but evidence has accumulated that it also contributes to the development of type 2 diabetes (Bhatnagar 2009; Rajagopalan and Brook 2012). Based on current estimates suggesting that exposure to air pollution represents a modifiable risk factor (on the societal level) that causes 3.7 million deaths every year (WHO 2013), a better understanding of the mechanisms that link air pollution and morbidity as well as premature mortality is important.

In comparison, data supporting a role for air pollution in the pathophysiology of type 2 diabetes is fairly novel. Animal studies demonstrate that exposure to particulate matter (PM) induces proinflammatory processes and insulin resistance in mice

(Sun et al. 2009). Epidemiological studies described the role of air pollution (PM and gaseous compounds) initially in ecological and cross-sectional studies, but in recent years, data from prospective studies became available that corroborated these previous findings (Krämer et al. 2010; Rajagopalan and Brook 2012; Liu et al. 2013c). The link between air pollution and type 2 diabetes appears biologically plausible (Rajagopalan and Brook 2012; Liu et al. 2013c), and inflammatory processes are among the mechanisms that may mediate this association.

The objective of this chapter is to review data on the interplay between air pollution, inflammation and type 2 diabetes. We aim to (i) provide an overview of epidemiological data on air pollution and risk of type 2 diabetes, (ii) discuss the role of inflammatory processes in the development of this disease and (iii) appraise the evidence that inflammation represents a mechanistic link between the exposure to air pollution and diabetes risk.

11.2 Air Pollution and Risk of Type 2 Diabetes

Air pollution is a global burden and represents a health hazard both in outdoor and indoor environments. Especially developing countries have reported rapidly increasing problems with air pollution-related diseases. The World Health Organization (WHO) reviewed the associations between air pollution and adverse health effects and proposed global air quality guidelines in 2005. In more recent statements, the WHO designates air pollution as the world's leading cause of disease (WHO 2014a; Lippmann 2014) in part because of its cancerogenic potential in addition to its connection to chronic obstructive pulmonary disease, hypertension, stroke, ischaemic heart diseases and mortality.

11.2.1 Classification and Sources of Air Pollution

Pollutants can be classified into solid and gaseous compounds. The common group of solid compounds suspended in the atmosphere is particulate matter (PM), which is furthermore characterised by its particle size. According to the US Environmental Protection Agency (EPA 2014), there are four groups of PM:

1. PM with a diameter below 10 μm: PM_{10}
2. PM with diameter between 10 and 2.5 μm: PM_{coarse}
3. Fine PM with a diameter below 2.5 μm: $PM_{2.5}$
4. Ultrafine PM with a diameter below 0.1 μm: $PM_{0.1}$, UFP

PM originates from incomplete combustion processes in motor vehicles, from industrial processes or from burning of wood and coal. The larger particles are more often based on natural phenomena and are brought into the atmosphere by dusted soils, storms or volcanic activity (Valavanidis et al. 2008). Monitoring projects on a national level have previously been established and allow the user to follow worldwide PM monitoring stations on a live broadcast basis (http://aqicn.org/).

Important gaseous pollutants, like the most commonly monitored nitrogen oxides (NO_x), sulfur oxides (SO_x), ozone (O_3) and carbon monoxide (CO), were extensively examined in several studies (Rückerl et al. 2011). These compounds may have the same man-made origin in combustion processes, but they can also arise from photochemical reactions in the atmosphere.

Indoor air pollution is a more neglected research topic but is predominant in cultures and countries with simple in-house solid fuel burning for heating or cooking. Around three billion people depend on these systems and are thereby exposed to varying levels of household air pollution. Its evaluation is necessary to assess cumulative and synergistic effects with outdoor air pollution on our health status (WHO 2014a).

11.2.2 Analysing the Impact of Air Pollution

In order to quantify the toxicological potential of ambient air pollution, several parameters should be considered. Ambient PM levels can be measured by either local monitoring stations with data provided by governmental institutions or individual exposure monitors. The efficiency of monitoring long-term exposure data and a validated measurement protocol lend an advantage to the first method, which yields useful results that can be fitted to the exposure level of each study participant by appropriate statistical methods (Eeftens et al. 2012; Beelen et al. 2013a). In contrast, personal monitors collect exact exposure data over the observation period and are able to detect even indoor pollutant exposure, which is not possible by governmental outdoor monitoring stations. The high costs to date and the reliance on the accurate use by the study participant currently restrict this system to short-term exposure studies.

11.2.3 Air Pollution and Health Hazards

Data for adverse health effects of exposure to air pollution is most detailed for North America (Pope and Dockery 2006). The European Study of Cohorts for Air Pollution Effects (ESCAPE) consortium was initiated to assess long-term exposure-related effects of air pollutants on public health in Europe. Table 11.1 summarises the key findings related to the respiratory tract of this recent collaborative multinational effort (Gehring et al. 2013; Liu et al. 2013a; MacIntyre et al. 2013; Raaschou-Nielsen et al. 2013; Dimakopoulou et al. 2014; Schikowski et al. 2014). There is convincing evidence for a plethora of adverse effects on the respiratory tract in both children and adolescents as well as adults after long-term exposure to ambient air pollutants representing a complex mixture of hazardous substances. The impact of a particular pollutant varies for the analysed health outcome. Synergistic effects of the exposure to different air pollutants cannot be excluded, but detailed data on the impact of the composition of air pollutants on health is currently not available. Differences of composition and concentration can be expected between rural and urban areas (Valavanidis et al. 2008).

Additional results of the ESCAPE consortium, related to health outcomes to other sides than the pulmonary system, are presented in Table 11.2 (Gruzieva et al.

Table 11.1 Key findings from the ESCAPE consortium on the association between air pollution and pulmonary health outcomes

Outcome	Findings	Reference
Eosinophilic airway inflammation	Short-term increase of exhaled NO after NO_2 exposure in schoolchildren	Liu et al. (2013a)
Respiratory infections	Increased risk of pneumonia in children with increasing concentrations of several pollutants	MacIntyre et al. (2013)
Lung function	Decreased lung function in children after exposure to solid and gaseous pollutants	Gehring et al. (2013)
COPD	No associations between NO_2 or PM_{10} levels with incidence of COPD	Schikowski et al. (2014)
Lung cancer	HR (95 % CI) for lung cancer of 1.22 (1.03–1.45) for 10 μg/m³ increases in PM_{10} levels	Raaschou-Nielsen et al. (2013)
Respiratory mortality	No significant links between exposure to air pollution and nonmalignant respiratory mortality	Dimakopoulou et al. (2014)

CI confidence interval, *COPD* chronic obstructive pulmonary disease, *HR* hazard ratio, *NO* nitrogen monoxide, NO_2 nitrogen dioxide

Table 11.2 Key findings on the association between air pollution and health outcomes related to other sides than the lung from the ESCAPE consortium

Outcome	Findings	Reference
Allergic sensitisation	No associations between air pollution exposure and allergic sensitisation in early childhood	Gruzieva et al. (2013)
Blood pressure	No associations between ambient air pollutants and blood pressure in schoolchildren	Liu et al. (2013b)
Coronary events	HR (95 % CI) of 1.12 (1.01–1.25) for coronary events (myocardial infarction, unstable angina) for 10 μg/m³ increases in PM_{10}	Cesaroni et al. (2014)
Low birth weight	OR (95 % CI) of 1.18 (1.06–1.33) for low birth weight at term for 5 μg/m³ increase in $PM_{2.5}$	Pedersen et al. (2013)
Natural-cause mortality	HR (95 % CI) of 1.07 (1.02–1.13) for mortality for 5 μg/m³ increase in $PM_{2.5}$	Beelen et al. (2013b)
Cardiovascular mortality	No association between air pollution and cardiovascular mortality	Wang et al. (2014)

CI confidence interval, *HR* hazard ratio, *OR* odds ratio

2013; Liu et al. 2013b; Pedersen et al. 2013; Beelen et al. 2013b; Cesaroni et al. 2014; Wang et al. 2014). According to the results of the studies from the ESCAPE consortium, an increase in ambient PM exposure was neither associated with allergic sensitisation (Gruzieva et al. 2013) nor with increased blood pressure (Liu et al. 2013b) in schoolchildren. In contrast, low birth weight (Pedersen et al. 2013) and natural-cause mortality were related to increased levels of $PM_{2.5}$, even at low ambient concentrations (Beelen et al. 2013b). These studies are important because they help to better quantify the effect sizes of associations between air pollution and health outcomes at low and high levels of exposure. Further studies are desirable to better characterise the susceptibility of different subgroups of the population to air pollutants, e.g. with respect to age, sex, obesity, type 2 diabetes or pre-existing cardiovascular disease.

11.2.4 Air Pollution and Type 2 Diabetes

The notion that exposure to air pollution contributes to the development of type 2 diabetes is substantiated by several lines of evidence. First, the impact of PM with a diameter below 2.5 μm on glucose metabolism was repeatedly studied in mice (Sun et al. 2009; Xu et al. 2011a, 2012; Zheng et al. 2013; Liu et al. 2014). These studies are consistent in their findings and report an enhanced insulin resistance after exposing mice to particulate matter (Table 11.3). Furthermore, some of the studies summarised in Table 11.3 also addressed possible mechanisms linking the exposure to PM with the development of impaired glucose metabolism. One study (Zheng et al. 2013) explained impaired glucose metabolism by alterations in the liver which were similar to non-alcoholic steatohepatitis. Enhanced proinflammatory processes (Sun et al. 2009; Xu et al. 2011a), altered phosphorylation of 5′-adenosine monophosphate-activated protein kinase (AMPK) and modified gene expression profiles related to insulin signaling pathways are additional contributors to insulin resistance and impaired glucose metabolism (Xu et al. 2011a, 2012).

A second line of evidence regarding a potential role of air pollutants for the development of type 2 diabetes comes from a range of epidemiological studies investigating associations between exposure to air pollutants and type 2 diabetes as well as related traits using a cross-sectional design (Table 11.4) (Lockwood 2002; Brook et al. 2008; Kelishadi et al. 2009; Pearson et al. 2010; Chuang et al. 2011;

Table 11.3 Experimental studies in mice demonstrating associations between particulate matter and impaired glucose regulation

Author	Mouse model	Findings
Sun et al. (2009)	C57BL/6 mice	$PM_{2.5}$ exposure exacerbated insulin resistance and inflammation in adipose tissue
Xu et al. (2011a)	C57BL/6 mice	10-month exposure to $PM_{2.5}$ caused multiple-tissue insulin resistance and inflammatory responses
Xu et al. (2012)	Male $ApoE^{-/-}$ mice	Nickel and concentrated $PM_{2.5}$ induced insulin resistance and decreased AMPK phosphorylation
Zheng et al. (2013)	C57BL/6 mice	$PM_{2.5}$ exposure triggered a NASH-like phenotype and caused insulin resistance
Liu et al. (2014)	C57BL/6 mice	$PM_{2.5}$ exposure exacerbated insulin resistance, adipose tissue inflammation and accumulation of lipids in the liver

AMPK 5′ adenosine monophosphate-activated protein kinase, *ApoE* apolipoprotein E, *C57BL/6* inbred strain of laboratory mice, *NASH* non-alcoholic steatohepatitis

Table 11.4 Cross-sectional studies on associations between air pollution exposure and the prevalence of type 2 diabetes and related phenotypes

Author	Cohort (country)	Findings
Lockwood (2002)	184,450 participants (USA)	Total air toxicant levels (assessed on state level) were correlated with T2D prevalence ($r=0.54$; $p<0.0001$)
Brook et al. (2008)	7,364 participants (Canada)	Increase by 1 µg/m³ of ambient NO_2 level was associated with an increase of 4 % in T2D prevalence
Kelishadi et al. (2009)	374 children (Iran)	HOMA-IR was increased after exposure to PM_{10}
Pearson et al. (2010)	Cross-sectional data of >2,700 counties (USA)	Increase of 10 µg/m³ in $PM_{2.5}$ concentration was associated with 1 % higher T2D prevalence
Dijkema et al. (2011)	8,018 participants (the Netherlands)	No association between traffic-related air pollution and prevalence of T2D
Chuang et al. (2011)	1,023 elderly adults (Taiwan)	Exposure to PM_{10}, $PM_{2.5}$, NO_2 and O_3 was associated with higher levels of fasting glucose and HbA1c
Teichert et al. (2013)	360 elderly women (Germany)	Long-term exposure to NO_2 was related to impaired fasting glucose
Thiering et al. (2013)	397 schoolchildren (Germany)	Insulin resistance (HOMA-IR) was increased by 17 % and 19 % in children for every elevation of NO_2 and PM_{10} levels by 10 µg/m³ and 6 µg/m³, respectively
Tamayo et al. (2014a)	9,120 patients with T2D (Germany)	Significantly lower HbA_{1c} levels in the lowest quartile of PM_{10} exposure

HbA1c glycated haemoglobin, *HOMA-IR* homeostasis model assessment insulin resistance, *T2D* type 2 diabetes

Dijkema et al. 2011; Teichert et al. 2013; Thiering et al. 2013; Tamayo et al. 2014a). Interestingly, the positive associations were not restricted to $PM_{2.5}$-mediated effects, which have been tested in mice, but several studies also extended the findings from mouse models to nitrogen dioxide as a harmful component in polluted air. Overall, findings are fairly consistent and are supported by the results of prospective studies presented in Table 11.5 representing the third line of evidence. The particular value of prospective studies lies in their study design which requires that the measurement of the exposure to air pollution as potential risk factor precedes the development of type 2 diabetes as a health outcome of interest. Therefore, this study type possesses a higher degree of evidence for causal relationships than cross-sectional studies. Accumulating data indicates that traffic-related nitrogen dioxide may have deleterious impacts on public health and may contribute to the development of type 2 diabetes in addition to PM. Despite the relatively consistent observations, it has to be noted that pollutant composition, exposure levels, exposure time, community settings, personal lifestyle, anthropometric confounders and genetic susceptibility limit the comparability between the studies summarised in Table 11.5.

Table 11.5 Epidemiological studies investigating associations between air pollutants and the risk or incidence of type 2 diabetes

Author	Cohort (abbreviation, country)	Findings
Krämer et al. (2010)	1,776 elderly women (SALIA; Germany)	Increase of 15 μg/m³ NO_2 leads to an 42 % increased risk in T2D incidence
Puett et al. (2011)	89,450 men and women from two studies (NHS and HPFS; USA)	*Short distance* (<50 m versus ≥200 m) to a street with high traffic load increased the incidence of T2D by 14 %
Coogan et al. (2012)	3,992 women (BWHS; USA)	Increases in NO_2 by 12.4 ppb were related to a 25 % higher T2D incidence rate
Andersen et al. (2012)	51,818 nondiabetic participants (DCH; Denmark)	Increases in NO_2 by 4.9 μg/m³ increased the risk of confirmed T2D by 4 %
Chen et al. (2013)	62,012 participants (Canada)	An increase of 10 μg/m³ in $PM_{2.5}$ levels resulted in an 11 % increased risk of diabetes

BWHS Black Women's Health Study, *DCH* Danish Diet, Cancer and Health Study, *HPFS* Health Professionals Follow-Up Study, *NHS* Nurses' Health Study, *SALIA* Study on the influence of Air pollution on Lung function, Inflammation and Aging

11.2.5 Air Pollution and Diabetes-Related Co-morbidities

There is evidence that patients with type 2 diabetes are more susceptible to harmful effects of ambient air pollutants than nondiabetic individuals. Research related to the cardiovascular system indicated that diabetic subjects show attenuated heart rate variability after exposure to $PM_{2.5}$ and PM_{10} (Park et al. 2005; Whitsel et al. 2009). This association appeared insulin dependent, as subjects with increased insulin resistance showed greater alterations (Whitsel et al. 2009). Besides heart rate variability, many other factors contribute to cardiovascular events. After exposure to $PM_{2.5}$, both increased blood pressure (Hoffmann et al. 2012) and impaired vascular reactivity, in terms of nitroglycerin-mediated and flow-mediated reactivity (O'Neill et al. 2005), were more pronounced in diabetic subjects than in healthy controls. Furthermore, the risk of cardiovascular admission to hospitals was doubled for diabetes patients after exposure to PM_{10} compared to nondiabetic individuals (Zanobetti and Schwartz 2002).

It is well established that inflammatory processes contribute to the development and progression of cardiovascular diseases. Therefore, proinflammatory biomarkers, like cytokines or soluble adhesion molecules, are thought to be key factors in this process. In line with this concept and the aforementioned observations, studies reported elevated levels of the acute-phase protein C-reactive protein (CRP) (Dubowsky et al. 2006) and white blood cell count (Khafaie et al. 2013) in patients with diabetes after exposure to PM. These findings may help to explain the increased susceptibility of type 2 diabetes patients for vascular complications after exposure to air pollution and point towards inflammatory processes as potential mediators in the association between air pollution and cardiometabolic health. A higher susceptibility to cardiovascular disease may eventually lead to premature mortality. Indeed, higher mortality rates of patients with type 2 diabetes were related to exposure to ambient air pollution (Jerrett et al. 2005; Goldberg et al. 2006; Raaschou-Nielsen et al. 2012; Brook et al. 2013). One study described a 49 % raised mortality rate in patients suffering from type 2 diabetes for an increase of 10 μg/m^3 in ambient $PM_{2.5}$ concentration. This rate was relatively high compared to pooled results from 22 longitudinal cohort studies across Europe which found an increase of 7 % in natural mortality with every 5 μg/m^3 increase in $PM_{2.5}$ concentration (Beelen et al. 2013b).

The enhanced susceptibility of T2D patients to air pollution is not restricted to mortality and cardiovascular events. Moreover, respiratory diseases, i.e. reduced lung function (Berclaz et al. 2009), tuberculosis (Jeon and Murray 2008), COPD (Seshasai et al. 2011) and lung cancer (Cesaroni et al. 2013), occurred more often in diabetic patients compared to nondiabetic controls.

11.2.6 Conclusion

Taken together, there is a growing body of evidence from animal models to large epidemiological studies that supports the association between air pollution and impaired glucose regulation, type 2 diabetes and co-morbidities in patients with diabetes. While the mechanistic background remains unknown, several studies addressed the potential role of inflammation in the increased susceptibility of diabetes patients to air pollution. The following section will illustrate that type 2 diabetes has a proinflammatory component and that subclinical inflammation represents an important risk factor, with air pollution as one of its determinants.

11.3 Subclinical Inflammation as a Risk Factor for Type 2 Diabetes

We outlined in the preceding section that there is substantial evidence from epidemiological studies that exposure to air pollutants increases the risk for type 2 diabetes. However, most of these studies did not investigate which mechanisms are responsible for this association. As reviewed recently, several mechanisms appear biologically plausible and are supported by experimental data (Rajagopalan and Brook 2012; Liu et al. 2013c). These include endothelial dysfunction, insulin resistance, mitochondrial dysfunction, alterations in adipose tissue and local and systemic subclinical inflammation (Fig. 11.1). Before discussing the interplay of air pollution, subclinical inflammation and type 2 diabetes in more detail in the next section, we will first illustrate how inflammatory processes contribute to the development of type 2 diabetes.

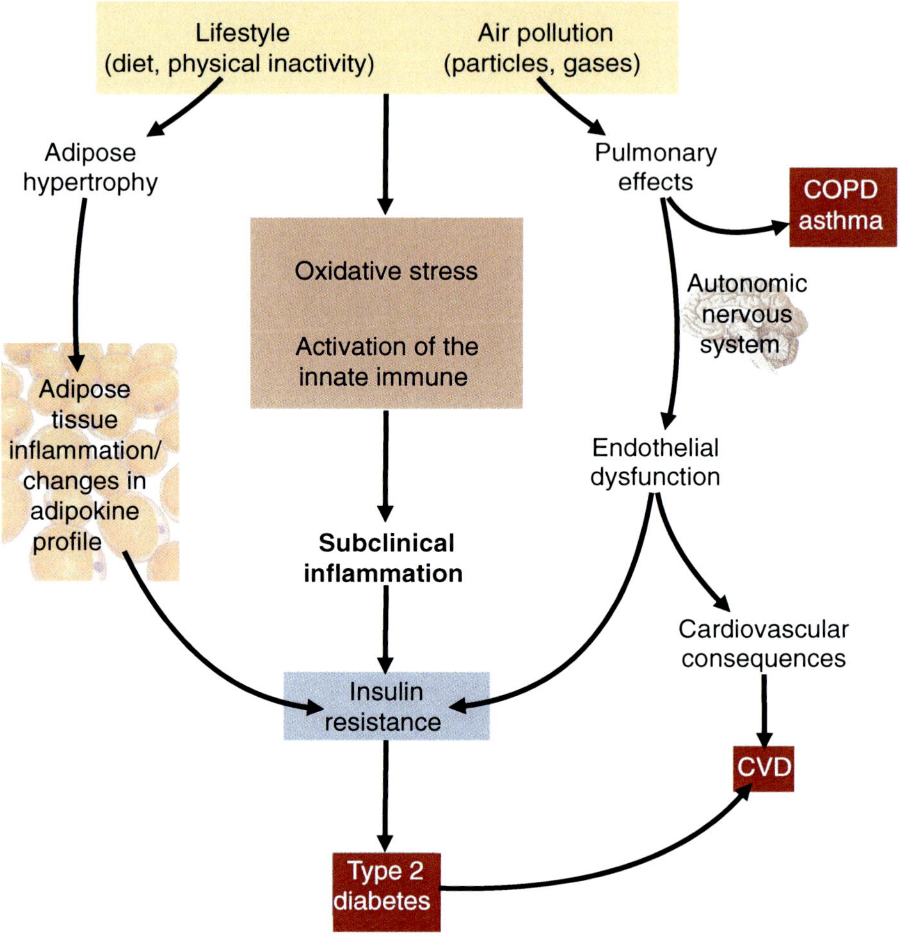

Fig. 11.1 Pathways linking risk factors with the development of type 2 diabetes. Schematic presentation of the impact of risk factors on the development of subclinical inflammation and type 2 diabetes. Figure modified from Sun et al. 2009. *COPD* chronic obstructive pulmonary disease, *CVD* cardiovascular disease

11.3.1 Subclinical Inflammation and Incident Type 2 Diabetes: Epidemiological Studies

In contrast to classical inflammation which is characterised by the clinical symptoms of pain, heat, redness, swelling and loss of function, subclinical inflammation is asymptomatic. Although there is no universal definition, individuals with subclinical inflammation exhibit increased serum or plasma concentrations of proinflammatory immune mediators resulting in a state of chronic, low-grade systemic inflammation (Kolb and Mandrup-Poulsen 2005). In the first prospective studies investigating the role of inflammation, elevated baseline levels of C-reactive protein (CRP) and the cytokine interleukin-6 (IL-6) were found in individuals who developed type 2 diabetes during the follow-up period compared to individuals who remained diabetes-free (Schmidt et al. 1999; Barzilay et al. 2001; Pradhan et al. 2001; Festa et al. 2002). Subsequent studies corroborated these findings and reported associations for further proinflammatory cytokines (e.g. IL-1β, IL-18), chemokines (e.g. monocyte chemotactic protein-1 (MCP-1)/CCL2, IL-8/CXCL8), adipokines (e.g. leptin) and soluble adhesion molecules (e.g. soluble E-selectin, soluble intercellular adhesion molecule-1 (sICAM-1) with incident type 2 diabetes (Spranger et al. 2003; Thorand et al. 2005; Herder et al. 2006; Meigs et al. 2004). In addition to altered concentrations of proinflammatory immune mediators in the circulation, individuals at high risk of type 2 diabetes also showed an increase in systemic levels of the anti-inflammatory protein IL-1 receptor antagonist (IL-1RA) and decreased levels of the adipocyte-derived, anti-inflammatory protein adiponectin (Herder et al. 2009a; Lindsay et al. 2002). Increased IL-1RA levels most likely reflect a counter-regulatory response to metabolic and immunological disturbances in the prediabetic phase, which, however, are not sufficient to protect against the onset of type 2 diabetes (Herder et al. 2009a; Carstensen et al. 2010). Adiponectin clearly has insulin-sensitising effects in rodents, and a decrease in circulating adiponectin may be causally related to the development of insulin resistance and the manifestation of type 2 diabetes (Turer and Scherer 2012; Li et al. 2009; Herder et al. 2013). It is important to note that subclinical inflammation is not only associated with risk for type 2 diabetes but also with increased risk to develop myocardial infarction and stroke and with higher risk of general frailty and all-cause mortality (Hansson and Hermansson 2011; Lee et al. 2012). Further studies suggest that an activation of the immune system may additionally contribute to diabetic polyneuropathy, cognitive decline in old age, depression, dementia and several types of cancer (Herder et al. 2009b; Xu et al. 2009; Khandekar et al. 2011) which emphasises the relevance of research in this field in the context of public health.

The established data from cohort studies that subclinical inflammation precedes the onset of type 2 diabetes raise three important questions: (i) Do we have evidence from mechanistic studies that subclinical inflammation may be causal for type 2 diabetes rather than representing an epiphenomenon in the development of the disease? (ii) Which factors trigger subclinical inflammation – or, in other words, does subclinical inflammation represent a modifiable risk factor? (iii) What is the evidence that anti-inflammatory therapy has beneficial metabolic effects and can reduce hyperglycaemia?

11.3.2 Causal Role of Subclinical Inflammation in the Development of Type 2 Diabetes

The first question has mainly been addressed in a range of mouse models in which the effects of genetic manipulations of the immune system on glucose metabolism have been investigated. A complete discussion of these experiments is beyond the scope of this overview, but important results can be summarised as follows: A shift in the balance between pro- and anti-inflammatory processes by genetic overexpression of proinflammatory immune mediators, their receptors or essential components of their signal transduction cascades in the whole animal or in specific tissues (e.g. leukocytes, adipocytes, hepatocytes) or by genetic deletion of anti-inflammatory proteins frequently leads to insulin resistance, impaired glucose tolerance and diabetes. In contrast, genetic deletions of proinflammatory proteins or overexpression of anti-inflammatory components usually have the opposite effect and improve glucose tolerance or protect against diet-induced obesity and concomitant insulin resistance (Kolb and Mandrup-Poulsen 2005; Shoelson et al. 2007; Donath and Shoelson 2011). Molecular and cellular mechanisms linking inflammation with both insulin resistance and beta-cell dysfunction in mouse models and humans as hallmarks of type 2 diabetes have been reviewed recently in more detail elsewhere (Donath and Shoelson 2011; Gregor and Hotamisligil 2011; Osborn and Olefsky 2012; Sell et al. 2012). Thus, changes in the immune system are sufficient to have profound effects on the regulation of glucose metabolism and the onset of diabetes in mice.

11.3.3 Triggers of Subclinical Inflammation

The second question is important because knowledge of factors that trigger subclinical inflammation may be used for therapeutic approaches to attenuate inflammatory processes in order to prevent or improve inflammation-related conditions. Among the aforementioned non-modifiable risk factors, old age can be considered a contributor to low-grade inflammation. It has been hypothesised that this may be at least in part attributable to impairments in anti-inflammatory regulatory processes that counteract the activation of the immune system. Moreover, genes encoding components of the immune system harbour many polymorphisms that have an impact on their expression and protein levels so that the genetic background of every individual represents a determinant of immune functions. However, it should be noted that no genetic variant has been identified so far that increases the risk for type 2 diabetes by mainly proinflammatory mechanisms. In addition, most lifestyle or environmental factors that predispose to type 2 diabetes also have a proinflammatory component (Kolb and Mandrup-Poulsen 2010). This has clearly been shown for obesity, poor diet, physical inactivity, psychosocial stress factors and environmental toxins including air pollution (the latter will be discussed in detail in the following section) (Guilherme et al. 2008; Kolb and Mandrup-Poulsen 2010; Sidawi and Al-Hairi 2012; Nimmo et al. 2013; Howren et al. 2009; Carpenter 2008). A removal or reversal of these factors is not always possible in experimental settings, but weight loss, dietary changes and an increase in physical activity have frequently been demonstrated to reduce systemic levels of proinflammatory cytokines and to increase adiponectin levels (Ziccardi et al. 2002; Herder et al. 2009c; Beavers et al. 2010; Gleeson et al. 2011). Therefore, many different factors representing primary causes in the aetiology of diabetes are linked with subclinical inflammation that may be considered an integrating mechanism and potential therapeutic target.

11.3.4 Subclinical Inflammation as Potential Therapeutic Target

The third question has been addressed in several intervention studies (Donath and Shoelson 2011; Donath Marc et al. 2013). A proof-of-principle study demonstrated that subcutaneous administration of recombinant IL-1RA in order to block the action of the proinflammatory cytokine IL-1β attenuated systemic levels of proinflammatory immune mediators and improved hyperglycaemia in patients with type 2 diabetes (Larsen et al. 2007). Several subsequent studies attempted to reduce IL-1β reactivity by monoclonal anti-IL-1β antibodies with modest improvements in hyperglycaemia (Cavelti-Weder et al. 2012). In addition to targeting cytokines, it is also possible to interfere with proinflammatory signal transduction pathways. Salsalate, a non-acetylated salicylate with a better safety profile than aspirin, inhibits the transcription factor nuclear factor-kappa B (NF-κB) and thereby attenuates inflammatory processes. Interestingly, treatment of patients with type 2 diabetes with salsalate also improved hyperglycaemia (Goldfine et al. 2010). Although ongoing studies will have to show to what extent anti-inflammatory drugs can be used in the future to treat type 2 diabetes or to prevent/delay its onset (Ridker et al. 2012), the findings that anti-inflammatory treatment has beneficial effects on glucose homeostasis emphasise the tight interactions between immune system and metabolic regulation (Donath and Shoelson 2011; Donath et al. 2013),

11.3.5 Conclusion

Taken together, subclinical inflammation as evident by increased circulating levels of mainly proinflammatory immune mediators represents a risk factor of type 2 diabetes. Subclinical inflammation is triggered by diabetes-related lifestyle and environmental factors that include air pollution, and removal of these factors or targeted pharmacological approaches can be used to attenuate subclinical inflammation and improve glucose metabolism.

11.4 Link between Air Pollution, Local Inflammation in the Airways, Systemic Inflammation and Type 2 Diabetes

Ambient air pollution exposure has been associated with a range of diseases and increased morbidity and mortality. Among the pollutants, NO_2 and $PM_{2.5}$ appear to be key contributors in the development of type 2 diabetes and related co-morbidities. Diabetes and its co-morbidities are multifactorial diseases with a range of risk factors, but it is interesting to note that they are all characterised by low-grade systemic inflammation, which could be one unifying factor for their relationship with exposure to air pollutants. It is recognised that dose-dependent exposure to air pollution activates the adaptive and innate immune system, followed by proinflammatory cascades in the respiratory tract (Miyata and van Eeden 2011). Beyond the respiratory tract, air pollution has also been repeatedly linked to increased levels of IL-6, IL-8, TNF-α and CRP in the circulation (Eeden et al. 2001; Donaldson et al. 2005; Rückerl et al. 2006; Agustí et al. 2012). With respect to the inflammatory response, individuals already suffering from metabolic diseases appear to be more susceptible to the effects of air pollution than healthy subjects (Schneider et al. 2011).

11.4.1 Air Pollution and Airway Inflammation

The respiratory tract is in direct contact with ambient air and thus possibly the most vulnerable and sensitive organ against air pollution (Fig. 11.2). Common methods to analyse the inflammatory levels are based on cytological measurements in bronchoalveolar lavage fluid (BALF) and sputum. Inhaled particles are deposited to a certain degree, whereas 25 % are removed by macrophages and mucociliary clearance (Geiser and Kreyling 2010). The deposition is more efficient for UFP than for coarse particles (Devlin et al. 2014). Therefore, it has been hypothesised that UFP possess a greater toxicological potential as the contact time with the tissue appears to be longer (Devlin et al. 2014).

After PM inhalation, the lung tissue releases locally pro-oxidative (e.g. reactive oxygen species, ROS) and proinflammatory mediators in a dose-dependent manner (Michael et al. 2013). Increased levels of IL-8, IL-1β and granulocyte-macrophage colony-stimulating factor (GM-CSF) were reported in supernatant of human primary bronchial epithelial cells after incubation with PM_{10} (Fujii et al. 2001). Higher IL-8 mRNA transcription rates were observed in epithelial A549 cells from rats after exposure to particles with different diameters (Brown et al. 2001). Interestingly, particles with a smaller diameter provoked a less intense response of the cultivated

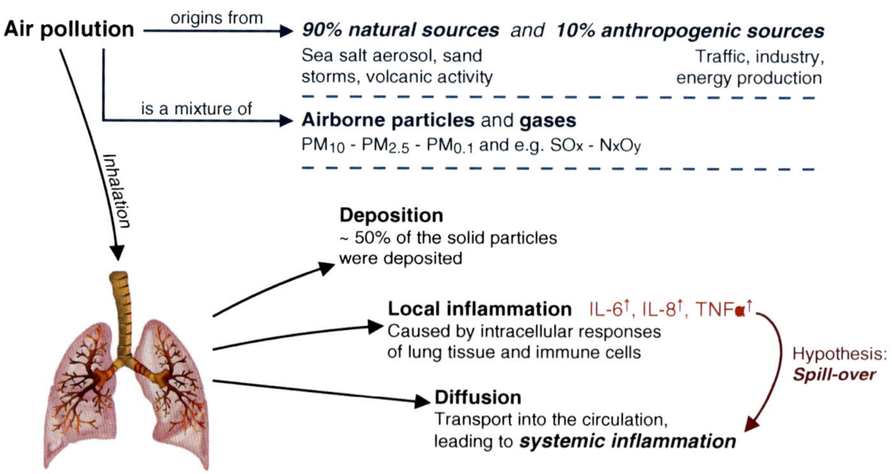

Fig. 11.2 Brief overview of effects of air pollution on lung tissue. Origin and characteristics of air pollutants and their pathological potential after inhalation. *IL* interleukin, *NO* nitrite oxides, *PM* particulate matter, *SO* sulfoxides, *TNFα* tumor necrosis factor alpha

cells than larger-sized particles in one study. This result is contrary to the general hypothesis regarding an inverse relationship between particle diameter and toxicological potential (Gilmour et al. 2004). However, this study was based on an in vitro cell system and thus may not reflect the harmful potential of UFP in vivo after inhalation.

Short-term exposure to black carbon and diesel exhaust particles increased leukocyte cell numbers (Gilmour et al. 2004) and IL-1β levels (Provoost et al. 2010) in BALF in animal studies. IL-1β activates proinflammatory cascades and is associated with the progression of lung injury and cystic fibrosis (Kolb et al. 2001; Levy et al. 2009). In addition, cell activation results in ROS generation and oxidative burst in the respiratory tract and may promote the development and progression of respiratory diseases (Dutta et al. 2013) in line with proinflammatory reactions (Valavanidis et al. 2008).

Adverse effects from experimental studies in cell cultures and animals were confirmed by increased immune cell counts in nasal and airway lavages in man after exposure to $PM_{2.5}$ (Chen et al. 2012a) and diesel exhaust particles (Salvi et al. 1999). Exposure to indoor PM_{10} was also associated with increased counts of a variety of immune cells (Dutta et al. 2013) as well as significantly higher concentrations of IL-6, IL-8 and TNF-α in lavages (Eeden et al. 2001; Dutta et al. 2013). In another study, indoor air pollution led to partial depletion of the antioxidative enzyme superoxide dismutase (SOD) and increased levels of ROS (Dutta et al. 2013). Biopsies revealed increased IL-8 gene transcription, elevated immune cell counts and higher levels of the adhesion molecules ICAM-1 and VCAM-1 in bronchial epithelial cells (Salvi et al. 1999, 2000). These molecular mechanisms in connection with increasing oxidative stress may play a key role in the development of several inflammatory diseases like acute lung injury, COPD, bronchitis and asthma (Valavanidis et al. 2008). Due to the translocation of UFP, adverse health effects are not exclusive to the lung (Terzano et al. 2010).

11.4.2 Air Pollution and Proinflammatory Effects on Other Organs

Once particles have entered the circulation, UFPs are translocated to other organs including the liver, the spleen, the kidneys, the heart and the brain, where they may be deposited (Nemmar et al. 2002). This distribution of PM throughout the body entails the potential for pathological consequences of exposure to PM in many different tissues (Bai et al. 2007; Rückerl et al. 2011).

First of all, inhaled particles will be translocated along the olfactory nerve into the olfactory bulb and permeate directly into the central nervous system (Oberdörster et al. 2004) where they promote hippocampal proinflammatory cytokine production (Fonken et al. 2011) and neuroinflammation (Block and Calderón-Garcidueñas 2009). Increased levels of IL-1α and TNF-α (Campbell et al. 2005; Fonken et al. 2011) were observed in the cytoplasmic fraction of mouse brains, whereas the concentrations of IL-6 remained unaffected (Fonken et al. 2011).

Second, particles can be translocated via the vagus nerve into the lung followed by the distribution to the connecting parasympathetic nerves (Belvisi 2003). Particles entering the lung can pass the lung-blood barrier and enter the circulation. This will ultimately lead to the distribution to different tissues throughout the body and possible infiltration of target organs, followed by an immune response reflected by a local inflammation (Sinden and Stockley 2010). Nemmar et al. showed in hamsters (Nemmar et al. 2001) and man (Nemmar et al. 2002) that inhaled radioactively labelled (^{99m}Tc) particles entered the circulation within minutes after exposure. As a consequence, distributed particles may reach the liver (Oberdörster and Utell 2002), adipose tissue (Xu et al. 2011b), the stomach (Péry et al. 2009) and lymph nodes by particle-loaded macrophages (Lippmann et al. 1980). Kinetic studies estimated that 12.7 % of the inhaled particles were translocated (Péry et al. 2009).

Among the aforementioned organs, activated proinflammatory mechanisms were found in the liver, adipose tissue and bladder of mice. After the entry of UFP into lung tissue and after direct contact with hepatocytes, the production of proinflammatory cytokines was triggered (Bai et al. 2007), causing liver fibrosis and non-alcoholic steatohepatitis-like phenotypes (Zheng et al. 2013), and may thereby have contributed to the impairment of hepatic glucose control (Zheng et al. 2013; Xu et al. 2011a; Liu et al. 2014).

Alterations in adipose tissue were present after long-term exposure to $PM_{2.5}$ in an extended study of mice exposed to air pollution (Xu et al. 2011b; Liu et al. 2013c). Visceral adipose tissue was infiltrated by activated F4/80$^+$ macrophages, which were also present in activated lung tissue (Sun et al. 2009). Independently of diet, the exposure to $PM_{2.5}$ doubled the number of monocytes adhering to endothelial cells in adipose tissue depots and led to an increase in adipocyte size, thereby increasing proinflammatory susceptibility (Liu et al. 2013c). $PM_{2.5}$ exposure resulted in notable changes in the size of the mitochondria, but not in their number in brown adipose tissue, and increased oxidative and nitrous stress. In line with these findings, altered levels of the adipokines leptin and adiponectin were observed in serum (Xu et al. 2011b).

Another targeted tissue in rat studies was the bladder, where enzymatic activities important for oxidative stress regulation were altered after exposure to diesel exhaust particles (Luo et al. 2013).

11.4.3 Air Pollution and Systemic Inflammation as Risk Factors of Type 2 Diabetes

Proinflammatory effects of inhaled particles are not restricted to the respiratory tract or infiltrated tissues. Long-term exposure to $PM_{2.5}$ also results in the upregulation of circulating CRP levels (Salvi et al. 1999; Zhao et al. 2013), which was observed as a dose-dependent reaction to exposure in man (Kelishadi et al. 2009; Hoffmann et al. 2009; Hertel et al. 2010; Khafaie et al. 2013). Even short-term exposure may affect circulating hsCRP levels in exposed subjects (Rückerl et al. 2006; Dubowsky et al. 2006; Seaton et al. 1999; Delfino et al. 2008). Besides activation of acute-phase proteins, increased serum levels of TNF-α, IL-6 and plasma fibrinogen (Schwartz 2001; Bai et al. 2007; Sun et al. 2009; Wang et al. 2013a) were present in rodents exposed to PM, whereas findings in epidemiological studies were more controversial (Törnqvist et al. 2007; Dubowsky et al. 2006; Rückerl et al. 2007; Ljungman et al. 2009; Seaton et al. 1999; Hoffmann et al. 2009; Delfino et al. 2009). This increase was dependent on the PM dose (Wang et al. 2013b) as well as particle diameter (Gilmour et al. 2004) and was accompanied by increased peripheral blood monocyte (Kampfrath et al. 2011) and circulating leukocyte (Gilmour et al. 2004) numbers. Histological analyses suggested that $PM_{2.5}$ exposure activated Toll-like receptor (TLR)-4-dependent pathways (Kampfrath et al. 2011) and chemoattractant molecules (Xu et al. 2013). It is important to note that there are other studies including some with improved exposure assessment (Sullivan et al. 2007), and large population cohorts (Steinvil et al. 2008; Diez Roux et al. 2006), which have not found a relationship between PM exposure and inflammation.

Several studies analysed gene expression profiles of leukocytes in human subjects and reported increased peripheral blood monocyte expression of CD18, CD54 (Frampton et al. 2006), CD80, CD40, CD86, HLA-DR and CD23 (Schneider et al. 2011) after exposure to $PM_{2.5}$ from 2 (Frampton et al. 2006) to 24 h (Huang et al. 2010; Schneider et al. 2011). Additional pathway analyses revealed that differentially expressed genes were involved in host defence, insulin-growth-factor (IGF)-I and insulin receptor signalling (Huang et al. 2010). It is possible that these alterations in insulin-dependent pathways contribute to the development of type 2 diabetes, but the long-term effects remain to be evaluated (Huang et al. 2010). Circulating cell recruitment and platelet activation after exposure to a variety of ambient air pollutants were confirmed by studies in man (Salvi et al. 1999; Gilmour et al. 2004; Brook et al. 2013; Chen et al. 2008). Platelet activation may increase the vulnerability to cardiovascular diseases (CVD) and risk of mortality by cardiovascular events (Salvi et al. 1999).

The aforementioned systemic proinflammatory effects of PM may not only be attributable to translocation of PM throughout the body but also consequences of local inflammatory processes in the airways. Due to its small diameter, PM can penetrate deeply into the lung tissue and trigger local inflammation involving macrophage activation. After repeated exposure, the local inflammatory response may intensify, and subsequently proinflammatory mediators have been hypothesised to *spill over* into the circulation resulting in systemic inflammation (Sinden and Stockley 2010). A direct correlation between airway inflammatory markers and circulating markers would support this hypothesis. However, results from different studies are inconsistent as several research groups observed significant correlations between sputum and serum markers in COPD patients (Vernooy et al. 2002; Broekhuizen et al. 2005; Hurst et al. 2006; Singh et al. 2010), whereas others did not report such associations (Sapey et al. 2009; Zeng et al. 2009; Röpcke et al. 2012). A more detailed discussion of the spillover hypothesis, the mechanisms linking airway and systemic inflammation and methodological challenges in this context can be found in a recent review (Sinden and Stockley 2010).

It is important to note that although it is biologically plausible based on the studies discussed above that proinflammatory changes resulting from exposure to air pollution represent one link between air pollution and the risk of type 2 diabetes, CVD and other co-morbidities, studies corroborating this hypothesis are still scarce.

The role of inflammation as potential intermediary in this context has been shown in mouse models, but convincing studies in human subjects are still lacking.

11.4.4 Conclusion

Inhaled ambient PM causes airway inflammation and thus may promote the development and progression of respiratory diseases. Local inflammatory responses in the lungs may depend on particle dose and size, which are determinants of macrophage activation and oxidative stress production. Translocation of particles throughout the body also leads to proinflammatory responses in other organs including the brain, liver and adipose tissue. A growing number of studies demonstrated a relationship between the exposure to air pollution and systemic inflammation, mainly measured as CRP, IL-6, TNF-α and fibrinogen in the blood. Several reports suggested a relationship between local airway inflammation and systemic inflammation based on a spillover of proinflammatory mediators into the circulation, but data were controversial. Overall, the mechanisms linking on the one hand air pollution-induced airway inflammation and systemic inflammatory responses as potential risk factors for type 2 diabetes and on the other hand chronic diseases require further studies.

11.5 Future Directions

Based on the projections of increasing prevalences of type 2 diabetes in most countries worldwide during the next decades and an incomplete understanding of risk factors and pathogenesis of this disease, recent observations regarding the role of air pollution in this context are timely and important. Although our knowledge of air pollution as a modifiable risk factor for type 2 diabetes has improved during the last 5–10 years, a range of open questions remains that require further research. These relate to the measurement of exposure, to the analysis of pathophysiological mechanisms and to the assessment of metabolic health as an outcome.

First, current exposure measurements usually rely on estimated exposures for individuals' home addresses. These estimations are based on data from monitoring stations and different statistical models. The assessment of the individual exposure to air pollutants is expensive and often not feasible in larger or long-term studies. However, a higher precision in the assessment of individual exposure, the measurement of both outdoor and indoor air pollution and repeated measurements to better assess the temporal and spatial variation of exposure to air pollution are required for a more accurate quantification of air pollution-related health hazards. From the analytical point of view, most current studies concentrate on effects of $PM_{2.5}$, PM_{10} or NO_x, whereas studies on further components (e.g. UFP) or taking into account the composition of PM are virtually absent in general and in particular in the context of type 2 diabetes.

Second, the identification of mechanisms linking air pollution and health outcomes such as type 2 diabetes requires a more detailed characterisation of potentially pathophysiological pathways. The measurement of systemic levels of biomarkers reflecting not only systemic subclinical inflammation but also endothelial dysfunction, coagulation, platelet activation, oxidative stress and lipid metabolism represents one approach. In addition, effects of exposure to air pollution should also be assessed at the level of tissues that are crucial for the regulation of glucose metabolism. As summarised in Table 11.3, the impact of PM on adipose tissue and liver function was assessed in mice, but comparable data for humans or studies regarding the potential effect on beta-cell function are not available.

Third, we have increasing evidence that different components of air pollution affect the risk of type 2 diabetes. Most data are based on epidemiological studies in which type 2 diabetes was not the main outcome, so that the presence of type 2 diabetes was often assessed using questionnaires which implicates the risk of misclassification due to the presence of unknown (previously undiagnosed) diabetes. Ideally, the diagnosis of type 2 diabetes should rely on laboratory tests which include fasting glucose, 2 h post-load glucose after an oral glucose tolerance test and HbA1c. One important advantage of these tests is the fact that they also allow to investigate the impact of air pollution on different stages of prediabetes which are already associated with increased risk of micro- and macrovascular complications (Tabák et al. 2012). In addition to the early stages of diabetes, a detailed analysis of air pollutants and their impact on the incidence of diabetic complications and co-morbidities is required to capture the full impact on diabetes-related morbidity and mortality.

Despite a number of publications on the relationships between (i) air pollution and proinflammatory processes, (ii) air pollution and risk of type 2 diabetes and (iii) subclinical inflammation and type 2 diabetes, the lack of studies analysing these associations in a combined approach is striking. Therefore, it is currently not possible to estimate to what extent air pollution and inflammatory processes contribute to the risk of type 2 diabetes in areas with different levels of exposure. Furthermore, more prospective studies are needed to quantify effect sizes in subgroups of the population because the susceptibility to detrimental effects of air pollution may be modified by genetic predisposition, age, sex, obesity and pre-existing cardiometabolic disease. A better understanding of these questions is not only of purely academic interest, but these data will be needed for improved evidence-based regulations to reduce the burden of air pollution and to improve public health.

Acknowledgements This study was supported by research grants from the Deutsche Forschungsgemeinschaft (DFG; HE-4510/2-1, KR 1938/3-1, LU 691/4-1). This work was supported by the Ministry of Innovation, Science and Research of the State of North Rhine-Westphalia (MIWF NRW) and the German Federal Ministry of Health (BMG). This study was supported in part by a grant from the German Federal Ministry of Education and Research (BMBF) to the German Center for Diabetes Research (DZD e.V.). We thank Kirti Kaul (German Diabetes Center) for critically reading the manuscript and for helpful discussions.

References

Agustí A, Edwards LD, Rennard SI, MacNee W, Tal-Singer R, Miller BE, Vestbo J, Lomas DA, Calverley PMA, Wouters E, Crim C, Yates JC, Silverman EK, Coxson HO, Bakke P, Mayer RJ, Celli B (2012) Persistent systemic inflammation is associated with poor clinical outcomes in COPD: a novel phenotype. PLoS One 7(5):e37483

Andersen ZJ, Raaschou-Nielsen O, Ketzel M, Jensen SS, Hvidberg M, Loft S, Tjonneland A, Overvad K, Sorensen M (2012) Diabetes incidence and long-term exposure to air pollution: a cohort study. Diabetes Care 35(1):92–98

Bai N, Khazaei M, van Eeden SF, Laher I (2007) The pharmacology of particulate matter air pollution-induced cardiovascular dysfunction. Pharmacol Ther 113(1):16–29

Barzilay J, Abraham L, Heckbert S, Cushman M, Kuller L, Resnick H, Tracy R (2001) The relation of markers of inflammation to the development of glucose disorders in the elderly the Cardiovascular Health Study. Diabetes 50(10):2384–2389

Beavers KM, Brinkley TE, Nicklas BJ (2010) Effect of exercise training on chronic inflammation. Clin Chim Acta 411(11–12):785–793

Beelen R, Hoek G, Vienneau D, Eeftens M, Dimakopoulou K, Pedeli X, Tsai M-Y, Künzli N, Schikowski T, Marcon A, Eriksen KT, Raaschou-Nielsen O, Stephanou E, Patelarou E, Lanki T, Yli-Tuomi T, Declercq C, Falq G, Stempfelet M, Birk M, Cyrys J, Sv K, Nádor G, Varró MJ, Dėdelė A, Gražulevičienė R, Mölter A, Lindley S, Madsen C, Cesaroni G, Ranzi A, Badaloni C, Hoffmann B, Nonnemacher M, Krämer U, Kuhlbusch T, Cirach M, Nazelle AD, Nieuwenhuijsen M, Bellander T, Korek M, Olsson D, Strömgren M, Dons E, Jerrett M, Fischer P, Wang M, Brunekreef B, Hoogh KD (2013a) Development of NO2 and NOx land use regression models for estimating air pollution exposure in 36 study areas in Europe – The ESCAPE project. Atmos Environ 72(C):10–23

Beelen R, Raaschou-Nielsen O, Stafoggia M, Andersen Z, Weinmayr G, Hoffmann B, Wolf K, Samoli E, Fischer P, Nieuwenhuijsen M, Vineis P, Xun W, Katsouyanni K, Dimakopoulou K, Oudin A, Forsberg B, Modig L, Havulinna A, Lanki T, Turunen A, Oftedal B, Nystad W, Nafstad P, De Faire U, Pedersen N, Östenson C, Fratiglioni L, Penell J, Korek M, Pershagen G, Eriksen K, Overvad K, Ellermann T, Eeftens M, Peeters P, Meliefste K, Wang M, Bueno-de-Mesquita B, Sugiri D, Krämer U, Heinrich J, de Hoogh K, Key T, Peters A, Hampel R, Concin H, Nagel G, Ineichen A, Schaffner E, Probst-Hensch N, Künzli N, Schindler C, Schikowski T, Adam M, Phuleria H, Vilier A, Clavel-Chapelon F, Declercq C, Grioni S, Krogh V, Tsai M, Ricceri F, Sacerdote C, Galassi C, Migliore E, Ranzi A, Cesaroni G, Badaloni C, Forastiere F, Tamayo I, Amiano P, Dorronsoro M, Katsoulis M, Trichopoulou A, Brunekreef B, Hoek G (2013b) Effects of long-term exposure to air pollution on natural-cause mortality: an analysis of 22 European cohorts within the multicentre ESCAPE project. Lancet 383(9919):785–795

Belvisi M (2003) Sensory nerves and airway inflammation: role of Aδ and C-fibres. Pulm Pharmacol Ther 16(1):1–7

Berclaz P-Y, Gao H, Tobian J, Swanson D, Webb D, Crapo R, Jensen R (2009) The impact of diabetes and age on pulmonary function: data from the National Health and Nutrition Examination Survey. Diabetes Res Clin Pract 83(1):e1–e3

Bhatnagar A (2009) Could dirty air cause diabetes? Circulation 119(4):492–494

Block ML, Calderón-Garcidueñas L (2009) Air pollution: mechanisms of neuroinflammation and CNS disease. Trends Neurosci 32(9):506–516

Broekhuizen R, Vernooy J, Schols A, Dentener M, Wouters E (2005) Leptin as local inflammatory marker in COPD. Respir Med 99(1):70–74

Brook RD, Jerrett M, Brook JR, Bard RL, Finkelstein MM (2008) The relationship between diabetes mellitus and traffic-related air pollution. J Occup Environ Med 50(1):32–38

Brook RD, Rajagopalan S, Pope CA, Brook JR, Bhatnagar A, Diez-Roux AV, Holguin F, Hong Y, Luepker RV, Mittleman MA, Peters A, Siscovick D, Smith SC, Whitsel L, Kaufman JD (2010) Particulate matter air pollution and cardiovascular disease: an update to the scientific statement from the American Heart Association. Circulation 121(21):2331–2378

Brook R, Xu X, Bard R, Dvonch J, Morishita M, Kaciroti N, Sun Q, Harkema J, Rajagopalan S (2013) Reduced metabolic insulin sensitivity following sub-acute exposures to low levels of ambient fine particulate matter air pollution. Sci Total Environ 448:66–71

Brown D, Wilson M, MacNee W, Stone V, Donaldson K (2001) Size-dependent proinflammatory effects of ultrafine polystyrene particles: a role for surface area and oxidative stress in the enhanced activity of ultrafines. Toxicol Appl Pharmacol 175(3):191–199

Campbell A, Oldham M, Becaria A, Bondy S, Meacher D, Sioutas C, Misra C, Mendez L, Kleinman M (2005) Particulate matter in polluted air may increase biomarkers of inflammation in mouse brain. Neurotoxicology 26(1):133–140

Carpenter D (2008) Environmental contaminants as risk factors for developing diabetes. Rev Environ Health 23(1):59–74

Carstensen M, Herder C, Kivimäki M, Jokela M, Roden M, Shipley MJ, Witte DR, Brunner EJ, Tabák AG (2010) Accelerated increase in serum interleukin-1 receptor antagonist starts 6 years before diagnosis of type 2 diabetes Whitehall II prospective cohort study. Diabetes 59(5):1222–1227

Cavelti-Weder C, Babians-Brunner A, Keller C, Stahel MA, Kurz-Levin M, Zayed H, Solinger AM, Mandrup-Poulsen T, Dinarello CA, Donath MY (2012) Effects of gevokizumab on glycemia and inflammatory markers in type 2 diabetes. Diabetes Care 35(8):1654–1662

Cesaroni G, Badaloni C, Gariazzo C, Stafoggia M, Sozzi R, Davoli M, Forastiere F (2013) Long-term exposure to urban air pollution and mortality in a cohort of more than a million adults in Rome. Environ Health Perspect 121(3):324–331

Cesaroni G, Forastiere F, Stafoggia M, Andersen Z, Badaloni C, Beelen R, Caracciolo B, de Faire U, Erbel R, Eriksen K (2014) Long term exposure to ambient air pollution and incidence of acute coronary events: prospective cohort study and meta-analysis in 11 European cohorts from the ESCAPE Project. Br Med J 348:f7412

Chen H, Goldberg MS, Villeneuve PJ (2008) A systematic review of the relation between long-term exposure to ambient air pollution and chronic diseases. Rev Environ Health 23(4):243–298

Chen B-Y, Chan C-C, Lee C-T, Cheng T-J, Huang W-C, Jhou J-C, Han Y-Y, Chen C-C, Guo Y (2012a) The association of ambient air pollution with airway inflammation in schoolchildren. Am J Epidemiol 175(8):764–774

Chen L, Magliano D, Zimmet P (2012b) The worldwide epidemiology of type 2 diabetes mellitus – present and future perspectives. Nat Rev Endocrinol 8(4):228–236

Chen H, Burnett R, Kwong C, Villeneuve P, Goldberg M, Brook R, van Donkelaar A, Jerrett M, Martin R, Brook J, Copes R (2013) Risk of incident diabetes in relation to long-term exposure to fine particulate matter in Ontario, Canada. Environ Health Perspect 121(7):804–810

Chuang K-J, Yan Y-H, Chiu S-Y, Cheng T-J (2011) Long-term air pollution exposure and risk factors for cardiovascular diseases among the elderly in Taiwan. Occup Environ Med 68(1):64–68

Coogan PF, White LF, Jerrett M, Brook RD, Su JG, Seto E, Burnett R, Palmer JR, Rosenberg L (2012) Air pollution and incidence of hypertension and diabetes mellitus in black women living in Los Angeles. Circulation 125(6):767–772

Delfino RJ, Staimer N, Tjoa T, Polidori A, Arhami M, Gillen DL, Kleinman MT, Vaziri ND, Longhurst J, Zaldivar F, Sioutas C (2008) Circulating biomarkers of inflammation, antioxidant activity, and platelet activation are associated with primary combustion aerosols in subjects with coronary artery disease. Environ Health Perspect 116(7):898–906

Delfino RJ, Staimer N, Tjoa T, Gillen DL, Polidori A, Arhami M, Kleinman MT, Vaziri ND, Longhurst J, Sioutas C (2009) Air pollution exposures and circulating biomarkers of effect in a susceptible population: clues to potential causal component mixtures and mechanisms. Environ Health Perspect 117(8):1232–1238

Devlin R, Smith C, Schmitt M, Rappold A, Hinderliter A, Graff D, Carraway M (2014) Controlled exposure of humans with metabolic syndrome to concentrated ultrafine ambient particulate matter causes cardiovascular effects. Toxicol Sci 140(1):61–72

Diez Roux AV, Auchincloss AH, Astor B, Barr R, Cushman M, Dvonch T, Jacobs D, Kaufman J, Lin X, Samson P (2006) Recent exposure to particulate matter and C-reactive protein concentration in the multi-ethnic study of atherosclerosis. Am J Epidemiol 164(5):437–448

Dijkema MB, Mallant SF, Gehring U, Kvd H, Alssema M, van Strien RT, Fischer PH, Nijpels G, Stehouwer CD, Hoek G, Dekker JM, Brunekreef B (2011) Long-term exposure to traffic-related air pollution and type 2 diabetes prevalence in a cross-sectional screening-study in the Netherlands. Environ Health 10:76

Dimakopoulou K, Samoli E, Beelen R, Stafoggia M, Andersen Z, Hoffmann B, Fischer P, Nieuwenhuijsen M, Vineis P, Xun W, Hoek G, Raaschou-Nielsen O, Oudin A, Forsberg B, Modig L, Jousilahti P, Lanki T, Turunen A, Oftedal B, Nafstad P, Schwarze P, Penell J, Fratiglioni L, Andersson N, Pedersen N, Korek M, De Faire U, Eriksen K, Tjønneland A, Becker T, Wang M, Bueno-de-Mesquita B, Tsai M, Eeftens M, Peeters P, Meliefste K, Marcon A, Krämer U, Kuhlbusch T, Vossoughi M, Key T, de Hoogh K, Hampel R, Peters A, Heinrich J, Weinmayr G, Concin H, Nagel G, Ineichen A, Jacquemin B, Stempfelet M, Vilier A, Ricceri F, Sacerdote C, Pedeli X, Katsoulis M, Trichopoulou A, Brunekreef B, Katsouyanni K (2014) Air pollution and nonmalignant respiratory mortality in 16 cohorts within the ESCAPE Project. Am J Respir Crit Care Med 189(6):684–696

Donaldson G, Seemungal T, Patel I, Bhowmik A, Wilkinson T, Hurst J, MacCallum P, Wedzicha J (2005) Airway and systemic inflammation and decline in lung function in patients with COPD. Chest 128(4):1995–2004

Donath MY, Dalmas É, Sauter Nadine S, Böni-Schnetzler M (2013) Inflammation in obesity and diabetes: islet dysfunction and therapeutic opportunity. Cell Metab 17(6):860–872

Donath MY, Shoelson SE (2011) Type 2 diabetes as an inflammatory disease. Nat Rev Immunol 11(2):98–107

Dubowsky SD, Suh H, Schwartz J, Coull BA, Gold DR (2006) Diabetes, obesity, and hypertension may enhance associations between air pollution and markers of systemic inflammation. Environ Health Perspect 114(7):992–998

Dutta A, Roychoudhury S, Chowdhury S, Ray MR (2013) Changes in sputum cytology, airway inflammation and oxidative stress due to chronic inhalation of biomass smoke during cooking in premenopausal rural Indian women. Int J Hyg Environ Health 216(3):301–308

Echouffo-Tcheugui J, Dagogo-Jack S (2012) Preventing diabetes mellitus in developing countries. Nat Rev Endocrinol 8(9):557–562

Eeden SV, Tan W, Suwa T, Mukae H, Terashima T, Fujii T, Qui D, Vincent R, Hogg J (2001) Cytokines involved in the systemic inflammatory response induced by exposure to particulate matter air pollutants (PM10). Am J Respir Crit Care Med 164(5):826–830

Eeftens M, Beelen R, de Hoogh K, Bellander T, Cesaroni G, Cirach M, Declercq C, Dedele A, Dons E, de Nazelle A (2012) Development of land use regression models for PM2.5, PM2.5 absorbance, PM10 and PMcoarse in 20 European study areas; results of the ESCAPE project. Environ Sci Technol 46(20):11195–11205

Festa A, D'Agostino R, Tracy R, Haffner S (2002) Elevated levels of acute-phase proteins and plasminogen activator inhibitor-1 predict the development of type 2 diabetes the insulin resistance atherosclerosis study. Diabetes 51(4):1131–1137

Fonken LK, Xu X, Weil ZM, Chen G, Sun Q, Rajagopalan S, Nelson RJ (2011) Air pollution impairs cognition, provokes depressive-like behaviors and alters hippocampal cytokine expression and morphology. Mol Psychiatry 16(10):987–995

Frampton MW, Stewart JC, Oberdörster G, Morrow PE, Chalupa D, Pietropaoli AP, Frasier LM, Speers DM, Cox C, Huang L-S, Utell MJ (2006) Inhalation of ultrafine particles alters blood leukocyte expression of adhesion molecules in humans. Environ Health Perspect 114(1):51–58

Fujii T, Hayashi S, Hogg J, Vincent R, van Eeden S (2001) Particulate matter induces cytokine expression in human bronchial epithelial cells. Am J Respir Cell Mol Biol 25(3):265–271

Gehring U, Gruzieva O, Agius R, Beelen R, Custovic A, Cyrys J, Eeftens M, Flexeder C, Fuertes E, Heinrich J (2013) Air pollution exposure and lung function in children: the ESCAPE Project. Environ Health Perspect 121(11–12):1357–1364

Geiser M, Kreyling WG (2010) Deposition and biokinetics of inhaled nanoparticles. Part Fibre Toxicol 7:2

Gilmour P, Ziesenis A, Morrison E, Vickers M, Drost E, Ford I, Karg E, Mossa C, Schroeppel A, Ferron G (2004) Pulmonary and systemic effects of short-term inhalation exposure to ultrafine carbon black particles. Toxicol Appl Pharmacol 195(1):35–44

Gleeson M, Bishop NC, Stensel DJ, Lindley MR, Mastana SS, Nimmo MA (2011) The anti-inflammatory effects of exercise: mechanisms and implications for the prevention and treatment of disease. Nat Rev Immunol 11(9):607–615

Gold DR, Mittleman MA (2013) New insights into pollution and the cardiovascular system: 2010 to 2012. Circulation 127(18):1903–1913

Goldberg M, Burnett R, Yale J-F, Valois M-F, Brook J (2006) Associations between ambient air pollution and daily mortality among persons with diabetes and cardiovascular disease. Environ Res 100(2):255–267

Goldfine A, Fonseca V, Jablonski K, Pyle L, Staten M, Shoelson S (2010) The effects of salsalate on glycemic control in patients with type 2 diabetes: a randomized trial. Ann Intern Med 152(6):346–357

Gregor MF, Hotamisligil GS (2011) Inflammatory mechanisms in obesity. Annu Rev Immunol 29(1):415–445

Gruzieva O, Gehring U, Aalberse R, Agius R, Beelen R, Behrendt H, Bellander T, Birk M, de Jongste J, Fuertes E (2013) Meta-analysis of air pollution exposure association with allergic sensitization in European birth cohorts. J Allergy Clin Immunol 133(3):767–776

Guilherme A, Virbasius JV, Puri V, Czech MP (2008) Adipocyte dysfunctions linking obesity to insulin resistance and type 2 diabetes. Nat Rev Mol Cell Bio 9(5):367–377

Hansson GK, Hermansson A (2011) The immune system in atherosclerosis. Nat Immunol 12(3):204–212

Hectors TLM, Vanparys C, van der Ven K, Martens G, Jorens P, Van Gaal L, Covaci A, De Coen W, Blust R (2011) Environmental pollutants and type 2 diabetes: a review of mechanisms that can disrupt beta cell function. Diabetologia 54(6):1273–1290

Herder C, Baumert J, Thorand B, Koenig W, Jager W, Meisinger C, Illig T, Martin S, Kolb H (2006) Chemokines as risk factors for type 2 diabetes: results from the MONICA/KORA Augsburg study, 1984–2002. Diabetologia 49(5):921–929

Herder C, Brunner E, Rathmann W, Strassburger K, Tabák A, Schloot N, Witte D (2009a) Elevated levels of the anti-inflammatory interleukin-1 receptor antagonist precede the onset of type 2 diabetes. The Whitehall II study. Diabetes Care 32(3):421–423

Herder C, Lankisch M, Ziegler D, Rathmann W, Koenig W, Illig T, Doring A, Thorand B, Holle R, Giani G, Martin S, Meisinger C (2009b) Subclinical inflammation and diabetic polyneuropathy: MONICA/KORA Survey F3 (Augsburg, Germany). Diabetes Care 32(4):680–682

Herder C, Peltonen M, Koenig W, Sütfels K, Lindström J, Martin S, Ilanne-Parikka P, Eriksson JG, Aunola S, Keinänen-Kiukaanniemi S, Valle TT, Uusitupa M, Kolb H, Tuomilehto J (2009c) Anti-inflammatory effect of lifestyle changes in the Finnish Diabetes Prevention Study. Diabetologia 52(3):433–442

Herder C, Carstensen M, Ouwens D (2013) Anti-inflammatory cytokines and risk of type 2 diabetes. Diabetes Obes Metab 15(s3):39–50

Herder C, Kowall B, Tabak AG, Rathmann W (2014) The potential of novel biomarkers to improve risk prediction of type 2 diabetes. Diabetologia 57(1):16–29

Hertel S, Viehmann A, Moebus S, Mann K, Bröcker-Preuss M, Möhlenkamp S, Nonnemacher M, Erbel R, Jakobs H, Memmesheimer M, Jöckel K-H, Hoffmann B (2010) Influence of short-term exposure to ultrafine and fine particles on systemic inflammation. Eur J Epidemiol 25(8):581–592

Hoffmann B, Moebus S, Dragano N, Stang A, Möhlenkamp S, Schmermund A, Memmesheimer M, Bröcker-Preuss M, Mann K, Erbel R, Jöckel K-H (2009) Chronic residential exposure to particulate matter air pollution and systemic inflammatory markers. Environ Health Perspect 117(8):1302–1308

Hoffmann B, Luttmann-Gibson H, Cohen A, Zanobetti A, de Souza C, Foley C, Suh H, Coull B, Schwartz J, Mittleman M (2012) Opposing effects of particle pollution, ozone, and ambient temperature on arterial blood pressure. Environ Health Perspect 120(2):241–246

Howren MB, Lamkin DM, Suls J (2009) Associations of depression with C-reactive protein, IL-1, and IL-6: a meta-analysis. Psychosom Med 71(2):171–186

Huang Y-CT, Schmitt M, Yang Z, Que LG, Stewart JC, Frampton MW, Devlin RB (2010) Gene expression profile in circulating mononuclear cells after exposure to ultrafine carbon particles. Inhal Toxicol 22(10):835–846

Hurst JR, Perera WR, Wilkinson TMA, Donaldson GC, Wedzicha JA (2006) Systemic and upper and lower airway inflammation at exacerbation of chronic obstructive pulmonary disease. Am J Respir Crit Care Med 173(1):71–78

IDF Diabetes Atlas Group (2013) IDF diabetes atlas, 6th edn. International Diabetes Federation, Brussels

Jeon CY, Murray MB (2008) Diabetes mellitus increases the risk of active tuberculosis: a systematic review of 13 observational studies. PLoS Med 5(7), e152

Jerrett M, Burnett RT, Ma R, Pope CA, Krewski D, Newbold KB, Thurston G, Shi Y, Finkelstein N, Calle EE, Thun MJ (2005) Spatial analysis of air pollution and mortality in Los Angeles. Epidemiology 16(6):727–736

Kahn SE, Cooper ME, Del Prato S (2014) Pathophysiology and treatment of type 2 diabetes: perspectives on the past, present, and future. Lancet 383(9922):1068–1083

Kampfrath T, Maiseyeu A, Ying Z, Shah Z, Deiuliis JA, Xu X, Kherada N, Brook RD, Reddy KM, Padture NP, Parthasarathy S, Chen LC, Moffatt-Bruce S, Sun Q, Morawietz H, Rajagopalan S (2011) Chronic fine particulate matter exposure induces systemic vascular dysfunction via NADPH oxidase and TLR4 pathways. Circ Res 108(6):716–726

Kelishadi R, Mirghaffari N, Poursafa P, Gidding SS (2009) Lifestyle and environmental factors associated with inflammation, oxidative stress and insulin resistance in children. Atherosclerosis 203(1):311–319

Khafaie MA, Salvi SS, Ojha A, Khafaie B, Gore SS, Yajnik CS (2013) Systemic inflammation (C-reactive protein) in type 2 diabetic patients is associated with ambient air pollution in Pune city, India. Diabetes Care 36(3):625–630

Khandekar M, Cohen P, Spiegelman B (2011) Molecular mechanisms of cancer development in obesity. Nat Rev Cancer 11(12):886–895

Kolb H, Mandrup-Poulsen T (2005) An immune origin of type 2 diabetes? Diabetologia 48(6):1038–1050

Kolb H, Mandrup-Poulsen T (2010) The global diabetes epidemic as a consequence of lifestyle-induced low-grade inflammation. Diabetologia 53(1):10–20

Kolb M, Margetts P, Anthony D, Pitossi F, Gauldie J (2001) Transient expression of IL-1β induces acute lung injury and chronic repair leading to pulmonary fibrosis. J Clin Invest 107(12):1529–1536

Krämer U, Herder C, Sugiri D, Strassburger K, Schikowski T, Ranft U, Rathmann W (2010) Traffic-Related air pollution and incident type 2 diabetes: results from the SALIA cohort study. Environ Health Perspect 118(9):1273–1279

Larsen C, Faulenbach M, Vaag A, Vølund A, Ehses J, Seifert B, Mandrup-Poulsen T, Donath M (2007) Interleukin-1–receptor antagonist in type 2 diabetes mellitus. N Engl J Med 356(15):1517–1526

Lee JK, Bettencourt R, Brenner D, Le T-A, Barrett-Connor E, Loomba R (2012) Association between serum interleukin-6 concentrations and mortality in older adults: the Rancho Bernardo Study. PLoS One 7(4):e34218

Levy H, Murphy A, Zou F, Gerard C, Klanderman B, Schuemann B, Lazarus R, García K, Celedón J, Drumm M (2009) IL1B polymorphisms modulate cystic fibrosis lung disease. Pediatr Pulmonol 44(6):580–593

Li S, Shin H, Ding E, van Dam R (2009) Adiponectin levels and risk of type 2 diabetes: a systematic review and meta-analysis. JAMA 302(2):179–188

Lindsay R, Funahashi T, Hanson R, Matsuzawa Y, Tanaka S, Tataranni P, Knowler W, Krakoff J (2002) Adiponectin and development of type 2 diabetes in the Pima Indian population. Lancet 360(9326):57–58

Lippmann M (2014) Toxicological and epidemiological studies of cardiovascular effects of ambient air fine particulate matter (PM 2.5) and its chemical components: coherence and public health implications. Crit Rev Toxicol 44(4):299–347

Lippmann M, Yeates D, Albert R (1980) Deposition, retention, and clearance of inhaled particles. Br J Ind Med 37(4):337–362

Liu C, Flexeder C, Fuertes E, Cyrys J, Bauer C-P, Koletzko S, Hoffmann B, Von Berg A, Heinrich J (2013a) Effects of air pollution on exhaled nitric oxide in children: results from the GINIplus and LISAplus studies. Int J Hyg Environ Health 217(4–5):483–491

Liu C, Fuertes E, Tiesler C, Birk M, Babisch W, Bauer C-P, Koletzko S, Von Berg A, Hoffmann B, Heinrich J (2013b) The associations between traffic-related air pollution and noise with blood pressure in children: results from the GINIplus and LISAplus studies. Int J Hyg Environ Health 217(4–5):499–505

Liu C, Ying Z, Harkema J, Sun Q, Rajagopalan S (2013c) Epidemiological and experimental links between air pollution and type 2 diabetes. Toxicol Pathol 41(2):361–373

Liu C, Xu X, Bai Y, Wang T, Rao X, Wang A, Sun L, Ying Z, Gushchina L, Maiseyeu A, Morishita M, Sun Q, Harkema J, Rajagopalan S (2014) Air pollution–mediated susceptibility to inflammation and insulin resistance: influence of CCR2 pathways in mice. Environ Health Perspect 122(1):17–26

Ljungman P, Bellander T, Schneider A, Breitner S, Forastiere F, Hampel R, Illig T, Jacquemin B, Katsouyanni K, Von Klot S, Koenig W, Lanki T, Nyberg F, Pekkanen J, Pistelli R, Pitsavos C, Rosenqvist M, Sunyer J, Peters A (2009) Modification of the interleukin-6 response to air pollution by interleukin-6 and fibrinogen polymorphisms. Environ Health Perspect 117(9):1373–1379

Lockwood A (2002) Diabetes and air pollution. Diabetes Care 25(8):1487–1488

Luo L, Hong X, Chen C, Brooks SP, Song Y (2013) Identification of pathology from diesel exhaust particles in the bladder in a rat model by aspiration of particles from the pharynx. Environ Toxicol Pharmacol 35(3):380–387

MacIntyre E, Gehring U, Mölter A, Fuertes E, Klümper C, Krämer U, Quass U, Hoffmann B, Gascon M, Brunekreef B (2013) Air pollution and respiratory infections during early childhood: an analysis of 10 European birth cohorts within the ESCAPE project. Environ Health Perspect 122(1):107

Meigs J, Hu F, Rifai N, Manson J (2004) Biomarkers of endothelial dysfunction and risk of type 2 diabetes mellitus. JAMA 291(16):1978–1986

Michael S, Montag M, Dott W (2013) Pro-inflammatory effects and oxidative stress in lung macrophages and epithelial cells induced by ambient particulate matter. Environ Pollut 183(c):19–29

Miyata R, van Eeden SF (2011) The innate and adaptive immune response induced by alveolar macrophages exposed to ambient particulate matter. Toxicol Appl Pharmacol 257(2):209–226

Nemmar A, Vanbilloen H, Hoylaerts M, Hoet P, Verbruggen A, Nemery B (2001) Passage of intratracheally instilled ultrafine particles from the lung into the systemic circulation in hamster. Am J Respir Crit Care Med 164(9):1665–1668

Nemmar A, Hoet PHM, Vanquickenborne B, Dinsdale D, Thomeer M, Hoylaerts M, Vanbilloen H, Mortelmans L, Nemery B (2002) Passage of inhaled particles into the blood circulation in humans. Circulation 105(4):411–414

Nimmo M, Leggate M, Viana J, King J (2013) The effect of physical activity on mediators of inflammation. Diabetes Obes Metab 15(s3):51–60

O'Neill MS, Veves A, Zanobetti A, Sarnat JA, Gold DR, Economides PA, Horton ES, Schwartz J (2005) Diabetes enhances vulnerability to particulate air pollution–associated impairment in vascular reactivity and endothelial function. Circulation 111(22):2913–2920

Oberdörster G, Utell M (2002) Ultrafine particles in the urban air: to the respiratory tract – and beyond? Environ Health Perspect 110(8):A440–A441

Oberdörster G, Sharp Z, Atudorei V, Elder A, Gelein R, Kreyling W, Cox C (2004) Translocation of inhaled ultrafine particles to the brain. Inhal Toxicol 16(6–7):437–445

Osborn O, Olefsky JM (2012) The cellular and signaling networks linking the immune system and metabolism in disease. Nat Med 18(3):363–374

Park S, O'Neill M, Vokonas P, Sparrow D, Schwartz J (2005) Effects of air pollution on heart rate variability: the VA Normative Aging Study. Environ Health Perspect 113(3):304–309

Pearson JF, Bachireddy C, Shyamprasad S, Goldfine AB, Brownstein JS (2010) Association between fine particulate matter and diabetes prevalence in the U.S. Diabetes Care 33(10):2196–2201

Pedersen M, Giorgis-Allemand L, Bernard C, Aguilera I, Andersen A-MN, Ballester F, Beelen RMJ, Chatzi L, Cirach M, Danileviciute A, Dedele A, Eijsden MV, Estarlich M, Fernández-Somoano A, Fernández MF, Forastiere F, Gehring U, Grazuleviciene R, Gruzieva O, Heude B, Hoek G, Hoogh KD, Van Den Hooven EH, Håberg SE, Jaddoe VWV, Klümper C, Korek M, Krämer U, Lerchundi A, Lepeule J, Nafstad P, Nystad W, Patelarou E, Porta D, Postma D, Raaschou-Nielsen O, Rudnai P, Sunyer J, Stephanou E, Sørensen M, Thiering E, Tuffnell D, Varró MJ, Vrijkotte TGM, Wijga A, Wilhelm M, Wright J, Nieuwenhuijsen MJ, Pershagen G, Brunekreef B, Kogevinas M, Slama R (2013) Ambient air pollution and low birthweight: a European cohort study (ESCAPE). Lancet Respir Med 1(9):695–704

Péry ARR, Brochot C, Hoet PHM, Nemmar A, Bois FY (2009) Development of a physiologically based kinetic model for 99m-technetium-labelled carbon nanoparticles inhaled by humans. Inhal Toxicol 21(13):1099–1107

Pope CA, Dockery DW (2006) Health effects of fine particulate air pollution: lines that connect. J Air Waste Manage Assoc 56(6):709–742

Pouwer F, Kupper N, Adriaanse M (2010) Does emotional stress cause type 2 diabetes mellitus? A review from the European Depression in Diabetes (EDID) research consortium. Discov Med 9(45):112–118

Pradhan A, Manson J, Rifai N, Buring J, Ridker P (2001) C-reactive protein, interleukin 6, and risk of developing type 2 diabetes mellitus. JAMA 286(3):327–334

Provoost S, Maes T, Willart M, Joos G, Lambrecht B, Tournoy K (2010) Diesel exhaust particles stimulate adaptive immunity by acting on pulmonary dendritic cells. J Immunol 184(1):426–432

Puett R, Hart J, Schwartz J, Hu F, Liese A, Laden F (2011) Are particulate matter exposures associated with risk of type 2 diabetes? Environ Health Perspect 119(3):384–389

Raaschou-Nielsen O, Sørensen M, Ketzel M, Hertel O, Loft S, Tjønneland A, Overvad K, Andersen ZJ (2012) Long-term exposure to traffic-related air pollution and diabetes-associated mortality: a cohort study. Diabetologia 56(1):36–46

Raaschou-Nielsen O, Andersen Z, Beelen R, Samoli E, Stafoggia M, Weinmayr G, Hoffmann B, Fischer P, Nieuwenhuijsen M, Brunekreef B (2013) Air pollution and lung cancer incidence in 17 European cohorts: prospective analyses from the European Study of Cohorts for Air Pollution Effects (ESCAPE). Lancet Oncol 14(9):813–822

Rajagopalan S, Brook RD (2012) Air pollution and type 2 diabetes: mechanistic insights. Diabetes 61(12):3037–3045

Ridker P, Howard C, Walter V, Everett B, Libby P, Hensen J, Thuren T (2012) Effects of interleukin-1β inhibition with canakinumab on hemoglobin A1c, lipids, C-reactive protein, interleukin-6, and fibrinogen: a phase IIb randomized, placebo-controlled trial. Circulation 126(23):2739–2748

Röpcke S, Holz O, Lauer G, Müller M, Rittinghausen S, Ernst P, Lahu G, Elmlinger M, Krug N, Hohlfeld JM (2012) Repeatability of and relationship between potential COPD biomarkers in bronchoalveolar lavage, bronchial biopsies, serum, and induced sputum. PLoS One 7(10):e46207

Rückerl R, Ibald-Mulli A, Koenig W, Schneider A, Woelke G, Cyrys J, Heinrich J, Marder V, Frampton M, Wichmann H (2006) Air pollution and markers of inflammation and coagulation in patients with coronary heart disease. Am J Respir Crit Care Med 173(4):432–441

Rückerl R, Greven S, Ljungman P, Aalto P, Antoniades C, Bellander T, Berglind N, Chrysohoou C, Forastiere F, Jacquemin B, Von Klot S, Koenig W, Küchenhoff H, Lanki T, Pekkanen J, Perucci CA, Schneider A, Sunyer J, Peters A (2007) Air pollution and inflammation (interleukin-6, C-reactive protein, fibrinogen) in myocardial infarction survivors. Environ Health Perspect 115(7):1072–1080

Rückerl R, Schneider A, Breitner S, Cyrys J, Peters A (2011) Health effects of particulate air pollution: a review of epidemiological evidence. Inhal Toxicol 23(10):555–592

Salvi S, Blomberg A, Rudell B, Kelly F, Sandstrom T, Holgate S, Frew A (1999) Acute inflammatory responses in the airways and peripheral blood after short-term exposure to diesel exhaust in healthy human volunteers. Am J Respir Crit Care Med 159(3):702–709

Salvi S, Nordenhall C, Blomberg A, Rudell B, Pourazar J, Kelly F, Wilson S, Sandstrom T, Holgate S, Frew A (2000) Acute exposure to diesel exhaust increases IL-8 and GRO-α production in healthy human airways. Am J Respir Crit Care Med 161(2):550–557

Sapey E, Ahmad A, Bayley D, Newbold P, Snell N, Rugman P, Stockley RA (2009) Imbalances between interleukin-1 and tumor necrosis factor agonists and antagonists in stable COPD. J Clin Immunol 29(4):508–516

Schikowski T, Adam M, Marcon A, Cai Y, Vierkötter A, Carsin A, Jacquemin B, Al Kanani Z, Beelen R, Birk M (2014) Association of ambient air pollution with the prevalence and incidence of COPD. Eur Respir J 44(3):614–626

Schmidt M, Duncan B, Sharrett A, Lindberg G, Savage P, Offenbacher S, Azambuja M, Tracy R, Heiss G (1999) Markers of inflammation and prediction of diabetes mellitus in adults (Atherosclerosis Risk in Communities study): a cohort study. Lancet 353(9165):1649–1652

Schneider A, Alexis N, Diaz-Sanchez D, Neas L, Harder S, Herbst M, Cascio W, Buse J, Peters A, Devlin R (2011) Ambient PM2.5 exposure up-regulates the expression of costimulatory receptors on circulating monocytes in diabetic individuals. Environ Health Perspect 119(6):778–783

Schwartz J (2001) Air pollution and blood markers of cardiovascular risk. Environ Health Perspect 109(Suppl 3):405–409

Seaton A, Soutar A, Crawford V, Elton R, McNerlan S, Cherrie J, Watt M, Agius R, Stout R (1999) Particulate air pollution and the blood. Thorax 54(11):1027–1032

Sell H, Habich C, Eckel J (2012) Adaptive immunity in obesity and insulin resistance. Nat Rev Endocrinol 8(12):709–716

Seshasai S, Kaptoge S, Thompson A, Di Angelantonio E, Gao P, Sarwar N, Whincup P, Mukamal K, Gillum R, Holme I (2011) Diabetes mellitus, fasting glucose, and risk of cause-specific death. N Engl J Med 364(9):829–841

Shoelson S, Herrero L, Naaz A (2007) Obesity, inflammation, and insulin resistance. Gastroenterology 132(6):2169–2180

Sidawi B, Al-Hairi MTA (2012) The impact of built environment on diabetic patients: the case of Eastern Province, Kingdom of Saudi Arabia. Glob J Health Sci 4(4):126–138

Sima C, Glogauer M (2013) Diabetes mellitus and periodontal diseases. Curr Diab Rep 13(3):445–452

Sinden NJ, Stockley RA (2010) Systemic inflammation and comorbidity in COPD: a result of 'overspill' of inflammatory mediators from the lungs? Review of the evidence. Thorax 65(10):930–936

Singh D, Edwards L, Tal-Singer R, Rennard S (2010) Sputum neutrophils as a biomarker in COPD: findings from the ECLIPSE study. Respir Res 11:77

Spranger J, Kroke A, Möhlig M, Hoffmann K, Bergmann M, Ristow M, Boeing H, Pfeiffer A (2003) Inflammatory cytokines and the risk to develop type 2 diabetes results of the prospective population-based European Prospective Investigation into Cancer and Nutrition (EPIC)-Potsdam Study. Diabetes 52(3):812–817

Steinvil A, Kordova-Biezuner L, Shapira I, Berliner S, Rogowski O (2008) Short-term exposure to air pollution and inflammation-sensitive biomarkers. Environ Res 106(1):51–61

Stringhini S, Batty GD, Bovet P, Shipley MJ, Marmot MG, Kumari M, Tabak AG, Kivimäki M (2013) Association of lifecourse socioeconomic status with chronic inflammation and type 2 diabetes risk: the Whitehall II prospective cohort study. PLoS Med 10(7):e1001479

Sullivan J, Hubbard R, Liu SL-J, Shepherd K, Trenga C, Koenig J, Chandler W, Kaufman J (2007) A community study of the effect of particulate matter on blood measures of inflammation and thrombosis in an elderly population. Environ Health 6:3

Sun Q, Yue P, Deiuliis JA, Lumeng CN, Kampfrath T, Mikolaj MB, Cai Y, Ostrowski MC, Lu B, Parthasarathy S, Brook RD, Moffatt-Bruce SD, Chen LC, Rajagopalan S (2009) Ambient air pollution exaggerates adipose inflammation and insulin resistance in a mouse model of diet-induced obesity. Circulation 119(4):538–546

Tabák AG, Herder C, Rathmann W, Brunner EJ, Kivimäki M (2012) Prediabetes: a high-risk state for diabetes development. Lancet 379(9833):2279–2290

Tamayo T, Rathmann W, Krämer U, Sugiri D, Grabert M, Holl R (2014a) Is particle pollution in outdoor air associated with metabolic control in type 2 diabetes? PLoS One 9(3):e91639

Tamayo T, Rosenbauer J, Wild SH, Spijkerman AMW, Baan C, Forouhi NG, Herder C, Rathmann W (2014b) Diabetes in Europe: an update. Diabetes Res Clin Pract 103(2):206–217

Taylor KW, Novak RF, Anderson HA, Birnbaum LS, Blystone C, Devito M, Jacobs D, Köhrle J, Lee D-H, Rylander L, Rignell-Hydbom A, Tornero-Velez R, Turyk ME, Boyles AL, Thayer KA, Lind L (2013) Evaluation of the association between persistent organic pollutants (POPs) and diabetes in epidemiological studies: a national toxicology program workshop review. Environ Health Perspect 121(7):774–783

Teichert T, Vossoughi M, Vierkötter A, Sugiri D, Schikowski T, Schulte T, Roden M, Luckhaus C, Herder C, Krämer U (2013) Association between traffic-related air pollution, subclinical inflammation and impaired glucose metabolism: results from the SALIA study. PLoS One 8(12):e83042

Terzano C, Di Stefano F, Conti V, Graziani E, Petroianni A (2010) Air pollution ultrafine particles: toxicity beyond the lung. Eur Rev Med Pharmacol Sci 14(10):809–821

Thiering E, Cyrys J, Kratzsch J, Meisinger C, Hoffmann B, Berdel D, Berg A, Koletzko S, Bauer C-P, Heinrich J (2013) Long-term exposure to traffic-related air pollution and insulin resistance in children: results from the GINIplus and LISAplus birth cohorts. Diabetologia 56(8):1696–1704

Thorand B, Kolb H, Baumert J, Koenig W, Chambless L, Meisinger C, Illig T, Martin S, Herder C (2005) Elevated levels of interleukin-18 predict the development of type 2 diabetes results from the MONICA/KORA Augsburg study, 1984–2002. Diabetes 54(10):2932–2938

Törnqvist H, Mills NL, Gonzalez M, Miller MR, Robinson SD, Megson IL, MacNee W, Donaldson K, Söderberg S, Newby DE, Sandström T, Blomberg A (2007) Persistent endothelial dysfunction in humans after diesel exhaust inhalation. Am J Respir Crit Care Med 176(4):395–400

Turer AT, Scherer PE (2012) Adiponectin: mechanistic insights and clinical implications. Diabetologia 55(9):2319–2326

United States Environmental Protection Agency (2014) Particulate matter. http://www.epa.gov/ncer/science/pm/. Accessed 7 Apr 2014

Valavanidis A, Fiotakis K, Vlachogianni T (2008) Airborne particulate matter and human health: toxicological assessment and importance of size and composition of particles for oxidative damage and carcinogenic mechanisms. J Environ Sci Health C Environ Carcinog Ecotoxicol Rev 26(4):339–362

Vernooy JH, Küçükaycan M, Jacobs JA, Chavannes NH, Buurman WA, Dentener MA, Wouters EF (2002) Local and systemic inflammation in patients with chronic obstructive pulmonary disease. Am J Respir Crit Care Med 166(9):1218–1224

Wang G, Jiang R, Zhao Z, Song W (2013a) Effects of ozone and fine particulate matter (PM2.5) on rat system inflammation and cardiac function. Toxicol Lett 217(1):23–33

Wang Y-B, Watts AB, Peters JI, Iii ROW (2013b) The impact of pulmonary diseases on the fate of inhaled medicines – a review. Int J Pharm 461(1–2):112–128

Wang M, Beelen R, Stafoggia M, Raaschou-Nielsen O, Andersen ZJ, Hoffmann B, Fischer P, Houthuijs D, Nieuwenhuijsen M, Weinmayr G, Vineis P, Xun WW, Dimakopoulou K, Samoli

E, Laatikainen T, Lanki T, Turunen AW, Oftedal B, Schwarze P, Aamodt G, Penell J, Faire UD, Korek M, Leander K, Pershagen G, Pedersen NL, Östenson C-G, Fratiglioni L, Eriksen KT, Sørensen M, Tjønneland A, Bueno-De-Mesquita B, Eeftens M, Bots ML, Meliefste K, Krämer U, Heinrich J, Sugiri D, Key T, Kd H, Wolf K, Peters A, Cyrys J, Jaensch A, Concin H, Nagel G, Tsai M-Y, Phuleria H, Ineichen A, Künzli N, Probst-Hensch N, Schaffner E, Vilier A, Clavel-Chapelon F, Declerq C, Ricceri F, Sacerdote C, Marcon A, Galassi C, Migliore E, Ranzi A, Cesaroni G, Badaloni C, Forastiere F, Katsoulis M, Trichopoulou A, Keuken M, Jedynska A, Kooter IM, Kukkonen J, Sokhi RS, Brunekreef B, Katsouyanni K, Hoek G (2014) Long-term exposure to elemental constituents of particulate matter and cardiovascular mortality in 19 European cohorts: results from the ESCAPE and TRANSPHORM projects. Environ Int 66(C):97–106

Whitsel EA, Quibrera PM, Christ SL, Liao D, Prineas RJ, Anderson GL, Heiss G (2009) Heart rate variability, ambient particulate matter air pollution, and glucose homeostasis: the environmental epidemiology of arrhythmogenesis in the Women's Health Initiative. Am J Epidemiol 169(6):693–703

World Health Organisation (2013) Burden of disease: deaths data by region. http://apps.who.int/gho/data/node.main.156?lang=en. Accessed 7 Apr 2014

World Health Organisation (2014a) 7 million premature deaths annually linked to air pollution. http://www.who.int/mediacentre/news/releases/2014/air-pollution/en/. Accessed 7 Apr 2014

World Health Organisation (2014b) Global Health Observatory (GHO). http://www.who.int/gho/phe/outdoor_air_pollution/exposure/en/. Accessed 7 Apr 2014

Xu G, Zhou Z, Zhu W, Fan X, Liu X (2009) Plasma C-reactive protein is related to cognitive deterioration and dementia in patients with mild cognitive impairment. J Neurol Sci 284(1–2):77–80

Xu X, Liu C, Xu Z, Tzan K, Zhong M, Wang A, Lippmann M, Chen L-C, Rajagopalan S, Sun Q (2011a) Long-term exposure to ambient fine particulate pollution induces insulin resistance and mitochondrial alteration in adipose tissue. Toxicol Sci 124(1):88–98

Xu Z, Xu X, Zhong M, Hotchkiss IP, Lewandowski RP, Wagner JG, Bramble LA, Yang Y, Wang A, Harkema JR, Lippmann M, Rajagopalan S, Chen L-C, Sun Q (2011b) Ambient particulate air pollution induces oxidative stress and alterations of mitochondria and gene expression in brown and white adipose tissues. Part Fibre Toxicol 8:20

Xu X, Rao X, Wang T-Y, Jiang S, Ying Z, Liu C, Wang A, Zhong M, Deiuliis J, Maiseyeu A (2012) Effect of co-exposure to nickel and particulate matter on insulin resistance and mitochondrial dysfunction in a mouse model. Part Fibre Toxicol 9:40

Xu X, Jiang SY, Wang T-Y, Bai Y, Zhong M, Wang A, Lippmann M, Chen L-C, Rajagopalan S, Sun Q (2013) Inflammatory response to fine particulate air pollution exposure: neutrophil versus monocyte. PLoS One 8(8):e71414

Zanobetti A, Schwartz J (2002) Cardiovascular damage by airborne particles: are diabetics more susceptible? Epidemiology 13(5):588–592

Zeng M, Wen Y, Liu L-Y, Wang H, Guan K-P, Huang X (2009) Role of TNF-α, sTNF-R55 and sTNF-R75 in inflammation of acute exacerbations of chronic obstructive pulmonary disease. Respiration 78(4):399–403

Zhao J, Gao Z, Tian Z, Xie Y, Xin F, Jiang R, Kan H, Song W (2013) The biological effects of individual-level PM2. 5 exposure on systemic immunity and inflammatory response in traffic policemen. Occup Environ Med 70(6):426–431

Zheng Z, Xu X, Zhang X, Wang A, Zhang C, Hüttemann M, Grossman LI, Chen LC, Rajagopalan S, Sun Q, Zhang K (2013) Exposure to ambient particulate matter induces a NASH-like phenotype and impairs hepatic glucose metabolism in an animal model. J Hepatol 58(1):148–154

Ziccardi P, Nappo F, Giugliano G, Esposito K, Marfella R, Cioffi M, D'Andrea F, Molinari AM, Giugliano D (2002) Reduction of inflammatory cytokine concentrations and improvement of endothelial functions in obese women after weight loss over one year. Circulation 105(7):804–809

Immunotoxic Effects of Mercury

12

Renee M. Gardner and Jennifer F. Nyland

Contents

R.M. Gardner, PhD
Department of Public Health Sciences, Karolinska Institutet, Stockholm, Sweden
e-mail: renee.gardner@ki.se

J.F. Nyland, PhD (✉)
Department of Pathology, Microbiology and Immunology, University of South Carolina School of Medicine, Columbia, SC, USA
e-mail: jnyland@uscmed.sc.edu

C. Esser (ed.), *Environmental Influences on the Immune System*,
DOI 10.1007/978-3-7091-1890-0_12

12.1 Human Exposure to Mercury Species

Human populations are primarily exposed occupationally to inorganic or elemental mercury as a result of dental amalgam or through other manufacturing processes and to methylmercury (MeHg) as a result of contaminated fish consumption (National Research Council (U.S.) Board on Environmental Studies and Toxicology 2000). Most human exposure to mercury occurs in the form of MeHg. Human exposure to and toxicokinetic handling of the different species of mercury are reviewed below.

12.1.1 Elemental or Metallic Mercury (Hg^0)

The greatest source of human exposure to elemental mercury (Hg^0) is dental amalgam, a combination of elemental mercury and various metal alloys used in the restorations of dental caries (WHO 1991; Berry et al. 1998). While animal studies indicate that less than 1 % of orally ingested elemental mercury is absorbed in the gastrointestinal (GI) tract, both animal and human studies have shown that the mercury from dental amalgam fillings can cross epithelial barriers in the lung and be dissolved in tissue fluids and blood, such that 74 % of the inhaled dose is absorbed and retained in the body (Hursh et al. 1976). It is rapidly transported to other areas of the body, where it can cross the blood-brain and placental barriers and also enter breast milk (Maas et al. 1996; Oskarsson et al. 1996; Hultman et al. 1998; Takahashi et al. 2003). Elemental mercury is oxidized *in vivo* to mercuric mercury (Hg^{2+}) by the catalase–hydrogen peroxide pathway; once this oxidation has occurred, mercury can no longer pass through the blood-brain barrier or placental barrier.

Significant exposures to metallic mercury can occur occupationally as well. Mercury is used in a wide variety of manufacturing processes, most notably the manufacture of caustic soda and chlorine using the chloralkali process, measurement devices such as thermometers and manometers, and most recently an increasing number of compact fluorescent light bulbs. Manufacture and use of compact fluorescent light bulbs containing mercury may have the potential to increase exposure in the general population as well by releasing up to 4 t of mercury per year into the environment in the United States alone, once breakage and disposal are taken into account (Aucott et al. 2003).

Many occupational exposures to mercury are well controlled in the developed world, with airborne mercury vapor concentration standards put into place by organizations such as NIOSH in order to protect employee health. However, the practice of small-scale artisanal gold mining is widespread globally, and the resultant exposures to miners and those in surrounding areas to mercury from this process are quite high (Castilhos et al. 1998; Crompton et al. 2002; Limbong et al. 2003; Silva et al. 2004; Taylor et al. 2005; Donkor et al. 2006; Gammons et al. 2006; Gardner et al. 2010b). Mercury use is integral to artisanal gold mining, a process which involves hydraulic excavation of riverbeds to expose gold-containing placer deposits. Elemental mercury is used to dissolve gold particles by kneading several grams of liquid mercury through the slurry, creating a mercury–gold amalgam. The amalgam is then heated to evaporate the mercury, leaving gold behind. This process usually occurs without the benefit of personal protective equipment or measures to reduce release into the environment, leading to contaminated watersheds. This activity initiates the environmental cycle of mercury biotransformation and bioaccumulation and results in high methylmercury (MeHg) exposures to populations living downstream of mining camps and consuming fish from local rivers and lakes (Alves et al. 2006; Dominique et al. 2007; de Andrade Lima et al. 2008).

12.1.2 Inorganic Mercury: Mercuric and Mercurous Species

Human exposures to both mercuric and mercurous species of mercury are generally low. Both forms were used in medical preparations up until the 1950s, but this use has been almost entirely discontinued. The one notable exception to this rule is the use of skin lightening creams that contain mercuric chloride and are still distributed and used in some countries (Dyall-Smith and Scurry 1990; Weldon et al. 2000).

Inorganic mercury salts are not well absorbed in the GI tract, with less than 10 % of the ingested dose being retained. Of the mercurous mercury that is absorbed, the Hg–Hg^{2+} ion quickly dissociates to the mercuric ion (Hg^{2+}) and an atom of uncharged mercury (Hg^0), which is then oxidized to Hg^{2+} (Hand et al. 1943). As noted below, MeHg is also converted to Hg^{2+} in the body. Inorganic mercury is complexed with reduced glutathione (GSH) in the liver and transported to the kidney, where it is secreted by proximal tubular cells into the tubular lumen. The enzyme gamma-glutamyl transpeptidase degrades GSH, releasing cysteine–mercury complex that is reabsorbed by renal cells (Tanaka et al. 1990; Tanaka-Kagawa et al. 1993; Wei et al. 1999). This cycle leads to the accumulation of mercury in the kidneys and resultant damage to these tissues.

12.1.3 Organic Mercury: Methylmercury

Geochemical cycling and biological conversion cause elemental and inorganic mercury to be released into the environment where they are converted to methylmercury (MeHg) by microorganisms in the sediment of aquatic ecosystems (Mason et al. 2005; Harris et al. 2007). MeHg accumulates in the biological tissues of organisms that feed on MeHg-contaminated organisms lower in the trophic food chain, causing bioaccumulation and biomagnification of mercury in aquatic food chains (Watras et al. 1998; Braga et al. 2000). Because this biomagnification occurs in aquatic environments, the major source of human exposure to MeHg is through the consumption of contaminated fish and water mammals, especially predatory fish and mammals. While contaminated fish remains the major source of MeHg for human exposures, recent studies have demonstrated that rice, because it is grown in partially aquatic environment, can also be a source of MeHg exposure (Feng et al. 2008; Zhang et al. 2010; Li et al. 2012)

The United Nation Environmental Programme estimates that the global daily average intake of MeHg is 2.4 μg/person, though this amount can vary greatly depending on the dietary composition of different populations (UNEP 2002). Allen et al. conducted a physiologically based pharmacokinetic modeling study, using Bayesian methods to estimate the posterior distribution of exposure to MeHg in the US populations based on MeHg data (Allen et al. 2007). Less than 1% of the US population of women of childbearing age were estimated to have mercury exposures greater than the EPA reference dose (RfD) for MeHg of 0.1 μg MeHg/kg body weight (bw)/day, and exposures greater than the ATSDR minimal risk levels (MRL) of 0.3 μg MeHg/kg bw/day were estimated to be very few.

Approximately 95 % of the MeHg in fish muscle that is consumed by humans is absorbed in the GI tract (WHO 1990). MeHg can rapidly cross biological barriers, including the gut epithelium, the blood-brain barrier, and the placental barrier, by complexing with the amino acid cysteine to form a structure that closely resembles the amino acid methionine. This complex is transported via the large neutral amino acid transporter across biological membranes (Leaner and Mason 2002; Simmons-Willis et al. 2002). MeHg is demethylated to inorganic mercury (Hg^{2+}) *in situ* in mammalian tissues such as the liver, brain, and phagocytic cell populations by an unknown mechanism (Suda and Hirayama 1992; Suda et al. 1992). Once metabolized, the presence of inorganic mercury in the brain and other body tissues is extremely persistent (Suda and Hirayama 1992; Suda et al. 1992; Vahter et al. 1994).

12.2 Experimental Evidence of Mercury Immunotoxicity

12.2.1 Immunosuppressive Effects of Mercury

Until relatively recently, all forms of mercury were considered to be immunosuppressive, given their potent cytotoxicity in culture and high-dose effects in animals. Both inorganic mercury ($HgCl_2$) and MeHg prevented mitogen-induced proliferation in B cells and immunoglobulin (IgG and IgM) production *in vitro*, with MeHg being 10 times more potent than $HgCl_2$ (Shenker et al. 1993a). B cells were more resistant to these effects if stimulated with mitogen before mercury exposure, suggesting that the activation state of the immune cells could modify their sensitivity to mercury. T-cell proliferation was also prevented, though only if monocytes were also present in the culture (Shenker et al. 1992b). Shenker et al. have reported that T cells are specifically sensitive to mercury-induced apoptosis as a result of GSH depletion and increases in oxidative stress (Shenker et al. 1997, 1998, 2000, 2002). The doses used in these studies were extremely high (1000 ng/ml to 10 μg/ml, or 3.6 to 36 μM), inducing lymphocyte death within 24 h and monocyte death within 4 h (Shenker et al. 1992a, b, 1993a).

Similar effects for MeHg were seen in animal studies. Short-term exposure to extremely high and acutely toxic doses of MeHg (3–9 mg Hg/kg bw/day) caused severe immunosuppression in mice in terms of proliferative response to mitogen

(Ohi et al. 1976; Hirokawa and Hayashi 1980; Brown et al. 1988). Subcutaneous injections of 1–4 mg MeHg/kg bw every 3 days (280–600 μg Hg/kg bw/day) were observed to reduce the number of splenocytes in mice, mostly by depleting B cells, within 5 days of commencement of treatment (Haggqvist et al. 2005). MeHg treatment (58–180 μg Hg/kg bw/day) has also been shown to reduce calcium (Ca^{2+}) signaling at cellular membranes in lymphocytes after mitogen challenge, indicating that Hg could be effecting transmembrane signaling in splenocytes with potential immunosuppressive effects (Thompson et al. 1998). Extremely high doses of $HgCl_2$ administered in water (100 mg $HgCl_2$/L for 28 days, with an estimated internal dose of 1.8 mg Hg/kg bw/day) did not show similar effects on immune cell depletion in treated mice (Brunet et al. 1993). $HgCl_2$ treatment (37 μg Hg/kg bw/day) in Lewis rats induces a general suppression of T-cell function (Pelletier et al. 1990).

12.2.2 Immune Stimulation by Mercury: Experimental Evidence

The immune-stimulating effects of mercury were described in early experimental models of mercury-induced autoimmune kidney disease in susceptible strains of rats that were dosed with $HgCl_2$ (25–37 μg Hg/kg bw/day) (Bariety et al. 1971). High doses of mercury damaged the kidney, leading to production of autoantibodies against laminin, a component of the glomerular basement membrane (Sapin et al. 1977; Icard et al. 1993). The resulting pathology is immune complex mediated, with inflammation and thickening of the glomerular basement membrane, reduced filtration capacity of the kidney, and proteinuria (Bigazzi 1999). Immune complex was also reported to be deposited in other tissues of the body, without major pathological consequences (Bernaudin et al. 1979). In terms of pathophysiology, the disease in rats reflected the nephrotic syndrome that had been recently described in patients who were exposed to high levels of mercury occupationally or therapeutically (Cameron and Trounce 1965; Turk and Baker 1968; Tubbs et al. 1982).

Certain mouse strains injected with $HgCl_2$ (395 μg Hg/kg bw/day) were also shown to be susceptible to mercury-induced autoimmunity, but the induced pathophysiology resembled the autoimmune disease lupus more than the autoimmune kidney disease described in rats and humans (Hultman and Enestrom 1988; Hultman et al. 1989). This lupus-like syndrome is characterized by lymphoproliferation, hyperproduction of immunoglobulin, systemic autoantibody circulation, and immune complex-mediated glomerulonephritis and vasculitis (reviewed by (Lawrence and McCabe 2002; Pollard et al. 2005)). Mercury-induced autoimmunity can be observed whether the exposure was through subcutaneous injection or oral ingestion of $HgCl_2$ (generally between 20 and 450 μg Hg/kg bw/day) or peritoneal implantation of dental amalgam (8–100 mg/mouse, or about 320 mg/kg bw to 4000 mg/kg bw) (Hultman et al. 1994; Nielsen and Hultman 2002; Havarinasab et al. 2007a).

Despite initial findings that organic species of mercury were immunosuppressive, MeHg (280–600 μg Hg/kg bw/day) can induce autoimmune disease in susceptible strains of mice (Hultman and Hansson-Georgiadis 1999; Haggqvist et al. 2005;

Havarinasab et al. 2005; Havarinasab and Hultman 2005). The response to mercury is slower and somewhat attenuated when organic species of mercury are used. The initial decline in splenocytes observed in MeHg-treated mice was reversed after 9 days, leading to a delayed onset of lymphoproliferation (Haggqvist et al. 2005; Havarinasab et al. 2005). These results suggest that organic species of mercury can also stimulate autoimmunity in mice, but the delayed onset of effects suggests that metabolism of the organic species to inorganic mercury may be required. Studies have shown that treatment with MeHg leads to uptake into lymphoid tissue, followed by progressive demethylation of MeHg within the lymph nodes, likely due to the activity of phagocytic cells (Suda et al. 1992; Havarinasab et al. 2007b).

This model of murine mercury-induced autoimmunity, with supporting *in vitro* work, has been extensively characterized. As a mechanism of murine mercury-induced autoimmunity, Pollard et al. suggest that $HgCl_2$ induces cell death, which exposes and modifies self-antigens (Pollard et al. 2005). These antigens are processed and presented to CD4+ T helper cells by monocytes, macrophages, and dendritic cells (all antigen-presenting cells, or APCs). CD4+ T helper cells then drive autoantibody production by B cells, leading to the observed increases in autoantibody production and subsequent pathologies such as immune complex deposition.

12.2.3 T-Cell Populations in Mercury-Induced Immune Dysfunction

Mercury exposure causes an initial activation of T cells (Jiang and Moller 1996), followed by a period of T-cell proliferation in mouse models (Johansson et al. 1997; Havarinasab et al. 2007a). This proliferation is more pronounced in CD4+ T cells and is dependent upon the presence of adherent macrophages and the presence of IL-1 *in vitro* (at 10 μM $HgCl_2$) (Reardon and Lucas 1987; Jiang and Moller 1995; Pollard and Landberg 2001). Mercury-induced T-cell proliferation is prevented by concurrent treatment with anti-CD4 antibodies in this model (Jiang and Moller 1995). T-cell activation is required in models of mercury-induced autoimmunity (Pelletier et al. 1987, 1988; Stiller-Winkler et al. 1988; Kubicka-Muranyi et al. 1993; Hultman et al. 1995). T cells have been suggested to be the primary effector cells in mercury-mediated autoimmunity, as transfer of T cells from mercury-treated brown Norway rats was sufficient to induce autoimmune disease in non-treated rats (Pelletier et al. 1988). Brown Norway rats treated neonatally with $HgCl_2$ are resistant to mercury-induced autoimmunity later in life, though continuous exposure to $HgCl_2$ is required throughout life to maintain tolerance (Field et al. 2000). Depletion of CD8+ T cells breaks tolerance in this model, suggesting that CD8+ T cells may suppress mercury-induced autoimmunity (Field et al. 2003).

In an *in vitro* model using an immortalized T-cell line (Jurkat cells (Schneider et al. 1977)), 5 μM $HgCl_2$ disrupted intracellular signaling in the CD95/Fas apoptotic pathway, allowing cells which were stimulated with CD95 to resist apoptosis (Whitekus et al. 1999; McCabe et al. 2003, 2005). $HgCl_2$ treatment caused the dissociation of preassembled Fas receptors in this model, preventing the formation of the death-inducing signaling complex (DISC) (Whitekus et al. 1999; Ziemba et al. 2005). McCabe et al. have suggested that T-cell resistance to apoptosis as a result of mercury exposure could be the underlying biological mechanism that allows self-tolerance to be broken in mercury-induced autoimmunity (McCabe et al. 2003, 2005), though studies by Shenker et al. show that human T cells are sensitive to mercury-induced apoptosis at the same treatment concentrations (Shenker et al. 1992a; Guo et al. 1998). Laiosa et al. reported that exposure of mice to high levels of mercury in drinking water (10 mg/L, 133.2 μg Hg/kg bw/day) led to reduced activation of apoptotic caspases in T cells after superantigen exposure, supporting the notion that mercury treatment may allow T cells to escape apoptosis (Laiosa et al. 2007).

12.2.4 B Cells in Mercury-Induced Immune Dysfunction

B-cell activation occurs subsequently to and is dependent upon T-cell activation, although B-cell activation is directly responsible for many of the pathological outcomes of mercury-induced autoimmunity including elevated levels of circulating immunoglobulins IgG and IgE (Johansson et al. 1998; Nielsen and Hultman 2002). Included in the increased immunoglobulin production are antibodies specific to self-antigens, most notably directed at nuclear and nucleolar proteins. Antinuclear autoantibodies (ANA) and anti-nucleolar autoantibodies (ANoA) are frequently observed in patients with systemic autoimmune diseases directed toward connective tissues, such as lupus and scleroderma (Gonzalez and Rothfield 1966; Ho and Reveille 2003). ANA and ANoA increase in a dose-response manner with $HgCl_2$ treatment (Havarinasab et al. 2007a). In mice, antibodies directed against the nucleolar protein fibrillarin predominate, similar to what is observed in scleroderma (Hultman et al. 1989; Pollard et al. 1989). This anti-fibrillarin response is dependent upon T cells and likely arises due to modification of fibrillarin by mercury (Hultman et al. 1995; Pollard et al. 1997); however, similar anti-fibrillarin antibodies were not observed in mercury-exposed mining populations with elevated ANA and ANoA (Silva et al. 2004) (discussed in more detail below). In a B-cell lymphoma cell line, McCabe et al. have shown that 5–10 μM $HgCl_2$ causes dysregulation of B-cell receptor (BCR) signaling pathways, including reduced activation of Erk–MAPK signaling transduction pathways downstream of BCR antigen-induced activation (McCabe et al. 1999, 2007). This suggests that mercury may interact with signal transduction pathways in B cells as well as T cells.

12.2.5 Requirement for Cellular Interactions in Mercury-Induced Immune Dysfunction

As described above, much of the experimental literature on mercury-induced immune dysfunction focuses on the actions and requirements of the effector T and B cells that are responsible for directly mediating pathology, with much less emphasis on the actions of APCs which may be driving the autoimmune responses. However, evidence that cell–cell interactions are required suggests that APCs may be necessary.

In terms of cell–cell communication, mercury-induced autoimmunity in mice is absolutely dependent upon dysregulation of cytokine signaling. Given the importance of T cells in this model, the majority of cytokine studies have focused on the role of cytokines produced by the T_H1 and T_H2 T helper cell subsets. While diseases that are dominated by pathological B-cell responsiveness are generally thought to be skewed toward a T_H2-type response (characterized by increases in IL-4 and IL-5 production), IL-4 is not required for mercury-induced autoimmunity in mice, as evidenced by the development of mercury-induced autoimmunity in IL-4 knockout mice. On the other hand, the signature T_H1 cytokine, interferon-γ (IFN-γ), is required for disease onset and progression (Bagenstose et al. 1998b, 1999; Kono et al. 1998; Hu et al. 1999). Surprisingly, IL-12, which induces IFN-γ expression and drives T_H1-type responses, downregulates autoantibody production in mice, though the renal deposition of immune complex is not affected (Bagenstose et al. 1998a). Treatment with IL-10, which suppresses T_H1 cytokine responses, suppresses mercury-induced autoimmunity, but does not completely prevent the disease (Haggqvist and Hultman 2005).

In vitro studies of cytokine production have also focused on the balance of T_H1 and T_H2 cytokines. In human peripheral blood mononuclear cells (PBMCs) activated with monoclonal antibodies to the T-cell receptor, IFN-γ and IL-12 production were decreased while IL-4 was increased over a wide range of $HgCl_2$ concentrations, mostly in the subcytotoxic range (Hemdan et al. 2007a, b; Hemdan 2008). Changes in T_H1/T_H2 cytokine expression in response to $HgCl_2$ treatment were observed to be dependent upon the developmental stage at which splenocytes were harvested from mice and did not affect lymph node cells or thymocytes (Silva et al. 2005).

In addition to T helper cell-related cytokines, proinflammatory cytokines have also been reported to be dysregulated as a result of mercury treatment. Zdolsek et al. reported that macrophages isolated from mercury-sensitive mouse strains (SJL and DBA) and cultured with 1 μM $HgCl_2$ significantly increased IL-1 production (though IL-1 was not further classified) (Zdolsek et al. 1994). The authors noted that treatment at concentrations higher than 10 μM significantly reduced the production of IL-1, due to decreases in cell viability. *In vivo*, mercury treatment has been reported to cause systemic, early increases in TNF-α. As mentioned, T-cell proliferation was reported to be dependent upon adherent macrophages *in vitro* and the production of IL-1 (Pollard and Landberg 2001). In *in vitro* modeling of human PBMC responses to subcytotoxic $HgCl_2$ treatments, there have been mixed results. In one report, TNF-α release was decreased in response to subcytotoxic $HgCl_2$ treatments (Hemdan et al. 2007b). In two other more recent reports, subcytotoxic $HgCl_2$ (Gardner et al. 2009) and MeHg (Gardner et al. 2010a) concentrations caused a dose-dependent increase in the release of proinflammatory cytokines with a concomitant decrease in anti-inflammatory cytokines.

In addition to cytokine regulation, direct cell–cell interactions have been shown to be necessary for the induction of autoimmunity by mercury treatment. Costimulatory interactions between APCs, including B cells, and T cells are required for the induction of autoimmunity by mercury treatment in mice (Biancone et al. 1996; Pollard et al. 2004). Disruption of CD40/CD40L interactions or CD28/CD80 or CD86 interactions prevents the development of autoimmunity in mercury-treated mice, whether the interruption was induced genetically or via culture conditions.

12.2.6 Experimental Literature in the Context of Human Exposure

In the preceding sections, the doses used in animal models and the concentrations used for *in vitro* models have been emphasized. If these levels are compared with human daily exposures (Fig. 12.1) or blood mercury levels (Fig. 12.2), respectively, it becomes clear that most of these studies have occurred outside an exposure range that is relevant to humans, even when highly exposed populations are considered.

Figure 12.1 compares doses used in animal models (standardized to daily exposure) to the daily exposure levels reported in human populations. Passos et al. reported daily exposures based on detailed food-frequency questionnaires and analysis of MeHg in fish captured and eaten by participants (by species and region of the river) in a Brazilian population living along the Tapajos River (Passos et al. 2008). This river is heavily polluted by mercury release that occurs upstream as a result of small-scale artisanal gold mining, and the reported MeHg levels in fish were elevated. In the most affected communities, noncarnivorous fish had mercury levels which were as high as those levels reported for tuna consumed within the United States, and carnivorous fish had levels which approached levels seen in swordfish (Mahaffey et al. 2004). The high level of fish consumption in these populations leads to the elevated daily intake of MeHg compared to other populations. The animal studies of mercury immunotoxicity reported in the literature and discussed so far had daily exposures to mercury that were above this level. Two more recent studies (Via et al. 2003; Nyland et al. 2012) used the lowest doses of mercury exposure in comparison to other studies. These experiments are discussed in detail below.

The most common concentration range used in *in vitro* studies is 5–10 μM, exemplified in Fig. 12.2 by the concentration used by both McCabe et al. (1999) and Whitekus et al. (1999), and the higher concentration used by Mattingly et al. (2001). This corresponds to a blood mercury level that is higher than the cord blood level that has been reported by Harada to be the threshold for the development of congenital Minamata disease in infants (Harada 1995). This dose range is also close to the blood mercury levels that are observed in persons who have died as a result of mercury intoxication, as reported by Bakir et al. after analysis of the mass MeHg poisoning incident which occurred in Iraq in 1972 (Bakir et al. 1973). While Whitekus et al. were careful to show that these doses did not induce loss of viability in cultures of Jurkat cells, concentrations of this magnitude have been convincingly

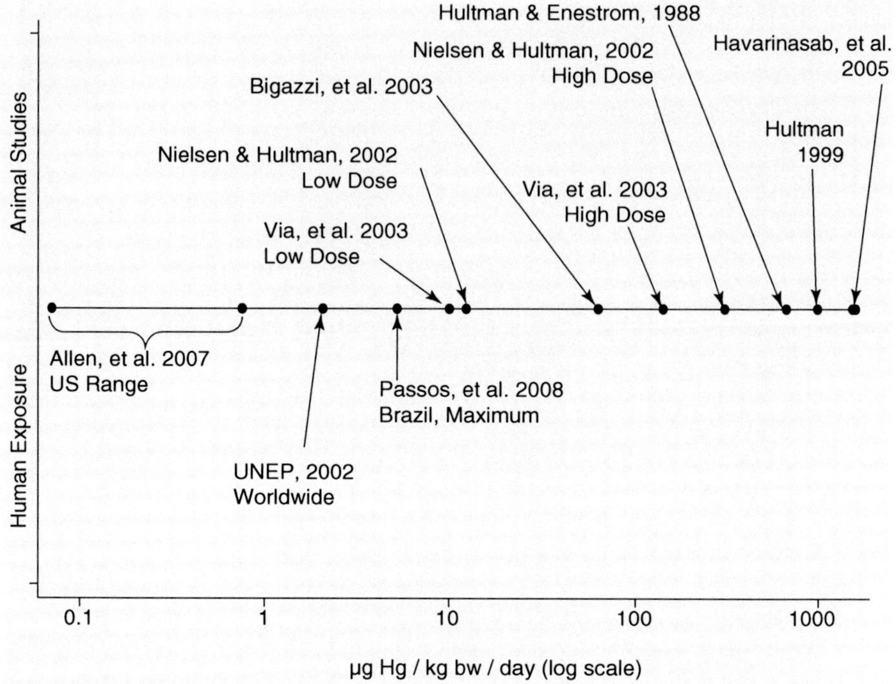

Fig. 12.1 Animal model doses as compared to human exposures. Reported doses from the studies of mercury immunotoxicity in animal models were standardized to represent the average amount of mercury the animals receive per day in terms of μg Hg/kg bw/day. In order to allow these studies to fit on the same scale as the range of human exposures shown, doses are shown here on the log scale. Via et al. (2003) and Hultman and Enestrom (1988) represent studies using subcutaneous injections of $HgCl_2$ in mice. Nielsen and Hultman (2002) represent a study in which mice were dosed with $HgCl_2$ in drinking water. Hultman and Hansson-Georgiadis (1999) used subcutaneous injections of MeHg in mice. Bigazzi et al. (2003) represent a standard dose of $HgCl_2$ given subcutaneously to brown Norway rats in models of mercury-induced kidney autoimmunity. These doses are compared with log-transformed reported exposures in human populations. Allen et al. (2007) used Bayesian methods to estimate the parameters of MeHg exposure in a physiologically based pharmacokinetic modeling study based on NHANES data gathered in US populations (as reported by Schober et al. (2003); see Fig. 2). The UNEP reported the average daily worldwide exposure to MeHg (UNEP 2002). Passos et al. (2008) reported daily exposures based on detailed food-frequency questionnaires and analysis of MeHg in fish (by species and region of the river) in a Brazilian population living along the Tapajos River. This river is heavily polluted by mercury release that occurs upstream as a result of small-scale gold mining, and the point represents the maximum estimated daily dose of mercury in this highly exposed population

shown to induce apoptosis in normal human lymphocytes *in vitro* (Shenker et al. 1992a, 2000; Guo et al. 1998). In the study by Walker al. represented in Fig. 12.2, heat shock protein genes were significantly upregulated in human lymphocytes after incubation with 20 μM ethylmercury (another organic mercury compound like MeHg) (Walker et al. 2006). When human lymphocytes are cultured with such high

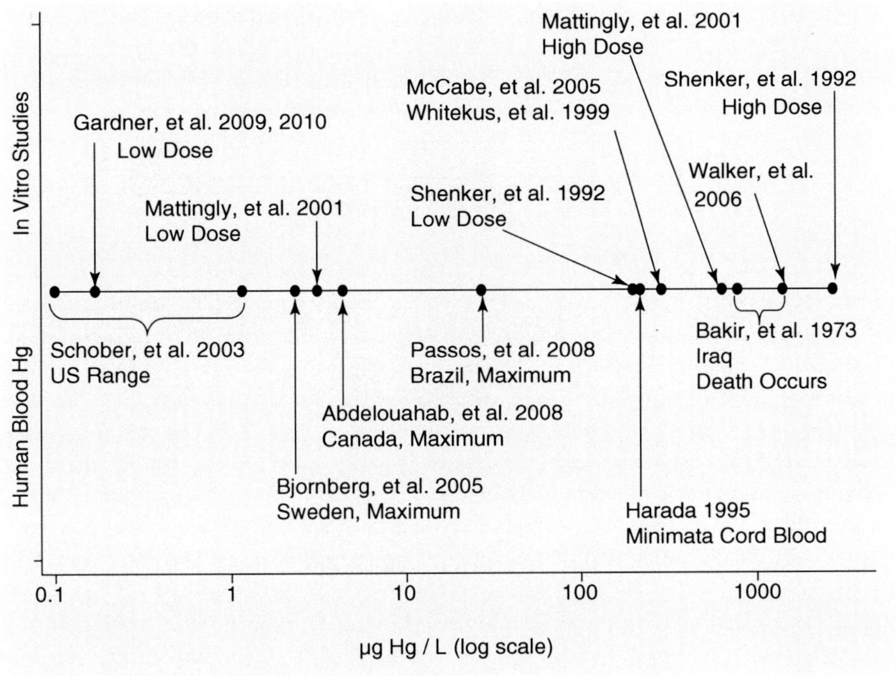

Fig. 12.2 In vitro model doses as compared to human blood mercury levels. Reported concentrations from the studies of mercury immunotoxicity in cell culture models were standardized to represent the concentration of mercury in terms of µg Hg/L. In order to allow these studies to fit on the same scale as the range of human exposures shown, concentrations are shown here on the log scale. Mattingly et al. (2001) and Whitekus et al. (1999) used Jurkat cells, a T-cell-like leukemia cell line. McCabe et al. (2007) used WEHI-231 cells, derived from a B-cell lymphoma. Shenker et al. (1992a, b, 1993a, b), Walker et al. (2006), and Gardner et al. (2009, 2010a) used human peripheral blood cells. These concentrations are compared with log-transformed reported blood mercury levels in human populations. Schober et al. (2003) reported the range of blood mercury levels in adult women in the United States using NHANES data. Passos et al. (2008) measured blood mercury levels in the population for which exposure was represented in Fig. 12.1. This point represents the maximum blood mercury level observed in this highly exposed population. Similarly, the data point from Abdelouahab et al. (2008) represents the maximum blood mercury level reported in a Canadian population with consumption of highly MeHg-contaminated fish species, and the data point for Björnberg et al. (2005) represents the maximum blood mercury level reported in a Swedish population with frequent fish consumption. Harada (1995) reported the cord blood mercury level above which most exposed infants developed congenital Minamata disease in Japan in the 1960s. Finally, Bakir et al. (1973) reported the range at which death of exposed persons occurred after consumption of MeHg-contaminated grain during the mass poisoning incident in Iraq in 1972

levels of mercury, cell defense mechanisms such as heat shock proteins come into play. Heat shock proteins function to preserve the basic functions of the cell in the presence of extreme stress (Beere 2004; Arya et al. 2007). In such a stressed state, specialized functions of cells are downregulated, especially energy-intensive processes like cytokine or antibody peptide synthesis and secretion which are critical to immune function. If the stress is not relieved, cells will undergo apoptosis. At the levels required to induce heat shock protein genes and cell death, immunotoxicity is not likely to be of relevance in exposed humans as the other acutely toxic aspects of mercury exposure will likely supersede any effect on the immune system.

12.3 Immunotoxic Effects of Hg in Humans

There is limited epidemiological data on the immunological effects of mercury in human populations. There have been no prospective studies on the immunological consequences of mercury exposure in humans. In a case-control study, Cooper et al. reported that past self-reported occupational exposure to mercury significantly increased the risk of lupus (Cooper et al. 2004). Arnett et al., in another case-control study, revealed a correlation between urine mercury levels and the severity of scleroderma, though disease-related alterations in renal functions could not be ruled out as a cause of this trend (Arnett et al. 1996).

Silva et al. report that gold miners exposed to elemental mercury and fish consumers exposed to MeHg had increased prevalence and titers of ANA and ANoA in serum compared to a nearby referent population with less exposure (Silva et al. 2004). In another study in the same elemental mercury-exposed mining population compared to referent mining populations with no mercury exposure, a positive correlation between elevated mercury exposure and increased ANA/ANoA titers was also documented (Gardner et al. 2010b). Given the similarities in ANA/ANoA response in these studies compared to murine models of mercury-induced autoimmunity, anti-fibrillarin antibodies were measured by Silva et al. Of 40 subjects with elevated ANoA titers, only 3 had detectable anti-fibrillarin antibodies, suggesting that the development of an anti-fibrillarin response may not be necessary to the development of mercury-induced autoimmunity in humans as it is in murine models (Hultman et al. 1989). Other studies have also confirmed the increased likelihood of elevated ANA/ANoA in MeHg-exposed populations (Alves et al. 2006; Nyland et al. 2011a), although not in a study of paired maternal–cord blood samples from the same region of Amazonian Brazil (Nyland et al. 2011b).

The body of evidence on immunological effects of mercury in occupationally exposed individuals is otherwise inconclusive. Cardenas et al. report that workers exposed to mercury show an increase in anti-DNA antibodies and increased IgE in serum compared to non-exposed referents (Cardenas et al. 1993). Dantas and Queiroz report an increase in serum IgE, but no detectable increase in anti-DNA antibodies or ANA was observed (Dantas and Queiroz 1997). Moszczynski et al. reported evidence of lymphoproliferation in mercury-exposed subjects, but Queiroz et al. found reduced numbers of T and B cells in exposed workers (Moszczynski et al. 1996; Queiroz and Dantas 1997a, b). Queiroz et al. also reported evidence of hyperimmunoglobulin production in mercury-exposed individuals, with detectable increases in IgG, IgM, and IgA (Queiroz et al. 1994). Studies by Barregard et al., Langworth et al., and Ellingsen et al. found no differences in immunoglobulin, immune complexes, or ANA levels in exposed workers. The equivocal findings of these studies may be due to the relatively small sample sizes used or to differences in the underlying genetic susceptibility to mercury-induced immune dysfunction or both.

12.4 Gene–Environment Interactions in Immune Responses

12.4.1 Genetic Basis of Susceptibility to Mercury Immunotoxicity: Experimental Evidence

In rodent models of mercury-induced autoimmunity, the effects of mercury on immune function are highly dependent on the genetic background of the mercury-challenged mouse or rat (Pelletier et al. 1990; Kosuda et al. 1994; Abedi-Valugerdi and Moller 2000; Hultman and Nielsen 2001). Susceptibility and resistance to mercury-induced autoimmunity are clearly complex traits. Hu et al. showed that resistance to mercury-induced autoimmunity was not due to general immunosuppression as a result of mercury exposure (Hu et al. 1998). Extensive analyses of inbred strains and backcrosses of mice have confirmed the importance of the H-2 locus of the major histocompatability complex (MHC) (Pelletier et al. 1990; Hultman et al. 1992; Kosuda et al. 1994; Hansson and Abedi-Valugerdi 2003). Mouse strains with the major histocompatibility complex H-2^S genotype are the most sensitive to mercury-induced autoimmune disease. In a study of 15 different mouse strains representing seven H-2 genotypes, at least one strain of mice from every H-2 genotype was shown to be susceptible to mercury-induced changes in immunoglobulin production (Abedi-Valugerdi and Moller 2000). MHC class II molecules were required for cytokine induction and lymphoproliferation induced by mercury exposure *in vitro* (Hu et al. 1997).

Abedi-Valugerdi et al. determined that resistance to mercury-induced autoimmunity was a dominant trait in mice (2001). Some aspects of the resistant phenotype appeared to involve a limited set of genes, while others appeared to involve more extensive loci including both H-2 and non-H-2 loci. Kono et al. (2001) conducted two genome-wide scans for genes related to mercury-induced autoimmunity using intercrosses of two susceptible mouse strains. They found a single major quantitative trait locus on chromosome 1, designated as Hmr1, which was common to both crosses. Interestingly, this locus encompasses a region containing several identified lupus susceptibility loci.

When mice that are genetically predisposed to the development of lupus are treated with mercury, including organic species, autoimmune pathology is accelerated and exacerbated in the mercury-exposed animals compared to non-treated animals of the same strain (al-Balaghi et al. 1996; Abedi-Valugerdi et al. 1997; Pollard et al. 1999, 2001; Havarinasab and Hultman 2006; Hultman et al. 2006). Lower doses of mercury are required for this effect, compared to those necessary to induce autoimmune disease in H-2^S mice strains.

12.4.2 Mercury and the Epigenome

Not only is genetic composition (DNA sequence) a factor in susceptibility to the immunotoxic effects of mercury, but epigenetic changes can also be induced by environmental contaminant exposure. Epigenetic changes, those stable and heritable changes in gene expression patterns that do not involve DNA sequence alterations, come in many flavors. DNA methylation and demethylation, histone hypoacetylation and hyperacetylation, and expression of certain noncoding RNAs are all types of epigenetic changes that can affect gene expression and regulation. These changes in gene expression can result in altered immune responses and immune dysfunction.

In a study utilizing murine embryonic stem cells *in vitro*, Gadhia et al. found that mercury altered histone modification pathways (Gadhia et al. 2012). In a study examining the effects of mercury exposure in wildlife, MeHg exposure was associated with DNA hypomethylation in the brains of polar bears (Pilsner et al. 2010). Finally, a small study in dental professionals found an association between MeHg exposure and hypomethylation of a particular gene promoter region (Goodrich et al. 2013).

Noncoding miRNAs, small 21 to 23 base pair RNAs that function as posttranslational regulators, are believed to regulate approximately one third of the "transcriptome" (Brooks et al. 2010). Multiple different miRNAs have been found to be differentially modulated in some patients with autoimmune disease, including miR-146a (Luo et al. 2011). miRNA levels can change for a variety of reasons, and many of the underlying triggers are unclear; however, a recent study demonstrated that, at least in plants, Hg exposure can modulate miRNA expression (Zhou et al. 2012).

12.4.3 Autoimmune Disease "Triggers" and Mercury Modulation of Autoimmunity

Environmental exposure to xenobiotics, infectious disease exposure, and genetic background all interact to influence susceptibility to and severity of autoimmune disease in humans (Rioux and Abbas 2005). Antigen exposure is considered to be a "trigger" in autoimmune disease, as exposure to common pathogens has been associated with increased risk of autoimmune disease (Bach 2005; Cooke et al. 2008). Inappropriate early innate responses to antigens may condition exaggerated immune responses, excessive inflammation, and tissue damage, which later results in autoimmune pathophysiology. In this context, it is interesting to consider murine models which include both mercury exposure and antigenic exposure.

Abedi-Valugerdi et al. reported that the antigenic stimulant lipopolysaccharide (LPS) influenced susceptibility to mercury-induced autoimmunity (Abedi-Valugerdi et al. 2005). When mice were treated with LPS and $HgCl_2$ (395 μg Hg/kg bw/day), autoimmunity was exacerbated in susceptible H-2^S mice. These mice showed elevated ANA and ANoA titers and increased glomerular immune complex deposition compared to mice treated with mercury only. In H-2^D mice that are otherwise highly resistant to mercury-induced autoimmunity (Abedi-Valugerdi and Moller 2000; Abedi-Valugerdi et al. 2001), concomitant exposure to LPS and $HgCl_2$ induced autoimmunity characterized by excess production of immunoglobulins and glomerular immune complex deposition, without an increase in ANA or ANoA titers.

Via et al. (2003) also reported that $HgCl_2$ treatment at much lower doses (7.4 or 74 μg Hg/kg bw/day) accelerated a lupus-like disease in a murine model of acquired autoimmune disease, chronic graft-versus-host disease (cGVHD). cGVHD is widely used to study mechanisms of autoimmune disease progression and to test novel pharmacotherapies (Gleichmann and Gleichmann 1985; Portanova et al. 1988; Via and Shearer 1988). Exposure to $HgCl_2$ before induction of cGVHD caused premature mortality, increased proteinuria, and more severe glomerulonephritis. The nephropathic changes were similar to those observed in the non-mercury-treated cGVHD mice and are distinct from those observed in rats with high-dose mercury exposure (Bigazzi et al. 2003).

Silbergeld et al. have reported that the same low doses of $HgCl_2$ (7.4 or 74 μg Hg/kg bw/day) exacerbate cardiac myosin antigen-induced autoimmune myocarditis (experimental autoimmune myocarditis, EAM) in mice (Silbergeld et al. 2005). Similarly, $HgCl_2$ treatment prior to infection with coxsackievirus B3 (CVB3) increased the severity of myocardial inflammation and the prevalence of dilated cardiomyopathy in mice (Nyland et al. 2012). Viral load in the heart was not affected in the CVB3-induced model, indicating that this effect did not depend on immune suppression, which can occur in this model at higher doses (Ilback et al. 1995, 1996). However, timing of the mercury exposure was important as mercury treatment after CVB3 infection was not sufficient to modulate the disease pathology. Interestingly, the number of macrophages was increased in the hearts of mercury-treated mice, suggesting a role for the activation of the innate immune response and proinflammatory cytokines in the mercury-exacerbated autoimmune disease.

12.4.4 Diversity in Human Immunogenetics: Implications for Gene–Environment Interactions

One reason for the apparent discrepancies between experimental and epidemiological studies on mercury-induced immunotoxicity described above may relate to unexamined differences in individual susceptibility to mercury-induced immunotoxicity as a result of the inherent variability in the human immune system.

As noted above, there is a clear role for genotype in murine models of mercury-induced autoimmune diseases and some suggestion that some of these genotypes may be important in modulating risk of autoimmune disease in human populations. However, humans are hypervariable in the genes which together control immune responses (Rodey and Fuller 1987). In humans, human leukocyte antigen (HLA) molecules on antigen-presenting cells (APCs) restrict immune responses by binding antigenic peptides and presenting them to T cells, analogous to the MHC in mice. HLA molecules greatly influence susceptibility and resistance to infectious disease based on variable affinity for antigenic peptides (Krensky 1997; Ghodke et al.

2005). The hypervariability present in HLA loci is driven by the evolutionary advantage conferred in infectious disease resistance. Heterozygotes (people who have two types of antigen-presenting molecule encoded per HLA locus, rather than one) can present a wider range of pathogenic antigens, increasing the likelihood that they will be able to effectively respond to and survive infection (Doherty and Zinkernagel 1975; Martin and Carrington 2005). Diversity in HLA molecules between people also confers a selective advantage to populations, enhancing the likelihood that at least some members can survive a disease outbreak (Singh et al. 1997). Despite the variability in HLA molecules in humans and its selective advantage, most genetic components of heritable difference in disease susceptibility map to loci outside of the HLA, including cytokine promoter regions and genes for their receptors (Hill 1998, 2001).

In addition to controlling susceptibility to infectious diseases, HLA and non-HLA polymorphisms in humans have been linked with risk of autoimmune disease (Braun 1992; Ebringer and Wilson 2000). Autoimmune diseases cluster in families, although not every family member has same disease and many family members go unaffected (Cooper et al. 1999). Alternately, for those autoimmune diseases which have been associated with exposure to some risk factor, only a small minority of those exposed go on to develop autoimmune disease (Inadera 2006). This suggests that the etiology of autoimmune diseases involves complex interactions between genetic and environmental risk factors.

In addition to genetic variability in genes controlling immune response, there is evidence that responses to environmental toxins can also vary within populations as a result of gene polymorphisms. There is relatively little data on the direct genetic modulation of responses to mercury in humans, except for a recent study reporting a modest association between brain-derived neurotrophic factor (BDNF) polymorphisms and neurotoxic symptoms in persons exposed to mercury in dental medicine (Heyer et al. 2004). Toxicokinetic handling of mercury can be modulated in humans as a result of polymorphisms in the GSH pathway, given its prominent role in the excretion of mercury from the body. Reports have indicated that polymorphisms in genes related to GSH synthesis and conjugation may result in increased body burden in exposed humans (Ballatori et al. 1998; Zalups et al. 1999; Custodio et al. 2005; James et al. 2005; Clarkson et al. 2007). Until recently, a separate biomarker of mercury was not used in population studies to assess the possibility that increased mercury levels may be a result of increased exposure. Custodio et al. and Schlawicke-Engstrom et al. recently demonstrated that certain polymorphisms in glutathione-S-transferase $\pi 1$ were correlated with higher mercury levels in erythrocytes once fish consumption was controlled for by the measurement of a highly specific exposure marker for fish, long-chain n-3 polyunsaturated fatty acids in plasma (Custodio et al. 2004; Schlawicke Engstrom et al. 2008).

12.5 Immunological Interactions in Other Mercury-Induced Pathologies

Mercury exposure had wide-ranging consequences for human health. Many toxic end points that result from mercury exposure have been extensively characterized, though little is known about whether these pathologies also have some aspects that are immune mediated. Here, we evaluate similarities between other end points of mercury-mediated toxicity and the immunotoxic effects of mercury that are observed in experimental models.

12.5.1 Kidney Disease

Mercury-induced kidney pathology provides the most direct evidence for an underlying dysregulation of immune response. Mercury, especially in elemental and inorganic forms, is directly toxic to kidney tissue as a result of mercury toxicodynamics discussed above. The accumulation of mercury in the kidneys damages tubular cells and causes the release of enzymes into the urine that are often used as biomarkers of damage to the kidney. At lower exposure levels, an increase in the lysosomal enzyme N-acetyl-β-d-glucosaminidase (NAG) is the most consistently observed biomarker in the urine of exposed subjects, an effect that is reversible once exposure has ceased (Ellingsen et al. 2000; Frumkin et al. 2001; Mandic et al. 2002). At higher exposures levels, intracellular enzymes including lactate dehydrogenase, aspartate aminotransferase, and acid phosphatase are released into the urine along with proteins such as albumin (Cardenas et al. 1993; Van Vleet and Schnellmann 2003; Barbier et al. 2005).

However, there is considerable evidence that damage to the kidney can result in autoimmune pathology directed against kidney tissue in susceptible individuals. As described above, mercury-induced glomerulonephritis has been characterized in both occupationally exposed human populations and animal models (Tubbs et al. 1982; Cardenas et al. 1993; Bigazzi 1999). This disease has also been characterized in populations exposed to mercury via use of skin lightening creams containing inorganic mercury (Barr et al. 1972, 1973).

Glomerulonephritis is characterized by an infiltration of inflammatory cells, including macrophages and monocytes. In a rat model of mercury-induced kidney autoimmunity, blocking the accumulation of glomerular macrophages and excessive TNF-α secretion by treatment with anti-TNF-α polyclonal antibodies blocked the development of nephritis in $HgCl_2$-exposed animals (Molina et al. 1995). As reviewed above, experimental evidence suggests that mercury may be able to modulate both proinflammatory cytokine signaling and the actions of APCs including macrophages.

12.5.2 The Nervous System

The developing nervous system and the immune system are in many ways conceptually and biologically linked: neuronal migration involves intercellular interactions that utilize cytokines and chemokines, the same signaling molecules as the immune system, in addition to other signaling molecules. The developing nervous system is considered to be the most sensitive target organ of MeHg toxicity, and the effects of MeHg have been extensively reviewed elsewhere (National Research Council (U.S.) Board on Environmental Studies and Toxicology. 2000; Clarkson and Magos 2006).

Several experimental findings in the studies of the neurological effects of mercury exposure are relevant to the discussion of the immunotoxicity of mercury. Charleston et al. reported an increase in reactive gliosis, or an influx of activated glial cells, in brains of nonhuman primates exposed to both organic and inorganic mercury (Charleston et al. 1994). Sass et al. reported that *in vitro* studies of microglia-directed neuronal migration showed that mercury decreased both the number of migrating neurons and the distance traveled by migrating neurons (Sass et al. 2001). Microglia are macrophage lineage cells resident in the CNS that perform important roles in neuronal migration and that synthesize, release, and respond to immunological signals. They function in the mature CNS to respond to infection and cell injury (Gehrmann et al. 1995). Decreases in neuronal migration observed by Sass et al. were related to disrupted cytokine signaling in the microglia. These results suggest that a possible mechanism of mercury-induced CNS damage may be the disruption of the complex neuro-immune signaling essential for neuronal migration and brain development.

12.5.3 Dermal Effects of Mercury

Toxic effects of mercury on the skin (Pink disease or acrodynia, mercury exanthem, and contact dermatitis) are also mediated by the immune system (Boyd et al. 2000). Acrodynia normally occurs in young children or infants, historically as a result of exposure to calomel (mercurous chloride) used as teething powder until the 1950s or mercurial antibacterial agents used in diaper laundering (Warkany and Hubbard 1953; Warkany 1966). In addition to general listlessness, anorexia, and irritability, affected children experience pain and swelling in their hands and feet, the skin of which become inflamed and pink, along with the skin of the nose. Skin biopsies reveal infiltrate of inflammatory cells. Given the inflammatory nature of the disease and the occurrence of the disease in only 1 in about every 500 exposed children, acrodynia is considered to be a hypersensitivity reaction that will only occur in susceptible individuals.

12.5.4 Cardiovascular Disease

The most recently characterized toxic end point of mercury exposure is effects on cardiovascular health. Several studies have reported increased risk of acute coronary events and cardiovascular disease as a result of mercury exposure from fish consumption (Salonen et al. 1995; Virtanen et al. 2005). Guallar et al. presented evidence that mercury body burdens resulting from fish intake may directly counteract the cardioprotective benefits of n-3 fatty acids in fish tissue and confirmed that mercury exposure increased the risk of myocardial infarction (Guallar et al. 2002). Interestingly, the studies in the Faroe and Seychelles Islands cohorts have shown that higher prenatal exposures to mercury are associated with higher blood pressure in children at age seven (Faroe Islands) and fifteen (Seychelles Islands, in males only) (Sorensen et al. 1999; Thurston et al. 2007). The relationship between mercury exposure and increased blood pressure was also observed in adult populations exposed to mercury via fish consumption in the Brazilian Amazon, though analysis in a US population (NHANES data) failed to find a correlation between blood mercury levels and blood pressure (Vupputuri et al. 2005; Fillion et al. 2006). Population studies of mercury effects on cardiovascular disease are difficult to conduct given the negative confounding relationship between the health effects of fish oils and the toxic effects of mercury exposure (Choi et al. 2008).

The inflammatory basis of cardiovascular disease has been extensively characterized in human populations, including evidence for the role of macrophage involvement and elevated proinflammatory cytokine levels in disease etiology and progression (Dinarello 1996; Boyle 2005; Apostolakis et al. 2008). As reviewed above, experimental evidence suggests that mercury may be able to modulate both proinflammatory cytokine signaling and the actions of APCs including macrophages. Given that mercury exposure is ubiquitous and that cardiovascular disease is the leading cause of death worldwide (Mathers et al. 2008), investigations of the direct links between mercury exposure, inflammation, and cardiovascular disease risk should be a high priority in future studies.

12.6 Conclusions

Animal models strongly suggest that mercury can disrupt immune processes, with pathological consequences for the exposed animal. These effects occur at exposures that are above the range of human exposure to mercury. The ability of inbred strains of mice to appropriately model the range of susceptibility to mercury-induced immune dysfunction in humans is questionable, given the inherent variability in human immune responses and the complex gene–environment interactions that are likely to be involved.

In order to evaluate the risk that mercury immunotoxicity poses to human health, studies that specifically examine the effects of mercury on the human immune system are required. Work that directly examines the impact of mercury on immune function in population-based studies is of high priority. The evidence presented in this review suggests that future experimental work ought to be conducted at low concentrations that accurately reflect the range of human exposures and ought to be conducted with human cells. Given the importance of cell–cell interactions in animal models of mercury immunotoxicity, monotypic cell lines should not be considered an adequate model to evaluate the effects of mercury in human cells. Instead, mixed culture models, such as PBMCs, ought to be used. If the appropriate study design is used, the use of primary cells would also allow for the evaluation of natural variability in response to mercury in human immune cells from different volunteers, presumably arising from intrinsic (genetic) and acquired factors such as previous infections. Finally evidence in animal models suggests that the activation state of mercury-exposed immune cells may play a significant role in determining the nature or magnitude of the impact of mercury, so the immunotoxic effects of mercury, including those effects on the epigenome, ought to be evaluated in both the presence and absence of antigenic stimulation or triggers of immune response.

References

Abdelouahab N, Vanier C, Baldwin M, Garceau S, Lucotte M, Mergler D (2008) Ecosystem matters: fish consumption, mercury intake and exposure among fluvial lake fish-eaters. Sci Total Environ 407(1):154–164

Abedi-Valugerdi M, Moller G (2000) Contribution of H-2 and non-H-2 genes in the control of mercury-induced autoimmunity. Int Immunol 12(10):1425–1430

Abedi-Valugerdi M, Hu H, Moller G (1997) Mercury-induced renal immune complex deposits in young (NZB x NZW)F1 mice: characterization of antibodies/autoantibodies. Clin Exp Immunol 110(1):86–91

Abedi-Valugerdi M, Hansson M, Moller G (2001) Genetic control of resistance to mercury-induced immune/autoimmune activation. Scand J Immunol 54(1–2):190–197

Abedi-Valugerdi M, Nilsson C, Zargari A, Gharibdoost F, DePierre JW, Hassan M (2005) Bacterial lipopolysaccharide both renders resistant mice susceptible to mercury-induced autoimmunity and exacerbates such autoimmunity in susceptible mice. Clin Exp Immunol 141(2):238–247

al-Balaghi S, Moller E, Moller G, Abedi-Valugerdi M (1996) Mercury induces polyclonal B cell activation, autoantibody production and renal immune complex deposits in young (NZB x NZW)F1 hybrids. Eur J Immunol 26(7):1519–1526

Allen BC, Hack CE, Clewell HJ (2007) Use of Markov chain Monte Carlo analysis with a physiologically-based pharmacokinetic model of methylmercury to estimate exposures in US women of childbearing age. Risk Anal 27(4):947–959

Alves MF, Fraiji NA, Barbosa AC, De Lima DS, Souza JR, Dorea JG, Cordeiro GW (2006) Fish consumption, mercury exposure and serum antinuclear antibody in Amazonians. Int J Environ Health Res 16(4):255–262

Apostolakis S, Vogiatzi K, Krambovitis E, Spandidos DA (2008) IL-1 cytokines in cardiovascular disease: diagnostic, prognostic and therapeutic implications. Cardiovasc Hematol Agents Med Chem 6(2):150–158

Arnett FC, Reveille JD, Goldstein R, Pollard KM, Leaird K, Smith EA, Leroy EC, Fritzler MJ (1996) Autoantibodies to fibrillarin in systemic sclerosis (scleroderma). An immunogenetic, serologic, and clinical analysis. Arthritis Rheum 39(7):1151–1160

Arya R, Mallik M, Lakhotia SC (2007) Heat shock genes – integrating cell survival and death. J Biosci 32(3):595–610

Aucott M, McLinden M, Winka M (2003) Release of mercury from broken fluorescent bulbs. J Air Waste Manag Assoc 53(2):143–151

Bach JF (2005) Infections and autoimmune diseases. J Autoimmun 25(Suppl):74–80

Bagenstose LM, Salgame P, Monestier M (1998a) IL-12 down-regulates autoantibody production in mercury-induced autoimmunity. J Immunol 160(4):1612–1617

Bagenstose LM, Salgame P, Monestier M (1998b) Mercury-induced autoimmunity in the absence of IL-4. Clin Exp Immunol 114(1):9–12

Bagenstose LM, Salgame P, Monestier M (1999) Cytokine regulation of a rodent model of mercuric chloride-induced autoimmunity. Environ Health Perspect 107(Suppl 5):807–810

Bakir F, Damluji SF, Amin-Zaki L, Murtadha M, Khalidi A, al-Rawi NY, Tikriti S, Dahahir HI, Clarkson TW, Smith JC, Doherty RA (1973) Methylmercury poisoning in Iraq. Science 181(96):230–241

Ballatori N, Wang W, Lieberman MW (1998) Accelerated methylmercury elimination in gamma-glutamyl transpeptidase-deficient mice. Am J Pathol 152(4):1049–1055

Barbier O, Jacquillet G, Tauc M, Cougnon M, Poujeol P (2005) Effect of heavy metals on, and handling by, the kidney. Nephron Physiol 99(4):105–110

Bariety J, Druet P, Laliberte F, Sapin C (1971) Glomerulonephritis with – and 1C-globulin deposits induced in rats by mercuric chloride. Am J Pathol 65(2):293–302

Barr RD, Rees PH, Cordy PE, Kungu A, Woodger BA, Cameron HM (1972) Nephrotic syndrome in adult Africans in Nairobi. Br Med J 2(5806):131–134

Barr RD, Woodger BA, Rees PH (1973) Levels of mercury in urine correlated with the use of skin lightening creams. Am J Clin Pathol 59(1):36–40

Beere HM (2004) "The stress of dying": the role of heat shock proteins in the regulation of apoptosis. J Cell Sci 117(Pt 13):2641–2651

Bernaudin JF, Druet E, Belair MF, Pinchon MC, Sapin C, Druet P (1979) Extrarenal immune complex type deposits induced by mercuric chloride in the Brown Norway rat. Clin Exp Immunol 38(2):265–273

Berry TG, Summitt JB, Chung AK, Osborne JW (1998) Amalgam at the new millennium. J Am Dent Assoc 129(11):1547–1556

Biancone L, Andres G, Ahn H, Lim A, Dai C, Noelle R, Yagita H, De Martino C, Stamenkovic I (1996) Distinct regulatory roles of lymphocyte costimulatory pathways on T helper type-2 mediated autoimmune disease. J Exp Med 183(4):1473–1481

Bigazzi PE (1999) Metals and kidney autoimmunity. Environ Health Perspect 107(Suppl 5):753–765

Bigazzi PE, Kosuda LL, Hannigan MO, Whalen B, Greiner DL (2003) Lack of graft-versus-host-like pathology in mercury-induced autoimmunity of Brown Norway rats. Clin Immunol 109(2):229–237

Bjornberg KA, Vahter M, Grawe KP, Berglund M (2005) Methyl mercury exposure in Swedish women with high fish consumption. Sci Total Environ 341(1–3):45–52

Boyd AS, Seger D, Vannucci S, Langley M, Abraham JL, King LE Jr (2000) Mercury exposure and cutaneous disease. J Am Acad Dermatol 43(1 Pt 1):81–90

Boyle JJ (2005) Macrophage activation in atherosclerosis: pathogenesis and pharmacology of plaque rupture. Curr Vasc Pharmacol 3(1):63–68

Braga MC, Shaw G, Lester JN (2000) Mercury modeling to predict contamination and bioaccumulation in aquatic ecosystems. Rev Environ Contam Toxicol 164:69–92

Braun WE (1992) HLA molecules in autoimmune diseases. Clin Biochem 25(3):187–191

Brooks WH, Le Dantec C, Pers JO, Youinou P, Renaudineau Y (2010) Epigenetics and autoimmunity. J Autoimmun 34(3):J207–J219

Brown DL, Reuhl KR, Bormann S, Little JE (1988) Effects of methyl mercury on the microtubule system of mouse lymphocytes. Toxicol Appl Pharmacol 94(1):66–75

Brunet S, Guertin F, Flipo D, Fournier M, Krzystyniak K (1993) Cytometric profiles of bone marrow and spleen lymphoid cells after mercury exposure in mice. Int J Immunopharmacol 15(7):811–819

Cameron JS, Trounce JR (1965) Membranous glomerulonephritis and the nephrotic syndrome appearing during mersalyl therapy. Guys Hosp Rep 114:101–107

Cardenas A, Roels H, Bernard AM, Barbon R, Buchet JP, Lauwerys RR, Rosello J, Hotter G, Mutti A, Franchini I et al (1993) Markers of early renal changes induced by industrial pollutants. I. Application to workers exposed to mercury vapour. Br J Ind Med 50(1):17–27

Castilhos ZC, Bidone ED, Lacerda LD (1998) Increase of the background human exposure to mercury through fish consumption due to gold mining at the Tapajos River region, Para State, Amazon. Bull Environ Contam Toxicol 61(2):202–209

Charleston JS, Bolender RP, Mottet NK, Body RL, Vahter ME, Burbacher TM (1994) Increases in the number of reactive glia in the visual cortex of Macaca fascicularis following subclinical long-term methyl mercury exposure. Toxicol Appl Pharmacol 129(2):196–206

Choi AL, Cordier S, Weihe P, Grandjean P (2008) Negative confounding in the evaluation of toxicity: the case of methylmercury in fish and seafood. Crit Rev Toxicol 38(10):877–893

Clarkson TW, Magos L (2006) The toxicology of mercury and its chemical compounds. Crit Rev Toxicol 36(8):609–662

Clarkson TW, Vyas JB, Ballatori N (2007) Mechanisms of mercury disposition in the body. Am J Ind Med 50(10):757–764

Cooke A, Ferraccioli GF, Herrmann M, Romani L, Schulze C, Zampieri S, Doria A (2008) Induction and protection of autoimmune rheumatic diseases. The role of infections. Clin Exp Rheumatol 26(1 Suppl 48):S1–S7

Cooper GS, Miller FW, Pandey JP (1999) The role of genetic factors in autoimmune disease: implications for environmental research. Environ Health Perspect 107(Suppl 5):693–700

Cooper GS, Parks CG, Treadwell EL, St Clair EW, Gilkeson GS, Dooley MA (2004) Occupational risk factors for the development of systemic lupus erythematosus. J Rheumatol 31(10):1928–1933

Crompton P, Ventura AM, de Souza JM, Santos E, Strickland GT, Silbergeld E (2002) Assessment of mercury exposure and malaria in a Brazilian Amazon riverine community. Environ Res 90(2):69–75

Custodio HM, Broberg K, Wennberg M, Jansson JH, Vessby B, Hallmans G, Stegmayr B, Skerfving S (2004) Polymorphisms in glutathione-related genes affect methylmercury retention. Arch Environ Health 59(11):588–595

Custodio HM, Harari R, Gerhardsson L, Skerfving S, Broberg K (2005) Genetic influences on the retention of inorganic mercury. Arch Environ Occup Health 60(1):17–23

Dantas DC, Queiroz ML (1997) Immunoglobulin E and autoantibodies in mercury-exposed workers. Immunopharmacol Immunotoxicol 19(3):383–392

de Andrade Lima LR, Bernardez LA, Barbosa LA (2008) Characterization and treatment of artisanal gold mine tailings. J Hazard Mater 150(3):747–753

Dinarello CA (1996) Biologic basis for interleukin-1 in disease. Blood 87(6):2095–2147

Doherty PC, Zinkernagel RM (1975) Enhanced immunological surveillance in mice heterozygous at the H-2 gene complex. Nature 256(5512):50–52

Dominique Y, Muresan B, Duran R, Richard S, Boudou A (2007) Simulation of the chemical fate and bioavailability of liquid elemental mercury drops from gold mining in Amazonian freshwater systems. Environ Sci Technol 41(21):7322–7329

Donkor AK, Bonzongo JC, Nartey VK, Adotey DK (2006) Mercury in different environmental compartments of the Pra River Basin, Ghana. Sci Total Environ 368(1):164–176

Dyall-Smith DJ, Scurry JP (1990) Mercury pigmentation and high mercury levels from the use of a cosmetic cream. Med J Aust 153(7):409–410, 414–415

Ebringer A, Wilson C (2000) HLA molecules, bacteria and autoimmunity. J Med Microbiol 49(4):305–311

Ellingsen DG, Efskind J, Berg KJ, Gaarder PI, Thomassen Y (2000) Renal and immunologic markers for chloralkali workers with low exposure to mercury vapor. Scand J Work Environ Health 26(5):427–435

Feng X, Li P, Qiu G, Wang S, Li G, Shang L, Meng B, Jiang H, Bai W, Li Z, Fu X (2008) Human exposure to methylmercury through rice intake in mercury mining areas, Guizhou province, China. Environ Sci Technol 42(1):326–332

Field AC, Caccavelli L, Fillion J, Kuhn J, Mandet C, Druet P, Bellon B (2000) Neonatal induction of tolerance to T(h)2-mediated autoimmunity in rats. Int Immunol 12(10):1467–1477

Field AC, Caccavelli L, Bloch MF, Bellon B (2003) Regulatory CD8+ T cells control neonatal tolerance to a Th2-mediated autoimmunity. J Immunol 170(5):2508–2515

Fillion M, Mergler D, Sousa Passos CJ, Larribe F, Lemire M, Guimaraes JR (2006) A preliminary study of mercury exposure and blood pressure in the Brazilian Amazon. Environ Health 5:29

Frumkin H, Letz R, Williams PL, Gerr F, Pierce M, Sanders A, Elon L, Manning CC, Woods JS, Hertzberg VS, Mueller P, Taylor BB (2001) Health effects of long-term mercury exposure among chloralkali plant workers. Am J Ind Med 39(1):1–18

Gadhia SR, Calabro AR, Barile FA (2012) Trace metals alter DNA repair and histone modification pathways concurrently in mouse embryonic stem cells. Toxicol Lett 212(2):169–179

Gammons CH, Slotton DG, Gerbrandt B, Weight W, Young CA, McNearny RL, Camac E, Calderon R, Tapia H (2006) Mercury concentrations of fish, river water, and sediment in the Rio Ramis-Lake Titicaca watershed, Peru. Sci Total Environ 368(2–3):637–648

Gardner RM, Nyland JF, Evans SL, Wang SB, Doyle KM, Crainiceanu CM, Silbergeld EK (2009) Mercury induces an unopposed inflammatory response in human peripheral blood mononuclear cells in vitro. Environ Health Perspect 117(12):1932–1938

Gardner RM, Nyland JF, Silbergeld EK (2010a) Differential immunotoxic effects of inorganic and organic mercury species in vitro. Toxicol Lett 198(2):182–190

Gardner RM, Nyland JF, Silva IA, Ventura AM, de Souza JM, Silbergeld EK (2010b) Mercury exposure, serum antinuclear/antinucleolar antibodies, and serum cytokine levels in mining populations in Amazonian Brazil: a cross-sectional study. Environ Res 110(4):345–354

Gehrmann J, Matsumoto Y, Kreutzberg GW (1995) Microglia: intrinsic immuneffector cell of the brain. Brain Res Brain Res Rev 20(3):269–287

Ghodke Y, Joshi K, Chopra A, Patwardhan B (2005) HLA and disease. Eur J Epidemiol 20(6):475–488

Gleichmann E, Gleichmann H (1985) Pathogenesis of graft-versus-host reactions (GVHR) and GVH-like diseases. J Invest Dermatol 85(1 Suppl):115s–120s

Gonzalez EN, Rothfield NF (1966) Immunoglobulin class and pattern of nuclear fluorescence in systemic lupus erythematosus. N Engl J Med 274(24):1333–1338

Goodrich JM, Basu N, Franzblau A, Dolinoy DC (2013) Mercury biomarkers and DNA methylation among Michigan dental professionals. Environ Mol Mutagen 54(3):195–203

Guallar E, Sanz-Gallardo MI, van't Veer P, Bode P, Aro A, Gomez-Aracena J, Kark JD, Riemersma RA, Martin-Moreno JM, Kok FJ (2002) Mercury, fish oils, and the risk of myocardial infarction. N Engl J Med 347(22):1747–1754

Guo TL, Miller MA, Shapiro IM, Shenker BJ (1998) Mercuric chloride induces apoptosis in human T lymphocytes: evidence of mitochondrial dysfunction. Toxicol Appl Pharmacol 153(2):250–257

Haggqvist B, Hultman P (2005) Interleukin-10 in murine metal-induced systemic autoimmunity. Clin Exp Immunol 141(3):422–431

Haggqvist B, Havarinasab S, Bjorn E, Hultman P (2005) The immunosuppressive effect of methylmercury does not preclude development of autoimmunity in genetically susceptible mice. Toxicology 208(1):149–164

Hand WC, Edwards BB, Caify ER (1943) Studies in the pathology of mercury III Histochemical demonstration and differentiation of metallic mercury, mercurous mercury, and mercuric mercury. J Lab Clin Med 28:1835–1841

Hansson M, Abedi-Valugerdi M (2003) Xenobiotic metal-induced autoimmunity: mercury and silver differentially induce antinucleolar autoantibody production in susceptible H-2s, H-2q and H-2f mice. Clin Exp Immunol 131(3):405–414

Harada M (1995) Minamata disease: methylmercury poisoning in Japan caused by environmental pollution. Crit Rev Toxicol 25(1):1–24

Harris RC, Rudd JW, Amyot M, Babiarz CL, Beaty KG, Blanchfield PJ, Bodaly RA, Branfireun BA, Gilmour CC, Graydon JA, Heyes A, Hintelmann H, Hurley JP, Kelly CA, Krabbenhoft DP, Lindberg SE, Mason RP, Paterson MJ, Podemski CL, Robinson A, Sandilands KA, Southworth GR, St Louis VL, Tate MT (2007) Whole-ecosystem study shows rapid fish-mercury response to changes in mercury deposition. Proc Natl Acad Sci U S A 104(42):16586–16591

Havarinasab S, Hultman P (2005) Organic mercury compounds and autoimmunity. Autoimmun Rev 4(5):270–275

Havarinasab S, Hultman P (2006) Alteration of the spontaneous systemic autoimmune disease in (NZB x NZW)F1 mice by treatment with thimerosal (ethyl mercury). Toxicol Appl Pharmacol 214(1):43–54

Havarinasab S, Haggqvist B, Bjorn E, Pollard KM, Hultman P (2005) Immunosuppressive and autoimmune effects of thimerosal in mice. Toxicol Appl Pharmacol 204(2):109–121

Havarinasab S, Bjorn E, Ekstrand J, Hultman P (2007a) Dose and Hg species determine the T-helper cell activation in murine autoimmunity. Toxicology 229(1–2):23–32

Havarinasab S, Bjorn E, Nielsen JB, Hultman P (2007b) Mercury species in lymphoid and non-lymphoid tissues after exposure to methyl mercury: correlation with autoimmune parameters during and after treatment in susceptible mice. Toxicol Appl Pharmacol 221(1):21–28

Hemdan NY (2008) The role of interleukin-12 in the heavy metal-elicited immunomodulation: relevance of various evaluation methods. J Occup Med Toxicol 3(1):25

Hemdan NY, Emmrich F, Faber S, Lehmann J, Sack U (2007a) Alterations of TH1/TH2 reactivity by heavy metals: possible consequences include induction of autoimmune diseases. Ann N Y Acad Sci 1109:129–137

Hemdan NY, Lehmann I, Wichmann G, Lehmann J, Emmrich F, Sack U (2007b) Immunomodulation by mercuric chloride in vitro: application of different cell activation pathways. Clin Exp Immunol 148(2):325–337

Heyer NJ, Echeverria D, Bittner AC Jr, Farin FM, Garabedian CC, Woods JS (2004) Chronic low-level mercury exposure, BDNF polymorphism, and associations with self-reported symptoms and mood. Toxicol Sci 81(2):354–363

Hill AV (1998) The immunogenetics of human infectious diseases. Annu Rev Immunol 16:593–617

Hill AV (2001) Immunogenetics and genomics. Lancet 357(9273):2037–2041

Hirokawa K, Hayashi Y (1980) Acute methyl mercury intoxication in mice – effect on the immune system. Acta Pathol Jpn 30(1):23–32

Ho KT, Reveille JD (2003) The clinical relevance of autoantibodies in scleroderma. Arthritis Res Ther 5(2):80–93

Hu H, Moller G, Abedi-Valugerdi M (1997) Major histocompatibility complex class II antigens are required for both cytokine production and proliferation induced by mercuric chloride in vitro. J Autoimmun 10(5):441–446

Hu H, Moller G, Abedi-Valugerdi M (1998) Non-responsiveness to mercury-induced autoimmunity in resistant DBA/2 mice is not due to immunosuppression or biased Th1-type response. Scand J Immunol 48(5):515–521

Hu H, Moller G, Abedi-Valugerdi M (1999) Mechanism of mercury-induced autoimmunity: both T helper 1- and T helper 2-type responses are involved. Immunology 96(3):348–357

Hultman P, Enestrom S (1988) Mercury induced antinuclear antibodies in mice: characterization and correlation with renal immune complex deposits. Clin Exp Immunol 71(2):269–274

Hultman P, Hansson-Georgiadis H (1999) Methyl mercury-induced autoimmunity in mice. Toxicol Appl Pharmacol 154(3):203–211

Hultman P, Nielsen JB (2001) The effect of dose, gender, and non-H-2 genes in murine mercury-induced autoimmunity. J Autoimmun 17(1):27–37

Hultman P, Enestrom S, Pollard KM, Tan EM (1989) Anti-fibrillarin autoantibodies in mercury-treated mice. Clin Exp Immunol 78(3):470–477

Hultman P, Bell LJ, Enestrom S, Pollard KM (1992) Murine susceptibility to mercury. I. Autoantibody profiles and systemic immune deposits in inbred, congenic, and intra-H-2 recombinant strains. Clin Immunol Immunopathol 65(2):98–109

Hultman P, Johansson U, Turley SJ, Lindh U, Enestrom S, Pollard KM (1994) Adverse immunological effects and autoimmunity induced by dental amalgam and alloy in mice. FASEB J 8(14):1183–1190

Hultman P, Johansson U, Dagnaes-Hansen F (1995) Murine mercury-induced autoimmunity: the role of T-helper cells. J Autoimmun 8(6):809–823

Hultman P, Lindh U, Horsted-Bindslev P (1998) Activation of the immune system and systemic immune-complex deposits in Brown Norway rats with dental amalgam restorations. J Dent Res 77(6):1415–1425

Hultman P, Taylor A, Yang JM, Pollard KM (2006) The effect of xenobiotic exposure on spontaneous autoimmunity in (SWR x SJL)F1 hybrid mice. J Toxicol Environ Health A 69(6):505–523

Hursh JB, Cherian MG, Clarkson TW, Vostal JJ, Mallie RV (1976) Clearance of mercury (HG-197, HG-203) vapor inhaled by human subjects. Arch Environ Health 31(6):302–309

Icard P, Pelletier L, Vial MC, Mandet C, Pasquier R, Michel A, Druet P (1993) Evidence for a role of antilaminin-producing B cell clones that escape tolerance in the pathogenesis of HgCl2-induced membranous glomerulopathy. Nephrol Dial Transplant 8(2):122–127

Ilback NG, Lindh U, Fohlman J, Friman G (1995) New aspects of murine coxsackie B3 myocarditis – focus on heavy metals. Eur Heart J 16 Suppl O:20–24

Ilback NG, Wesslen L, Fohlman J, Friman G (1996) Effects of methyl mercury on cytokines, inflammation and virus clearance in a common infection (coxsackie B3 myocarditis). Toxicol Lett 89(1):19–28

Inadera H (2006) The immune system as a target for environmental chemicals: xenoestrogens and other compounds. Toxicol Lett 164(3):191–206

James SJ, Slikker W 3rd, Melnyk S, New E, Pogribna M, Jernigan S (2005) Thimerosal neurotoxicity is associated with glutathione depletion: protection with glutathione precursors. Neurotoxicology 26(1):1–8

Jiang Y, Moller G (1995) In vitro effects of HgCl2 on murine lymphocytes. I. Preferable activation of CD4+ T cells in a responder strain. J Immunol 154(7):3138–3146

Jiang Y, Moller G (1996) Unresponsiveness of CD4+ T cells from a non-responder strain to HgCl2 is not due to CD8(+)-mediated immunosuppression: an analysis of the very early activation antigen CD69. Scand J Immunol 44(6):565–570

Johansson U, Sander B, Hultman P (1997) Effects of the murine genotype on T cell activation and cytokine production in murine mercury-induced autoimmunity. J Autoimmun 10(4):347–355

Johansson U, Hansson-Georgiadis H, Hultman P (1998) The genotype determines the B cell response in mercury-treated mice. Int Arch Allergy Immunol 116(4):295–305

Kono DH, Balomenos D, Pearson DL, Park MS, Hildebrandt B, Hultman P, Pollard KM (1998) The prototypic Th2 autoimmunity induced by mercury is dependent on IFN-gamma and not Th1/Th2 imbalance. J Immunol 161(1):234–240

Kono DH, Park MS, Szydlik A, Haraldsson KM, Kuan JD, Pearson DL, Hultman P, Pollard KM (2001) Resistance to xenobiotic-induced autoimmunity maps to chromosome 1. J Immunol 167(4):2396–2403

Kosuda LL, Greiner DL, Bigazzi PE (1994) Mercury-induced renal autoimmunity in BN → LEW.1N chimeric rats. Cell Immunol 155(1):77–94

Krensky AM (1997) The HLA system, antigen processing and presentation. Kidney Int Suppl 58:S2–S7

Kubicka-Muranyi M, Behmer O, Uhrberg M, Klonowski H, Bister J, Gleichmann E (1993) Murine systemic autoimmune disease induced by mercuric chloride (HgCl2): Hg-specific helper T-cells react to antigen stored in macrophages. Int J Immunopharmacol 15(2):151–161

Laiosa MD, Eckles KG, Langdon M, Rosenspire AJ, McCabe MJ Jr (2007) Exposure to inorganic mercury in vivo attenuates extrinsic apoptotic signaling in Staphylococcal aureus enterotoxin B stimulated T-cells. Toxicol Appl Pharmacol 225(3):238–250

Lawrence DA, McCabe MJ Jr (2002) Immunomodulation by metals. Int Immunopharmacol 2(2–3):293–302

Leaner JJ, Mason RP (2002) Methylmercury accumulation and fluxes across the intestine of channel catfish, Ictalurus punctatus. Comp Biochem Physiol C Toxicol Pharmacol 132(2):247–259

Li P, Feng X, Yuan X, Chan HM, Qiu G, Sun GX, Zhu YG (2012) Rice consumption contributes to low level methylmercury exposure in southern China. Environ Int 49:18–23

Limbong D, Kumampung J, Rimper J, Arai T, Miyazaki N (2003) Emissions and environmental implications of mercury from artisanal gold mining in North Sulawesi, Indonesia. Sci Total Environ 302(1–3):227–236

Luo X, Yang W, Ye DQ, Cui H, Zhang Y, Hirankarn N, Qian X, Tang Y, Lau YL, de Vries N, Tak PP, Tsao BP, Shen N (2011) A functional variant in microRNA-146a promoter modulates its expression and confers disease risk for systemic lupus erythematosus. PLoS Genet 7(6):e1002128

Maas C, Bruck W, Haffner HT, Schweinsberg F (1996) Study on the significance of mercury accumulation in the brain from dental amalgam fillings through direct mouth-nose-brain transport. Zentralbl Hyg Umweltmed 198(3):275–291

Mahaffey KR, Clickner RP, Bodurow CC (2004) Blood organic mercury and dietary mercury intake: National Health and Nutrition Examination Survey, 1999 and 2000. Environ Health Perspect 112(5):562–570

Mandic L, Radmila M, Jelena A, Dubravka D (2002) Change in the iso-enzyme profiles of urinary N-acetyl-beta-D-glucosaminidase in workers exposed to mercury. Toxicol Ind Health 18(5):207–214

Martin MP, Carrington M (2005) Immunogenetics of viral infections. Curr Opin Immunol 17(5):510–516

Mason RP, Abbott ML, Bodaly RA, Bullock OR Jr, Driscoll CT, Evers D, Lindberg SE, Murray M, Swain EB (2005) Monitoring the response to changing mercury deposition. Environ Sci Technol 39(1):14A–22A

Mathers C, Fat DM, Boerma JT, World Health Organization (2008) The global burden of disease: 2004 update. World Health Organization, Geneva

Mattingly RR, Felczak A, Chen CC, McCabe MJ Jr, Rosenspire AJ (2001) Low concentrations of inorganic mercury inhibit Ras activation during T cell receptor-mediated signal transduction. Toxicol Appl Pharmacol 176(3):162–168

McCabe MJ Jr, Santini RP, Rosenspire AJ (1999) Low and nontoxic levels of ionic mercury interfere with the regulation of cell growth in the WEHI-231 B-cell lymphoma. Scand J Immunol 50(3):233–241

McCabe MJ Jr, Whitekus MJ, Hyun J, Eckles KG, McCollum G, Rosenspire AJ (2003) Inorganic mercury attenuates CD95-mediated apoptosis by interfering with formation of the death inducing signaling complex. Toxicol Appl Pharmacol 190(2):146–156

McCabe MJ Jr, Eckles KG, Langdon M, Clarkson TW, Whitekus MJ, Rosenspire AJ (2005) Attenuation of CD95-induced apoptosis by inorganic mercury: caspase-3 is not a direct target of low levels of Hg2+. Toxicol Lett 155(1):161–170

McCabe MJ Jr, Laiosa MD, Li L, Menard SL, Mattingly RR, Rosenspire AJ (2007) Low and nontoxic inorganic mercury burdens attenuate BCR-mediated signal transduction. Toxicol Sci 99(2):512–521

Molina A, Sanchez-Madrid F, Bricio T, Martin A, Escudero E, Alvarez V, Mampaso F (1995) Abrogation of mercuric chloride-induced nephritis in the Brown Norway rat by treatment with antibodies against TNFalpha. Mediators Inflamm 4(6):444–451

Moszczynski P, Rutowski J, Slowinski S, Bem S, Jakus-Stoga D (1996) Effects of occupational exposure to mercury vapors on T-cell and NK-cell populations. Arch Med Res 27(4):503–507

National Research Council (U.S.) Board on Environmental Studies and Toxicology (2000) Toxicological effects of methylmercury. National Academy Press, Washington, DC

Nielsen JB, Hultman P (2002) Mercury-induced autoimmunity in mice. Environ Health Perspect 110(Suppl 5):877–881

Nyland JF, Fillion M, Barbosa F Jr, Shirley DL, Chine C, Lemire M, Mergler D, Silbergeld EK (2011a) Biomarkers of methylmercury exposure immunotoxicity among fish consumers in Amazonian Brazil. Environ Health Perspect 119(12):1733–1738

Nyland JF, Wang SB, Shirley DL, Santos EO, Ventura AM, de Souza JM, Silbergeld EK (2011b) Fetal and maternal immune responses to methylmercury exposure: a cross-sectional study. Environ Res 111(4):584–589

Nyland JF, Fairweather D, Shirley DL, Davis SE, Rose NR, Silbergeld EK (2012) Low-dose inorganic mercury increases severity and frequency of chronic coxsackievirus-induced autoimmune myocarditis in mice. Toxicol Sci 125(1):134–143

Ohi G, Fukuda M, Seto H, Yagyu H (1976) Effect of methylmercury on humoral immune responses in mice under conditions simulated to practical situations. Bull Environ Contam Toxicol 15(2):175–180

Oskarsson A, Schultz A, Skerfving S, Hallen IP, Ohlin B, Lagerkvist BJ (1996) Total and inorganic mercury in breast milk in relation to fish consumption and amalgam in lactating women. Arch Environ Health 51(3):234–241

Passos CJ, Da Silva DS, Lemire M, Fillion M, Guimaraes JR, Lucotte M, Mergler D (2008) Daily mercury intake in fish-eating populations in the Brazilian Amazon. J Expo Sci Environ Epidemiol 18(1):76–87

Pelletier L, Pasquier R, Vial MC, Mandet C, Moutier R, Salomon JC, Druet P (1987) Mercury-induced autoimmune glomerulonephritis: requirement for T-cells. Nephrol Dial Transplant 1(4):211–218

Pelletier L, Pasquier R, Rossert J, Vial MC, Mandet C, Druet P (1988) Autoreactive T cells in mercury-induced autoimmunity. Ability to induce the autoimmune disease. J Immunol 140(3):750–754

Pelletier L, Rossert J, Pasquier R, Vial MC, Druet P (1990) Role of CD8+ T cells in mercury-induced autoimmunity or immunosuppression in the rat. Scand J Immunol 31(1):65–74

Pilsner JR, Lazarus AL, Nam DH, Letcher RJ, Sonne C, Dietz R, Basu N (2010) Mercury-associated DNA hypomethylation in polar bear brains via the LUminometric Methylation Assay: a sensitive method to study epigenetics in wildlife. Mol Ecol 19(2):307–314

Pollard KM, Landberg GP (2001) The in vitro proliferation of murine lymphocytes to mercuric chloride is restricted to mature T cells and is interleukin 1 dependent. Int Immunopharmacol 1(3):581–593

Pollard KM, Reimer G, Tan EM (1989) Autoantibodies in scleroderma. Clin Exp Rheumatol 7(Suppl 3):S57–S62

Pollard KM, Lee DK, Casiano CA, Bluthner M, Johnston MM, Tan EM (1997) The autoimmunity-inducing xenobiotic mercury interacts with the autoantigen fibrillarin and modifies its molecular and antigenic properties. J Immunol 158(7):3521–3528

Pollard KM, Pearson DL, Hultman P, Hildebrandt B, Kono DH (1999) Lupus-prone mice as models to study xenobiotic-induced acceleration of systemic autoimmunity. Environ Health Perspect 107(Suppl 5):729–735

Pollard KM, Pearson DL, Hultman P, Deane TN, Lindh U, Kono DH (2001) Xenobiotic acceleration of idiopathic systemic autoimmunity in lupus-prone bxsb mice. Environ Health Perspect 109(1):27–33

Pollard KM, Arnush M, Hultman P, Kono DH (2004) Costimulation requirements of induced murine systemic autoimmune disease. J Immunol 173(9):5880–5887

Pollard KM, Hultman P, Kono DH (2005) Immunology and genetics of induced systemic autoimmunity. Autoimmun Rev 4(5):282–288

Portanova JP, Arndt RE, Kotzin BL (1988) Selective production of autoantibodies in graft-vs-host-induced and spontaneous murine lupus. Predominant reactivity with histone regions accessible in chromatin. J Immunol 140(3):755–760

Queiroz ML, Dantas DC (1997a) B lymphocytes in mercury-exposed workers. Pharmacol Toxicol 81(3):130–133

Queiroz ML, Dantas DC (1997b) T lymphocytes in mercury-exposed workers. Immunopharmacol Immunotoxicol 19(4):499–510

Queiroz ML, Perlingeiro RC, Dantas DC, Bizzacchi JM, De Capitani EM (1994) Immunoglobulin levels in workers exposed to inorganic mercury. Pharmacol Toxicol 74(2):72–75

Reardon CL, Lucas DO (1987) Heavy-metal mitogenesis: Zn++ and Hg++ induce cellular cytotoxicity and interferon production in murine T lymphocytes. Immunobiology 175(5):455–469

Rioux JD, Abbas AK (2005) Paths to understanding the genetic basis of autoimmune disease. Nature 435(7042):584–589

Rodey GE, Fuller TC (1987) Public epitopes and the antigenic structure of the HLA molecules. Crit Rev Immunol 7(3):229–267

Salonen JT, Seppanen K, Nyyssonen K, Korpela H, Kauhanen J, Kantola M, Tuomilehto J, Esterbauer H, Tatzber F, Salonen R (1995) Intake of mercury from fish, lipid peroxidation, and the risk of myocardial infarction and coronary, cardiovascular, and any death in eastern Finnish men. Circulation 91(3):645–655

Sapin C, Druet E, Druet P (1977) Induction of anti-glomerular basement membrane antibodies in the Brown-Norway rat by mercuric chloride. Clin Exp Immunol 28(1):173–179

Sass JB, Haselow DT, Silbergeld EK (2001) Methylmercury-induced decrement in neuronal migration may involve cytokine-dependent mechanisms: a novel method to assess neuronal movement in vitro. Toxicol Sci 63(1):74–81

Schlawicke Engstrom K, Stromberg U, Lundh T, Johansson I, Vessby B, Hallmans G, Skerfving S, Broberg K (2008) Genetic variation in glutathione-related genes and body burden of methylmercury. Environ Health Perspect 116(6):734–739

Schneider U, Schwenk HU, Bornkamm G (1977) Characterization of EBV-genome negative "null" and "T" cell lines derived from children with acute lymphoblastic leukemia and leukemic transformed non-Hodgkin lymphoma. Int J Cancer 19(5):621–626

Schober SE, Sinks TH, Jones RL, Bolger PM, McDowell M, Osterloh J, Garrett ES, Canady RA, Dillon CF, Sun Y, Joseph CB, Mahaffey KR (2003) Blood mercury levels in US children and women of childbearing age, 1999–2000. JAMA 289(13):1667–1674

Shenker BJ, Berthold P, Decker S, Mayro J, Rooney C, Vitale L, Shapiro IM (1992a) Immunotoxic effects of mercuric compounds on human lymphocytes and monocytes. II. Alterations in cell viability. Immunopharmacol Immunotoxicol 14(3):555–577

Shenker BJ, Rooney C, Vitale L, Shapiro IM (1992b) Immunotoxic effects of mercuric compounds on human lymphocytes and monocytes. I. Suppression of T-cell activation. Immunopharmacol Immunotoxicol 14(3):539–553

Shenker BJ, Berthold P, Rooney C, Vitale L, DeBolt K, Shapiro IM (1993a) Immunotoxic effects of mercuric compounds on human lymphocytes and monocytes. III. Alterations in B-cell function and viability. Immunopharmacol Immunotoxicol 15(1):87–112

Shenker BJ, Mayro JS, Rooney C, Vitale L, Shapiro IM (1993b) Immunotoxic effects of mercuric compounds on human lymphocytes and monocytes. IV. Alterations in cellular glutathione content. Immunopharmacol Immunotoxicol 15(2–3):273–290

Shenker BJ, Datar S, Mansfield K, Shapiro IM (1997) Induction of apoptosis in human T-cells by organomercuric compounds: a flow cytometric analysis. Toxicol Appl Pharmacol 143(2):397–406

Shenker BJ, Guo TL, Shapiro IM (1998) Low-level methylmercury exposure causes human T-cells to undergo apoptosis: evidence of mitochondrial dysfunction. Environ Res 77(2):149–159

Shenker BJ, Guo TL, Shapiro IM (2000) Mercury-induced apoptosis in human lymphoid cells: evidence that the apoptotic pathway is mercurial species dependent. Environ Res 84(2):89–99

Shenker BJ, Pankoski L, Zekavat A, Shapiro IM (2002) Mercury-induced apoptosis in human lymphocytes: caspase activation is linked to redox status. Antioxid Redox Signal 4(3):379–389

Silbergeld EK, Silva IA, Nyland JF (2005) Mercury and autoimmunity: implications for occupational and environmental health. Toxicol Appl Pharmacol 207(2 Suppl):282–292

Silva IA, Nyland JF, Gorman A, Perisse A, Ventura AM, Santos EC, Souza JM, Burek CL, Rose NR, Silbergeld EK (2004) Mercury exposure, malaria, and serum antinuclear/antinucleolar antibodies in Amazon populations in Brazil: a cross-sectional study. Environ Health 3(1):11

Silva IA, Graber J, Nyland JF, Silbergeld EK (2005) In vitro HgCl2 exposure of immune cells at different stages of maturation: effects on phenotype and function. Environ Res 98(3):341–348

Simmons-Willis TA, Koh AS, Clarkson TW, Ballatori N (2002) Transport of a neurotoxicant by molecular mimicry: the methylmercury-L-cysteine complex is a substrate for human L-type large neutral amino acid transporter (LAT) 1 and LAT2. Biochem J 367(Pt 1):239–246

Singh N, Agrawal S, Rastogi AK (1997) Infectious diseases and immunity: special reference to major histocompatibility complex. Emerg Infect Dis 3(1):41–49

Sorensen N, Murata K, Budtz-Jorgensen E, Weihe P, Grandjean P (1999) Prenatal methylmercury exposure as a cardiovascular risk factor at seven years of age. Epidemiology 10(4):370–375

Stiller-Winkler R, Radaszkiewicz T, Gleichmann E (1988) Immunopathological signs in mice treated with mercury compounds – I. Identification by the popliteal lymph node assay of responder and nonresponder strains. Int J Immunopharmacol 10(4):475–484

Suda I, Hirayama K (1992) Degradation of methyl and ethyl mercury into inorganic mercury by hydroxyl radical produced from rat liver microsomes. Arch Toxicol 66(6):398–402

Suda I, Totoki S, Uchida T, Takahashi H (1992) Degradation of methyl and ethyl mercury into inorganic mercury by various phagocytic cells. Arch Toxicol 66(1):40–44

Takahashi Y, Tsuruta S, Arimoto M, Tanaka H, Yoshida M (2003) Placental transfer of mercury in pregnant rats which received dental amalgam restorations. Toxicology 185(1–2):23–33

Tanaka T, Naganuma A, Imura N (1990) Role of gamma-glutamyltranspeptidase in renal uptake and toxicity of inorganic mercury in mice. Toxicology 60(3):187–198

Tanaka-Kagawa T, Naganuma A, Imura N (1993) Tubular secretion and reabsorption of mercury compounds in mouse kidney. J Pharmacol Exp Ther 264(2):776–782

Taylor H, Appleton JD, Lister R, Smith B, Chitamweba D, Mkumbo O, Machiwa JF, Tesha AL, Beinhoff C (2005) Environmental assessment of mercury contamination from the Rwamagasa artisanal gold mining centre, Geita District, Tanzania. Sci Total Environ 343(1–3):111–133

Thompson SA, Roellich KL, Grossmann A, Gilbert SG, Kavanagh TJ (1998) Alterations in immune parameters associated with low level methylmercury exposure in mice. Immunopharmacol Immunotoxicol 20(2):299–314

Thurston SW, Bovet P, Myers GJ, Davidson PW, Georger LA, Shamlaye C, Clarkson TW (2007) Does prenatal methylmercury exposure from fish consumption affect blood pressure in childhood? Neurotoxicology 28(5):924–930

Tubbs RR, Gephardt GN, McMahon JT, Pohl MC, Vidt DG, Barenberg SA, Valenzuela R (1982) Membranous glomerulonephritis associated with industrial mercury exposure. Study of pathogenetic mechanisms. Am J Clin Pathol 77(4):409–413

Turk JL, Baker H (1968) Nephrotic syndrome due to ammoniated mercury. Br J Dermatol 80(9):623–624

UNEP (2002) Global mercury assessment. United Nations Environment Programme, Geneva

Vahter M, Mottet NK, Friberg L, Lind B, Shen DD, Burbacher T (1994) Speciation of mercury in the primate blood and brain following long-term exposure to methyl mercury. Toxicol Appl Pharmacol 124(2):221–229

Van Vleet TR, Schnellmann RG (2003) Toxic nephropathy: environmental chemicals. Semin Nephrol 23(5):500–508

Via CS, Shearer GM (1988) Murine graft-versus-host disease as a model for the development of autoimmunity. Relevance of cytotoxic T lymphocytes. Ann N Y Acad Sci 532:44–50

Via CS, Nguyen P, Niculescu F, Papadimitriou J, Hoover D, Silbergeld EK (2003) Low-dose exposure to inorganic mercury accelerates disease and mortality in acquired murine lupus. Environ Health Perspect 111(10):1273–1277

Virtanen JK, Voutilainen S, Rissanen TH, Mursu J, Tuomainen TP, Korhonen MJ, Valkonen VP, Seppanen K, Laukkanen JA, Salonen JT (2005) Mercury, fish oils, and risk of acute coronary events and cardiovascular disease, coronary heart disease, and all-cause mortality in men in eastern Finland. Arterioscler Thromb Vasc Biol 25(1):228–233

Vupputuri S, Longnecker MP, Daniels JL, Guo X, Sandler DP (2005) Blood mercury level and blood pressure among US women: results from the National Health and Nutrition Examination Survey 1999–2000. Environ Res 97(2):195–200

Walker SJ, Segal J, Aschner M (2006) Cultured lymphocytes from autistic children and non-autistic siblings up-regulate heat shock protein RNA in response to thimerosal challenge. Neurotoxicology 27(5):685–692

Warkany J (1966) Acrodynia – postmortem of a disease. Am J Dis Child 112(2):147–156

Warkany J, Hubbard DM (1953) Acrodynia and mercury. J Pediatr 42(3):365–386

Watras CJ, Back RC, Halvorsen S, Hudson RJ, Morrison KA, Wente SP (1998) Bioaccumulation of mercury in pelagic freshwater food webs. Sci Total Environ 219(2–3):183–208

Wei H, Qiu L, Divine KK, Ashbaugh MD, McIntyre LC Jr, Fernando Q, Gandolfi AJ (1999) Toxicity and transport of three synthesized mercury-thiol-complexes in isolated rabbit renal proximal tubule suspensions. Drug Chem Toxicol 22(2):323–341

Weldon MM, Smolinski MS, Maroufi A, Hasty BW, Gilliss DL, Boulanger LL, Balluz LS, Dutton RJ (2000) Mercury poisoning associated with a Mexican beauty cream. West J Med 173(1):15–18; discussion 19

Whitekus MJ, Santini RP, Rosenspire AJ, McCabe MJ Jr (1999) Protection against CD95-mediated apoptosis by inorganic mercury in Jurkat T cells. J Immunol 162(12):7162–7170

WHO (1990) Environmental health criteria 101. Methylmercury. International Program on Chemical Safety, World Health Organization, Geneva

WHO (1991) Environmental health criteria 118. Inorganic mercury. International Programme on Chemical Safety, World Health Organization, Geneva

Zalups RK, Barfuss DW, Lash LH (1999) Relationships between alterations in glutathione metabolism and the disposition of inorganic mercury in rats: effects of biliary ligation and chemically induced modulation of glutathione status. Chem Biol Interact 123(3):171–195

Zdolsek JM, Soder O, Hultman P (1994) Mercury induces in vivo and in vitro secretion of interleukin-1 in mice. Immunopharmacology 28(3):201–208

Zhang H, Feng X, Larssen T, Qiu G, Vogt RD (2010) In inland China, rice, rather than fish, is the major pathway for methylmercury exposure. Environ Health Perspect 118(9):1183–1188

Zhou ZS, Zeng HQ, Liu ZP, Yang ZM (2012) Genome-wide identification of Medicago truncatula microRNAs and their targets reveals their differential regulation by heavy metal. Plant Cell Environ 35(1):86–99

Ziemba SE, McCabe MJ Jr, Rosenspire AJ (2005) Inorganic mercury dissociates preassembled Fas/CD95 receptor oligomers in T lymphocytes. Toxicol Appl Pharmacol 206(3):334–342

Part III

Challenges

Environment and Autoimmunity: Facts and Gaps

13

Angela Ceribelli, Elena Generali, and Carlo Selmi

Contents

13.1 Introduction

The etiology of autoimmune diseases remains unknown; even if many studies have investigated it in the past and are still exploring it, until now no unique genetic or environmental risk factor has been identified to be responsible for the onset of

A. Ceribelli
Division of Rheumatology and Clinical Immunology, Humanitas Research Hospital, Rozzano, Milan, Italy

BIOMETRA Department, University of Milan, Milan, Italy

E. Generali
Division of Rheumatology and Clinical Immunology, Humanitas Research Hospital, Rozzano, Milan, Italy

C. Selmi, MD, PhD (✉)
Division of Rheumatology and Clinical Immunology, Humanitas Clinical and Research Center, via A. Manzoni 56, Rozzano, Milan 20089, Italy

BIOMETRA Department, University of Milan, Milan, Italy
e-mail: carlo.selmi@unimi.it

C. Esser (ed.), *Environmental Influences on the Immune System*,
DOI 10.1007/978-3-7091-1890-0_13

autoimmune diseases. We now consider autoimmune diseases to be induced by multiple factors, genetic and environmental, and this is well supported by the fact that no direct genetic transmission of the same autoimmune disease has yet been found in a familiar group, even if different expressions of autoimmunity (organ specific or systemic) are found in families with affected individuals. An important observation comes from the incomplete concordance rate for autoimmune diseases in couples of monozygotic twins, in which only one twin expresses the rheumatic disease, despite the identical genomic sequences (Miller et al. 2012b; Selmi et al. 2012a). Twin studies are indicated in the study of autoimmune diseases as these allow also the possibility to calculate the genetic heritability of each condition, i.e., the proportion of observable differences in a phenotype between individuals that is due to genetic differences (Bogdanos et al. 2012; Christensen et al. 2011; De Santis and Selmi 2012). Since heritability is a proportion, its value ranges from 0 (when genes do not contribute to phenotypic differences) to 1 (when the environment does not contribute to phenotypic differences). This estimate depends on numerous variables, including the prevalence of the disease in the general population, and does not reflect the risk of getting a disease but the variance between twins. The term "heritability" defines the proportion of the phenotypic variance attributable to genetics (A + D), while the less known term "environmentability" represents the proportion of phenotypic variance attributable to environmental features (or 1-heritability). Of particular importance in the discussion of environmental factors is the fact that "environmentability" includes environmental influences that are the sum of common/shared environmental factors and individual environmental variance, while also including the differences due to measurement errors or observation bias. In specific autoimmune diseases, genetic heritability estimates are summarized in Table 13.1 and demonstrate that the weight of genetic influences is mostly observed in some conditions (as for Crohn's disease or ankylosing spondylitis) while being almost negligible in others (as for systemic sclerosis) (Selmi et al. 2004, 2011). To tackle this complex view in which causality is difficult to prove, the National Institute of Environmental Health Sciences (NIEHS) organized an expert workshop in September 2010 to provide a consensus on the definition of environmentally induced autoimmune diseases (Parks et al. 2014). The results of the expert panel were reported in three comprehensive articles on the mechanisms, epidemiology, and animal models, while a consensus statement was later reported (Miller et al. 2012a, b; Parks et al. 2014; Selmi et al. 2012a). The Delphi exercise results are summarized for immune or other mechanisms (Table 13.2), animal studies (Table 13.3), and factor-specific mechanisms (Table 13.4).

Based on these observations, it is now commonly accepted that multiple genetic predisposition factors must interact with epigenetic and environmental triggers to induce the clinical expression of autoimmune diseases, such as rheumatic ones (Ehrenfeld 2010). This could explain why some diseases develop mainly in some geographic areas (usually industrial zones), and they seem to have a seasonal concentration, maybe related to viral infections, as in the case of dermatomyositis (Tanaka and Takikawa 2013; Invernizzi 2010; Shapira et al. 2010; Chandran and

Table 13.1 Genetic heritability based on twin concordance rates for specific autoimmune diseases (Selmi et al. 2012b)

	Genetic heritability	Reference
Acute rheumatic fever	0.60 (0.41–0.81)	Engel et al. (2011)
Ankylosing spondylitis	0.97 (0.92–0.99)	Brown et al. (1997)
Celiac disease	0.57 (0.32–0.93) *if 1/1000 prevalence* 0.87 (0.49–1.00) *if 1/91 prevalence*	Nistico et al. (2006)
Crohn's disease	1.00 (0.34–1.00) 0.55 (*)	Tysk et al. (1988) So et al. (2011)
Multiple sclerosis	0.25 (0–0.88) 0.76 (0.33–0.88)	Hawkes and Macgregor (2009)
Psoriasis	0.66 (0.52–0.77)	Grjibovski et al. (2007)
Psoriatic arthritis	0.65 (0.22–1.00)	Pedersen et al. (2008)
Rheumatoid arthritis	0.68 (0.55–0.79) *ACPA positive* 0.66 (0.21–0.82) *ACPA negative*	van der Woude et al. (2009)
Sarcoidosis	0.66 (0.52–0.80)	Sverrild et al. (2008)
Systemic lupus erythematosus	0.66 (*)	So et al. (2011)
Systemic sclerosis	0.008	Feghali-Bostwick et al. (2003)
Type 1 diabetes	0.88 (0.78–0.94) 0.80 (*)	Hyttinen et al. (2003) So et al. (2011)

* prospective longitudinal study
When available, 95 % confidence intervals are provided

Raychaudhuri 2010; Zeki et al. 2010; Moroni et al. 2012; Selmi and Tsuneyama 2010; Prieto and Grau 2010). Despite the increasing interest about environmental risk factors, the recommendations and the ongoing growth of pathophysiological data about autoimmune diseases, there are still numerous gaps in our knowledge. The aim of this chapter is to describe the environmental factors and the general and specific mechanisms that are considered to play a role in the onset of autoimmune diseases.

13.2 Environment and Autoimmune Diseases: The Role of Chemicals, Xenobiotics, and Adjuvants

Every day an increasing number of factors responsible for autoimmune diseases are under evaluation, and in particular growing evidence for a pathogenic role is associated to chemicals, xenobiotics, and adjuvants (Chandran and Raychaudhuri 2010; Costenbader et al. 2012; Hemminki et al. 2010; Miller et al. 2012a, b). In the case of chemicals, some studies have demonstrated that aryl hydrocarbon receptor (AhR) can induce Th17 cell to differentiate with their activity exacerbating autoimmune diseases in animal models (Veldhoen et al. 2008), while others have shown that AhR promotes the expansion of regulatory T-cell (Treg) populations, decreases Th17

Table 13.2 The 2010 NIEHS expert workshop and panel findings on mechanisms involved in the role of environmental factors and development of autoimmune disease (Parks et al. 2014)

We are confident of the following	We consider the following likely, but requiring confirmation	Broad themes to be pursued in future investigations
B cells		
Dysfunctions of B-cell tolerance checkpoints are directly correlated with autoimmune disease in murine models B cells modulate autoimmunity positively and negatively as secretors of antibodies and inflammatory cytokines, as antigen-presenting cells to autoreactive T cells, and secretors of anti-inflammatory cytokines such as IL-10 Follicular B cells (B2) are a major source of autoreactive pathogenic antibodies B cells secreting pathogenic autoantibodies can emerge when somatic hypermutation occurs outside of germinal centers Sex hormones like estrogen and prolactin can differentially activate autoreactive B-cell populations from different subsets (e.g., B2)	B1 cells and marginal zone B cells can modulate autoimmunity by exacerbating it through secretion of autoreactive antibodies and/or by down-modulating it through secretion of anti-inflammatory cytokines B10 cells secrete IL-10 may be functionally specialized to carry out a negative regulatory role in inflammation and autoimmunity	The roles of B1 and marginal zone B cells in autoimmunity The role of the recently discovered B10 cell population in autoimmunity The survival/apoptotic pathways that when deregulated lead to expansion and survival of autoreactive B cells (such as the BAFF/BlyS receptor system and CD40) Tolerance checkpoint mechanisms regulating the formation of high-affinity autoreactive B2 cells both in and outside the germinal center Environmental agents with the potential to disrupt B-cell function
T-helper 17 (TH17) cells		
Deregulated Th17 cell activity can lead to pathology, as in chronic inflammatory diseases such as asthma or inflammatory bowel disease Th17 cells are involved in MS, RA, Crohn's disease, and psoriasis, where they seem to be involved in disease development and relapse	Smoking is an important risk factor for RA; and nicotine exerts effects via Th17 cells Aryl-hydrocarbon receptor (AhR) binding by aromatic hydrocarbons and non-halogenated polycyclic aromatic hydrocarbons favors differentiation of Th17 cells and can exacerbate autoimmunity	The involvement of environmental agents and exacerbation of autoimmune disease through Th17 cells Therapeutic modulation of Th17 cells

Innate immunity		
1. The interaction between xenobiotics and TLR is a major mechanism involved in the interaction of environmental factors with autoimmunity development 2. Innate immune activation TLR predisposes to toxic-induced inflammation 3. Adjuvants activate both innate and adaptive immunities, inducing the release of chemokines and inflammatory cytokines 4. Immunization must be accompanied by a strong adjuvant, such as complete Freund's adjuvant, including the mycobacterium component 5. Incomplete Freund's adjuvant results in production of antibodies, but without the occurrence of autoimmune diseases	Altered innate immune responses and deregulated TLR signaling are a key step in triggering autoimmune diseases, as in virus-induced animal models of type I diabetes TLR activation in macrophages may predispose cells to toxin-induced inflammatory cytokine production Active infection or microbial products of infection can provide the adjuvant effect necessary for the induction of many autoimmune disorders	Allergenicity, functional mimicry of environmental contaminants, and physical/chemical elements resembling TLR ligands Deregulation of the regulatory B cell (IL-10 producing, CD5+ B cells) through modulation of TLR signaling Molecular motifs of adjuvants and their physiological receptors that are associated with clinical manifestation of autoimmunity Genomic predisposition to innate immunity dysfunctions
T regulatory (Treg) cells		
Quantitative and qualitative Treg changes are culprit for tolerance breakdown The AhR ligand dioxin TCDD induces immunosuppressive T cells expressing specific Treg markers AhR ligands also affect skewing of the T-cell repertoire toward Treg cells indirectly via antigen-presenting cells TCDD induces IDO transcription to skew the T-cell repertoire toward FoxP3+ Tregs Activation of PPAR promotes Treg induction from naive cells	Most studies suggest that AhR activation in T cells or in antigen-presenting cells may increase Treg production and therefore decrease autoimmunity, but the opposite outcome is also likely and possibly ligand specific Context-specific activation of the AhR by specific ligands may result in either increased or decreased Treg activity Sex hormones play an important role in Treg development and may underlie female predominance of autoimmune diseases	Specific chemical, infectious, or physical agents capable of modulating Tregs Environmental modulators of AhR stimulation Mechanisms of sex-specific Treg changes

(continued)

Table 13.2 (continued)

We are confident of the following	We consider the following likely, but requiring confirmation	Broad themes to be pursued in future investigations
Modification of self-antigens		
The majority of human proteins undergo posttranslational modification (PTM), and these modifications or lack thereof may lead to tolerance breakdown PTM may explain the tissue specificity of autoimmune diseases MS pathogenesis includes PTM that increase the complexity of myelin proteins through the autoimmune response or neurodegenerative processes In RA, citrullination is an apoptotic PTM that seems to be helpful in opening protein conformation and favoring cleavage processes In PBC, cholangiocytes do not covalently link glutathione to lysine-lipoyl groups during apoptosis leading to accumulation and exposure to potentially self-reactive antigens, accounting for bile duct-specific pathology	Multiple self-protein modifications (phosphorylation, glycosylation, acetylation, deamidation) can lead to either T- or B-cell responses to self-antigens Serum autoantibodies to modified self-antigens may bind either modified or unmodified forms and thus be crucial to effector immune reaction in target tissues Mercury-induced cell death results in the formation of a unique and more immunogenic 19 kDa cleavage fragment of fibrillarin	Mechanisms by which citrullination and glutathionylation lead to tolerance breakdown in susceptible individuals The role of glucosylation in MS and other autoimmune diseases Experimental models to prove that autoantigens can be modified to increase their immunogenicity Technologies to reverse or induce PTM in animal models of autoimmunity
Modification of DNA methylation		
DNA methylation profiles are associated with environmental factors including prenatal tobacco smoke, alcohol, and environmental pollutants The importance of DNA methylation in regulating immune function is suggested by two rare congenital diseases, Silver-Russell and Beckwith-Wiedemann syndromes Changes in DNA methylation in specific peripheral immune cell types are associated with autoimmune diseases	Phenotypic differences are increased with age in twins in a trend coined as "epigenetic drift," due to different environmental exposures, and may explain late-onset autoimmunity Specific impairments in epigenetic regulation in immune cells may be responsible for immune-tolerance breakdown through hypo-methylation of genes or involvement of transcription repressors Recent genome-wide association studies demonstrate that genomics significantly predispose to SLE onset, but experimental studies indicate that epigenetic mechanisms, especially impaired T- and B-cell DNA methylation, may be one of these factors	The functional effects in vivo of DNA methylation changes under different environmental and genomic conditions The development of new therapeutic molecules capable to prevent or counteract DNA methylation changes in a cell-specific manner The DNA methylation changes in the target cells and not only in the rapidly accessible effector immune cells

Table 13.3 The 2010 NIEHS expert workshop and panel finding studies of animal models in the role of environmental factors and development of autoimmune disease (Parks et al. 2014)

We are confident of the following	We consider the following likely, but requiring confirmation	Broad themes to be pursued in future investigations
Chemical factors		
1. Autoimmune responses are influenced by species and strain of animal model 2. There is no preferred species or strain, but studies with rats and mice predominate 3. Forms of inorganic mercury (HgCl2, vapor, amalgam) induce systemic autoimmune disease in rats (transient) and mice 4. Inorganic mercury (HgCl2) exacerbates or accelerates systemic autoimmune disease in lupus-prone mice 5. Gold causes (transient) nephropathy in rats 6. Several mineral oil components and certain other hydrocarbons can induced an acute inflammatory arthritis in some rat strains 7. The mineral oil component 2,6,10,14-tetra-methylpentadecane (TMPD or pristane) induces lupus-like disease and inflammatory arthritis in several strains of mice 8. 2,3,7,8-Tetrachlorodibenzo-p-dioxin (TCDD) suppresses models of autoimmune disease in adult mice	Gold and silver cause autoimmune responses in mice; effects of other metals (e.g., organic mercury, cadmium, lead, arsenic) require additional studies Silica/asbestos (at lower doses) stimulate autoimmune disease, but more studies are needed using more species/strains and a wider range of doses and exposure routes High-dose silica (≥100 mg) suppresses organ-specific autoimmune disease, but mechanisms are not known Trichloroethylene (TCE) exacerbates systemic autoimmunity although responses are often limited and transient. Studies of autoimmune liver disease are needed with additional species/strains and in developmental studies Dimethyl sulfoxide (DMSO) reduces some aspects of autoimmunity, but effectiveness depends on timing and mode of exposure. Mechanisms and side effects are not known Hydrocarbons may induce lupus-like disease in mice or inflammatory arthritis in rats Silicon may enhance autoimmunity in strains of mice susceptible to arthritis or lupus PBC-like disease can be induced by immunization with xenobiotically modified lipoyl hapten-carrier conjugates TCDD exposure during fetal or early neonatal development may promote autoimmunity Hexachlorobenzene (an organochlorine pesticide) suppresses adjuvant arthritis and exacerbates EAE Organochlorine pesticides may enhance lupus-like disease in a predisposed mouse strain	Epidemiological and clinical studies should be "shaped by what is observed in humans, not by what is possible in mice" Studies should not be restricted to a "gold standard" animal model Multiple models should be investigated to reflect human genetic heterogeneity When using spontaneous disease models, it is important to consider whether environmental exposures exacerbate/accelerate idiopathic autoimmunity or reflect environmental factor-specific autoimmunity Genetic and biological markers should be sought in easily obtained biological fluids to enhance comparison with human studies Biomarkers of stress exposure are needed to identify contributing or confounding roles of stress in autoimmune disease outcomes Determining whether an environmental agent or compound affects autoimmune diseases should include screening autoimmune-prone and non-autoimmune species and strains If necessary, animal models should be tailored (e.g., humanized) to verify findings from human epidemiological data of suspected modifiers of autoimmune disease More studies on the effects of environmental factor exposure on expression of autoimmunity during different stages of life (gestational to adulthood) are needed

(continued)

Table 13.3 (continued)

We are confident of the following	We consider the following likely, but requiring confirmation	Broad themes to be pursued in future investigations
Physical factors		
Sunlight/UV exposure exacerbates lupus in genetically prone mice Emotional stress (e.g., noise) can modify disease incidence, onset, or severity	Stressful life events interact with other risk factors (e.g., chemicals, infections), thereby confounding indices of autoimmune diseases Common underlying mechanisms of chemical, biological, and physical stressor effects on autoimmune disease are likely and may provide a better framework for predicting the role of oxidative stress and cytotoxicity and cell clearance in autoimmune disease	
Biological factors		
For a limited number of pathogens (i.e., Streptococcal group A, coxsackie B virus), there is a clear association with development of autoimmune diseases Caloric restriction in adults protects against a wide range of autoimmune diseases Restricting various dietary components, such as protein and especially fats, can be more protective than limiting calories Excess iodine increases the incidence of autoimmune thyroiditis in genetically predisposed animal models	For many pathogens, evidence from animal models suggests associations with specific autoimmune diseases, but corroborative human data is not always available Data on their mode of action and strong evidence in animal models suggest at least some vaccine additives are capable of inducing autoimmunity Infections may protect (hygiene hypothesis) against some types of autoimmune diseases, as animal models show reduced exposure to infections increases autoimmune disease risk Some dietary and nutritional supplements (vitamin D and antioxidants) protect against specific autoimmune diseases	

Table 13.4 The 2010 NIEHS expert workshop and panel findings on mechanisms involved in the role of environmental factors and development of autoimmune disease (Parks et al. 2014)

We are confident of the following	We consider the following likely, but requiring confirmation	Broad themes to be pursued in future investigations
Chemicals		
Crystalline silica (quartz) contributes to the development of several systemic autoimmune diseases, including RA, SSc, SLE, and antineutrophil cytoplasmic antibody (ANCA)-related vasculitis Solvents contribute to the development SSc Smoking contributes to the development of ACPA-positive and antirheumatoid factor (RF)-positive RA (with an interaction with the shared epitope genetic susceptibility factor)	Solvents contribute to the development of MS Smoking contributes to the development of seronegative rheumatoid arthritis (RA), multiple sclerosis (MS), systemic lupus, Hashimoto's (HT) thyroiditis, Graves' disease (GD), Crohn's disease (CD) Current smoking protects against the development of ulcerative colitis	There is insufficient evidence on the role of metals, including those associated with animal models of autoimmunity, e.g., mercury The identification of single causal agents within groups of exposures is needed (e.g., specific solvents or pesticides contributing to increased risk for the group) Studies are needed on plasticizers (e.g., phthalates and bisphenol A), some of which may be endocrine or immune disruptors, and have been associated with other immune-mediated diseases There is insufficient evidence on the role of cosmetics in autoimmune diseases
Physical factors		
An inverse association exists between increased ultraviolet radiation exposure and risk of developing MS	Ionizing radiation contributes to the development of HT and GD	There is insufficient evidence on a possible protective role of ultraviolet radiation on TID Prospective data are needed on sun exposure as a risk factor for SLE (prior to early clinical symptoms) and dermatomyositis
Biological agents		
Ingestion of gluten contributes to the development of gluten-sensitive enteropathy (GSE) Ingestion of certain lots of L-tryptophan contributes to the development of eosinophilia-myalgia syndrome Dietary intake of 1,2-di-oleyl ester (DEPAP) and oleic anilide-contaminated rapeseed oil contributes to the development of toxic oil syndrome	EBV infection contributes to MS development Early introduction of complex foods contributes to the development of T1D and GSE Low dietary vitamin D intake and blood levels contribute to the development of MS	Studies are needed on MS and vitamin D in racial/ethnic groups with darker skin (associated with UV-associated vitamin D deficiency) and examining dose effects Prospective data are needed on vitamin D and other autoimmune diseases Additional studies are needed on associations of food chemicals, dyes, or additives Prospective studies are needed on nitrates/nitrosamines and T1D

frequency, and limits the clinical symptoms of disease (Quintana et al. 2008). Mechanisms linking nicotine, inflammation, and interleukin-17 (IL-17) production were studied in a rat adjuvant-induced arthritis model of human rheumatoid arthritis (RA). In this model, nicotine pretreatment aggravates arthritis increasing interferon (IFN) and IL-17 production, whereas posttreatment nicotine suppressed the disease (Yu et al. 2011).

Moreover, toxicological factors, such as silica, have been recently considered as triggering factors for autoimmune diseases based on both epidemiological and experimental evidences. The observation of a possible link between chemicals and disease onset is based on epidemiological studies that associate chemical exposure with biological markers of autoimmunity and also by laboratory studies that identify plausible biological mechanisms through which environmental agents can influence autoimmunity (Miller et al. 2012a, b). As for the role of xenobiotics in autoimmune diseases, recent works showed that the activation of toll-like receptors (TLR) in macrophages predisposes the cells to produce toxin-induced inflammatory cytokines. In fact, co-exposure to nickel and TLR2 agonists induces the release of IL-6 by lung fibroblasts in a protein kinase-dependent pathway (Gao et al. 2010). Because TLR signaling and IL-6 production are key elements in autoimmune diseases, they may be involved in the onset and perpetuation of an autoimmune response in the long term. In mouse models undergoing mercury exposure later developing an autoimmune disease, lipopolysaccharide (LPS) exposure has been shown to trigger the disease onset. Similarly, studies performed in vitro on human peripheral blood mononuclear cells stimulated with mercuric chloride showed that proinflammatory cytokines were induced only when the cells were co-exposed to LPS (Gardner et al. 2009). Similarly to chemicals and xenobiotics, adjuvants can stimulate the immune system even without having an antigenic effect per se. Adjuvants such as pristane, squalene, and mineral oil are capable of activating the immune response and can induce the release of chemokines and proinflammatory cytokines. Some studies, mainly performed on animal models, showed that adjuvants can activate the innate immune system mainly by binding to TLR and inducing dendritic cells or macrophage function, and moreover, they can modulate the release of chemokines and the recruitment of immune cells. Based on these data, we assume that environmental adjuvants can stimulate the innate immune response and then activate the adaptive response, which finally leads to chronic arthritis and production of autoantibodies such as lupus-related anti-Sm/RNP or Su antibodies (Rose 2008; Meroni 2011). Beside adjuvants that can be traced because of their use in experimental models, no specific autoimmune-associated biomarker of adjuvant exposure is currently used to identify and design studies of specific biomarkers of adjuvant exposure.

Another mechanism through which dietary components and environmental toxins can influence autoimmune disease onset is through the modification of the Th17 response in susceptible individuals. This has been evaluated and supported by scientific evidence in autoimmune diseases such as RA, Crohn's disease, and psoriasis, where Th17 cells seem to be involved in the development and in the relapse of the diseases (Di Cesare et al. 2009; Sarkar and Fox 2010; Segal 2010). In the last decade, several reports showed that vitamin A and vitamin D can exert an immune modulatory effect by controlling the Th17 and Treg balance, and also they play a hormonal role linked not only to bone health but also to immune homeostasis (Quintana and Weiner 2009).

Several studies have indicated that mercury-induced cell death leads to the formation of a unique 19 kDa cleavage fragment of fibrillarin, which cannot be detected in cells died from other causes. These mercury modifications of fibrillarin appear to increase its immunogenicity, and it is unclear whether this process is limited to fibrillarin itself or whether nontargeted cellular proteins are left intact following mercury exposure (Havarinasab and Hultman 2005; Pollard et al. 1997).

Recent reports suggest that environmental chemicals could influence the autoimmune response through the alteration of Treg production or function mediated by multiple intracellular receptors. But new studies are necessary to analyze the assumption that exposure to environmental chemicals, capable of modulating intracellular receptor signaling, is associated with the risk or severity of autoimmune diseases in humans.

13.3 Environment and Autoimmune Diseases: The Role of Physical Elements

Among many environmental factors that can play a role in autoimmune diseases, receptor-independent stressor-mediated environmental effects have been reported. Among these factors, ultraviolet B (UVB) light is able to induce Treg cell differentiation to produce IL-10 for antigen-specific immunosuppression (Maeda et al. 2008; Shintani et al. 2008). This process mediated by UV light may represent an immunosuppressive response to UV-mediated epithelial cell death and autoantigen presentation by Langerhans cells (Lehmann and Homey 2009). Ionizing radiations are also likely to contribute to the onset of autoimmune diseases, especially thyroid diseases, such as Hashimoto thyroiditis and Graves' disease, even if research bias (i.e., few studies of medical radiation therapy and inconsistency in findings from nuclear testing fallout and accidental radiation contamination) is still present. Another important field of investigation that deserves further development is the role of UV exposure as risk factor for dermatomyositis and systemic lupus erythematosus (SLE), as suggested by prospectively collected and preclinical data on sun sensitivity.

13.4 Environment and Autoimmune Disease: The Role of Infectious Agents

Infectious agents are one of the most studied environmental factors in the etiopathogenesis of autoimmune diseases. A well-known example is represented by rheumatic fever, caused by streptococcal antigens that can induce systemic symptoms such as fever, arthritis, and also heart disease (Malkiel et al. 2000). Other infections, mainly mediated by viruses such as Epstein-Barr virus (EBV), are linked to the development of autoimmune diseases such as SLE (Barzilai et al. 2007) but also RA (Balandraud et al. 2004), and Sjögren's syndrome (Padalko and Bossuyt 2001). Besides EBV, a potential role for cytomegalovirus (CMV) in the development of SLE has been suggested (Su et al. 2007), similarly to antiphospholipid syndrome (Blank et al. 2004; Blank and Shoenfeld 2004; Shoenfeld and Blank 2004).

The mainly accepted hypothesis assumes that genetically predisposed individuals with a normal immune system who develop a viral infection (probably with other concomitant environmental factors) activate autoimmunity through the action of viral superantigens, molecular mimicry, polyclonal activation, epitope spreading, and bystander activation (Barzilai et al. 2007). This uncontrolled activation leads to autoimmunity, which we can detect through sera autoantibodies, even before the onset of the clinical symptoms of the disease. Whether these are stochastic associations or significant pathogenetic links remains to be clarified in most cases; however, the time lapse between induction of infection and clinical manifestations will make a direct proof poorly feasible in humans, maybe in animal models.

13.5 Conclusions

The data reported in this chapter show that multiple agents are under investigation for their capacity to lead to multiple modifications (i.e., phosphorylation, glycosylation, acetylation, deamidation) in response to self-antigens, with the breakdown of tolerance and the onset of autoimmune reactions and diseases. Beside cellular mechanisms of innate and adaptive immunity, also autoantibodies to modified self-antigens can be crucial to the effector immune reaction against target tissues, as well represented in RA.

The lack of epidemiological data reflects that main difficulties and gaps are still present in this field, such as quantifying the exposure to environmental chemicals and analyzing multiple cellular subsets and their functions in human populations. A few recommendations for future research include (i) analysis of multiple chemical mixtures, reflecting the real-life complexity of human life in contact with the environment; (ii) exposure-related risks within specific disease phenotypes and in the context of genetic risk factors, such as the smoking associated with RA-defined anti-citrullinated peptide antibody positivity; and (iii) the definition of critical "windows of opportunity" in the timing of exposure and latency relative to age/developmental stage, understanding dose-response relationship, and identifying mechanisms that lead to autoimmunity.

More "translational" epidemiological studies of environmental autoimmunity are needed and should be guided by mechanisms defined in animal model systems and vice versa. An integrated, multidisciplinary approach is critical, and programs should be established to provide opportunities for collaboration and improve communication between epidemiologists, exposure scientists, and basic cellular/molecular biologists, i.e., fostering of interdisciplinary research through forums, funding, and training. Moreover, funding opportunities need to be specifically addressed toward autoimmunity and environmental factor research studies. In fact, a better coordination across the different disciplines and agencies conducting autoimmune research may help to encourage collaborations. Such coordinated efforts may also promote a more cohesive body of knowledge through studies of multiple autoimmune diseases with similar underlying mechanisms and shared genetic or environmental risk factors.

An important need for human autoimmune research is, for example, the availability of high-quality, validated measurement tools. As to the efforts to characterize the genome, new technologies should be harnessed to address the critical need to characterize human environmental exposures. An environment-wide association (exposure/EWAS) study database (complementing PHENX) would facilitate future epidemiological studies. More data are also needed to establish the contribution of psychosocial factors, infections, complex mixtures, and susceptibility factors to the development and severity of autoimmune diseases. Biomarkers identified by mechanistic studies should be applied to epidemiological research in the context of relevant exposure measures. Investments in high-quality exposure measures and biological markers will increase the ability to identify environmental contributions to the etiopathogenesis of autoimmune diseases.

Finally, a consensus-based approach should be established to define autoimmune phenotypes (rather than diseases), which may improve comparability between human studies and animal models. The focus on studying diseases defined by classification criteria may limit interpretation of animal model data and the ability to identify human exposure cohorts using the broadest disease definitions. Conversely, there is a need for animal models to better represent phenotypes that occur in human diseases (e.g., CNS-SLE). Some environmental exposures may cause diseases characterized by a mixture of outcomes or multiple phenotypes that do not fit the standard diagnostic criteria. Outbreak investigations should collect data to characterize the emerging phenotypes and include the preservation and archiving of biological specimens. Long-term follow-up of affected individuals is critical to assess phenotypes that might develop with long latency.

References

Balandraud N, Roudier J, Roudier C (2004) Epstein-Barr virus and rheumatoid arthritis. Autoimmun Rev 3:362–367

Barzilai O, Ram M, Shoenfeld Y (2007) Viral infection can induce the production of autoantibodies. Curr Opin Rheumatol 19:636–643

Blank M, Shoenfeld Y (2004) Beta-2-glycoprotein-I, infections, antiphospholipid syndrome and therapeutic considerations. Clin Immunol 112:190–199

Blank M, Asherson RA, Cervera R, Shoenfeld Y (2004) Antiphospholipid syndrome infectious origin. J Clin Immunol 24:12–23

Bogdanos DP, Smyk DS, Rigopoulou EI, Mytilinaiou MG, Heneghan MA, Selmi C, Gershwin ME (2012) Twin studies in autoimmune disease: genetics, gender and environment. J Autoimmun 38:J156–J169

Brown MA, Kennedy LG, MacGregor AJ, Darke C, Duncan E, Shatford JL, Taylor A, Calin A, Wordsworth P (1997) Susceptibility to ankylosing spondylitis in twins: the role of genes, HLA, and the environment. Arthritis Rheum 40:1823–1828

Chandran V, Raychaudhuri SP (2010) Geoepidemiology and environmental factors of psoriasis and psoriatic arthritis. J Autoimmun 34:J314–J321

Christensen K, Kyvik KO, Holm NV, Skytthe A (2011) Register-based research on twins. Scand J Public Health 39:185–190

Costenbader KH, Gay S, Alarcon-Riquelme ME, Iaccarino L, Doria A (2012) Genes, epigenetic regulation and environmental factors: which is the most relevant in developing autoimmune diseases? Autoimmun Rev 11:604–609

De Santis M, Selmi C (2012) The therapeutic potential of epigenetics in autoimmune diseases. Clin Rev Allergy Immunol 42:92–101

Di Cesare A, Di Meglio P, Nestle FO (2009) The IL-23/Th17 axis in the immunopathogenesis of psoriasis. J Invest Dermatol 129:1339–1350

Ehrenfeld M (2010) Geoepidemiology: the environment and spondyloarthropathies. Autoimmun Rev 9:A325–A329

Engel ME, Stander R, Vogel J, Adeyemo AA, Mayosi BM (2011) Genetic susceptibility to acute rheumatic fever: a systematic review and meta-analysis of twin studies. PLoS One 6:e25326

Feghali-Bostwick C, Medsger TA Jr, Wright TM (2003) Analysis of systemic sclerosis in twins reveals low concordance for disease and high concordance for the presence of antinuclear antibodies. Arthritis Rheum 48:1956–1963

Gao F, Brant KA, Ward RM, Cattley RT, Barchowsky A, Fabisiak JP (2010) Multiple protein kinase pathways mediate amplified IL-6 release by human lung fibroblasts co-exposed to nickel and TLR-2 agonist, MALP-2. Toxicol Appl Pharmacol 247:146–157

Gardner RM, Nyland JF, Evans SL, Wang SB, Doyle KM, Crainiceanu CM, Silbergeld EK (2009) Mercury induces an unopposed inflammatory response in human peripheral blood mononuclear cells in vitro. Environ Health Perspect 117:1932–1938

Grjibovski AM, Olsen AO, Magnus P, Harris JR (2007) Psoriasis in Norwegian twins: contribution of genetic and environmental effects. J Eur Acad Dermatol Venereol 21:1337–1343

Havarinasab S, Hultman P (2005) Organic mercury compounds and autoimmunity. Autoimmun Rev 4:270–275

Hawkes CH, Macgregor AJ (2009) Twin studies and the heritability of MS: a conclusion. Mult Scler 15:661–667

Hemminki K, Li X, Sundquist J, Sundquist K (2010) The epidemiology of Graves' disease: evidence of a genetic and an environmental contribution. J Autoimmun 34:J307–J313

Hyttinen V, Kaprio J, Kinnunen L, Koskenvuo M, Tuomilehto J (2003) Genetic liability of type 1 diabetes and the onset age among 22,650 young Finnish twin pairs: a nationwide follow-up study. Diabetes 52:1052–1055

Invernizzi P (2010) Geoepidemiology of autoimmune liver diseases. J Autoimmun 34:J300–J306

Lehmann P, Homey B (2009) Clinic and pathophysiology of photosensitivity in lupus erythematosus. Autoimmun Rev 8:456–461

Maeda A, Beissert S, Schwarz T, Schwarz A (2008) Phenotypic and functional characterization of ultraviolet radiation-induced regulatory T cells. J Immunol 180:3065–3071

Malkiel S, Liao L, Cunningham MW, Diamond B (2000) T-Cell-dependent antibody response to the dominant epitope of streptococcal polysaccharide, N-acetyl-glucosamine, is cross-reactive with cardiac myosin. Infect Immun 68:5803–5808

Meroni PL (2011) Autoimmune or auto-inflammatory syndrome induced by adjuvants (ASIA): old truths and a new syndrome? J Autoimmun 36:1–3

Miller FW, Alfredsson L, Costenbader KH, Kamen DL, Nelson LM, Norris JM, De Roos AJ (2012a) Epidemiology of environmental exposures and human autoimmune diseases: findings from a National Institute of Environmental Health Sciences Expert Panel Workshop. J Autoimmun 39:259–271

Miller FW, Pollard KM, Parks CG, Germolec DR, Leung PS, Selmi C, Humble MC, Rose NR (2012b) Criteria for environmentally associated autoimmune diseases. J Autoimmun 39:253–258

Moroni L, Bianchi I, Lleo A (2012) Geoepidemiology, gender and autoimmune disease. Autoimmun Rev 11:A386–A392

Nistico L, Fagnani C, Coto I, Percopo S, Cotichini R, Limongelli MG, Paparo F, D'Alfonso S, Giordano M, Sferlazzas C, Magazzu G, Momigliano-Richiardi P, Greco L, Stazi MA (2006) Concordance, disease progression, and heritability of coeliac disease in Italian twins. Gut 55:803–808

Padalko EY, Bossuyt X (2001) Anti-dsDNA antibodies associated with acute EBV infection in Sjogren's syndrome. Ann Rheum Dis 60:992

Parks CG, Miller FW, Pollard KM, Selmi C, Germolec D, Joyce K, Rose NR, Humble MC (2014) Expert panel workshop consensus statement on the role of the environment in the development of autoimmune disease. Int J Mol Sci 15:14269–14297

Pedersen OB, Svendsen AJ, Ejstrup L, Skytthe A, Junker P (2008) On the heritability of psoriatic arthritis. Disease concordance among monozygotic and dizygotic twins. Ann Rheum Dis 67:1417–1421

Pollard KM, Lee DK, Casiano CA, Bluthner M, Johnston MM, Tan EM (1997) The autoimmunity-inducing xenobiotic mercury interacts with the autoantigen fibrillarin and modifies its molecular and antigenic properties. J Immunol 158:3521–3528

Prieto S, Grau JM (2010) The geoepidemiology of autoimmune muscle disease. Autoimmun Rev 9:A330–A334

Quintana FJ, Weiner HL (2009) Environmental control of Th17 differentiation. Eur J Immunol 39:655–657

Quintana FJ, Basso AS, Iglesias AH, Korn T, Farez MF, Bettelli E, Caccamo M, Oukka M, Weiner HL (2008) Control of T(reg) and T(H)17 cell differentiation by the aryl hydrocarbon receptor. Nature 453:65–71

Rose NR (2008) The adjuvant effect in infection and autoimmunity. Clin Rev Allergy Immunol 34:279–282

Sarkar S, Fox DA (2010) Targeting IL-17 and Th17 cells in rheumatoid arthritis. Rheum Dis Clin North Am 36:345–366

Segal BM (2010) Th17 cells in autoimmune demyelinating disease. Semin Immunopathol 32:71–77

Selmi C, Tsuneyama K (2010) Nutrition, geoepidemiology, and autoimmunity. Autoimmun Rev 9:A267–A270

Selmi C, Mayo MJ, Bach N, Ishibashi H, Invernizzi P, Gish RG, Gordon SC, Wright HI, Zweiban B, Podda M, Gershwin ME (2004) Primary biliary cirrhosis in monozygotic and dizygotic twins: genetics, epigenetics, and environment. Gastroenterology 127:485–492

Selmi C, Maria Papini A, Pugliese P, Claudia Alcaro M, Gershwin ME (2011) Environmental pathways to autoimmune diseases: the cases of primary biliary cirrhosis and multiple sclerosis. Arch Med Sci 7:368–380

Selmi C, Leung PS, Sherr DH, Diaz M, Nyland JF, Monestier M, Rose NR, Gershwin ME (2012a) Mechanisms of environmental influence on human autoimmunity: a National Institute of Environmental Health Sciences expert panel workshop. J Autoimmun 39:272–284

Selmi C, Lu Q, Humble MC (2012b) Heritability versus the role of the environment in autoimmunity. J Autoimmun 39:249–252

Shapira Y, Agmon-Levin N, Shoenfeld Y (2010) Defining and analyzing geoepidemiology and human autoimmunity. J Autoimmun 34:J168–J177

Shintani Y, Yasuda Y, Kobayashi K, Maeda A, Morita A (2008) Narrowband ultraviolet B radiation suppresses contact hypersensitivity. Photodermatol Photoimmunol Photomed 24:32–37

Shoenfeld Y, Blank M (2004) The infectious etiology of the antiphospholipid syndrome (APS). Autoimmun Rev 3(Suppl 1):S32–S34

So HC, Gui AH, Cherny SS, Sham PC (2011) Evaluating the heritability explained by known susceptibility variants: a survey of ten complex diseases. Genet Epidemiol 35:310–317

Su BY, Su CY, Yu SF, Chen CJ (2007) Incidental discovery of high systemic lupus erythematosus disease activity associated with cytomegalovirus viral activity. Med Microbiol Immunol 196:165–170

Sverrild A, Backer V, Kyvik KO, Kaprio J, Milman N, Svendsen CB, Thomsen SF (2008) Heredity in sarcoidosis: a registry-based twin study. Thorax 63:894–896

Tanaka A, Takikawa H (2013) Geoepidemiology of primary sclerosing cholangitis: a critical review. J Autoimmun 46:35–40

Tysk C, Lindberg E, Jarnerot G, Floderus-Myrhed B (1988) Ulcerative colitis and Crohn's disease in an unselected population of monozygotic and dizygotic twins. A study of heritability and the influence of smoking. Gut 29:990–996

van der Woude D, Houwing-Duistermaat JJ, Toes RE, Huizinga TW, Thomson W, Worthington J, van der Helm-van Mil AH, de Vries RR (2009) Quantitative heritability of anti-citrullinated protein antibody-positive and anti-citrullinated protein antibody-negative rheumatoid arthritis. Arthritis Rheum 60:916–923

Veldhoen M, Hirota K, Westendorf AM, Buer J, Dumoutier L, Renauld JC, Stockinger B (2008) The aryl hydrocarbon receptor links TH17-cell-mediated autoimmunity to environmental toxins. Nature 453:106–109

Yu H, Yang YH, Rajaiah R, Moudgil KD (2011) Nicotine-induced differential modulation of autoimmune arthritis in the Lewis rat involves changes in interleukin-17 and anti-cyclic citrullinated peptide antibodies. Arthritis Rheum 63:981–991

Zeki AA, Schivo M, Chan AL, Hardin KA, Kenyon NJ, Albertson TE, Rosenquist GL, Louie S (2010) Geoepidemiology of COPD and idiopathic pulmonary fibrosis. J Autoimmun 34:J327–J338

14 The Challenge of Predicting the Immunotoxic Potential of Chemicals

Hans-Werner Vohr

Contents

14.1 What Is Immunotoxicology?

14.1.1 Introduction

Well before the mechanisms were understood, pulmonary immune diseases had been associated with environmental chemicals, i.e., air contaminants. During the second half of the twentieth century, purpose and function of the different

H.-W. Vohr
Heinrich-Heine-University, Düsseldorf, Germany
e-mail: hwv.office@t-online.de

C. Esser (ed.), *Environmental Influences on the Immune System*,
DOI 10.1007/978-3-7091-1890-0_14

components of the immune system as well as their interactions with chemicals were extensively investigated and ultimately understood in more detail. In addition to a deeper understanding of immunological interactions and mechanisms, a series of accidents involving immunotoxic compounds pushed the development of immunotoxicological science in the last century.

In Europe in particular, the field of immunotoxicology came to the forefront following the accidental release of 2,3,7,8-TCDD near Seveso in 1976. This accident marked the starting point of public discussion around chemical-induced immune deficiencies which still continue to this day. Broad public attention led to a flood of activity by academics, authorities, and the industry. Immunotoxic effects induced by drugs had been observed and experimentally investigated prior to this incident. However, these studies were – in most cases – not conducted as special immunotoxicity investigations, i.e., based on understanding of immunological processes. Furthermore, they did not obtain wide publicity. Authorities started the process of drafting immunotoxicity guidelines, and several workshops were initiated to discuss possible testing strategies for immunotoxicological screening (e.g., by IPCS) as well as intra- and interlaboratory validation studies (ICICIS, BGA, NTP, etc.). Several proposals for such screening batteries had been published between 1982 and 1998 from various sites.

Development of guidelines for industrial chemicals and agrochemicals reached a milestone as a result of the findings published by M. Luster et al. in 1992 and 1993, in which the authors presented data from studies of 51 substances, 35 of which were declared immunotoxic, in a comprehensive test battery in mice to investigate changes in functional parameters after 28-day administration of the substance. The key finding in this study was that immunotoxic effects (immunosuppression of host resistance) could not be detected by the incorporation of a single immune parameter into the routine toxicological testing. A combination of two or three additional parameters was required. One of these additional parameters was a functional assay, the Plaque-Forming Cell Assay (PFCA; Fig. 14.1). Another test was the analysis of subpopulations of spleen cells by flow cytometry. However, from the very beginning, many discussions about whether advanced histopathology of lymphoid organs alone could be sufficient to pick up all of the immunotoxic effects of chemicals, a discussion which persists to this day.

14.1.2 Definitions

The first international seminar on the immune system as a target for toxic damage was held in Luxembourg in 1984 (UNEP, ILO, WHO: IPCS, CEC; 1987). During the meeting the definition for the term “immunotoxicology” was agreed upon for the first time by all the participants. The definition reads as follows:

> Immunotoxicity is defined as undesired effects as a result of interaction of xenobiotics with the immune system.

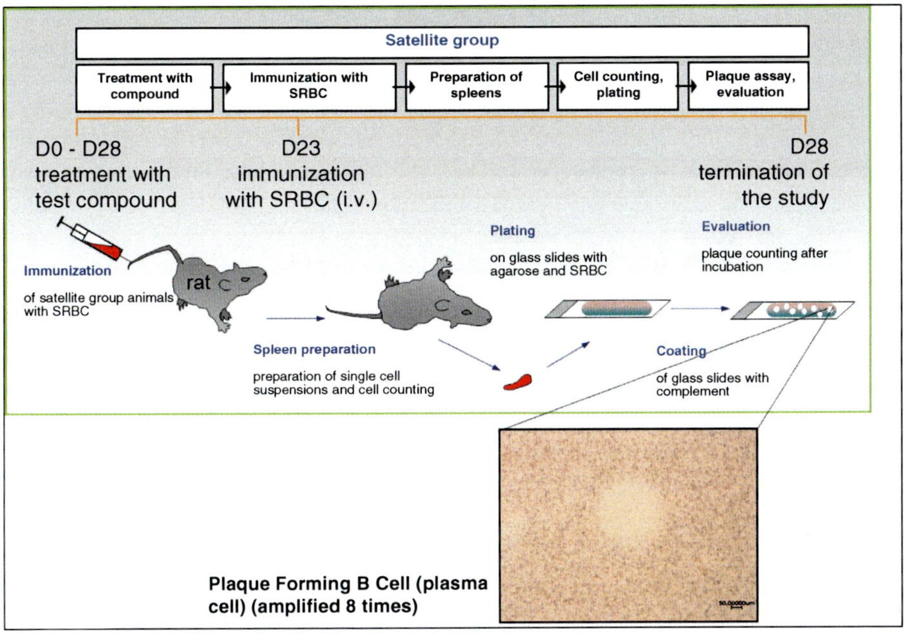

Fig. 14.1 Scheme of the Plaque-Forming Cell Assay (PFCA) as performed in accordance to relevant guidelines (Fischer, A, modified). In most cases, satellite animals are immunized with SRBC for the PFCA instead of animals in the main group to avoid any interference of the toxic effects by a strong immune reaction to SRBC

Such interactions of chemicals with the immune system can induce immunosuppression or enhancement (undesired immunostimulation). While immunosuppression may result in decreased resistance to infections or increased tumor development, overstimulation may increase the risk of autoimmune or allergic reactions. Although several additions and/or alterations have been made to the definition over the years, the end points of immunotoxicity are still valid. The meeting in Luxembourg was the starting point for several workshops, seminars, and symposia focusing on definitions and testing strategies for immunotoxic evaluation.

A workshop held in Hannover, Germany, in 1989 (IPCS 1990) was followed by a workshop in 1992 organized in Bilthoven, Netherlands (Vos and van Loveren 1995). Two special joint workshops of IPCS and the Norwegian National Institute of Public Health were held in Oslo in 1995 and 1996, of which the first was entitled "Environmental Chemicals and Respiratory Hypersensitisation" (Dybing et al. 1996). Finally, two additional scientific symposia were organized in Bilthoven, Netherlands, in 1997 (Van Loveren et al. 1999) and in 1999 in Berlin, Germany.

In course of this period, the most accepted definitions of immunotoxicology evolved significantly:

> Immunotoxicology is defined as the study of adverse effect on the immune system associated with exposure to environmental chemicals, pharmacologic agents, and biological.

> Generally these effects can be categorized as immunomodulation (immune suppression or potentiation), hypersensitivity (i.e., allergy), and autoimmunity.

While some authors added "chronic inflammation" and/or "flu-like reactions" to this enumeration of effects, Jacques Descotes published a further refined view on the possible end points of immunotoxicity. In 2005 he made precise distinctions between specific (immune-mediated) and nonspecific interactions, allergic and pseudoallergic reactions, etc. (Descotes 2005):

> The immunotoxic effects of drugs are divided into immunosuppression, immunostimulation, hypersensitivity and autoimmunity. The major adverse consequences of immunosuppression are infectious complications and virus-induced malignancies. Flu-like reactions, more frequent autoimmune diseases and hypersensitivity reactions to unrelated allergens, and inhibition of drug-metabolising enzymes are the adverse effects related to immunostimulation. Hypersensitivity reactions are the most frequent immunotoxic effects of drugs. They include immune-mediated ("allergic") and non-immune-mediated ("pseudoallergic") reactions. Drug-induced autoimmune reactions, either systemic or organ-specific, are seemingly rare.

Although Descotes' definition referred only to drug interactions with the immune system, the same holds true for environmental and industrial chemicals.

Immunotoxic effects can be a significant cause of morbidity or in some cases even mortality. Early immunotoxicological investigations in a regulatory environment were predominantly based on *in vivo* studies (28/90 days or short-term tests) with rats or mice. There were ongoing discussions regarding what parameters are essential for a sound immunotoxic assessment, and in parallel a set of relatively robust, standardized, and validated assays were established. However, with the exception of type IV investigations (guinea pig assays or LLNA), focus and experiences were mainly based on immunosuppression and not on immunostimulation. For non-clinical immunotoxicity, safety assessments of unexpected immunostimulation, like systemic or local hypersensitivity reactions, types I to III, or autoimmunity, the situation was unsatisfactory, because no validated and widely accepted assays for determining these end points were available.

Great efforts were made during the last few years to generate reliable assays for the prediction of immunogenicity of biologicals, but much less for the detection of immunostimulating properties of small molecules. Authorities and the public increased pressure on immunotoxicologists to develop additional *in vitro* assays and to extend immunotoxicity screening to animal species more relevant to humans (e.g., nonhuman primates (NHP)). While this effort has resulted in an increasing number of new models, protocols, and parameters, these are still a long way from standardization and validation. Few models and assays have so far been validated and used in preclinical safety assessment of undesired immunostimulation. The situation is more dire with respect to the prediction of hypersensitivity and autoimmune reactions. Last but not the least situation, industrial and environmental chemicals remain largely unstudied in relevant species. This is in contrast to therapeutic drug development where most studies are performed on rodents and other species such as dogs or monkey. In addition, side effects of drugs observed during preclinical development can shortly afterward be compared with findings in clinical investigations, i.e., in humans. This is unfortunately not the case for environmental chemicals. This will be the focus of the following chapters.

14.1.3 International Guidelines on Immunotoxicology

As described below, authorities started to think about immunotoxicity guidelines in the late 1970s and 1980s. In order to support the development of guidelines backed by a firm set of data, several national and international validation studies were initiated, one of which will be subsequently described in more detail.

The first guideline which was adapted to immunotoxicological end points was the OECD 407 (Repeated Dose 28-day Oral Toxicity Study in Rodents, 1981) for chemicals in 1992. Of note is that the immune parameters added to the existing guidelines were not the most sensitive ones as described by Luster et al. in 1992/1993 (Fig. 14.2). The history of the development of immunotoxicity guidelines is described in the following chapter.

14.1.4 Collaborative "BGA" Study

14.1.4.1 Cyclosporin A

In 1989, the German Federal Health Authority (Bundesgesundheitsamt, BGA) initiated a collaborative study in Europe to determine the most sensitive parameters for the detection of immunological side effects induced by chemicals after a 28-day treatment period. All five participating laboratories agreed to incorporate several additional investigations into the toxicological investigation in accordance with the OECD 407 guideline. Advanced histopathology of lymphoid organs was incorporated as well as a functional assay (PFCA), analyses of NK cell activity, subpopulations in the spleen, mitogen stimulation of splenic cells, and antibody titer in the sera of the treated animals. The "chemical" chosen for the first round was a well-known immunosuppressive standard, cyclosporin A. This drug was chosen not only because of its clear immunosuppressive potential but also because Cyclosporin A was used during the second international ring trial of ICICIS Dayan et al. (1998). Thus, the hope was to ultimately compare the two sets of data from these collaborative studies. Unfortunately, this has never been done, although the reason is not quite clear.

Although the additional immune parameters, especially flow cytometry and PFCA, turned out to be most sensitive for picking up significant effects on the immune system, the histopathology of the thymus was also more affected than other organs. This fact led to a discussion on whether the "new" additional immune parameters are valuable or not or if advanced histopathology would suffice. A final agreement about the necessity of additional immune parameters was also not reached by the participating labs of the BGA study (Richter-Reichhelm et al. 1995; Vohr 1995), and the discussion about this topic remains undecided to the present day. This is especially astonishing considering that Germolec et al. (2004), Lappin

	Plaque Forming Cells	NK Cell Activity	T Cell Mitogens	MLR	DHR	CTL	Surface Markers	Leukocyte Counts	Thymus/BW Ratio	Spleen/BW Ratio	Spleen Cellularity	LPS Response	
Plaque Forming Cells	**78 (45)**												**P<.0001**
NK Cell Activity	94 (34)	**69 (36)**											**P=.0014**
T Cell Mitogens	85 (40)	79 (34)	**67 (46)**										**P=.0003**
MLR	82 (34)	74 (31)	73 (37)	**56 (39)**									**P=.0458**
DHR	89 (27)	84 (19)	82 (28)	74 (23)	**57 (30)**								**P=.0348**
CTL	100 (8)	78 (9)	71 (7)	75 (8)	- (0)	**67 (9)**							**P=.2380**
Surface Markers	91 (23)	90 (21)	92 (24)	87 (23)	93 (14)	100 (5)	**83 (24)**						**P=.0017**
Leukocyte Counts	86 (28)	71 (24)	62 (29)	59 (27)	67 (18)	67 (6)	80 (20)	**43 (30)**					**P=.4490**
Thymus/BW Ratio	92 (38)	81 (31)	83 (36)	77 (30)	75 (24)	71 (7)	90 (21)	72 (29)	**68 (40)**				**P=.0009**
Spleen/BW Ratio	85 (39)	75 (32)	76 (37)	65 (31)	71 (24)	75 (8)	86 (22)	62 (29)	73 (40)	**61 (41)**			**P=.0395**
Spleen Cellularity	80 (35)	72 (29)	72 (32)	63 (30)	67 (21)	71 (7)	76 (21)	60 (25)	75 (32)	63 (32)	**56 (36)**		**P=.0694**
LPS Response	81 (37)	73 (30)	69 (39)	65 (31)	58 (24)	83 (6)	90 (20)	56 (27)	74 (34)	71 (35)	63 (27)	**50 (40)**	**P=0.2260**

Fig. 14.2 Individual and pairwise concordance to establish predictability using the immune panel. Values are presented as percentage concordance which is the sum of specificity (–/–) and sensitivity (+/+). Individual concordance values are shown in boldface on the diagonal of the matrix and combinations, using two tests on the off-diagonal element. Values in parenthesis are the number of chemicals tested for the assay. Since the individual tests were also used to establish the "immunotoxic classification," the frequency of concordance will obviously increase as the number of tests included for the analysis is increased (–). No overlapping studies were performed. P values are given for individual concordance only (Slightly modified after Luster et al. (1992))

and Black (2003), as well as Vohr and Rühl-Fehlert (2001) showed that only the combination of both advanced histopathology of lymphoid organs and determination of additional immune parameters are sufficient to pick up all immunotoxic effects of chemicals.

14.1.4.2 Hexachlorobenzene (HCB)

For the second round of the BGA ring study with nine participating labs, an "immunostimulating" chemical had to be chosen to investigate immunotoxic end points, immunosuppression, and immunostimulation. HCB had been selected for the second round on the basis of some reports, e.g., Vos et al. (1979), on the immunomodulating properties of this compound. However, during the evaluation of the histopathology, it became clear that an irritant property of the compound was primarily responsible for the observed changes. In accordance with this finding, the additional immunological parameters verified a secondary immunotoxic effect of HCB due to nonspecific activation of the immune system via irritation, i.e., inflammation. While a faint immunostimulating effect had been picked up for HCB, the additional immunological parameters clearly demonstrated the indirect effect of this reaction. As a result, there was no final agreement about the favorable additional immunological parameters to be included in the normal routine toxicology package to flag immunotoxicological changes.

Mike Luster from the National Toxicology Program (NTP) of the USEPA investigated a series of 52 chemicals, 39 of which were known immunosuppressives.

While some information on potential immunotoxic effects may be obtained from hematology, lymphoid organ weights, and histopathology, the data published by Mike Luster demonstrate that these end points alone are not sufficient to predict immunotoxicity (Luster et al. 1992, 1993).

14.1.5 Guidelines

The US Environmental Protection Agency (USEPA) was the first public authority to really push the development of guidelines on immunotoxicity. After a longlasting discussion about the optimal test battery for immunotoxicological screening, the USEPA developed its guideline, OPPTS 780.8700 (1998), which is exclusively based on the findings of Luster et al. mentioned above.

The USEPA further agreed and defined on the following definition: *"Immunotoxicity refers to the ability of a test substance to suppress immune responses that could enhance the risk of infectious or neoplastic disease, or to induce inappropriate stimulation of the immune system, thus contributing to allergic or autoimmune disease. This guideline only addresses potential immune suppression."* Nevertheless, the USEPA excluded all aspects of immunostimulation as this was also not part of Luster's analyses, and there were no validated widely accepted test methods available to pick up an allergic or autoimmune potential of chemicals. This is still true today for the prediction of allergic reactions of types I, II, and III and autoimmunity, but not for type IV reactions (contact allergy), which have been investigated for decades using a guinea pig assay (OECD TG 406) or the mouse local lymph node assay (OECD TG 429, 442A and 442B).

In principle most of the other guidelines on immunotoxicity published thereafter pursued similar concepts. Only the harmonized tripartite guideline ICH S6 "Preclinical Safety Evaluation of Biotechnologically Derived Pharmaceuticals" (1997; revision 2009) included immunostimulation as one of the end points to be determined. This is understandable because immunogenicity of biologicals is an issue during preclinical and clinical development of pharmaceuticals.

14.1.6 Summary

Based on the EPA guideline recommendations, it is vital to differentiate between primary and secondary immunotoxicities, the latter being a nonspecific sequela of toxicity to other organs. In our studies, we found examples for both mechanisms, where primary immunotoxic substances tend to be markedly more immunosuppressive, although primary effects on the whole occurred relatively seldom during toxicological screening of SMOL, i.e., in less than 10 % of the studies. In both cases, there is a strong correlation between cell analysis and functional parameters on the one hand and pathology on the other, thus ensuring that overt immunotoxicity would not remain undetected in routine studies with high dose levels. However, the higher predictivity of functional parameters and the analysis of special subpopulations are necessary for setting a correct no observed adverse effect level (NOAEL) and for fine differentiation during the screening of comparable immunotoxic compounds. As verified by the collaboration studies, an advanced histopathology of lymphoid organs, combined with flow cytometry of immune competent cells and a functional assay, is able to discriminate between primary and secondary effects as well as immunosuppression and immunostimulation and thus to identify an immunotoxic hazard.

14.1.7 Experiences in Screening Chemicals (Immunosuppression)

The development or selection of suitable tests for immunotoxicological screening and incorporation into guidelines presents a considerable problem. In the beginning, most of the tests which had been proposed for immunotoxicological investigations and the knowledge and experience in immunology were based on mouse models or on a few collaborative studies in rats (cf. above). Adaptation of suitable tests to rats was not always easy, partly because of lack of suitable reagents. The next problem was to find which tests could suitably be used for reliable identification of interactions with the immune system. As mentioned above, publications about collaborative studies as well as the investigations of Luster et al. were of major importance in this regard. Another question which still has not been answered yet was that of the dosages, kinetics, and changes in immunological parameters which are still tolerable over time, i.e., after short-term (28-day) or long-term (90-day) treatment. With respect to ideal time points to screen for immunotoxic effects, the vast majority of experts agree that short-term treatments (14 or 28 days) are of optimal length.

In order to put the discussion on a somewhat more sound footing, it was important to test the various models/parameters for the detection of immunotoxicological potential in practice. For this reason, Bayer AG (Bayer HealthCare AG) has not only been a collaborator in the trial mentioned above, but has also introduced a set of functional immunological tests into its routine toxicological testing of agrochemicals in rats to determine the informative value of these parameters in daily practice. Hence, we started to incorporate additional immunotoxicological parameters according to those used for the collaborative studies into each routine subacute study of agrochemicals as early as 1992. During the first few years, investigations were performed under GLP-like conditions before being subsequently changed to fully GLP compliant.

14.1.8 Results

During the screening of agrochemicals, we found that pesticides with primary immunotoxic effects were relatively rare (Vohr and Rühl-Fehlert 2001). To gain more experience with positive (immunotoxic) chemicals, we also screened closely related immunosuppressive drugs which showed an impact on the bone marrow. Histopathology revealed reduced hematopoiesis, with the affected dose level varying depending on the compound. The additional immune parameters showed higher sensitivity with respect to the affected dose level and confirmed primary immunotoxicity. Thus, the combination of histopathology of lymphoid organs and measurement of additional immune end points were of especial value in screening assays that can be adjusted according to the class of compound being tested.

It became evident that immunotoxic effects were detectable after only a few days of treatment by screening additional immune parameters. Histopathological changes of the lymphoid organs may only occur with a delay of some days to weeks. On the other hand, the immune system seemed to tolerate the test substance after longer exposure times (>90 days), and overall toxicity became most prominent with time. To check the correctness of these observations, we included an advanced screening battery at different time points of treatment during the development of an immunostimulating drug. As expected, the immunostimulating property of the compound was confirmed histopathologically by the increased number and size of germinal centers in the spleen in the high-dose group. There was no evidence of other organ toxicity that might have been causally related to this finding. As in the cyclosporin A collaborative study, the additional immune parameters were highly sensitive (mid-dose). However, these parameters detected the immunostimulating effect after only 2 weeks of exposure. At that time point, increased splenic germinal center formation was not yet detectable in histopathology. After 1 year of dosing with the test compound, the toxicological effects shifted from immunotoxicity to other organ toxicity.

14.2 What About Immunostimulation?

As mentioned at the end of the last chapter, determination of immunostimulating properties of chemicals is not as simple as it is for immunosuppression. The majority of clear cases of immunostimulation based on scientific evidence are due to pharmaceuticals developed for this purpose like vaccines or per se immunogenic pharmaceuticals like biologics. Especially the immunogenicity of biologics is one of the main concerns during preclinical and clinical development of such therapeutics. Immunogenic epitopes (e.g., OKT 3), directly T-cell-activating therapeutics (e.g., Tegenero antibody), and administration of high amounts of monoclonal antibodies can induce different kinds of inflammatory cytokines and chemokines. As a consequence patients suffer from flu-like reactions and, in severe cases, of "cytokine storm," i.e., the so-called cytokine release syndrome (CRS).

For environmental chemicals or small molecules (SMOL), such tremendous or comparable effects have not been described so far. However, after binding to carrier molecules, SMOL can act as haptens, i.e., they can elicit an immune response to "altered self" molecules. This immunological principle of anti-hapten antibody response was first described by Mitchison in 1971. The consequence of these hapten-carrier formations can be allergic responses of type I to type IV or autoimmune reactions depending on several additional factors like genetic background, duration of contact, route of exposure, preexisting condition, etc. Unfortunately, no validated and widely acknowledged prediction tools for these undesired health effects exist with the exception for type IV reactions. For the determination of cell-mediated hypersensitivity (type IV reactions), which results clinically in allergic contact dermatitis, there are well-established animal models available: guinea pig assays (Buehler or maximization assay) or a mouse assay (local lymph node assay, LLNA). Other hypersensitivities mediated by antibodies (types I to III) are difficult to predict, and most of our knowledge in this area comes from retrospective human data, i.e., from humans whose history of induction of sensitization remains unknown. It is therefore possible to investigate the specificity of reacting antibodies by different methods like the human prick test, but such investigations will not clarify the underlying mechanisms or intrinsic potential of the chemical to induce these reactions. Therefore, the risk assessment for these end points is often controversial.

The same holds true and is even worse for the prediction or determination of chemical-induced autoimmune reactions. Development of autoimmune responses is a longlasting process that it is in many cases impossible to narrow down and link to contact with a single chemical or class of chemicals.

Another hurdle of predicting chemical-specific immunostimulation is the fact that several environmental chemicals or SMOL exhibit different kinds of undesired properties to varying degrees. For example, cytotoxicity, severe irritation/corrosion, skin sensitization, photo irritation, photoallergy, or combinations of these belong to such properties. Therefore, it is hard to discriminate the nonspecific from chemical-specific immune reactions because nonspecific inflammatory reactions can show comparable characteristics as severe specific immune reactions (Hölzle et al. 1991).

Ketoprofen illustrates how numerous distinct properties can be expressed by a single chemical compound. Ketoprofen is not only irritating and sensitizing to the skin but is also known to induce photo irritation as well as photoallergy in humans. Although ketoprofen and other nonsteroidal anti-inflammatory drugs (NSAIDs) are not "classical" environmental chemicals due to their prevalence in plants, they are also part of our environment. This illuminates another aspect of the problem in assessing immunotoxicity of environmental chemicals. While drugs and most agrochemicals are intensively screened for immunotoxicological side effects, this is not the case for industrial or other environmental chemicals. With respect to risk assessment for environmental chemicals, we rely therefore mainly on experiences and data from non-environmental chemicals. There is still an enormous gap in the knowledge and understanding of undesired immunological side effects induced by molecules of natural origin or chemicals found ubiquitously in the environment.

14.3 REACH and Its Influence on Immunotoxicological Screening of Chemicals in Europe

Assessment of immunotoxic effects such as immunosuppression and undesired immunostimulation rely at present on several animal-based assays. The use of animals, however, faces a number of issues, e.g., ethical concerns and relevance to human risk assessment. There is a growing belief that non-animal approaches can eliminate these issues without impairing human safety, provided that biological markers are available to identify the immunotoxic potentials of chemicals to which humans may be exposed. As mentioned before, the growing knowledge that the immune system can be the target of many chemicals, resulting in a range of several adverse effects, has raised serious concerns from the public and within the regulatory agencies. In combination with the European REACH legislation (Regulation (EC) No. 1907/2006), immunotoxicological side effects such as skin hypersensitivity must be studied for preregistration. This new EU policy on chemicals has a strong impact on manufacturers, importers, distributors, and downstream users due to the underpinning principle: "no data, no market." Driven by the 7th Amendment to the EU Cosmetics Directive as well as the REACH act, animal-based testing for chemicals is to be reduced to an unavoidable minimum (REACH) or even prohibited (Cosmetics Directive). Hence, there is an enormous pressure on the industry to develop and establish batteries of *in vitro* methods for predicting general toxicity and immunotoxicological side effects. Such *in vitro* methods have to focus on immunosuppressive as well as on immunostimulative properties of chemicals. At present we are far away from predicting the toxicity of chemicals toward the immune system by simple, fast, and reliable cell-based immunotoxicity assays. Some new methods which may lead the way are described in more detail in the last chapter.

This dilemma between advanced animal welfare and the need to (re)evaluate chemicals for their toxic (and immunotoxic) properties has given rise to a series of new *in vitro* and *in silico* methods, many of which lack validation and general acceptance. These methods are nevertheless widely used due to the lack of alternatives. While it is relatively easy to establish and validate *in vitro* assays for simple end points like cytotoxicity (irritation/corrosion), it is much more difficult to develop assays for more complex end points, where metabolism, cell interaction, and induction of factors are all playing a role.

An example of the difference between simple and more complex end points is the *in vitro* testing of skin effects caused by chemicals. For years, skin irritation and corrosion are tested by *in vitro* methods on 3D human skin equivalents. These methods have been established, pre-validated, validated, and finally accepted by regulators over a period of more than a decade. All these efforts resulted in two OECD Guidelines, which came into force in 2004 (OECD TG 431; skin corrosion; update 2013) and 2008 (OECD TG 439; skin irritation; update 2013), respectively.

In contrast *in vitro* methods for predicting skin sensitization are still far away from proposal for an OECD Guideline. One assay, KeratinoSens™ Assay developed by Givaudan, which was published as a draft version in 2014, is the furthest along. However, skin sensitization is a multilayered process which includes penetration of the chemical through the *stratum corneum;* metabolism in the skin; cell interaction of keratinocytes, dendritic cells, and T cells; chemokine and cytokine induction; cell migration; etc. An interesting review about the complexity of skin sensitization testing has been published previously by Van der Veen et al. 2014.

The implications of the complex interactions mean that for *in vitro* skin sensitization analysis, a battery of tests for several end points has to be established, mimicking all the different steps necessary for interactions to develop contact dermatitis. Indeed, there are *in vitro* methods that are underdevelopment measuring skin penetration, hapten-carrier binding, dendritic cell activation, signal transduction, T-cell activation, or cytokine/chemokine release caused by chemicals. The validation status of these different methods is heterogeneous. While some are already almost accepted by regulators (e.g., KeratinoSens or Direct Peptide Reactivity Assay (DPRA)), others are still far away from any international validation or global acceptance (e.g., hCLAT or signal transduction, IL-18 production by 3D skin models, etc.).

To replace the *in vivo* evaluation of skin sensitization, a battery of at least three assays is necessary. It is clear that each of these models has a certain level of sensitivity and specificity and a certain percentage of false-negative and false-positive results. This is the primary disadvantage of such a composite of assays as the overall accuracy and/or variance will increase with the number of test models included. In addition, each *in vivo* or *in vitro* assay has a specific applicability domain (AD), i.e., the classes of chemicals which can be tested by the model are restricted to the intrinsic properties of the compounds like solubility in aqueous vehicles, cytotoxicity, etc. These ADs can be very specific for single test models and so will not always overlap 100 % in a test battery and thus will decrease the classes of chemicals which can be tested by such a battery. Therefore, the aim should be to develop test models with a broader applicability domain, e.g., gene expression analyses or use of organoid models like reconstructed human epidermis. However, these assays must be validated before they can be used in an *in vitro* test battery, which must itself also be validated.

A validated and widely accepted *in vitro* battery for skin sensitization will not be available in anytime soon. On the other hand, the cosmetic industry in Europe has to test new components exclusively in *in vitro* systems, and the same holds true for many chemicals to be re-evaluated under the REACH legislation. A relatively simple end point such as skin sensitization has taken decades to be developed; therefore, an *in vitro* alternative to testing the more complex toxicity effects of small molecules or environmental chemicals will likely require an enormous effort. Other ways to overcome this dilemma could be via biomonitoring and epidemiological investigations which is the topic of the next chapter.

14.4 Risk Assessment for the Immunotoxic Potential of (Environmental) Chemicals

14.4.1 Accidents and Biomonitoring

The history of immunotoxicity is closely related to accidents and pollution with small molecules showing immunomodulating properties like dioxin, heavy metals, polychlorinated biphenyls (PCBs), asbestos, latex, essential oils, pesticides, isocyanates, diesel engine emissions, etc. All of these molecules are ubiquitously available in the environment or are released into the environment by very different mechanisms. Some of these chemicals were released by accidental spillage where large amounts were released. It was during these accidents that the immunomodulating properties of these compounds became a focal point.

There are several examples of such accidents, be it the methylmercury contamination of fish and shellfish in Japanese Minamata Bay in the 1950s, the contamination of special feed supplement for lactating cows with polybrominated biphenyls in Michigan 1973 and 1974, or the accidental release of 2,3,7,8-TCDD near Seveso in 1976. All such cases of accidental or ignorantly released chemicals led to decades of investigations on the toxic mechanisms observed and raised the general public and regulatory awareness of immunotoxic screening.

Another approach for obtaining information about the possible immunotoxic side effects of (environmental) chemicals is through biomonitoring. The first investigation of workplace-related poisoning was that of chimney sweepers. As early as 1775, the dependency between intensive contact with soot and the development of testicular cancer was reported by Percivall Pott. This report is considered the starting point for the history of biomonitoring of workplace-specific body burden with chemicals. Today biomonitoring in the workplace spans the measurement of concentrations of relevant chemicals, the routes of possible application, as well as the detection of chemicals and their metabolites in a variety of biological fluids and tissues. The result is a sound risk assessment and determination of so-called maximum allowable concentrations (MACs). Many immunotoxic chemicals have been assessed by this manner, although in many cases, the discrimination between, e.g., cancerogenous and immunotoxic properties, was not made clear. Ultimately, whether a worker was protected from cancer or immunotoxicity was irrelevant.

While this is one way of providing a reasonable risk assessment for chemical exposures at work, biomonitoring is not well suited for a similar risk assessment in the overall population. This is the goal of epidemiological investigations as described in the next chapter.

14.4.2 Epidemiological Evaluation

Epidemiology is a scientific discipline dealing with prevalence, mechanisms, and consequences of health conditions and events of the entire population. One part of this discipline is epidemiological evaluation of possible toxic effects of environmental chemicals. Universities, private, and governmental institutions all over the world maintain several positions for this special part of epidemiology. One of the main and most efficient tools used in this discipline is surveys and networking. It is beyond the scope of this chapter to go into too much detail about different surveys. The most important surveys will be mentioned, and a closer look at one example will be undertaken, i.e., skin sensitization.

14.4.3 Surveys

As early as 1976, a comprehensive survey was initiated in the United States by the CDC (Centers for Disease Control and Prevention). The National Health and Nutrition Examination Survey (NHANES) program tested samples from the general population for lead and certain pesticides over a period of several years (Annest et al. 1983). The NHANES program had considerably been expanded over the years, not only to measure pollution in the general population with chemicals by biomonitoring but also to investigate the impact of this pollution on the general health status (Stokstad 2004). The aim in 2004 was to monitor nearly 1000 chemicals in persons from all over the United States. It is clear that this number of chemicals and all their varying interactions, metabolisms, distributions, and kinetics in the body would not simplify the interpretation and evaluation of risk assessment. In 2006 Dennis Paustenbach and David Galbraith critically discussed this aspect of biomonitoring in a much-noticed review.

Similar surveys were initiated in Germany, of which the German Environmental Survey (GerES) is one of the most important. It was started in 1985 with determination of chemicals in the urine of the general population grouped by age, gender, residential area, etc. This program developed stepwise to a more complex determination of different classes of environmental chemicals, in specific groups (ages and gender) or living areas (Angerer et al. 2007; Schulz et al. 2007a, b). For example, GerES IV survey was the first one in Germany to investigate body burden of pollutants in children and the exposure to pollutants in their homes at about 150 different locations in Germany. Due to the experience gained with a large amount of data, human biomonitoring has been subdivided into biological monitoring of exposure and biological effect monitoring. Accordingly the biomarker, organs, or body fluids tested became increasingly comprehensive. To further harmonize these, in order to maximize the value of the ever-increasing data set, the "Human Biomonitoring Commission of the German Federal Environment Agency" was established in 1992 (Schulz et al. 2007a, b).

There are a growing number of surveys across the globe, and due to more harmonized strategies, the focus of different initiatives is more and more refined. Also the existing methods were validated and/or harmonized with time to accurately measure biologically relevant concentrations of the chemicals investigated. However, due to the large amount of data produced and the enormous variety of interactions of the different parameters analyzed, controversial discussions about the reasonable selection of chemicals evaluated took place (Paustenbach and Galbraith 2006a, b) and objections about the reasonableness of such investigations were put forth (Boccia et al. 2007; Baker and Gibson 2014; Chang et al. 2014).

In spite of or maybe because of large amounts of data were generated by different groups, clear-cut and reliable effects of environmental chemicals on the immune system are rare and in most cases controversially discussed. This is often due to external or basic parameters of a survey which were not addressed or over-interpreted depending on the intention of an investigation. In many publications, the mere presence of a toxic compound around the detection limit in the environment or tissue is discussed as a major health problem. Realistic exposure scenarios or risk assessments based on relevant NOELs are simply not done. Even for experts in the field of immunology/immunotoxicology, it is sometimes hard to eliminate the noise. Therefore, data presented are often looked at with skepticism.

So it is really not clear at all whether huge amounts of date as would, for example, be generated by the American "National Children's Study" which should follow 100,000 children across the United States from birth until age 21, to address the effects of social, economic, and environmental factors on a child's health, would at the end help to understand all the factors influencing the development and the health status of children. Because of the expected extreme costs and the abovementioned shortcomings of such a study, the planning has lasted over 14 years (for details, http://www.nap.edu/catalog.php?record_id=18826).

14.4.4 German Survey on Skin Sensitization

The Information Network of Departments of Dermatology for recording and scientific analysis of contact allergies (IVDK) founded in 1988 cooperates with 55 dermatological hospitals in Germany, Austria, and Switzerland. As a multicentric project, the IVDK collects data about allergens and publishes lists at regular intervals on the prevalence of allergies in different regions (Schnuch et al. 2004; Uter et al. 2007, 2010). At present (2014) data of about 12,000 prick-tested patients are collected and analyzed each year.

The data is extrapolated into incidences of allergic contact dermatitis (ACD) in the general population between 3 (medium case scenario) and 7 (worst case scenario) cases per 1,000 persons a year. Thus far, no comprehensive studies exist to determine the actual incidences of such ACDs in the general populations. The data is based on extrapolation prick test data or interviews of patients by dermatologists (Hermann-Kunz 1999).

Beyond calculation of incidences and prevalence of allergic diseases, the IVDK also publishes ranking lists of the most frequent allergens. For years, nickel has always topped the charts, although it is a weak sensitizer (Geier et al. 2011). Thus, in contrast to the low potency of nickel to induce ACDs, the prevalence of nickel allergies is high due to the extreme frequency of contact in the general population. This discrepancy is one of the reasons for frequent discussions about potency of chemicals relative to their frequency of exposure. In addition, the different assay protocols used in dermatology are not yet fully standardized, and it is not always clear if the data from different sources can be pooled for evaluation of potency (Thyssen et al. 2012a, b, c).

This example illustrates the difficulty in evaluating or predicting the risk of an environmental chemical to induce contact allergy across the population, particularly as the majority of data analyzed is exclusively based on patients with a longer history of disease. If evaluation of a relatively simple side effect such as induction of contact dermatitis is already complex, the difficulty of measuring a more complex side effect of environment chemicals can be envisioned.

In conclusion, prediction of risk assessment of immunotoxic effects of environmental chemicals based solely on epidemiological data should be regarded with skepticism.

14.5 Future Developments

14.5.1 *In Vitro* Screenings

Although there are still several problems to be solved for "classical" immunotoxicity screening, such as autoimmunity or systemic hypersensitivity, the field of immunotoxicity is already expanding into new areas. Such new directions are *in vitro/in silico* immunotoxicology and developmental immunotoxicity. While the roots of the first are more common in Europe, those of the second field of interest are more prevalent in America.

As already mentioned before, there is increasing pressure in the European Union to develop *in vitro* screening methods and thus reduce the number of animals used in toxicological studies, including immunotoxicity screenings. However, a number of questions need to be addressed prior to embarking on validation studies: Which cell source should be used for these *in vitro* studies – human or mouse/rat cell lines or primary cells from lymphoid organs? How can we discriminate between overall cytotoxicity and immunotoxicity to cells? Which end points are to be measured – induction/inhibition of surface marker expression, and/or proliferation, and/or cytokine expression? Which activation stimuli should be used – T-cell mitogens, antigens, B cell, or macrophage stimuli? It is clear that a simple *in vitro* determination of the cytotoxicity of chemicals against cells of lymphoid organs like lymph nodes, bone marrow, thymus, blood, or spleen will not be sufficient to replace *in vivo* screening in immunotoxicology. The fact that the basic principle of immune cell activation is cell–cell interaction will make it absolutely necessary to develop an *in vitro* functional assay which includes coculture of various relevant cell types.

14.5.2 Mishell–Dutton Cultures (*In Vitro* PFCA)

An increasing aim in safety assessment of chemicals and drugs is to reduce, refine, and replace animal testing. Therefore, alternative methods for this purpose are highly desirable. Furthermore, importance of immunotoxicological studies for determining potential adverse effects of pesticides and pharmaceuticals is growing. However, *in vitro* alternatives for immunosuppression are available only at a research level, and up to now, no *in vitro* test for the prediction of immunotoxicity is fully validated or accepted by regulatory authorities.

The production of antigen-specific antibodies represents a major defense mechanism of humoral immune responses, and TDAR, like the PFCA, has been identified in a regulatory context as a main functional test for immunotoxicological investigations. The PFCA *in vitro* equivalent, also known as MD culture or MD test, represents a comprehensive evaluation of immune function based on the interaction of antigen-presenting cells, T cells, and B cells involved in the antigen-specific antibody response. Using MD cultures of mouse spleen cells treated with 11 different test items, we were able to both demonstrate immunosuppressive effects and clearly discriminate between specific immunosuppression and nonspecific cytotoxicity (Koeper and Vohr 2009). To compare the *in vitro* antibody responses of rats and mice, we performed a study with three standards using spleen cells as well as peripheral blood mononuclear cells (PBMC) from both species. Preliminary data showed an excellent concordance between species (including dog, monkey, and humans) as well as different cell sources (Fischer et al. 2011) (Fig. 14.3).

14.5.3 Developmental Immunotoxicity

Another challenge in the near future will be to screen for the impact of a chemical on the developing immune system, i.e., developmental immunotoxicity (DIT). There are different views and opinions with respect to a reasonable test strategy, but at present we are far away from a validated and fully accepted protocol. There are considerable differences in the time required for full development of the immune system across species and in the placental structure and transport mechanisms. Thus, caution is required both in planning the treatment period and dose in the species selected for developmental immunotoxicity studies and in translating the data from animals to humans (Holladay and Smialowicz 2000). Nevertheless, testing developmental immunotoxicity in rats instead of mice would have the advantage that such studies could easily be incorporated into existing test protocols (Luster et al. 2003; Barnett 2005). On the other hand, there is an ongoing discussion about the feasibility and value of incorporation of a DIT module in complex and longlasting rat studies like the extended one-generation reproductive toxicity study (EOGRTS) (Boverhof et al. 2014).

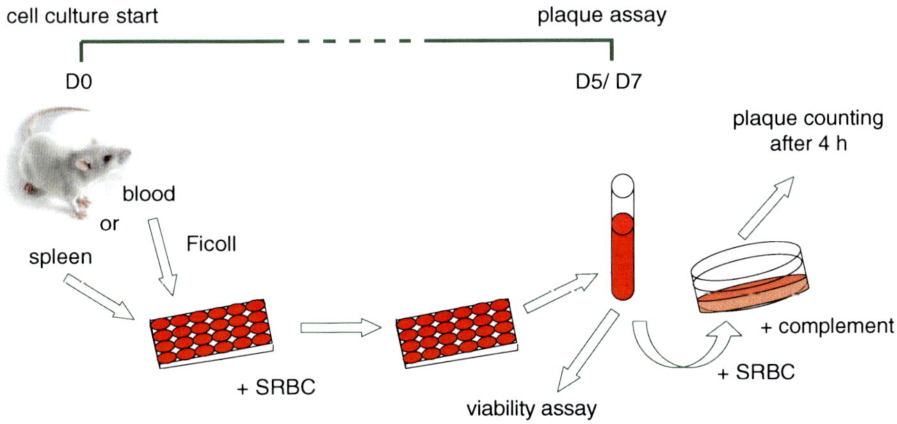

Fig. 14.3 Principle of the in vitro Plaque-Forming Cell Assay, the so-called Mishell–Dutton culture (MD test) as described previously by Koeper and Vohr 2009 and Fischer et al. 2011. The MD test can be performed with splenic or blood cells of rodents as well as other species like dog, monkey, or humans

14.6 Overall Summary

Current immunotoxicology testing approaches for pharmaceuticals, agrochemicals, or animal health products differ significantly from the testing of new and environmental chemicals. Whether comparable immunotoxicological screenings should also routinely be done for industrial, and to some extent also for environmental chemicals, has been discussed controversially for years. Such testing would subject these chemicals to the USEPA OPPTS 870.7800 guideline or the ICH S8 guidance by including a functional test (TDAR) as well as a flow cytometric analysis of blood or splenic cell populations. Especially controversial is whether environmental chemicals should also be subjected to such tests; a trigger-based approach would be necessary.

A trigger-based approach for environmental chemicals could be followed in accordance with the ICH S8 guidance, i.e., evaluation would start with a weight-of-evidence (WoE) assessment. This assessment should include all data available for the relevant chemical, i.e., data from standard toxicity studies, structural similarities to known toxicants, toxicokinetics data, intrinsic properties of a chemical class, and possible routes of exposure. This initial WoE assessment would then trigger an immunotoxicological screening as described in the abovementioned guidelines. Although such an approach sounds reasonable, it raises a number of key questions: Who will collect the data and prepare a WoE assessment? Are there sufficient and reliable data available for environmental chemicals as basis for such an approach? Who will sponsor a full-blown immunotoxicological screening if required? What will be the consequences of positive outcomes of such a study?

Routine toxicological investigations of agrochemicals have produced considerable amounts of data with respect to special immune parameters. These data show that a combination of advanced histopathology and some additional immunological investigations such as PFCA (TDAR) and/or flow cytometric analyses of subpopulations can be used not only to flag immunomodulating chemicals but also to discriminate whether such effects are due to the primary or secondary impact on the immune system. Although direct immunosuppressive as well as immunostimulating effects can be determined by such *in vivo*/*ex vivo* test batteries, there are as yet no robust and validated tests for the determination of other end points such as autoimmunity or type I allergy. The development of widely accepted models for such end points will necessitate much effort in the near future.

Nonetheless, there are already demands for additional new fields of immunotoxicology, i.e., *in vitro* immunotoxicity screenings or developmental immunotoxicity. This will likewise represent a significant challenge for the future.

We are still far from applying these investigations to environmental chemicals as described in this chapter. Our knowledge about the impact of environmental chemicals on the immune system is still fragmented and requires further study. For progress in the field, it will be critical to obtain consensus between the scientific and administrative communities about the path forward. A consolidated expert panel discussion and consultation to move the discussion forward is already being conducted by various associations like ILSI-HESI, ECVAM, and others who have recognized the importance of this alignment. Importantly, these efforts should go beyond publication by providing results in well-defined research projects.

Acknowledgement Thanks go to my esteemed colleague Fred Aswad for his assistance and thoroughly reviewing the manuscript. I like to thank Anna Fischer-Berenbein for providing the basis for two modified figures included here.

References

Angerer J, Ewers U, Wilhelm M (2007) Human biomonitoring: state of the art. Int J Hyg Environ Health 210:201–228

Annest JL, Pirkle JL, Makuc D, Neese JW, Bayse DD, Kovar MG (1983) Chronological trend in blood lead levels between 1976 and, 1980. N Engl J Med 308:1373–1377

Baker SR, Gibson BG (2014) Social oral epidemi(olog)2 y where next: one small step or one giant leap? Community Dent Oral Epidemiol. doi:10.1111/cdoe.12118 [Epub ahead of print]

Barnett J (2005) Developmental immunotoxicology. In: Vohr HW (ed) Immunotoxicology. Springer, Heidelberg, pp 201–203

Boccia S, La Torre G, Persiani R, D'Ugo D, van Duijn CM, Ricciardi G (2007) A critical appraisal of epidemiological studies comes from basic knowledge: a reader's guide to assess potential for biases. World J Emerg Surg 2:1–8

Boverhof DR, Ladics G, Luebke B, Botham J, Corsini E, Evans E, Germolec D, Holsapple M, Loveless SE, Lu H, van der Laan JW, White KL Jr, Yang Y (2014) Approaches and considerations for the assessment of immunotoxicity for environmental chemicals: a workshop summary. Regul Toxicol Pharmacol 68:96–107

Chang ET, Boffetta P, Adami HO, Cole P, Mandel JS (2014) A critical review of the epidemiology of Agent Orange/TCDD and prostate cancer. Eur J Epidemiol [Epub ahead of print]

Dayan AD, Kuper F, Madsen C, Smialowicz RJ, Smith E, Van Loveren H, Vos JC, White KL (1998) Report of validation study of assessment of direct immunotoxicity in the rat. The ICICIS group investigators. International collaborative immunotoxicity study. Toxicology 125(2–3):183

Descotes J (2005) Immunotoxicology: role in the safety assessment of drugs. Drug Saf 28(2):127–136

Dybing E, Schwarze PE, Løvik M, Magnus P (1996) [Air pollution and health]. Tidsskr Nor Laegeforen 116(18):2147–2148 (Norwegian)

Fischer A, Koeper LM, Vohr HW (2011) Specific antibody responses of primary cells from different cell sources are able to predict immunotoxicity in vitro. Toxicol In Vitro 25:1966–1973

Geier J, Uter W, Krautheim A, Lessmann H, Schnuch A (2011) Die häufigsten Kontaktallergene der Jahre 2007–2009. Aktuelle Daten aus dem Informationsverbund Dermatologischer Kliniken (IVDK). Allergo J 20:93–101. German

Germolec DR, Kashon M, Nyska A, Kuper CF, Portier C, Kommineni C, Johnson KA, Luster MI (2004) The accuracy of extended histopathology to detect immunotoxic chemicals. Toxicol Sci 82(2):504–514

Hermann-Kunz E (1999) Incidence of allergic diseases in East and West Germany. Gesundheitswesen 61:100–105. German

Holladay SD, Smialowicz RJ (2000) Development of the murine and human immune system: differential effects of immunotoxicants depend on time of exposure. Environ Health Perspect 108:463–473

Hölzle E, Neumann N, Hausen B, Przybilla B, Schauder S, Hönigsmann H, Bircher A, Plewig G (1991) Photopatch testing: the 5-year experience of the German, Austrian, and Swiss Photopatch Test Group. J Am Acad Dermatol 25:59–68

Koeper LM, Vohr HW (2009) Functional assays are mandatory for a correct prediction of immunosuppressant properties of compounds in vitro. Food Chem Toxicol 47:110–118 [Epub 2008]

Lappin PB, Black LE (2003) Immune modulator studies in primates: the utility of flow cytometry and immunohistochemistry in the identification and characterization of immunotoxicity. Toxicol Pathol 31(Suppl):111–118

Luster MI, Portier C, Pait DG et al (1992) Risk assessment in immunotoxicology. I. Sensitivity and predictability of immune tests. Fundam Appl Toxicol 18:200

Luster MI, Portier C, Pait DG et al (1993) Risk assessment in immunotoxicology. II. Relationship between immune and host resistance tests. Fundam Appl Toxicol 21:71–82
Luster MI, Dean JH, Germolec DR (2003) Consensus workshop on methods to evaluate developmental immunotoxicity. Environ Health Perspect 111:579–583
OECD, Organisation for Economic Cooperation and Development (1981) OECD guidelines for testing of chemicals no. 407 (Repeated Dose 28-day Oral Toxicity Study in Rodents), adopted 1992
Paustenbach D, Galbraith D (2006a) Biomonitoring: is body burden relevant to public health? Regul Toxicol Pharmacol 44:249–261
Paustenbach D, Galbraith D (2006b) Biomonitoring and biomarkers: exposure assessment will never be the same. Environ Health Perspect 114:1143–1149
Richter-Reichhelm H-B, Dasenbrock C, Descotes G, Emmendörfer A, Heinrich UE, Harlemann JH, Hilde¬brand B, Küttler K, Rühl-Fehlert CI, Schilling K, Schulte AE, Vohr H-W (1995) Validation of a modified 28-Day rat study to evidence effects of test compounds on the immune system. Regul Toxicol Pharmakol 22:54–56
Schnuch A, Uter W, Geier J, Lessmann H, Frosch PJ (2004) Contact allergy to farnesol in 2021 consecutively patch tested patients. Results of the IVDK. Contact Dermatitis 50:117–121
Schulz C, Angerer J, Ewers U, Kolossa-Gehring M (2007a) The German human biomonitoring commission. Int J Hyg Environ Health 210:373–382
Schulz C, Conrad A, Becker K, Kolossa-Gehring M, Seiwert M, Seifert B (2007b) Twenty years of the German Environmental Survey (GerES): human biomonitoring – Temporal and spatial (West Germany/East Germany) differences in population exposure. Int J Hyg Environ Health 210:271–297
Stokstad E (2004) Pollution gets personal. Science 304:1892–1894
Thyssen JP, Giménez-Arnau E, Lepoittevin J-P, Menné T, Boman A, Schnuch A (2012a) The critical review of methodologies and approaches to assess the inherent skin sensitization potential (skin allergies) of chemicals. Part I. Contact Dermatitis 66(Suppl 1):11–24
Thyssen JP, Giménez-Arnau E, Lepoittevin J-P, Menné T, Boman A, Schnuch A (2012b) The critical review of methodologies and approaches to assess the inherent skin sensitization potential (skin allergies) of chemicals. Part II. Contact Dermatitis 66(Suppl 1):25–52
Thyssen JP, Giménez-Arnau E, Lepoittevin J-P, Menné T, Boman A, Schnuch A (2012c) The critical review of methodologies and approaches to assess the inherent skin sensitization potential (skin allergies) of chemicals. Part III. Contact Dermatitis 66(Suppl 1):53–70
US-EPA, United States Environmental Protection Agency (1998) Health effects test guidelines: OPPTS 870.7800. Immunotoxicity
Uter W, Geier J, Schnuch A, Frosch PJ (2007) Patch test results with patients' own perfumes, deodorants and shaving lotions: results of the IVDK 1998–2002. J Eur Acade Dermatol Venereol 21:374–379
Uter W, Hegewald J, Pfahlberg A, Lessmann H, Schnuch A, Gefeller O (2010) Contact allergy to thiurams: multifactorial analysis of clinical surveillance data collected by the IVDK network. Int Arch Occup Environ Health 83:675–681
Van der Veen JW, Soeteman-Hernández LG, Ezendam J, Stierum R, Kuper FC, van Loveren H (2014) Anchoring molecular mechanisms to the adverse outcome pathway for skin sensitization: Analysis of existing data. Crit Rev Toxikol 44(7):590–599
Van Loveren H, Germolec D, Koren HS, Luster MI, Nolan C, Repetto R, Smith E, Vos JG, Vogt RF (1999) Report of the Bilthoven Symposium: Advancement of epidemiological studies in assessing the human health effects of immunotoxic agents in the environment and the workplace. Biomarkers 4:135–157
Vohr H-W (1995) Experiences with an advanced screening procedure for the identification of chemicals with an immunotoxic potential in routine toxicology. Toxicology 104:149–158
Vohr H-W, Rühl-Fehlert C (2001) Industry experience in the identification of the immunotoxic potential of agrochemicals. Sci Total Environ 270:123–134
Vos JG, van Loveren H (1995) Markers for immunotoxic effects in rodents and man. Toxicol Lett 82–83:385–394
Vos JG, Van Logten MJ, Kreeftenberg JG, Steerenberg PA, Kruizinga W (1979) Effect of hexachlorobenzene on the immune system of rats following combined pre- and post-natal exposure. Drug Chem Toxicol 2:61

Environmental Pollution: A Human Rights Perspective

15

Dinah Shelton

Contents

15.1 An Introduction to Human Rights Law

National constitutions throughout the world contain enumerated rights and freedoms for individuals residing within the state's territory or subject to its jurisdiction. In the twentieth century, the international community increasingly recognized that such constitutional guarantees sometimes prove inadequate or even illusory when military coups, armed conflicts, or repressive governments disrupt or deliberately ignore the rule of law, including constitutional limits on the exercise of power. Responding to this awareness, international and regional organizations created or reformed after the Second World War recognized that human rights must be considered a matter of international concern if individuals and groups are to be ensured

D. Shelton
Manatt/Ahn Professor of Law (emeritus), The George Washington University Law School, Washington, DC, USA
e-mail: dshelton@law.gwu.edu

C. Esser (ed.), *Environmental Influences on the Immune System*,
DOI 10.1007/978-3-7091-1890-0_15

their fundamental rights and freedoms. With the leadership of a group of states and strong advocacy from civil society groups, intergovernmental organizations began elaborating the international law of human rights.

15.1.1 The United Nations (UN)

At the global level, the United Nations Charter, a binding treaty, contains human rights obligations that are freely accepted by each state when it joins the organization, but the Charter does not contain a list of guaranteed rights. This lacuna led the UN to begin almost immediately to draft an International Bill of Rights. The first step was the adoption without dissent on December 10, 1948, of the Universal Declaration of Human Rights, a text cited in virtually every subsequent human rights instrument and incorporated into the constitutions of many new states. Although the Declaration was adopted as a nonbinding resolution of the UN General Assembly, it is today considered to define the term human rights as used in the binding UN Charter and is thus the standard by which the performance of each UN member state is judged.

The Declaration was further transformed into binding law through the adoption of two treaties in 1966: the International Covenant on Civil and Political Rights (ICCPR) and the International Covenant on Economic Social and Cultural Rights (ICESCR). Together, these three instruments are referred to as the International Bill of Rights (Box 15.1).

Box 15.1 The International Bill of Rights
Universal Declaration of Human Rights, 1948
International Covenant on Civil and Political Rights, 1966
International Covenant on Economic, Social, and Cultural Rights, 1966

Subsequent UN standard-setting has sought to protect particularly vulnerable groups (racial minorities, women, children, indigenous peoples, persons with disabilities) or to prevent and punish particularly egregious violations (slavery, torture, forced disappearances). The UN considers nine of its global treaties to be "core" agreements to which each member state should adhere (i.e., become a "state party"), although no treaty will bind a state without its consent.

Each UN core treaty establishes its own monitoring body, composed of independent experts elected by the participating states for a fixed term of office. The monitoring bodies receive periodic reports from states parties to the treaties and have (usually optional) jurisdiction to receive complaints by a state party or a victim against a state that has accepted the treaty and the complaints procedure. Some treaty bodies have additional powers of investigation. Reports on the work of each treaty body are submitted annually to the General Assembly.

Whether or not a state accepts to be bound by any or all of the core human rights treaties, the UN monitors human rights performance through a procedure known as

Universal Periodic Review, pursuant to which the UN Human Rights Council, an organ established by the General Assembly in 2006 to replace the former UN Commission on Human Rights, conducts a periodic peer review. The Council also appoints thematic working groups or rapporteurs to conduct studies or investigate particular human rights issues or problematic countries; and the Council maintains a complaints procedure that allows anyone to denounce a situation of gross and systematic violations of human rights within a state. The studies authorized by the Council in 2013 include the topic of human rights and the environment, undertaken by an independent expert (now called special rapporteur) appointed for a 3-year term.

15.1.2 Regional Systems

Regional organizations reinforce the human rights efforts of the United Nations and offer something that does not exist at the global level: courts with jurisdiction to render binding judgments and award redress to victims of violations.

15.1.2.1 Europe

In 1950, ten "like-minded" governments adopted the European Convention for the Protection of Human Rights and Fundamental Freedoms (ECHR) (Rome, Nov. 4, 1950) taking "the first steps for the collective enforcement of certain of the rights stated in the Universal Declaration" (ECHR, preamble). Those rights are largely civil and political rights; a separate European Social Charter (1961) contains economic and social rights. Both treaties have been amended and supplemented over time to add further guarantees and improve procedures.

Initially, the jurisdiction of the European Court of Human Rights was extremely limited, in that both the right of individuals to file petitions and the jurisdiction of the court were optional and need not be accepted by any of the small number of original states parties. Today, due to a series of reforms and geographic expansion, the Court has mandatory jurisdiction over the 47 member states of the Council of Europe, allowing more than 800 million people the possibility of "going to Strasbourg" once they have exhausted available local remedies.

15.1.2.2 The Americas

The Organization of American States (OAS) serves as the body of regional cooperation in the Americas, including on matters of human rights. From the late nineteenth century, regional conferences of the American states, which preceded the OAS, acted on issues concerning the rights of political asylum-seekers, aliens, women, and children.[1] The OAS adopted the American Declaration of the Rights and Duties of Man on May 2, 1948, simultaneously with concluding the constitutional Charter of the OAS, some 6 months before the adoption of the UN's Universal Declaration

[1] On the early history of the inter-American system, see ANNA P SCHREIBER, THE INTER-AMERICAN COMMISSION ON HUMAN RIGHTS (Sijthoff 1970).

of Human Rights. Notably, Latin American states were instrumental in promoting the inclusion of references to human rights in the UN Charter.[2]

The OAS did not immediately establish an institutional framework for monitoring compliance with the American Declaration, but in 1959, its General Assembly created the Inter-American Commission on Human Rights (IACHR), a body of seven independent experts who serve one or two four-year terms. A decade later, in 1969, the OAS member states adopted the American Convention on Human Rights (ACHR), which expanded the jurisdiction of the IACHR and created an Inter-American Court of Human Rights. For those persons residing within 23 states parties to the American Convention on Human Rights, it is possible to take a case from the domestic legal system to the Inter-American Commission on Human Rights and thereafter to the Inter-American Court, which has broad remedial powers. The other 11 OAS member states remain subject to the IACHR, but the cases cannot proceed to the Court because the states have not accepted the Court's jurisdiction. Protection of socioeconomic rights was added in 1988, with the adoption of the Protocol to the American Convention on Human Rights in the Area of Economic, Social, and Cultural Rights (Protocol of San Salvador, 17 Nov. 1988). This Protocol extends the petition or complaints system to include trade union rights and the right to education. The OAS has also adopted additional specialized treaties (Box 15.2) and important political declarations, for example, the Inter-American Democratic Charter of 2001.

Box 15.2 Inter-American Specialized Human Rights Treaties

Inter-American Convention to Prevent and Punish Torture (Cartagena de Indias, 9 Dec. 1985), OASTS No. 67

Protocol to the American Convention on Human Rights to Abolish the Death Penalty (Asuncion, 8 June 1990), OASTS No. 73

Inter-American Convention on Forced Disappearance of Persons (Belem do Para, 9 June 1994)

Inter-American Convention on the Prevention, Punishment, and Eradication of Violence against Women (Belem do Para, 9 June 1994), OASTS 5 March 1995

Inter-American Convention on the Elimination of All Forms of Discrimination against Persons with Disabilities (Guatemala City, 7 June 1999)

15.1.2.3 Africa

In 1981, the then-Organization of African Unity (OAU) adopted the African Charter on Human and Peoples' Rights (AfCHPR), now accepted by all 53 member states. In the Preamble to the Charter, the member states of the OAU reaffirmed their commitment to the human rights instruments of the United Nations. The sole

[2] Mary Ann Glendon, *The Forgotten Crucible: The Latin American Influence on the Universal Human Rights Idea* (2003) 16 Harv Hum Rts J 27.

supervisory body that the African Charter foresaw was the African Commission of Human and Peoples' Rights, which held its first session in 1987. The Charter explicitly mandates the African Commission to "draw inspiration from international law on human and peoples' rights" (African Charter, art. 60). The Protocol to the African Charter on Human and Peoples' Rights on the Rights of Women in Africa supplemented the Charter in 2003. A 1990 African Charter on the Rights and Welfare of the Child entered into force in 1999, following which the Committee of Experts on the Rights and Welfare of the Child was established in 2001. In 2002, the African Union (AU), whose Constitutive Act, Article 3 h, recognizes human rights as one of the Union's objectives, replaced the OAU. In 2004, a Protocol to the African Charter on Human and Peoples' Rights on the Establishment of an African Court on Human and Peoples' Rights (June 1988) entered into force; its first judges were elected in 2006.

15.1.2.4 Other Regions

Other regional and subregional intergovernmental organizations around the world have also addressed human rights in recent years. Some have created human rights institutions, if not fully equipped human rights systems. These include initiatives in Southeast Asia and the Arab-speaking world, as well as subregional bodies in Europe, the Americas, and Africa. These regional bodies provide platforms to states and civil society, where people can potentially make their voices heard in the global human rights discourse. The opportunities for participation offered by the regional systems can help bridge the gap between the universality of human rights norms, on the one hand, and the cultural and political particularities of each region, on the other.

15.2 The Environment as a Human Rights Issue

Environmental degradation became a matter of national and international concern beginning in the 1960s, some two decades after human rights emerged on the international agenda. Given the timing, there are few explicit references to environmental matters in the earlier-drafted international human rights instruments. The two most often-cited provisions are ICESCR article 12, where the right to health expressly calls on states parties to take steps for "the improvement of all aspects of environmental and industrial hygiene" and article 24 of the Convention on the Rights of the Child (20 Nov. 1989) which requires that states parties shall take appropriate measures to combat disease and malnutrition "through the provision of adequate nutritious foods and clean drinking water, taking into consideration the dangers and risks of environmental pollution."

Despite this lack of specific reference to the environment in human rights treaties, international awareness of the linkages between human rights and environmental protection has expanded considerably since the emergence of environmental protection as a legal issue. The definition of pollution in international and domestic law partly explains the linkage, establishing that only those substances that are harmful to human health or other interests constitute pollution (Box 15.3).

Box 15.3 International Treaty Definition of Pollution
Pollution of the marine environment is

> the introduction by man, directly or indirectly, of substances or energy into the marine environment, including estuaries, which results or is likely to result in such deleterious effects as harm to living resources and marine life, hazards to human health, hindrance to marine activities, including fishing and other legitimate uses of the sea, impairment of quality for use of sea water and reduction of amenities.
>
> United Nations Convention on the Law of the Sea (1982), Art. 1(4)

The links between human rights and environmental protection were apparent at least from the first international conference on the human environment, held in Stockholm in 1972. At the Stockholm concluding session, the preamble of the final declaration[3] proclaimed that

> Man is both creature and moulder of his environment, which gives him physical sustenance and affords him the opportunity for intellectual, moral, social and spiritual growth.... Both aspects of man's environment, the natural and the man-made, are essential to his well-being and to the enjoyment of basic human rights – even the right to life itself.

Principle 1 of the Stockholm Declaration established a further connection between human rights and environmental protection, declaring that

> Man has the fundamental right to freedom, equality and adequate conditions of life, in an environment of a quality that permits a life of dignity and well-being.

In resolution 45/94, the UN General Assembly recalled the language of Stockholm, stating that all individuals are entitled to live in an environment adequate for their health and well-being. The resolution called for enhanced efforts to ensure a better and healthier environment.

15.2.1 Environmental Quality as a Prerequisite to the Enjoyment of Human Rights

Many human rights tribunals and experts understand that environmental protection is a precondition to the enjoyment of several internationally guaranteed human rights, especially the rights to life and health. Environmental protection is thus seen as an essential instrument subsumed in or a prerequisite to the effort to secure the

[3] Stockholm Declaration of the United Nations Conference on the Human Environment, 16 June 1972, U.N. Doc. A/.CONF.48/14/Rev.1 at 3 (1973).

effective enjoyment of human rights. In this sense, the General Assembly has called the preservation of nature "a prerequisite for the normal life of man" (GA Res. 35/48 of 30 October 1980).

Several UN treaty bodies or experts have addressed the intersection of human rights and environmental protection from this perspective. The former United Nations Human Rights Commission appointed a Special Rapporteur on the adverse effects of the illicit movement and dumping of toxic and dangerous products and wastes on the enjoyment of human rights, conferring a mandate that included investigating complaints about such trade.[4] In its resolutions on this matter, the Commission consistently recognized that environmental law violations "constitute a serious threat to the human rights to life, good health and a sound environment for everyone" (Commission on Human Rights, Resolutions 199/23 and 2000/72).

The Commission also named a Special Rapporteur on the right to food whose mandate includes the issue of safe drinking water, subsequently proclaimed to be a right in itself. The Commission specifically linked the issue of the right to food with sound environmental policies and noted that problems related to food shortages "can generate additional pressures upon the environment in ecologically fragile areas." Other resolutions of the Commission referred explicitly to the right to a safe and healthy environment.[5] In recent years, the Human Rights Council has adopted resolutions on climate change as a human rights issue.

15.2.2 Human Rights Necessary for Environmental Protection

Another approach to the linkage of these issues considers certain human rights as essential elements to achieving sound environmental protection. This approach is well illustrated by the Rio Declaration on Environment and Development, adopted at the conclusion of the 1992 Conference of Rio de Janeiro on Environment and Development. Principle 10 formulates a link between human rights and environmental protection largely in procedural terms, declaring in Principle 10 that access to information, public participation, and access to effective judicial and administrative proceedings, including redress and remedy, should be guaranteed because "environmental issues are best handled with the participation of all concerned

[4]Resolution 2001/35, Adverse effects of the illicit movement and dumping of toxic and dangerous products and wastes on the enjoyment of human rights, E/CN.4/RES/2001/35. See the Report of the Special Rapporteur on the Adverse Effects of the Illicit Movement and Dumping of toxic and Dangerous Products and Wastes on the Enjoyment of Human Rights, Addendum, Commission on Human Rights, E/CN.4/2001/55/Add.1 (21 Dec. 2000), documenting *inter alia* damage to tissues from arsenic poisoning, risks to health from the dumping of heavy metals, illnesses from pesticide use at banana plantations, deaths from petrochemical dumping, and kidney failure in children due to contaminated pharmaceuticals.

[5]In Resolution 2001/65, entitled "Promotion of the Right to a Democratic and Equitable International Order", the Commission affirmed that "a democratic and equitable international order requires, *inter alia*, the realization of ... [t]he right to a healthy environment for everyone."

citizens, at the relevant level." These procedural rights, contained in all human rights instruments, are thus adopted in environmental texts in order to have better environmental decision-making and enforcement.

15.2.3 Environmental Quality as a Human Right

Still other legal texts proclaim the existence of a right to a safe and healthy environment as a human right. Two regional human rights treaties contain such a provision. The African Charter on Human and Peoples' Rights, Article 16, guarantees to every individual the right to enjoy the best attainable state of physical and mental health while Article 24 was the first international treaty to proclaim that "All peoples shall have the right to a general satisfactory environment favorable to their development." The 1988 Additional Protocol to the American Convention on Human Rights in the area of Economic, Social, and Cultural Rights,[6] in its Article 11, followed this precedent in proclaiming that "Everyone shall have the right to live in a healthy environment and to have access to basic public services" and that the states parties shall promote the protection, preservation, and improvement of the environment. Also at the regional level, the preambles of European Union legal texts often state their aim as being "to protect human health and the environment."[7]

On the national level, more than 100 constitutions throughout the world guarantee a right to a clean and healthy environment, impose a duty on the state to prevent environmental harm, or mention the protection of the environment or natural resources. Such provisions vary in the chosen description of the environmental quality that is protected. While many of the older provisions refer to a "healthy" or "healthful" environment, more recent formulations add references to ecology and/or biodiversity to the guarantee.

Box 15.4 Examples of Constitutional Guarantees for Environmental Quality

Angola: "all citizens shall have the right to live in a healthy and unpolluted environment" (art. 24–1).

Argentina: "all residents enjoy the right to a healthy, balanced environment which is fit for human development ..." (art. 41).

Azerbaijan: "everyone has the right to live in a healthy environment."

Brazil: "everyone has the right to an ecologically balanced environment, which is a public good for the people's use and is essential for a healthy life" (art. 225).

[6] Additional Protocol to the American Convention on Human Rights in the Area of Economic, Social, and Cultural Rights (San Salvador, Nov. 17, 1988, OAS T.S. 69).

[7] EC Council Directive No. 85/201 on Air Quality Standards for Nitrogen Dioxide, 7 Mar. 1985, L 87 O.J.E.C. (1985); EC Council Directive No. 80/779 on Air Quality Limit Values, 15 July 1980, L 229, O.J.E.C. 30 (1980).

Chile: Article 19 of the 1980 Constitution of Chile provides for a "right to life" and a "right to live in an environment free of contamination" and establishes that certain other individual rights may be restricted to protect the environment (CHILE CONST, art. 19 §§ 1, 8).

France: The right to live in a "balanced environment, favorable to human health" (Charter of the Environment, 2005).

Hungary: "Hungary recognizes and implements everyone's right to a healthy environment" (A MAGYAR KÜZTÁRSASÁG ALKOTMÁNYA [Constitution] art. 18 (Hung)).

Quebec: The provincial Charter provides "Every person has a right to live in a healthful environment in which biodiversity is preserved, to the extent and according to the standards provided by law" (Charter of Human Rights and Freedoms, R.S.Q., c. C-12, s. 46).

The US federal constitution does not mention the environment, but states in the USA have the power to provide their citizens with rights additional to those contained in the federal constitution and state constitutions revised or amended from 1970 to the present have added environmental protection among their provisions.[8] To mark the occasion of the first Earth Day in 1970, the Pennsylvania legislature adopted a proposed amendment to the state constitution,[9] subsequently approved overwhelmingly by voters in the state,[10] adding what is now Article I, section 27, to the state constitution:

> Section 27. Natural resources and the public estate.
>
> The people have a right to clean air, pure water, and to the preservation of the natural, scenic, historic and aesthetic values of the environment. Pennsylvania's public natural resources are the common property of all the people, including generations yet to come. As trustee of these resources, the Commonwealth shall conserve and maintain them for the benefit of all the people.

---

[8] See Ala. Const. art. VIII; Cal. Const. art. X, § 2; Fla. Const. art. II, § 7; Haw. Const. art. XI; Ill. Const. art. XI; La. Const. art. IX; Mass. Const. § 179; Mich. Const. art. IV, § 52; Mont. Const. art. IX, § 1; N.M. Const. art. XX, § 21; N.Y. Const. art. XIV; N.C. Const. art. XIV, § 5; Ohio Const. art. II, § 36; Pa. Const. art. I, § 27; R.I. Const. art. 1, § 17; Tex. Const. art. XVI, § 59; Utah Const. art. XVIII; and Va. Const. art. XI, § 1. For discussions of these provisions, see A. E. Dick Howard, *State Constitutions and the Environment*, 58 VA. L. REV. 193, 229 (1972); Roland M. Frye, Jr., *Environmental Provisions in State Constitutions*, 5 ENVTL. L. REP. 50028–29 (1975); Stewart G. Pollock, *State Constitutions, Land Use, and Public Resources: The Gift Outright*, 1984 ANN. SURV. AM. L. 13, 28–29; Robert A. McLaren, Comment, *Environmental Protection Based on State Constitutional Law: A Call for Reinterpretation*, 12 U. Haw. L. REV. 123, 126–27 (1990).; and Carole L. Gallagher, *The Movement to Create an Environmental Bill of Rights: From Earth Day 1970 to the Present*, 9 FORDHAM ENVTL. L.J. 107 (1997).

[9] Franklin L. Kury, *The Pennsylvania Environmental Protection Amendment*, PA. B. ASS'N Q., Apr. 1987, at 85, 87.

[10] The vote was more than 3–1 in favor of the amendment, with close to 2 million voters. See Franklin L. Kury, *The Environmental Amendment to the Pennsylvania Constitution: Twenty Years Later and Largely Untested*, 1 VILL. ENVTL. L.J. 123, 123–24 (1990).

The amendment and others like it were intended to elevate environmental protection as a fundamental value to a constitutional status above the states' legislative and regulatory norms and to protect the environment beyond issues of human health.[11] A second aim was to expand standing to sue to allow public interest litigation on behalf of the environment.[12] Illinois, Massachusetts, and Montana all amended their constitutions in 1972 to provide in similar fashion a right to a clean and healthful environment (Box 15.5).

Box 15.5 US State Constitutional Provisions on Environmental Rights

Hawai'i: "Each person has the right to a clean and healthful environment, as defined by law relating to environmental quality, including control of pollution and resources. Any person may enforce this right against any party, public or private, through appropriate legal proceedings" (Constitution, Article XI, section 9).

Massachusetts: Guarantees the right to clean air and water, freedom from excessive and unnecessary noise, and the natural scenic, historic, and aesthetic qualities of their environment (Mass. Const. art. XLIX).

Montana: "The people shall have the right to clean air and water, freedom from excessive and unnecessary noise, and the natural, scenic, historic, and esthetic qualities of their environment; and the protection of the people in their right to the conservation development and utilization of the agricultural, mineral, forest, water, air and other natural resources is hereby declared to be a public purpose" (Mont. Const. XLIX).

The Supreme Court of Montana has provided the most detail about the substantive implications of a right to a specified environmental quality. In *Montana Environmental Information Center et al. v. Department of Environmental Quality*,[13] the plaintiffs contended that the constitution's environmental protections were violated by the legislature when it amended state law to provide a blanket exception to requirements governing discharges from water well without regard to the degrading effect that the discharges would have on the surrounding

[11] The Pennsylvania Supreme Court has indicated that environmental litigants may sue for generalized harm because "[a]esthetic and environmental well-being are important aspects of the quality of life in our society" and because its constitution establishes a local government's duty to protect its citizen's "quality of life" (*Commonwealth, Pa. Game Comm'n v. Commonwealth, Dept. of Envtl. Resources*, 509 A.2d 877, 883–84 (Pa. Comm2. Ct. 1986), aff'd 555 A.2d 812 (Pa. 1989)).

[12] For example, see *Life of the Land v. Land Use Comm'n of the State of Hawai'i*, 623 P.2d 431 (Haw. 1981) (granting standing to an environmental organization which sought to challenge a reclassification of certain lands which were not owned by any of the organization's members. The Supreme Court held that the plaintiffs' "aesthetic and environmental interests" were "personal" rights guaranteed by Art. XI, Section. 9, of the Constitution). See also *Richard v. Metcalf*, 921 P.2d 122 (Haw. 1997); *Kahuna Sunset Owners Ass'n v. Mahui County Council*, 948 P.2d 122 (Haw. 1997).

[13] 296 Mont. 207, 988 P.2d 1236 (1999).

environment. The monitoring of well tests was also inadequate because it was done without regard to the harm caused by those tests. The Court concluded that "the right to a clean and healthful environment is a fundamental right because it is guaranteed by the Declaration of Rights found at Article II, Section 3 of Montana's Constitution, and that any statute or rule which implicates that right must be strictly scrutinized and can only survive scrutiny if the State establishes a compelling state interest and that its action is closely tailored to effectuate that interest and is the least onerous path that can be taken to achieve the State's objective." The Court examined the drafting history of the Constitutional provision and held as follows:

> The delegates did not intend to merely prohibit that degree of environmental degradation which can be conclusively linked to ill health or physical endangerment. Our constitution does not require that dead fish float on the surface of our state's rivers and streams before its farsighted environmental protections can be invoked....
>
> We conclude that the constitutional right to a clean and healthy environment and to be free from unreasonable degradation of that environment is implicated based on the Plaintiffs' demonstration that the pumping tests proposed by SPJV would have added a known carcinogen such as arsenic to the environment in concentrations greater than the concentrations present in the receiving water and that the DEQ or its predecessor after studying the issue and conducting hearings has concluded that discharges containing carcinogenic parameters greater than the concentrations of those parameters in the receiving water has a significant impact which requires review pursuant to Montana's policy of nondegradation....[14]

Other national courts have similarly given broad reading to constitutional guarantees and have done so through reference to national and international environmental standards. Article 24, the South African constitution, provides that:

> Everyone has a right to (a) to an environment that is not harmful to their health or well being; and (b) to have the environment protected, for the benefit of present and future generations, through reasonable legislative and other measures that (i) prevent pollution and ecological degradation; (ii) promote conservation; and (iii) secure ecologically sustainable development and use of natural resources while promoting justifiable economic and social development.[15]

---

[14] The Montana Supreme Court further applied its constitutional provision in the case *Cape-France Enterprises v. The Estate of Peed,* 305 Mont. 513, 29 P.3d 1011 (2001), in which it held that "the protections and mandates of this provision apply to private action – and thus to private parties – as well" as to state action. Thus, "it would be unlawful for Cape-France, a private business entity, to drill a well on its property in the face of substantial evidence that doing so may cause significant degradation of uncontaminated aquifers and pose serious public health risks." The court held that it would be a violation of the state's obligation under the constitution for it to grant specific performance of a contract for the sale of the land in question. See Chase Naber, *Murky Waters: Private Action and the Right to a Clean and Healthful Environment – An examination of Cape-France Enterprises v. Estate of Peed,* 64 MONT. L. REV. 357 (2003) and B. Thompson, *Constitutionalizing the Environment: The History and Future of Montana's Environmental Provisions*, 64 MONT.L.REV. 157 (2003).

[15] S. AFR. CONST. ch. IV, § 24.

The South African Constitutional Court has explicitly relied on international environmental principles in giving substantive content to this constitutional guarantee.[16] In the Court's view, the National Environmental Management Act, which was enacted to give effect to section 24 of the constitution, embraces the concept of sustainable development, defined to mean "the integration of social, economic and environmental factors into planning, implementation and decision-making for the benefit of present and future generations." In turn, this broad definition of sustainable development integrates environmental protection and socioeconomic development and incorporates the internationally recognized principle of intergenerational and intragenerational equity.[17] The Court thus set aside the decision of the environmental authorities and required reconsideration consistent with the judgment. As to the role of the courts in giving effect to environmental rights, the Court was clear:

> The role of the courts is especially important in the context of the protection of the environment and giving effect to the principle of sustainable development. The importance of the protection of the environment cannot be gainsaid. Its protection is vital to the enjoyment of the other rights contained in the Bill of Rights; indeed, it is vital to life itself. It must therefore be protected for the benefit of the present and future generations. The present generation holds the earth in trust for the next generation. This trusteeship position carries with it the responsibility to look after the environment. It is the duty of the court to ensure that this responsibility is carried out.[18]

---

[16] *Fuel Retailers Association of Southern Africa v Director-General Environmental Management, Department of Agriculture, Conservation and Environment, Mpumalanga Province, and Others,* Case no CCT 67/06; ILDC 783 (ZA 2007). The case arose out of a decision by a provincial Department of Agriculture, Conservation, and Environment to grant private parties permission to construct a filling station.

[17] *Id,* paras. 59. In addition, NEMA sets out some of the factors that are relevant to decisions on sustainable development. These factors largely reflect international experience. But as NEMA makes it clear, these factors are not exhaustive. The Court quoted the factors set forth in the domestic National Environmental Management Act, Section 2(4)(a):
"Sustainable development requires the consideration of all relevant factors including the following:

(i) That the disturbance of ecosystems and loss of biological diversity are avoided, or, where they cannot be altogether avoided, are minimised and remedied;
(ii) that pollution and degradation of the environment are avoided, or, where they cannot be altogether avoided, are minimised and remedied;
(iii) that the disturbance of landscapes and sites that constitute the nation's cultural heritage is avoided, or where it cannot be altogether avoided, is minimised and remedied;
(iv) that waste is avoided, or where it cannot be altogether avoided, minimised and re-used or recycled where possible and otherwise disposed of in a responsible manner;
(v) that the use and exploitation of non-renewable natural resources is responsible and equitable, and takes into account the consequences of the depletion of the resource;
(vi) that the development, use and exploitation of renewable resources and the ecosystems of which they are part do not exceed the level beyond which their integrity is jeopardised;
(vii) that a risk-averse and cautious approach is applied, which takes into account the limits of current knowledge about the consequences of decisions and actions; and
(viii) that negative impacts on the environment and on people's environmental rights be anticipated and prevented, and where they cannot be altogether prevented, are minimised and remedied."

[18] *Id.*, para. 102.

In India, a series of judgments between 1996 and 2000 responded to health concerns caused by industrial pollution in Delhi. In some instances, the courts issued orders to cease operations.[19] The Indian Supreme Court based closure orders on the principle that health is of primary importance and that residents are suffering health problems due to pollution. In Argentina, the right to environment is deemed a subjective right entitling any person to initiate an action for environmental protection.[20] Colombia also recognizes the enforceability of the right to environment.[21] In Costa Rica, a court stated that the rights to health and to the environment are necessary to ensure that the right to life is fully enjoyed.[22] The French Conseil Constitutionnel has used its Constitutional Charter to review legislative enactments, finding that the Charter constitutes a "fundamental freedom" of constitutional value allowing for the suspension of an administrative decision under French procedural law.

15.2.4 Why a "Rights-Based" Approach?

Many lawyers concerned either with the environment or with human rights prefer this "rights-based approach" to environmental protection in contrast to relying on environmental regulations, private litigation, or market-based incentives, because human rights are generally seen as maximum claims on society, elevating concern for the environment above a mere policy choice that may be modified or discarded at will. All legal systems establish a hierarchy of norms. Constitutional guarantees usually are at the apex and "trump" any conflicting norm of lower value. Thus, to include respect for the environment as a constitutional right ensures that it will be given precedence over other legal norms that are not constitutionally based. In addition, the moral weight afforded by the concept of rights as inherent attributes that must be respected in any well-ordered society exercises an important compliance pull. Finally, at the international level, enforcement of human rights law is more developed than are the procedures of international environmental law. The availability of individual complaints procedures has given rise to extensive jurisprudence from which the specific obligations of states to protect and preserve the environment are detailed. The danger of placing confidence in the regulatory process alone

[19] See, e.g., *M.C. Mehta v. Union of India & Others,* JT 1996, reprinted in 1 sat 631.

[20] *Kattan, Alberto and Others v. National Government, Juzgado Nacional de la Instancia en lo Contenciosoadministrativo Federal.* No. 2, Ruling of 10 May 1983, La Ley, 1983-D, 576; *Irazu Margarita v. Copetro S.A., Camara Civil y Comercial de la Plata,* Ruling of 10 May 1993 (available at www.eldial.com) ("The right to live in a healthy and balanced environment is a fundamental attribute of people. Any aggression to the environment ends up becoming a threat to life itself and to the psychological and physical integrity of the person.").

[21] *Fundepublico v. Mayor of Bugalagrande and Others,* Juzgado Primero superior, Interlocutorio # 032, Tulua, 19 Dec. 1991 ("It should be recognized that a healthy environment is a sina qua non condition for life itself and that no right could be exercised in a deeply altered environment.").

[22] *Presidente de la sociedad Marlene S.A. v. Municipalidad de Tibas,* Sala Constitucional de la corte Supreme de justicia. Decision No. 6918/94 of 25 Nov. 1994.

is illustrated by *Zander v. Sweden*,[23] where the applicants complained about contamination of their well water by cyanide from a neighboring dump site. The municipality initially furnished temporary water supplies, but later, adhering to the normal regulatory procedures, the town raised the permissible level of cyanide in the city water supply. The permit for the dump was later renewed and expanded while the applicant's request for safe drinking water was denied.[24] The European Court of Human Rights found in favor of the individual who had no redress before domestic courts.

Human rights, enshrined in international and constitutional law, thus set the limits of majority rule as well as provide protection against dictatorial repression. The scope and contours of substantive as well as procedural rights are sometimes detailed in legislation, but they are also given content through litigation. International human rights tribunals in particular elaborate on the often generally stated rights whose implementation they monitor.

15.2.5 Linking Human Rights to Obligations: What Human Rights Tribunals Say

Human rights tribunals have given effect to various human rights linked to environmental protection by reference to international environmental principles, standards, and norms. In addition, they have emphasized the importance of giving effect to national environmental rights provisions. As a general matter, the European Court of Human Rights has indicated that the scope of rights guaranteed by the European Convention is affected by the "growing and legitimate concern both in Europe and internationally about offenses against the environment."[25] In its *Öneryıldız v.*

[23] *Zander v. Sweden*, App. No. 14282/88, Eur. Ct. Hum. Rts [1993] Ser. A, No. 279B. Concededly, it was the denial of judicial review of this decision that formed the basis of Lander's successful claim before the European Court. The Court, finding that the applicants had a right to clean water under Swedish law, held that the lack of judicial review violated the European Convention, Article 6(1), because the applicants were entitled as of right to seek precautionary measures against water pollution.

[24] The European Court did not actually have to reach a conclusion on the substance of this decision, because it found that the applicant's procedural right of access to justice under Article 6 was violated. The applicants had been unable to obtain judicial review by Swedish courts of the board's permitting decision.

[25] See *Mangouras v. Spain*, no. 12050/04, 8 Jan. 2009, para. 41 (referred to a Grand Chamber 5 June 2009). Increased concern with the environment has also proved important in cases where states have taken measures to protect the environment and the actions are resisted on the grounds that they interfere with the right to property. In *Fredin v. Sweden*, the applicant argued that nature protection was an inadequate reason to revoke a license to extract gravel on his property and therefore was a violation of Article I, Protocol 1. The Court found no violation, noting that the protection of the environment is an increasingly important consideration (*Fredin v. Sweden*, No. 12033/86, 13 EHRR. 784 (1991)). The Court similarly found no violation of the same provision in *Pine Valley Developments Ltd v. Ireland*, where permission to carry out construction in a greenbelt area was revoked on grounds of environmental protection (*Pine Valley Devs. Ltd. v Ireland*, App. No. 12742/87, 14 EHRR. 319 (1992)). The most difficult and contentious cases in this respect have concerned travelers or gypsies, whose lifestyle may bring them into contact with modern land use

Turkey[26] judgment, the European Court referred to environmental laws, in particular the Convention on Civil Liability for Damage resulting from Activities Dangerous to the Environment (Lugano, 21 June 1993) and the Convention on the Protection of the Environment through Criminal Law (Strasbourg, 4 November 1998) in finding violations of the right to life.

Taşkin and Others v. Turkey involved challenges to the development and operation of gold mine, which the applicants alleged caused environmental damage to the detriment of people in the region. Appropriate procedures had been followed; the challenge was to the substance of the decision taken. Applicants litigated the issue and won in domestic courts. The Turkish Supreme Administrative Court repeatedly concluded that the mine's operating permit did not serve the public interest and that the safety measures which the company had taken did not suffice to eliminate the risks involved in such an activity. Before the European Court, the applicants alleged a violation of Article 8.

In reviewing the applicable legal framework, the Court referred to the procedural rights set forth in Rio Principle 10 and the subsequent Aarhus Convention elaborating on these rights. In addition, however, the Court also quoted from a Council of Europe Parliamentary Assembly resolution on environment and human rights that recommended that member states ensure appropriate protection of life, health, family and private life, physical integrity, and private property, taking particular account of the need for environmental protection, and that member states recognize a human right to a healthy, viable, and decent environment. Such right corresponds to an objective obligation for states to protect the environment in national laws, preferably at the constitutional level. Given this recommendation and the domestic Constitutional guarantees in Turkey, the Court found a violation despite the absence of any accidents or incidents at the mine. The mine was deemed to present an unacceptable risk.

The European Court's application of environmental standards reached a new level in a judgment delivered 27 January 2009 in the case of *Tatar v. Romania*. The case arose in the aftermath of an ecological disaster at a gold mine in Romania, which resulted in high levels of sodium cyanide and heavy metals being released into local freshwaters. The contaminated water passed into the Tisza River in Hungary and eventually into the Danube River, causing pollution as far as the Black Sea. After two Romanians, father and son, were unable to achieve any redress through Romanian administrative and penal procedures, they complained to the European Court, alleging violations of Convention articles 2 and 8. As in other cases, the Court made note of the right to a healthy and balanced environment under the Romanian Constitution and of the domestic law implementing this right. The Court considered the

planning. The European Court has repeatedly refused to override local zoning restrictions, especially the creation of green belts, in order to ensure a permanent home for this minority group. See *Buckley v. United Kingdom*, 1996-IV Eur. Ct. H.R. 1271 (1996) and the four recent cases: *Smith v. United Kingdom*, App. No. 25154/94, 33 EHRR 712 (2001); *Lee v. United Kingdom*, App. No. 25289/94, 33 EHRR 677 (2001); *Chapman v. United Kingdom*, App. No. 27238/94, 33 EHRR 399 (2001); and *Beard v. United Kingdom*, App. No. 24882/94, 33 EHRR (2001).

[26] *Oneryildiz v. Turkey* (GC), Reports 2004-VI (30 Nov.).

procedural rights to information, public participation, and redress, but it also assessed the substantive obligations of the government under international environmental standards. The Court relied on UN findings about the causes and consequences of the accident, as well as determinations of the World Health Organization about the health consequences of exposure to sodium cyanide, placing heavy reliance on them in the absence of adequate domestic fact-finding. The Court referred to international standards on best practices for the mining industry and, significantly, quoted extensively from the Stockholm Declaration on the Human Environment, the Rio Declaration on Environment and Development, and the Aarhus Convention.

Two of the Court's conclusions in the *Tatar* case were particularly important to the development of the law. First, the European Court declared in this case that the "precautionary principle" has become a legal norm with applicable content. This means the government must adopt reasonable and adequate measures capable of respecting the rights of individuals in the face of serious risks to their health and well-being, even where scientific certainty is lacking. Secondly, the Court recalled to Romania the obligation under Stockholm Principle 21 and Rio Principle 14 to prevent significant transboundary harm, in noting that both Hungary and Serbia were affected by the mining accident. These international norms, the Court found, should have been applied by the Romanian government.

15.2.6 The Environmental Quality Required

In the Western Hemisphere, the Inter-American Commission and Court have insisted on everyone's right to an environment at a quality that permits the enjoyment of all guaranteed human rights. In the cases presented to these institutions, applicants have asserted violations of the rights to life, health, property, culture, and access to justice, but some of them have also cited to guarantees of freedom of religion and respect for culture. The Commission's general approach to environmental protection has been to recognize that a basic level of environmental health is not linked to a single human right but is required by the very nature and purpose of human rights law:

> The American Convention on Human Rights is premised on the principle that rights inhere in the individual simply by virtue of being human. Respect for the inherent dignity of the person is the principle which underlies the fundamental protections of the right to life and to preservation of physical well-being. Conditions of severe environmental pollution, which may cause serious physical illness, impairment and suffering on the part of the local populace, are inconsistent with the right to be respected as a human being.[27]

Similarly, using environmental standards, the European Court has given some indications of the quality of environment required to comply with the Convention's substantive guarantees. In its first major decision[28] involving environmental harm as

[27] Inter-Am. Comm.H.R., *Report on the Situation of Human Rights in Ecuador*, OAS doc. OEA/Ser.L/V/II.96, doc. 10 rev. 1, April 24, 1997, at 92 [hereinafter Report on Ecuador].

[28] *Lopez Ostra v. Spain*, Eur. Ct. Hum. Rts [1994] Ser. A, No. 303C.

a breach of the right to private life and the home, guaranteed by Article 8 of the European Convention, the European Court held that severe environmental pollution may affect individuals' "well-being" to the extent that it constitutes a violation of Article 8. The pollution need not reach the point of affecting health, if the enjoyment of home, private, and family life is reduced and there is no fair balance struck between the community's economic well-being and the individual's effective enjoyment of guaranteed rights.[29]

The Court further explained this standard in *Fadayeva v. Russia*,[30] noting that the adverse effects of environmental pollution must attain a certain minimum level if they are to fall within the scope of Article 8. The requisite effects or interference need not reach the level of proven injury to health; it is enough if they pose serious risks. In *Fadayeva*, the applicant succeeded on her claim because she was made more vulnerable to various diseases, even though quantifiable harm to her health was deemed not proved; in addition, the Court found that her quality of life at her home was adversely affected.

15.2.7 Positive Obligations

Enforcement of environmental rights involves courts in not only determining the mandated environmental quality but also in assessing whether or not the government has taken the requisite actions to achieve that quality. Human rights tribunals have made clear that the state may be responsible whether pollution or other environmental harm is directly caused by the state or whether the state's responsibility arises from its failure to regulate adequately private sector activities.[31] Human rights instruments require states not only to respect the observance of rights and freedoms but also to guarantee their existence and the free exercise of all of them against private actors as well as the state. Any act *or omission* by a public authority which impairs guaranteed rights may violate a state's obligations.[32] This is particularly

[29] In *Powell & Raynor v. United Kingdom*, Eur. Ct. Hum. Rts [1990] Ser. A No. 172, the European Court found that aircraft noise from Heathrow Airport constituted a violation of Article 8 but was justified as "necessary in a democratic society" for the economic well-being of the country and was acceptable under the principle of proportionality because it did not "create an unreasonable burden for the person concerned." The latter text could be met by the State if the individual had "the possibility of moving elsewhere without substantial difficulties and losses." See also *Taşkin and Others v. Turkey*, App. No. 46117/99, 2004 Eur. Ct. Hum. Rts. 621 (10 Nov.).

[30] *Fadayeva v. Russia*, no. 55723/00, judgment of 9 June 2005, 2005/IV Eur. Ct.H.R. 255 (2005).

See also Leon and Agnieszka Kania v. Poland, no. 12605/03, 21 July 2009, para. 102; *Borysewicz v. Poland*, no. 71146/01, 1 July 2008, para. 55;. *Hatton and Others v. the United Kingdom*, (GC) no. 36022/97, Reports 2003-VIII.

[31] *See Mareno Gomez v. Spain*, no. 4143/02, 16 Nov. 2004, para. 55; *Giacomelli v. Italy*, paras. 78–79; *Surugiu v. Romania*, no. 48995/99, 20 April 2004.

[32] *Velasquez Rodriguez Case*, 4 Inter-Am. Ct. H.R. (ser. C) at 155 (Judgment of July 29, 1988) (concerning disappearance of civilians perpetrated by the Honduran army); *Godinez Cruz Case*, 5 Inter-Am. Ct. H.R. (ser. C) at 152–53 (Judgment of Jan. 20, 1989).

important in respect to the environment, where most activities causing harm are undertaken by the private sector.

In the inter-American system, positive obligations for the state to act derive not only from the generic obligations of Convention Article 1[33] but also from specific rights, including an individual's right to have his or her life respected and protected by law.[34] In the case of Yanomami v. Brazil,[35] the Inter-American Commission found that the government had violated the Yanomani rights to life, liberty, and personal security guaranteed by Article 1 of the Declaration, as well as the right of residence and movement (Article VIII) and the right to the preservation of health and well-being (Article XI)[36] because the government failed to implement measures of "prior and adequate protection for the safety and health of the Yanomami Indians."[37]

Other inter-American cases and country studies have specified that governments must enact appropriate laws and regulations and then fully enforce them. In a country report on Ecuador, the Commission referred generally to the obligation of the state to respect and ensure the rights of those within its territory and the responsibility of the government to implement the measures necessary to remedy existing pollution and to prevent future contamination which would threaten the lives and health of its people, including through addressing risks associated with hazardous development activities, such as mining.[38] Governments must regulate industrial and other activities that potentially could result in environmental conditions so detrimental that they create risks to health or life.[39] Furthermore, the government must enforce the laws that it enacts as well as any constitutional guarantee of a particular quality of environment.[40] The Commission was clear: "Where the right to life, to health and

---

[33] Article 1 provides: "The States Parties to this Convention undertake to respect the rights and freedoms recognized herein and to ensure to all persons subject to their jurisdiction the free and full exercise of those rights and freedoms…." (American Convention, art. 1).

[34] Art 4(1) reads: Every person has the right to have his life respected. This right shall be protected by law…. No one shall be arbitrarily deprived of his life.

[35] *Yanomami Case*, Res. No. 12/85, Case 7615 (Brazil), in *Annual Report of the IACHR 1984–1985*, OEA/Ser.L/V/II.66, doc. 10, rev. 1 (1985), 24.

[36] Id. at 33.

[37] Id. at 32.

[38] Report on Ecuador *supra* note 27 at 94.

[39] *Id*, p. v.

[40] In the Ecuador report, the Commission heard allegations that the Government had failed to ensure that oil exploitation activities were conducted in compliance with existing legal and policy requirements. The Commission's on-site delegation also heard that the Government of Ecuador had failed to enforce the inhabitants' constitutionally protected rights to life and to live in an environment free from contamination. The domestic law of Ecuador recognizes the relationship between the rights to life, physical security and integrity, and the physical environment in which the individual lives. The first protection accorded under Article 19 of the Constitution of Ecuador, the section which establishes the rights of persons, is of the right to life and personal integrity. The second protection establishes "the right to live in an environment free from contamination." Accordingly, the Constitution invests the State with responsibility for ensuring the enjoyment of this right and for establishing by law such restrictions on other rights and freedoms as are neces-

to live in a healthy environment is already protected by law, the Convention requires that the law be effectively applied and enforced."[41]

The state must also comply with and enforce the international agreements to which it is a signatory, whether these are human rights instruments or ones related to environmental protection. In the Ecuador report, the Commission noted that the state is party to or has supported a number of instruments "which recognize the critical connection between the sustenance of human life and the environment," including the Additional Protocol to the American Convention in the Area of Economic, Social, and Cultural Rights, the ICCPR and the ICESCR, the Stockholm Declaration, the Treaty for Amazonian Cooperation, the Amazon Declaration, the World Charter for Nature, the Convention on Nature Protection and Wildlife Preservation in the Western Hemisphere, the Rio Declaration on Environment and Development, and the Convention on Biological Diversity. Through the standard-setting and enforcement process, the state must "take the measures necessary to ensure that the acts of its agents … conform to its domestic and inter-American legal obligations."[42]

States thus are not exempt from human rights and environmental obligations in their development projects: "the absence of regulation, inappropriate regulation, or a lack of supervision in the application of extant norms may create serious problems with respect to the environment which translate into violations of human rights protected by the American Convention."[43] In the case of the *Saramaka People v. Suriname*,[44] the Inter-American Court set forth three safeguards it deemed essential to ensure that development is consistent with human rights and environmental protection: (1) the state must ensure the effective participation of the members of the Saramaka people, in conformity with their customs and traditions, regarding any development, investment, exploration, or extraction plan within Saramaka territory; (2) the state must guarantee that the Saramakas will receive a reasonable benefit from any such plan within their territory; and (3) the state must ensure that no concession will be issued within Saramaka territory unless and until independent and technically capable entities, with the state's supervision, perform a prior environmental and social impact assessment.[45] The Court viewed benefit sharing as inherent to the right of compensation recognized under Article 21(2) of the Inter-American Convention.[46]

sary to protect the environment. Thus, the Constitution establishes a hierarchy according to which protections which safeguard the right to a safe environment may have priority over other entitlements. *Id.* pp. 78–86.

[41] *Id.*

[42] Report on Ecuador, *supra* n. 27 at 92.

[43] *Id.* at 89.

[44] *Case of the Saramaka People v. Suriname*, Inter-Am. Ct. Hum. Rts, Ser. C No. 172 (28 Nov. 2007).

[45] *Id.* at para. 129.

[46] Article 21(2) provides that [n]o one shall be deprived of his property except upon payment of just compensation, for reasons of public utility or social interest, and in the cases and according to the forms established by law.

The European Court's jurisprudence is similar. The Court requires at a minimum that the state should have complied with its domestic environmental standards.[47] Noise pollution cases often turn on compliance with local environmental laws. When the state conducts inspections and finds that the activities do not exceed permissible noise levels established for the area, at least in the absence of evidence of serious and long-term health problems, the Court is unlikely to find that the state failed to take reasonable measures to ensure the enjoyment of Article 8 rights. In other words, where no specific environmental quality is guaranteed by the constitution or applicable human rights instrument, the courts accord considerable deference to the level of protection enacted by state or local authorities.

The issue of compliance with domestic law is particularly important when there is a domestic constitutional right to environmental protection. The European Court will review governmental actions in the light of the domestic law. *Okyay and Others v. Turkey*[48] concerned the failure of Turkish authorities to enforce constitutional rights and statutory environmental laws. The applicants had successfully challenged in domestic courts the operations of thermal power plants in Southwest Turkey, which they claimed would damage the environment and pose risks for the life and health of the Aegean region's population. They explicitly argued that Article 56 of the Turkish Constitution guaranteed them the right to life in a healthy and balanced environment. They did not argue that they had suffered any economic or other loss. The European Court agreed that they had a right under Turkish law to protection against damage to the environment and that their rights under Article 6(1) had been violated due to the failure of Turkish authorities to comply in practice and within a reasonable time with the domestic court's judgments.

The African Commission also has identified governmental obligations in this field by reference to environmental norms. In *SERAC v. Nigeria*, the African Commission held that African Charter Article 24 "imposes clear obligations upon a government to take reasonable and other measures to prevent pollution and ecological degradation, to promote conservation, and to secure an ecologically sustainable development and use of natural resources."[49] Compliance with these obligations includes ordering or permitting independent scientific monitoring of threatened environments, requiring environmental and social impact studies, monitoring hazardous materials and activities, as well as providing information and an opportunity for the public to participate in decision-making.[50] While the Commission did not cite to specific environmental agreements, the obligations it mentions are part of international environmental law.

---

[47] *See*, e.g., *Ashworth and Others v. the United Kingdom*, App. No. 39561/98, 20 Jan. 2004;, *Moreno Gomes v. Spain*, 2004-X Eur. Ct. H.R. 327 (2005).

[48] *Okyay and Others v. Turkey*, App. 36220/97, 2005 Eur. Ct. H.R. 476, 12 July 2005 at 57.

[49] *Social and Economic Rights Action Center/Center for Economic and Social Rights v. Nigeria*, Comm. 155/96, Case No. ACHPR/COMM/A044/1, May 27, 2002.

[50] *Id.* para. 53.

15.2.8 Reviewing National Actions and Decisions

Beyond ensuring that any domestic environmental rights are enforced, the European Court scrutinizes the adequacy of the domestic law, to see if the state has ensured a fair balance between the interests of the community and the rights of those affected. The Court accords each state considerable deference in this respect (known as the "margin of appreciation"), because national authorities "are in principle better placed than an international court to assess the requirements" in a particular local context and to determine the most appropriate environmental policies and individual measures while taking into account the needs of the local community,[51] especially in a technical sphere like environmental protection.[52] The Court will only find a violation if there is a "manifest error of appreciation" by the national authorities in striking a fair balance between the competing interests of the different private actors.[53] Only "in exceptional circumstances" will the Court look beyond the procedures followed to disallow the conclusions reached by domestic authorities on the environmental protection measures to be taken on the projects and activities allowed to proceed.[54] Even if it finds that the state decided wrongly, the Court will not determine exactly what should have been done to reduce the pollution in a more efficient way.[55]

Another human rights supervisory body has also found violations of substantive guarantees due to the failure of the government to legislate to protect the environment. The first European Social Charter complaint to concern environmental conditions, lodged April 4, 2005, claimed violations of the Charter's right to health provisions[56] because the state had not adequately prevented negative environmental impacts nor had it developed an appropriate strategy to prevent and respond to the health hazards stemming from lignite mining. The complaint also alleged that there was no legal framework guaranteeing security and safety of persons working in lignite mines. The European Committee of Social Rights concluded that the government had violated the Charter.[57] On the issue of the right to health (Article 11), the Committee examined the Greek National Action Plan for greenhouse gas emissions and found it inadequate in the light of the state's obligations under the Kyoto Protocol and the principle requiring use of the "best available techniques."[58] While

[51] *Giacomelli*, para. 80.

[52] *Fadayeva v. Russia*, supra note 30 at para 104, citing *Hatton*, *supra* n. 30 at para. 122.

[53] *Id.*

[54] *Id.* para. 105, citing *Taskin*.

[55] In particular, the Court says it would be going too far to assert that the State or the polluting undertaking was under an obligation to provide the applicant with free housing (Para. 133). It is enough to say that the situation called for a special treatment of those living near the plant.

[56] Complaint No. 30/2005 *Marangopoulos Foundation for Human Rights (MFHR) v. Greece.*

[57] The Committee transmitted its decision on the merits to the Committee of Ministers and to the Parties on 6 December 2006. The Committee of Ministers adopted its resolution on the matter on January 15, 2008.

[58] According to the Committee, "[t]he Greek National Action Plan for 2005–2007 (NAP1) provides for greenhouse gas emissions for the whole country and all sectors combined to rise by no more than 39.2 % until 2010, whereas Greece was committed, in the framework of the Kyoto Protocol,

the Committee found that Greek regulations on information and public participation were satisfactory, the evidence showed "that in practice the Greek authorities do not apply the relevant legislation satisfactorily" and very little had been done to organize systematic epidemiological monitoring of those concerned and no morbidity studies have been carried out.

In *Guerra v. Italy*,[59] the applicants alleged that the Italian authorities violated European Convention rights by failing to mitigate the risk of a major accident at a nearby chemical factory and by withholding information from local residents about the risks and emergency procedures. The "right to information" claim was dismissed, because the European Convention does not require the collection and dissemination of information about the environment, but the European Court effectively incorporated this requirement into the applicant's Article 8 claim as the "procedural dimension" of the obligation of states to secure effective respect for the applicants' right to family and home life.

15.2.9 Industrial Accidents and Natural Disasters

In *Oneryildiz v. Turkey*,[60] the European Court of Human Rights held the Turkish government responsible for the loss of life and property resulting from a methane explosion at a waste disposal site. It noted the duty under the Strasbourg Convention for authorities to establish criminal offenses for loss of life involving the disposal or treatment of hazardous wastes. The Court explained that the right to life provision, Article 2 of the European Convention, "must be construed as applying in the context of any activity, whether public or not, in which the right to life may be at stake, and a fortiori in the case of industrial activities, which by their very nature are dangerous, such as the operation of waste–collection sites." According to European standards, waste disposal is a hazardous activity; therefore, Article 2 applies. The resulting duty of care depends on several factors: the harmfulness of the phenomena inherent in the activity, the contingency of the risk to the applicant, the status of those involved in creating the risk, and whether or not the conduct was deliberate. The Court found that "particular emphasis" should be placed on the public's right to information concerning the risks to life and the duty to investigate when loss of life occurs.[61] Assessing the evidence, the Court found that the authorities must have known of the risk and of the need to take preventive measures "particularly as there

to an increase in these gases of no more than 25 % in 2010. When air quality measurements reveal that emission limit values have been exceeded, the penalties imposed are limited and have little dissuasive effect. Moreover, the initiatives taken by DEH (the public power corporation operating the Greek lignite mines) to adapt plant and mining equipment to the "best available techniques" have been slow."

59 *Guerra v. Italy*, App. No: 14967/89, Reports 1998-I, no. 64.

60 *Oneryildiz v. Turkey*, *supra* note 26.

61 *Id.* para. 90.

were specific regulations on the matter."[62] As such, they had an obligation under Convention Article 2, "to take such preventive measures as were necessary and sufficient to protect those individuals…."[63] The government failed in its duty.

Like *Oneryildiz*, the case of *Budayeva and Others v. Russia*[64] concerned governmental knowledge of a hazard and the failure to act on that knowledge. The difference was that *Budayeva* involved repeated natural disasters rather than hazards originating in human activities. The standard of care did not differ appreciably, however. Governmental authorities aware of mudslide hazards in a mining district failed to take reasonable precautions, with resulting deaths in a village and loss of property. The applicants pleaded violations of the right to life and the right to property. The Court held the government responsible for the loss of life but found that the causal link was not established in respect of the latter claims because the applicants could not demonstrate that "but for" the official failures to act, their property would have been safe. The July 2000 mudslide was of unprecedented severity.

Looking at the substantive aspect of the government's obligations respecting dangerous activities, the Court placed special emphasis on the adoption of regulations geared to the special features of the activity in question, particularly with regard to the level of the potential risk to human life.[65] Such regulations must govern the licensing, setting up, operation, security, and supervision of the activity and must make it compulsory for all those concerned to take practical measures to ensure the effective protection of citizens whose lives might be endangered by the inherent risks.[66] Supervision and monitoring are also required. The choice of particular practical measures is in principle a matter within the state's margin of appreciation and the Court will seek to avoid placing an impossible or disproportionate burden on authorities.[67]

The Court reviewed the measures taken by the government found that they were limited to a mud-retention dam and collector that were not adequately maintained. The Court held that there was no justification for the failure to act regarding foreseeable mortal risks to the residents of the town and there was a causal link between that failure and the death and injuries suffered in the mudslide. Accordingly, there was a violation of the right to life, even though the state's positive obligation is less in the context of natural disasters, "which are as such beyond human control," than in the sphere of dangerous activities of a man-made nature. The right to peaceful enjoyment of possessions, which is not absolute, requires only that the state do what is reasonable in the circumstances.[68] The standard of care is different and higher

[62] *Id.* para. 101.

[63] *Id.*

[64] *Budayeva and Others v. Russia*, App. No. 15339/02 & Ors (20 March 2008).

[65] *Id*, para. 132.

[66] *Id.*

[67] *Id.*, para. 135.

[68] *Id.*, at para. 174. While the Court found that the measures taken by the state were negligent, it found the causal link was not well established. The mudslide of 2000 being exceptionally strong, the Court said it was unclear whether a functioning warning system or proper maintenance of the defense infrastructure would have mitigated the damage.

when the risk involves potential loss of life. The state in this situation has a positive obligation to do everything within the authorities' power in the sphere of disaster relief for the protection of the right to life. The origin of the threat and the extent to which one or another risk is susceptible to mitigation are factors to be evaluated in determining the scope of the state's positive obligations.[69]

The Court found that the authorities had been given warnings about the risks, including the state of disrepair of the dam, and had failed to provide resources for strengthening the defense infrastructure – resources that became available immediately after the mudslide. Nor were any alternative land-planning policies being implemented or monitoring stations set up. The Court noted that the public's procedural right of information can only be implemented if the government obtains the relevant information, which in this case was indispensable for ensuring the residents' safety. The authorities' failure to ensure the functioning of an early warning system was thus also unjustified.

15.3 Causality, Evidence, and Precaution

Assessing risk is an important issue in litigating substantive environmental rights. Some human rights procedures limit standing to "victims" of violations and there must be a sufficient threat for the applicants or petitioners to qualify as victims.[70] The precautionary principle has begun to play a role in bringing more risks within the ambit of human rights litigation.

The *Taşkin* case described above was one based on risk, stemming from the use of cyanide in gold extraction. The Court referred to the various reports that had been done on site which highlighted the risks. Domestic judicial findings also demonstrated the threat to the environment and lives of the neighboring population. The Court found Article 8 to be applicable "where the dangerous effects of an activity to which the individuals *are likely to be exposed* have been determined as part of an environmental impact assessment procedure in such a way as to establish a sufficiently close link with private and family life for purposes of Article 8 of the Convention."[71] The Court held that this broad reading was necessary to ensure the effectiveness of Article 8.

The evidentiary basis of the *Taşkin* decision was the domestic court judgment. The Court also held that "in view of" the conclusion of the domestic court on the absence of a public interest in allowing the gold mine, it did not need to examine the case from the perspective of the normal wide margin of appreciation afforded governments in environmental matters. Therefore, there was a violation of Article 8.

[69] *Id.*, at para. 137.

[70] See *Bordes and Temeharo v. France*, Comm. No. 645/1995, CCPR/C/57/D/645/1995, 30 July 1996. The risk of harm from nuclear radiation due to nuclear testing by France in the South Pacific deemed too remote for the victims to qualify as victims.

[71] *Taşkin, supra* note 29 at para. 113 [emphasis added].

The problem of fact-finding and lack of expertise is frequently said to be a hurdle to giving substantive content to environmental rights. At the international level, this has not proved to be a high hurdle thus far, because in most of the cases, domestic fact-finding had already revealed the risks entailed or the consequent harm. This was the case in *Oneryildiz*, *Taşkin*,[72] and *Fadayeva*.[73] In the last-mentioned case, a government decree had recited statistics on the increases in respiratory and blood diseases linked to air pollution, as well as the increased number of deaths from cancer.[74] The government had also determined by legislation the safe levels of various polluting substances, many of which were exceeded in the security zone where the applicant lived. The mayor of the city said the steel plant was responsible for more than 95 % of industrial emissions into the town's air,[75] while a STATE REPORT ON THE ENVIRONMENT indicated that the plant in question was the largest contributor to air pollution of all metallurgical plants in Russia. The two statements reduced questions about causality.[76] In the end, both parties agreed that the applicant's place of residence was affected by industrial pollution caused by the steel plant, but they disagreed over the degree and effects of the pollution. The government claimed that the disturbance caused by the pollution was not so severe as to raise an issue under Article 8. The applicant and the European Court disagreed. The Court elaborated on its test for finding that environmental conditions are sufficiently severe to be encompassed within the guarantees of Article 8:

> The assessment of that minimum is relative and depends on all the circumstances of the case, such as the intensity and duration of the nuisance, and its physical or mental effects. The general context of the environment should also be taken into account. There would be no arguable claim under Article 8 if the detriment complained of was negligible in comparison to the environmental hazards inherent to life in every modern city.

Causality was an issue on the applicant's health claims. Her medical records indicated problems but did not attribute them to any specific causes. The doctors stated, however, that her problems would be exacerbated by working in conditions of vibration, toxic pollution, and an unfavorable climate.[77] The applicant also submitted an expert report[78] which linked the plant specifically to increased adverse

---

[72] *Id.*

[73] *Fadayeva v. Russia*, *supra* note 30. See also *Ledyayeva, Dobrokhotova, Zolotareva and Romashina v. Russia*, nos. 53157/99, 53247/99, 53695/00 and 56850/00, judgment of 26 Oct. 2006, also involving the same steel plant built during the Soviet era.

[74] Russia's Constitution, art. 42 guarantees as follows: "Everyone has the right to a favorable environment, to reliable information about its state, and to compensation for damage caused to his health or property by ecological disease." The provision was not invoked in the case.

[75] The Court noted that this made the case different from and more easily definable than other air pollution cases where multiple minor sources cumulate to produce the problem.

[76] The Court noted that the parties produced official documents containing generalized information on industrial pollution, because basic data on air pollution are not publicly available (para. 30).

[77] *Fadayeva v. Russia*, supra note 30 at para. 45.

[78] The court made it a point to recite the qualifications of the expert when discussing the report. See *id*, para. 46 n. 1.

health conditions of persons residing nearby. The Court found that the medical evidence did not establish a causal link between the pollution at her residence and her illnesses but accepted that the evidence, including submissions by the government, was clear about the unsafe excessive pollution around her home. The Court also made reference to the expert report and the findings of the domestic courts. The Court noted that Russian legislation defined the maximum permissible concentrations as "safe concentrations of toxic elements." Therefore, exceeding these limits produced a presumption of unsafe conditions potentially harmful to health and well-being of those exposed to it. This presumption, together with the evidence submitted, led the court to conclude that the applicant's health deteriorated as a result of her prolonged exposure to the industrial emissions from the steel plant. Alternatively, even if that harm could not be quantified, the pollution "inevitably made the applicant more vulnerable to various illnesses" and affected her quality of life at home.[79] Therefore Article 8 applied.

The analysis raises the question of what evidence is sufficient to raise the presumption the Court creates in the *Fadayeva* case. It should not be limited to legislative findings, because as *Zander v. Sweden* indicates safe levels may be changed to accommodate economic interests without necessarily being based on sound science. The World Health Organization (WHO) and other scientific bodies have determined through epidemiological studies what constitutes safe levels of concentration of toxic, carcinogenic, mutagenic, and other hazardous substances.[80] Reliable evidence from such studies can and should be introduced to demonstrate presumed harm when such levels are exceeded, even if local legislation permits higher concentrations. A petition to the Inter-American Commission, recently declared admissible, relies on such WHO standards to assert that the average sulfur dioxide levels from a metallurgical complex are detrimental to the lives and health of the nearby community in Peru.[81]

The European Court's standard of proof is high[82] but flexible and takes into account the fact that governments often are the sole repository of relevant evidence. Indeed, in the case of *Fägerskiöld v. Sweden*,[83] the Court cited to World Health Organization guidelines[84] on noise pollution, in rejecting the admissibility of an application concerning wind turbines constructed and operating near the applicants' property. The Court noted that the guidelines are set at the level of the lowest adverse

[79] *Id.* para. 88.

[80] The WHO has developed guidelines for safe and acceptable water quality and quantity. World Health Organization, "Guidelines for Drinking Water Quality" (3d ed. 2004). Independent surveillance of water quality, quantity, accessibility, affordability, and long-term availability is part of the WHO framework.

[81] Inter-American Commission on Human Rights, Report No. 76/09, Case 12.718, *Community of La Oroya, Peru,* admissibility decision of 5 August 2009, OAS/Ser/L/V/II.135, doc. 23.

[82] It has long demanded "proof beyond reasonable doubt." Fadayeva supra n. 29, para. 79 which can follow from the coexistence of sufficiently strong, clear, and concordant inferences or of similar unrebutted presumptions of fact.

[83] *Fägerskiöld v. Sweden*, no. 37664/04 (admissibility), 26 Feb. 2008.

[84] World Health Organization, "Guidelines for Community Noise" (Geneva 1999).

health effect associated with noise exposure. The Court also referred to even lower maximum levels adopted by most European countries. Applying these standards to the noise-level tests submitted in the case, the Court found that the levels of noise did not exceed the WHO guidelines and were minimally above the recommended maximum level in Sweden. Therefore, the environmental nuisance could not be found to reach the level of constituting severe environmental pollution. The Court also rejected the applicants' claims that their property rights were violated because the wind turbines decreased the value of their property. Assuming that there was an interference with property rights, the Court found that it was justified on several grounds, one of them being that the operation of the wind turbines was in the general interest as it is an environmentally friendly source of energy which contributes to the sustainable development of natural resources. The Court considered whether these beneficial environmental consequences were sufficient to outweigh the negative impact on the applicants. The Court reiterated its findings that the negative consequences were not severe while the availability of renewable energy is beneficial for both the environment and society. Moreover, the government had taken measures to mitigate the negative impacts on the applicants. In sum, the alleged interference was proportionate to the aims pursued and no violation of property rights occurred.

In the Greek case on lignite mining,[85] the European Social Charter Committee similarly relied on what it called "ample and unambiguous scientific evidence" that lignite-caused air pollution has a harmful effect on human health and life, without specifying the health risks. Despite the beneficial impacts of lignite use in providing energy independence, access to electricity at a reasonable cost, and economic growth, the Committee found that the government's actions violated the State's national and international obligations to combat pollution that caused health problems. It pointed to the right to environment in the Greek constitution, as well as national environmental protection legislation and regulations, noting that these were not applied and enforced in an effective manner. In sum, Greece had not stuck a reasonable balance between the interests of persons living in the lignite mining areas and the general interest and there was thus a violation of the right to protection of health under the Charter.

15.4 Conclusions

National and international tribunals increasingly are being asked to consider the link between environmental degradation and internationally guaranteed human rights. In some instances, the complaints brought have not been based upon a specific right to a safe and environmentally sound environment but rather upon rights to life, property, health, information, family, and home life. Underlying the complaints, however, are instances of pollution, deforestation, water pollution, and other types of

[85] *Marangopoulos Foundation for Human Rights v. Greece*, Complaint No. 30/2005, European Committee on Social Rights (2006).

environmental harm. International petition procedures thus allow those harmed to bring international pressure to bear when governments lack the will to prevent or halt severe pollution that threatens human well-being. Petitioners have been afforded redress and governments have taken measures to remedy the violation. Petition procedures at the least can help to identify problems and encourage a dialogue to resolve them. In addition, the emphasis given rights of information, participation, and access to justice can encourage an integration of democratic values and promotion of the rule of law in broad-based structures of governance. Even where there is a guaranteed right to environment, it still must be balanced against other rights should there be a conflict. Human rights exist to promote and protect human well-being, to allow the full development of each person and the maximization of the person's goals and interests, individually and in community with others. This cannot occur without basic healthy surroundings, which the state is to promote and protect.

Adjudicating cases under broadly-worded standards is not new for judges; however, nor is it uncommon for them to be faced with adjudicating highly technical matters. Courts must regularly, and on a case-by-case basis, define what constitutes "reasonable," "fair," or "equitable" conduct. With the adoption of constitutional environmental rights provisions and increasing acceptance of the links between environmental degradation and the violation of other human rights, national and international tribunals struggle to give substance to environmental rights without overstepping the judicial function. In general, courts have taken the view that such enactments serve to place environmental protection in a position superior to ordinary legislation. Over time, courts tend to create a balancing test to avoid too readily undoing the deliberative decisions reached by the political branches of government.

Human rights law is not about stopping all human activities, but about recognizing that they utilize scarce resources and produce emissions and waste that inevitably have individualized and cumulative environmental impacts. These impacts have to be considered, measured, and monitored, with the result that some activities will be limited or prohibited. Environmental science helps determine the causal links between the activities and the impacts, giving courts a set of data on which to base decisions about whether or not a proper balance of interests has been obtained, one which ensures an equitable outcome and minimizes the risk of harm to the environment and human rights. The substance of environmental rights involves evaluating ecological systems, determining the impacts that can be tolerated, and what is needed to maintain and protect the natural base on which life depends. Environmental quality standards, precaution, and principles of sustainability can establish the limits of environmental decision-making and continue to give specific content to environmental rights in law.

Both national and international courts have used environmental law and science to give content to the level of environmental protection required by human rights law. This approach can involve reference to World Health Organization standards on acceptable emissions levels, incorporation of the precautionary principle to judge the adequacy of measures taken by a government, or reference to environmental treaties and declarations. The breadth of the search for standards depends in part on whether or not there is a textual guarantee of environmental quality and if there is, on the descriptions of that quality.

There remain many questions to be addressed, including issues about the scope of the guaranteed rights, the scope of state responsibility, accountability of non-state actors, and procedural mechanisms to give effect to or monitor compliance with environmental rights. These issues will undoubtedly be raised in future litigation and debated in academic journals. In both contexts, contributions from scientists, especially the medical profession, and other relevant disciplines will be necessary to ensure that the law and policy reflect knowledge about the environment and the consequences of pollution.

Index

C. Esser (ed.), *Environmental Influences on the Immune System*,
DOI 10.1007/978-3-7091-1890-0

F